Springer Actuarial

Springer Actuarial Lecture Notes

This subseries of Springer Actuarial includes books with the character of lecture notes. Typically these are research monographs on new, cutting-edge developments in actuarial science; sometimes they may be a glimpse of a new field of research activity, or presentations of a new angle in a more classical field.

In the established tradition of Lecture Notes, the timeliness of a manuscript can be more important than its form, which may be informal, preliminary or tentative.

More information about this subseries at http://www.springer.com/series/15682

Michel Denuit · Donatien Hainaut ·
Julien Trufin

Effective Statistical Learning Methods for Actuaries I

GLMs and Extensions

 Springer

Michel Denuit
Institut de Statistique, Biostatistique et
Sciences Actuarielles
Université Catholique de Louvain
Louvain-la-Neuve, Belgium

Donatien Hainaut
Institut de Statistique, Biostatistique et
Sciences Actuarielles
Université Catholique de Louvain
Louvain-la-Neuve, Belgium

Julien Trufin
Département de Mathématiques
Université Libre de Bruxelles
Brussels, Belgium

ISSN 2523-3262 ISSN 2523-3270 (electronic)
Springer Actuarial
ISSN 2523-3289 ISSN 2523-3297 (electronic)
Springer Actuarial Lecture Notes
ISBN 978-3-030-25819-1 ISBN 978-3-030-25820-7 (eBook)
https://doi.org/10.1007/978-3-030-25820-7

Mathematics Subject Classification (2010): C1, C46, C23, C35, C8, 62P05, 62-XX, 68-XX

This Springer imprint is published by the registered company Springer Nature Switzerland AG
The registered company address is: Gewerbestrasse 11, 6330 Cham, Switzerland

Preface

The present material is written for students enrolled in actuarial master programs and practicing actuaries, who would like to gain a better understanding of insurance data analytics. It is built in three volumes, starting from the celebrated Generalized Linear Models, or GLMs and continuing with tree-based methods and neural networks.

After an introductory chapter, this first volume starts with a recap' of the basic statistical aspects of insurance data analytics and summarizes the state of the art using GLMs and their various extensions: GAMs, mixed models and credibility, and some nonlinear versions, or GNMs. Analytical tools from Extreme Value Theory are also presented to deal with tail events that arise in liability insurance or survival analysis. This book also goes beyond mean modeling, considering volatility modeling (double GLMs) and the general modeling of location, scale and shape parameters (GAMLSS).

Throughout this book, we alternate between methodological aspects and numerical illustrations or case studies to demonstrate practical applications of the proposed techniques. The numerous examples cover all areas of insurance, not only property and casualty but also life and health, being based on real data sets from the industry or collected by regulators.

The R statistical software has been found convenient to perform the analyses throughout this book. It is a free language and environment for statistical computing and graphics. In addition to our own R code, we have benefited from many R packages contributed by the members of the very active community of R-users. We provide the readers with information about the resources available in R throughout the text as well as in the closing section to each chapter. The open-source statistical software R is freely available from https://www.r-project.org/.

The technical requirements to understand the material are kept at a reasonable level so that this text is meant for a broad readership. We refrain from proving all results but rather favor an intuitive approach with supportive numerical illustrations, providing the reader with relevant references where all justifications can be found, as well as more advanced material. These references are gathered in a dedicated section at the end of each chapter.

The three authors are professors of actuarial mathematics at the universities of Brussels and Louvain-la-Neuve, Belgium. Together, they accumulate decades of teaching experience related to the topics treated in the three books, in Belgium and throughout Europe and Canada. They are also scientific directors at Detralytics, a consulting office based in Brussels.

Within Detralytics as well as on behalf of actuarial associations, the authors have had the opportunity to teach the material contained in the three volumes of "Effective Statistical Learning Methods for Actuaries" to various audiences of practitioners. The feedback received from the participants to these short courses greatly helped to improve the exposition of the topic. Throughout their contacts with the industry, the authors also implemented these techniques in a variety of consulting and R&D projects. This makes the three volumes of "Effective Statistical Learning Methods for Actuaries" the ideal support for teaching students and CPD events for professionals.

Louvain-la-Neuve, Belgium   Michel Denuit
Louvain-la-Neuve, Belgium   Donatien Hainaut
Brussels, Belgium   Julien Trufin
June 2019

Notation

Here are a few words on the notation and terminology used throughout the book. For the most part, the notation conforms to what is usual in mathematical statistics as well as insurance mathematics. Unless stated otherwise,

$n =$	Number of observations (or data points).
$y_i =$	The ith observed value of the response (or outcome, dependent variable), observed response for individual i, realization of the random variable $Y_i, i = 1, \ldots, n$.
$\widehat{y}_i =$	Fitted value, predicted response for the ith individual.
$\bar{y} =$	Average, or sample mean of the n observed responses $y_1, \ldots, y_n$.
$y =$	Column vector of all n response values. All vectors are column ones, by convention.
$p =$	Number of features (sometimes also called covariates, independent variables, regressors, as well as explanatory variables or predictors when they influence the response).
$x_{ij} =$	Value of the jth feature for the ith data point, $i = 1, \ldots, n$, $j = 1, \ldots, p$, realization of the random variable X_{ij}.
$x_i =$	Column vector of the p features for the ith data point.
$X =$	Matrix of the p features for all data points, known as the design matrix with n rows and p columns (x_i is the ith row of X)
	or random vector with components $X_j, j = 1, \ldots, p$ (clear from the context).
$X^\top =$	The transpose of the design matrix X, with p rows and n columns.
$I[\cdot] =$	Indicator function of an event (equal to 1 if the event appearing within the brackets is realized and to 0 otherwise).
$P[\cdot] =$	Probability of an event.
$E[\cdot] =$	Expectation of a random variable.
$\text{Var}[\cdot] =$	Variance of a random variable.
$\text{Cov}[\cdot, \cdot] =$	Covariance of a pair of random variables.

$\widehat{\theta} =$ Estimator or estimate (clear from the context) of the unknown parameter θ (parameters are denoted by Greek letters).

The real line is denoted as $(-\infty, \infty)$. We use the symbol $\sim$ to mean "is distributed as", $\approx$ to mean "is approximately equal to" or "is approximately distributed as" (clear from the context), and $\propto$ to mean "is proportional to".

Contents

Part I
Loss Models

Chapter 1
Insurance Risk Classification

1.1 Introduction

Contrarily to other industries, the insurance activity does not consist in selling products as such, but promises. An insurance policy is nothing else but a promise made by the insurer to the policyholder to defray the third parties or to pay for his or her own future losses due to the occurrence of adverse random events, in exchange of an upfront premium.

In case of an accident, illness or fatality, the policyholder or the beneficiary may file a claim against the insurance company asking for compensation in execution of the insurance contract. An insurance claim can be formally defined as the request to an insurance company for coverage or compensation for the financial consequences of a peril comprised in the insurance policy. The insurance company handles the claim from reporting to final settlement (or closure) and once approved, issues payment to the insured or to a third party on behalf of the insured.

Consequently, when the insurance policy is sold, the provider does not know the ultimate costs for this service, but relies on historical data and actuarial models to predict a sustainable price for their product. The central part of this price is the pure premium, that is, the best estimate or expected (present) value of future liabilities.

In this introductory chapter, we review the basic concepts of actuarial pricing that will be used throughout this book. We carefully set up the scene and introduce general principles behind insurance loss modeling and claim data analysis.

© Springer Nature Switzerland AG 2019
M. Denuit et al., *Effective Statistical Learning Methods for Actuaries I*,
Springer Actuarial, https://doi.org/10.1007/978-3-030-25820-7_1

1.2   Technical Versus Commercial Price List

1.2.1   Selling Price and Operating Cost

The technical tariff aims to evaluate as accurately as possible the pure premium for each policyholder and is the reference for internal risk evaluation. It incorporates all the available information about the risk.

Instead of considering observed claim costs that involve a large part of randomness, actuaries consider the combined operating cost for each policy as the technical price. The advantage of looking at the modeled cost instead of the actual cost is that random fluctuations in the observed experience are removed. Also, the vast majority of the observed costs are just null since no claims have been filed in relation to the corresponding policies, but it is clear that the price of the product must be positive. Notice that the modeled cost may include experience rating/credibility corrections if past claims history reveals future costs.

The combined operating cost is the sum of

- the modeled expected claim cost per policy;
- running expenses (staff, buildings, etc.);
- safety loading ensuring solvency and reinsurance costs;
- as well as the shareholders' profit.

The combined operating ratio corresponds to the combined operating cost divided by the net commercial premium (i.e. the selling price, taxes and intermediaries' commissions deducted). This differs from loss ratios, or actual over expected ratios, which are generally uninformative at individual policy level (but relevant for blocks of business, or in commercial insurance where a single policy may cover an entire firm with a large risk exposure).

There is thus a difference between the cost-based technical premium and the demand-based commercial one, actually charged to policyholders. The former needs to correctly evaluate the risk for each policyholder according to individual characteristics whereas the latter aims to optimize insurer's profit. Insurance is a heavily regulated business so that commercial premiums must comply with the rules imposed by the local supervisory authority. Such rules generally have two objectives: firstly ensuring the insurer's solvency (remember that insurers sell promises) and secondly protecting customers by banning unfair treatment (or treatment perceived as unfair).

Besides legal and regulatory compliance, IT limitations or sales channels may also severely constrain the commercial price list. Programming changes into an IT system has a cost, not to mention a certain degree of conservatism of people in charge. Also, if products are sold by brokers or agents, they must be understandable by these intermediaries and meet their concerns.

Commercial price lists also strongly depend on the organizational form of the insurance company, mutual insurer or stock/commercial one. Mutuals are owned by the policyholders and are generally comfortable with more solidarity among insureds. However, their technical pricing must be even finer to prevent any adverse selection effects induced by the competition with commercial insurers (whereas in practice, mutuals are often less advanced in developing accurate pricing models).

This book is entirely devoted to the evaluation of technical premiums. The simplifications leading to the commercial premiums are only briefly mentioned here and there. Actuaries resort nowadays to advanced statistical tools to be able to accurately assess the risk profile of the policyholders, as explained in the next chapters.

1.2.2 Risk Factor Versus Rating Factor

Risk classification is based on risk factors.[1] In motor insurance for instance, risk factors typically include policyholder's age, gender and occupation, the type and use of their car, the place where they reside and sometimes even the number of cars in the household or marital status. These observable risk characteristics are typically seen as non-random features and the analysis is conducted conditionally on their values (even if it is useful to recognize their random nature in some methodological developments).

As explained above, commercial premiums may differ from the cost-based technical premiums computed by actuaries because of regulation or because of the position of the company with respect to its market competitors. Commercial premiums depend on a set of rating factors. These rating factors form a subset of the risk factors, sometimes simplified. The next examples help to understand the difference between risk and rating factors.

Example 1.2.1 (*Gender in the European Union*) Gender is generally a relevant risk factor and therefore needs to be included in the technical price list. However, gender has to be excluded from the commercial price list in the European Union, in application of European directives banning every discrimination between male and female citizens. Thus, gender is a risk factor but not a rating factor in the European Union.

Example 1.2.2 (*Smoking status*) Smoking status is a risk factor in life insurance as it is well documented that expected remaining lifetime is significantly smaller for smokers compared to non-smokers. For death benefits, this risk factor is often used as a rating factor: smokers usually pay a higher premium for the same amount of death benefit in term insurance, for instance. For survival benefits (pure endowments or life annuities, for instance), however, smokers are not always awarded premium discounts despite their reduced remaining life expectancy.[2] Thus, smoking status may be a risk factor as well as a rating factor or only a risk factor, depending on the product and according to the commercial decisions made by the insurance provider.

---

[1] The term "factor" should not be confused with its statistical meaning for categorical features: an insurance risk factor may well be continuous, such as age or power of the car.

[2] Notice that enhanced annuities sold in the United States and United Kingdom offer higher benefits in case of impaired health status.

Example 1.2.3 (Geographic pricing) It is common in insurance of domestic property lines, such as householders' fire insurance, for instance, to let the risk premium per unit exposure vary with geographic area when all other risk factors are held constant. In motor insurance, most companies have adopted a risk classification according to the geographical zone where the policyholder lives (urban, sub-urban or rural area for instance). The spatial variation may be related to geographic factors (like traffic density or proximity to arterial roads in motor insurance) or to socio-demographic factors (perhaps affecting theft rates in homeowner's insurance). The technical premium reflects the spatial variation in claim experience given the geographical location contained in the postcode.[3] Thus, the geographic effect obtained from the technical price list is generally simplified before its integration to the commercial price list and this generates a difference between geographic location as a risk or as a rating factor.

In this book, we only discuss the determination of cost-based technical premiums incorporating all the relevant information in order to

– counteract adverse selection,
– monitor the premium transfers created inside the portfolio by the application of the commercial price list, and
– evaluate customer's value, by comparing technical premiums to commercial ones.

These are just three examples of the usefulness of keeping track of technical premiums besides commercial ones.

1.2.3   Top-Down Pricing

Actuaries usually combine two approaches:

– an analysis performed at individual policy level
– supplemented with a collective analysis at the portfolio level.

This is because an insurance policy can never be considered in isolation but always in relation with the portfolio in which it is comprised (the diversification gained by aggregating individual risks into portfolios being at the heart of insurance business). Reinsurance or coinsurance may provide substitutes to risk diversification for portfolios of limited volume.

Since insurers sell promises, the insurance activity has always been heavily regulated to ensure solvency, i.e. the insurer's capacity to pay for future claims. This is

---

[3]Technical pricing now tends to become even finer, with premiums sometimes varying by street or district within each postcode.

why regulators impose that insurers hold enough capital to fulfill their promises to policyholders. This capital can only be computed at the portfolio level, accounting for the risk diversification among policies.

Precisely, top-down pricing proceeds as follows. A forecast is first made at the portfolio level, to predict overall future claim costs given market cycles, claim-specific inflation, loss reserving dynamics, and so on. Using an appropriate risk measure, the amount of risk capital is then determined, leading to the portfolio premium. Notice that the portfolio is considered as a collective in this analysis and the detailed individual characteristics of policies are generally not taken into account.

An individual analysis is then performed to identify risky profiles. The output of this second stage is used to allocate the resulting total premium income among different policies: the relative riskiness of each policy (with respect to some reference policy) is determined and the resulting percentage is applied to the portfolio premium.

In this book, we only consider the individual analysis performed at policy level to identify the relevant risk factors, not the assessment of the appropriate level of capital needed to ensure solvency.

1.2.4 Prevention

Besides cost predictions, technical pricing is also helpful for prevention. By identifying the risk factors and the accident-prone behavior (behind the wheel for instance, with the help of telematics devices), insurers may inform policyholders and help them to diminish their claim frequency and/or severity.

In several countries, insurers now disclose part of their technical analysis to demonstrate that their price list appropriately reflects the underlying claim experience. This is sometimes imposed by regulatory authorities, or may be done to increase the transparency of the pricing system and ease its acceptance by customers, showing why a particular profile must pay more for the same coverage, compared to others.

It is worth to mention the GiveDataBack facility developed by AXA. Throughout a digital portal available at `https://givedataback.axa/`, consumers can access AXA's claims and policy data from Germany, Belgium, Spain, United Kingdom, Italy, Switzerland and France. By entering an address on the portal, consumers can find out the probability of theft or water damage at that location as well as expected frequencies and severities and the months in which they are most likely to happen. The portal also recommends how best to reduce risk exposure and suggests how to limit the impact of claims. This kind of initiative is certainly beneficial as it increases transparency and eases acceptance of risk classification.

1.3   The Risk Classification Problem

1.3.1   Insurance Risk Diversification

Insurance companies cover risks (that is, random financial losses) by collecting premiums. Premiums are generally paid in advance (hence their name). The pure premium is the amount collected by the insurance company, to be re-distributed as benefits among policyholders and third parties in execution of the contract, without loss nor profit. Under the conditions of validity of the law of large numbers, the pure premium is the expected amount of compensation to be paid by the insurer (sometimes discounted to policy issue in case of long-term liabilities).

This can easily be understood as follows. Consider independent and identically distributed random variables $Y_1, Y_2, Y_3, \ldots$ representing the total amount of benefits paid by the insurer in relation to individual policies, over one period. Then,

$$\overline{Y}_n = \frac{1}{n} \sum_{i=1}^{n} Y_i$$

represents the average payment per policy in a portfolio of size n.

Let F be the common distribution function of the risks Y_i, that is,

$$F(y) = P[Y_1 \leq y].$$

Recall that the mathematical expectation of Y_1 is computed from

$$E[Y_1] = \int_0^{\infty} y \, dF(y).$$

We assume that $E[Y_1]$ exists and is finite. The law of large numbers then ensures that

$$\overline{Y}_n \to E[Y_1] \text{ as } n \to \infty, \text{ with probability } 1.$$

Hence, the expected amount of benefit $E[Y_1]$ can be seen as the average cost per policy in an infinitely large homogeneous portfolio and thus appears to be the best candidate for calculating the pure premium.

Notice that the pure premiums are just re-distributed among policyholders to pay for their respective claims, without loss nor profit on average. Hence, they cannot be considered as insurance prices because loadings must be added to face operating cost, in order to ensure solvency, to cover general expenses, to pay commissions to intermediaries, to generate profit for stockholders, not to mention the taxes imposed by the local authorities.

1.3.2 Why Classifying Risks?

In practice, most portfolios are heterogeneous: they mix individuals with different risk levels. Some policyholders tend to report claims more often or to report more expensive claims, on average. In an heterogeneous portfolio with a uniform price list, the financial result of the insurance company depends on the composition of the portfolio as shown in the next example.

Example 1.3.1 If young policyholders are proved to report significantly smaller losses than older ones and if a company disregards this risk factor in its commercial pricing and charges an average premium to all policyholders regardless of age, most of its younger policyholders are tempted to move to another company recognizing this age difference by offering better premiums at young ages. The former company is then left with a disproportionate number of older policyholders and insufficient premium income to cover the reported losses (because premiums have been tailored to the previous mix between young and old policyholders so that older policyholders enjoy an undeserved premium reduction at the expenses of the younger ones). The premiums collected are no more sufficient to cover the losses. As a consequence, the tariff must be revised to reach the level of the competitor.

Notice that the recognition of the age difference in premiums can be

– explicit, in case the former insurer charges different premiums according to age, adopting the same risk classification than its competitor;
– or implicit, in case the former insurer raises its premium to the amount corresponding to older ages, discouraging younger policyholders to buy insurance from this provider.

In the latter situation, the market as a whole has also integrated the age difference in premiums but the former company now targets a specific segment of the insured population whereas the segmented tariff of the competitor allows it to cover different age profiles.

The modification in the composition of the portfolio may thus generate losses for the insurer charging a uniform premium to different risk profiles, when competitors distinguish premiums according to these profiles. Policyholders who are over-priced by the insurance tariff leave the insurer to enjoy premium discounts offered by the competitors whereas those who appear to have been under-priced remain with the insurer. This change in the portfolio composition generates systematic losses for the insurer applying the uniform tariff. In theory, each time a competitor uses an additional rating factor, the actuary has to refine the partition to avoid loosing the less risky policyholders with respect to this factor. This phenomenon is known as adverse selection, as policyholders are supposed to select the insurance provider offering them the best premium.

This (partly) explains why so many factors are used by insurance companies: insurance companies have to use a rating structure matching the premiums for the risks as closely as the rating structures used by competitors. If they do not, they become exposed to the risk of loosing the policyholders who are currently over-priced according to their tariff, breaking the equilibrium between expected losses and collected premiums. Of course things are more complicated in reality, because of limited information available to policyholders who also pay attention to other elements than just the amount of premium, such as the quality of services, the brand or the perceived financial solidity of the insurer, for instance, not to mention the natural tendency to remain with the same insurance provider.

To sum up, an insurance company may well charge the same amount of premiums to policyholders with different risk profile but it is then at risk to loose the over-priced one. This is one of the reasons why the technical price list must be as accurate as possible: it is only in this way that the insurer is able to manage its portfolio effectively, by knowing which profiles are over-priced and which ones subsidize the others. In other words, the insurer knows the value of each policy in the portfolio.

1.3.3   The Need for Regression Models

Considering that risk profiles differ inside insurance portfolios, it theoretically suffices to subdivide the entire portfolio into homogeneous risk classes, i.e. groups of policyholders sharing the same risk factors, and to determine an amount of pure premium specific to each risk class. However, if the data are subdivided into risk classes determined by many factors, actuaries often deal with sparsely populated groups of contracts. Even in a large portfolio, for instance of motor insurance policies it becomes clear that every policyholder becomes essentially unique when individual policies are partitioned by means of

– demographic factors like age or gender for instance;
– socio-economic factors, like occupation, place of residence or income for instance;
– insured risk characteristics, like power of the car or type of vehicle for instance;
– insurance package characteristics, like compulsory third-party liability only, or supplemented with other optional coverages such as material damage or theft, or choice of deductibles for instance;
– behavioral characteristics collected with the help of telematics devices installed in the vehicle for instance;
– external data basis recording credit scores or banking information, or the history of safety offenses for instance;
– past experience in relation with the different guarantees subscribed, sometimes combining all products at the household level.

Therefore, simple averages become useless and regression models are needed.

Regression models predict a response variable from a function of risk factors and parameters. This approach is also referred to as supervised learning. By connecting the different risk profiles, a regression analysis can deal with highly segmented problems resulting from the massive amount of information about the policyholders that has now become available to the insurers.

Notice that the question about the existence of insurance business at the "big data era" has been questioned. As it will be seen below, the increasing predictive accuracy of the mean loss amount impacts on the premium differentials among policyholders. But the insurer still covers the individual variations around this expected value and the volatility of individual risks remains huge in most insurance lines, even if a lot of information is integrated in the pricing model. Hence, the insurance business remains viable. This is especially the case when large claims sometimes occur, like in third-party liability coverages where bodily injuries often cause very expensive loss amounts. Another aspect making the insurance business attractive for policyholders is that the insurance company does not only pay for the claim, indemnifying the victims in place of the policyholders, but it also takes care of the claim settlement process. In a tort system, this service is certainly an important value added of the insurance activities. Last but not least, insurers also provide assistance to their policyholders, or other services (direct repair of covered damages, for instance).

1.3.4 Observable Versus Hidden Risk Factors

Some risk factors can easily be observed, such as the policyholder's age, gender, marital status or occupation, the type and use of the car, or the place of residence for instance. Other ones can be observed but subject to some effort or cost. This is typically the case with behavioral characteristics reflected in telematics data or information gathered in external databases that can be accessed by the insurer for a fee paid to the provider. But besides these observable factors, there always remain risk factors unknown to the insurer. In motor insurance for instance, these hidden risk factors typically include temper and skills, aggressiveness behind the wheel, respect of the highway code or swiftness of reflexes (even if telematics data now help insurers to figure out these behavioral traits, but only after contract inception).

Henceforth, we denote as

X = random vector gathering the observable risk factors used by the insurer
(not necessarily in causal relationship with the response Y)

X^+ = hidden risk factors influencing the risk, in addition to X
(in causal relationship with the response Y).

Notice that some components of X may become irrelevant once X^+ is available. The reason is as follows. Some components of X may not be directly related to Y but merely correlated to X^+. This means that X only brings some indirect information about Y.

Example 1.3.2 As an example, think about the number of children or marital status in motor insurance. Being engaged in a stable relationship or having one or several children does not improve nor worsen someone's driving abilities. But this generally increases his or her degree of risk aversion so that the individual starts driving more carefully. This means that the number of children enters X and because the degree of risk aversion is typically included in X^+ and impacts on Y, this feature appears to be indirectly related to Y. This kind of phenomenon has always to be kept in mind when analyzing insurance data.

1.3.5 Risk Sharing in Segmented Tariff: Random Versus Subsidizing Solidarity

Let Y be the total amount of benefit paid by the insurer in relation with a policyholder whose characteristics are (X, X^+). Pure premiums correspond to mathematical expectations under the assumptions of the law of large numbers, so that they must be conditional to the information used for pricing. Thus, the pure premium computed by the insurer is $E[Y|X]$ whereas the correct pure premium is $E[Y|X, X^+]$. In addition to insurance risk, the company is also subject to imperfect risk classification, quantified by the variation in the true premium $E[Y|X, X^+]$ not explained by X. This can be seen by decomposing the variance of Y as follows:

$$\text{Var}[Y] = E\big[\text{Var}[Y|X]\big] + \text{Var}\big[E[Y|X]\big].$$

The first term represents the part of the variance absorbed by the insurer: having created risk classes based on X only, the variations of Y around the premium $E[Y|X]$ inside each risk class (i.e. the variance of Y given X) are compensated by the insurer. As the premium varies according to X, the second term quantifies the variance borne by the policyholders who are charged different premium amounts according to their profile, i.e. they pay $E[Y|X]$ varying in function of X.

Now, the true premium depends on X^+, not only on X. The first term can therefore be further split to get

$$\text{Var}[Y] = \underbrace{E\Big[\text{Var}[Y|X, X^+]\Big]}_{\substack{\text{random solidarity} \\ \text{mutuality principle}}} + \underbrace{E\Big[\text{Var}\big[E[Y|X, X^+]\big|X\big]\Big]}_{\substack{\text{subsidizing solidarity} \\ \text{imperfect risk classification}}}$$

$$\xrightarrow{\text{insurance company}}$$

$$+ \underbrace{\text{Var}\Big[E[Y|X]\Big]}_{\rightarrow \text{policyholder}}.$$

It is important to realize that given (X, X^+), the variations in Y are purely random, i.e. only attributable to chance. They must therefore be compensated by the insurer according to the mutuality principle (which is the very essence of insurance activities). The mutuality principle is perfectly reflected in the motto "The contribution of the many to the misfortune of the few": the premiums collected among a large homogeneous group of policyholders are then re-distributed in favor of those few individuals who suffered some adverse events. As long as the group is homogeneous with respect to X^+ (or may be considered to be homogeneous, because of a lack of information for instance), the insurer just charges the same premium to every member. The mutuality principle is also sometimes referred to as "random solidarity" because the variability in loss amounts can only be attributed to chance, or pure randomness. This would be the case if the insurer would be able to use X^+.

As X^+ is not available, variations of Y given X partly reflect residual heterogeneity, i.e. differences in risk profiles despite sharing the same X. This means that some policyholders among those with observed features X are over-priced and thus subsidize other ones who are under-priced. This creates subsidizing solidarity due to imperfect risk classification. Precisely, the true insurance premium $E[Y|X, X^+]$ cannot be charged to the policyholder with risk profile (X, X^+), because X^+ is hidden, so that the insurer also absorbs the variations in the true premiums that cannot be explained by its rating structure based on X.

Charging $E[Y|X]$ instead of $E[Y|X, X^+]$ means that some policyholders with observed profile X pay too much, other ones paying not enough, depending on the value of $E[Y|X]$ compared to $E[Y|X, X^+]$. This is because policyholders sharing the same observed risk profile X may differ in terms of hidden risk factors X^+, as we just explained. This creates some subsidizing effect: part of the premium paid by those who are over-priced by the tariff based on X is transferred to cover the benefits granted to those who are under-priced. The global balance remains nevertheless in equilibrium as

$$E\big[E[Y|X, X^+]|X\big] = E[Y|X]$$

but this identity requires that the composition of the portfolio in terms of X (that is, the distribution of X in the portfolio under consideration) remains stable of time. This must be carefully monitored by the insurer. Problems arise when policyholders get knowledge about X^+, so that they gain an information advantage over the insurer, or when a competitor gets access to X^+ and targets some segments of the market that are over-priced when using X, only.

1.3.6 Insurance Ratemaking Versus Loss Prediction

Consider a response Y and a set of features $X_1, \ldots, X_p$ gathered in the vector X. Features are considered here as random variables so that they are denoted by capital letters. This means that we are working with a generic policyholder, taken at random from (and thus representative of) the portfolio under consideration. When it comes to pricing a specific contract, we work conditionally to the realized value of X, that is, given $X = x$.

The dependence structure inside the random vector $(Y, X_1, \ldots, X_p)$ is exploited to extract the information contained in X about Y. In actuarial pricing, the aim is to evaluate the pure premium as accurately as possible. This means that the target is the conditional expectation $\mu(X) = E[Y|X]$ of the response Y (claim number or claim amount) given the available information X. Henceforth, $\mu(X)$ is referred to as the true (pure) premium.

Notice that the function $x \mapsto \mu(x) = E[Y|X = x]$ is generally unknown to the actuary, and may exhibit a complex behavior in x. This is why this function is approximated by a (working, or actual) premium $x \mapsto \pi(x)$ with a relatively simple structure compared to the unknown regression function $x \mapsto \mu(x)$.

For many models considered in insurance studies, the features X are combined to form a score S (i.e. a real-valued function of the features $X_1, \ldots, X_p$) and the premium π is a monotonic, say increasing, function of S. Formally,

$$\pi(X) = h(S)$$

for some increasing function h.

Some models may include several scores, each one capturing a particular facet of the cost transferred to the insurer. This is the case for instance with zero-augmented regression models where the probability mass in zero and the expected cost when claims have been filed are both modeled by a specific score. This is the topic of Chap. 7.

1.3.7 A Priori Versus a Posteriori Variables

Actuaries must distinguish a priori and a posteriori variables entering their analysis. A priori information is available before (or prior to) the start of the yearly coverage period. Examples include policyholder's age and gender or the characteristics of the vehicle, that are known from the underwriting process in motor insurance. A posteriori information becomes gradually available during the coverage period, after the contract has been issued. Such variables include claims experience (numbers and cost of claims) as well as behavioral characteristics (revealed by connected objects or telematic devices installed in the insured vehicle), as well as their past values.

A priori information is available for all policyholders in the same, standardized way whereas a posteriori information is available in various quantities:

- not available for new risks (newly licensed drivers, for instance);
- available in abridged form for new contracts (certificate recording claims in tort over the past 5 years in motor third-party liability, for instance);
- cost of claims only available for claiming policyholders;
- behavior behind the wheel only measured when policyholders drive.

Since regression models target the conditional distribution of the response, given the a priori features, the latter are treated as known constants. The dynamics of a priori features is generally not modeled (if, when and where the policyholder moves, for instance) but some assumptions may sometimes be needed in that respect, such as in customer value calculation for example.

A posteriori variables are modeled dynamically, jointly with claim experience. Formally, this information is included in pricing by means of predictive distributions (i.e. credibility corrections, experience rating) which appear to be specific to each policy, accounting for the volume and reliability of own past experience.

As explained in Chap. 5, the unobservable risk characteristics X^+ are represented as latent variables (or random effects) in actuarial pricing models, in the vein of credibility theory. Observed values of the responses in the past are then used to refine a priori risk classification, beyond the differentials induced by X. Denoting as $Y^{\leftarrow}$ the past claims experience, the premium then becomes $E[Y|X, Y^{\leftarrow}]$. Of course, the information contained in $Y^{\leftarrow}$ varies from one policyholder to another: for new ones, no information is available, whereas those covered for several years have random vectors $Y^{\leftarrow}$ of different lengths according to the time spent in the portfolio (i.e. the seniority of the insurance policy). Thus, the format of $Y^{\leftarrow}$ differs from one policyholder to another. This is why $Y^{\leftarrow}$ is not treated as the other features but requires a specific treatment in a second stage of risk classification. This is the topic of credibility theory that can be implemented with the help of the mixed models developed in the statistical literature and available in statistical software.

1.4 Insurance Data

1.4.1 Claim Data

Because of adverse selection, most actuarial studies are based on insurance-specific data, generally consisting in claims data. It is important to realize that claims are made at the policyholder's discretion. Sometimes, an accident is not reported to the insurer because the policy conditions comprise an experience rating mechanism: if filing a claim increases future premiums then the policyholder may consider that it is cheaper to defray minor accidents to escape financial penalties imposed by the insurer (not to mention the possible cancellation of the policy). Sometimes, even if the claim is reported, it may be rejected by the company because of policy conditions. This is the case for instance when the claim amount falls below the deductible stated in the policy. Generally, such events are not recorded in the database (or may be recorded as a zero claim, but often not systematically so that such cases must be considered with great care if they are included in the analysis).

Dealing with claim data thus means that only limited information is available about events that actually occurred. Analyzing insurance data, the actuary is able to draw conclusions about the number of claims filed by policyholders subject to a specific ratemaking mechanism (bonus-malus rules or deductibles, for instance), not about the actual number of accidents. The conclusions of the actuarial analysis are valid only if the existing rules are kept unchanged. The effect of an extension of coverage (decreasing the amount of deductibles, for instance) is extremely difficult to assess.

Also, some policyholders may not report their claims immediately, for various reasons (for instance because they were not aware of the occurrence of the insured event), impacting on the available data. Because of late reporting, the observed number of claims may be smaller than the actual number of claims for recent observation periods. Once reported, claims require some time to be settled. This is especially the case in tort systems, for liability claims. This means that it may take years before the final claim cost is known to the insurer.

The information recorded in the data basis generally gathers one or several calendar years. The data are as seen from the date of extraction (6 months after the end of the observation period, say). Hence, most of the "small" claims are settled and their final cost is known. However, for the large claims, actuaries can only work with incurred losses (payments made plus reserve, the latter representing a forecast of the final cost still to be paid according to the evaluation made by the claim manager). Incurred losses are routinely used in practice but a better approach would consist in recognizing that actuaries only have partial knowledge about the claim amount.

1.4.2 Propensities

Propensity models, also called likelihood-to-buy models, aim to predict the likelihood of a certain type of customer purchasing behavior, like whether a customer browsing a website is likely to buy something. In insurance applications, propensity models help the actuaries to analyze the success rate in attracting new business. A file containing a record for each offer to new customers and whether or not the policy has been issued is used in that respect. Identifying the factors explaining this binary response helps the insurer to identify the segments where its premiums are not competitive.

Propensity models are also useful to measure retention rates. Analyzing renewal retention rates can provide an understanding of how price-sensitive (or elastic) the retention of a business can be. Here, the response is equal to 1 if the policy is renewed and to 0 if it is not. The analysis is based on a file containing one record for each renewal offered.

Throughout this book, we use the vocable propensity to refer to a binary 0–1 random variable with 0 coding failure or absence while 1 corresponds to success or presence. In insurance applications, such binary outcomes correspond to the occurrence or non-occurrence of certain events, such as

– whether the policyholder reports at least one claim during the coverage period,
– whether the policyholder dies during the coverage period,
– whether a submitted claim is fraudulent,

for instance. For such responses, the analyst is interested in the probability that the event will occur. Propensity models of this type are also sometimes called incidence models.

1.4.3 Frequency-Severity Decomposition

1.4.3.1 Claim Numbers

Even if the actuary wants to model the total claim amount Y generated by a policy of the portfolio over one period (typically, one year), this random variable is generally not the modeling target. Indeed, modeling Y does not allow to study the effect of per-claim deductibles nor bonus-malus rules, for instance. Rather, the total claim amount Y is decomposed into

$$Y = \sum_{k=1}^{N} C_k$$

where

$$N = \text{number of claims}$$
$$C_k = \text{cost (or severity) of the } k \text{ thclaim}$$
$$C_1, C_2, \dots \text{ identically distributed}$$

all these random variables being independent. By convention, the empty sum is zero, that is,

$$N = 0 \Rightarrow Y = 0.$$

Alternatively, Y can be represented as

$$Y = N \times \overline{C} \text{ with } \overline{C} = \frac{1}{N} \sum_{k=1}^{N} C_k,$$

with the convention that

$$Y = N = 0 \Rightarrow \overline{C} = 0.$$

The frequency component of Y refers to the number N of claims filed by each policyholder during one period. Considering the number of claims reported by a policyholder in Property and Casualty insurance, the Poisson model and its extensions are generally taken as a starting point. In life insurance, the Binomial model is useful to graduate general population death rates. These models are thoroughly presented in Chap. 2.

Generally, the different components of the yearly insurance losses Y are modeled separately. Costs may be of different magnitudes, depending on the type of the claim: standard, or attritional claims, with moderate costs versus large claims with much higher costs. If large claims may occur then the mix of these two types of claims is explicitly recognized by

$$C_k = \begin{cases} \text{large claim cost, with probability } p, \\ \text{attritional claim cost, with probability } 1 - p. \end{cases}$$

1.4.3.2 Claim Amounts

Having individual costs for each claim, the actuary often wishes to model their respective amounts (also called claim sizes or claim severities in the actuarial literature). Prior to the analysis, the actuary first needs to exclude possible large claims, keeping only the standard, or attritional ones. Traditionally, the famous Bar-le-Duc case occurred on March 18, 1976, in France, has been considered as the prototype for a large claim.

On that day, a goods train hit a 2CV Citroën at a railway crossing, at full speed, destroyed the car completely and came off the rails on a 100 m stretch. Fortunately, there were no fatalities as passengers of the car managed to leave the vehicle before the crash, but damages were a record 30 millions French francs. Claims were made

- by the French railway company SNCF for one engine type BB15011, 21 railroad cars and about 100 m of destroyed rails.
- by third parties for thousands of Kronenbourg beer cans and Knorr soup packs.
- and for business interruption: the rail connection between Paris and Strasbourg remained disrupted for 10 days, a detour of 200 km had to be made by 60 coaches and trucks every day. Also, the traffic on the nearby canal was laid off for some time.
- Even the fishermen's club "Bar-le-Duc" claimed the loss of hundreds of kilos of fish of various sorts because of overfeeding with Kronenbourg beer and Knorr soup.

Even if this event has been for decades considered as the prototype for large claim, nowadays, even larger claims with more dramatic consequences are faced by insurance companies. The way large claims are dealt with in the framework of Extreme Value Theory described in Chap. 9. Broadly speaking, attritional claims have probability density function that decay exponentially to 0 when their argument grows whereas those for large claims decrease polynomially (i.e. at a much slower pace) to 0. An example of the latter behavior is furnished by the Pareto distribution that is often used in actuarial studies.

Overall, the modeling of claim amounts is more difficult than claim frequencies. There are several reasons for that. First and foremost, claims sometimes need several years to be settled as explained before. Only estimates of the final cost appear in the insurer's records until the claim is closed. Moreover, the statistics available to fit a model for claim severities are much more scarce, since generally only 5–10% of the policies in the portfolio produced claims. Finally, the unexplained heterogeneity is sometimes more pronounced for costs than for frequencies. The cost of a traffic accident for instance is indeed for the most part beyond the control of a policyholder since the payments of the insurance company are determined by third-party characteristics. The degree of care exercised by a driver mostly influences the number of accidents, but in a much lesser way the cost of these accidents.

1.4.4 Observational Data

Statistical analyzes are conducted with data either from experimental or from observational studies. In the former case, random assignment of individual units (humans or animals, for instance) to the experimental treatments plays a fundamental role to draw conclusions about causal relationships (to demonstrate the usefulness of a new drug, for instance). This is however not the case with insurance data, which consist of observations recorded on past contracts issued by the insurer.

As an example, let us consider motor insurance. The policyholders covered by a given insurance company are generally not a random sample from the entire population of drivers in the country. Each company targets a specific segment of this population (with advertisement campaigns or specific product design, for instance) and attracts particular profiles. This may be due to consumers' perception of insurer's products, sales channels (brokers, agents or direct), not to mention the selection operated by the insurer, screening the applicants before accepting to cover their risks.

In insurance studies, we consider that the portfolio is representative of future policyholders, those who will stay insured by the company or later join the portfolio. The assumption that new policyholders conform with the profiles already in the portfolio needs to be carefully assessed as any change in coverage conditions or in competitors' price lists may attract new profiles with different risk levels (despite they are identical with respect to X, they may differ in X^+, due to adverse selection against the insurer).

The actuary has always to keep in mind the important difference existing between causal relationships and mere correlations existing among the risk factors and the number of claims or their severity. Such correlations may have been produced by a causal relationship, but could also result from confounding effects. The following simple example illustrates this situation.

Example 1.4.1 Consider motor insurance and assume that having children (observed feature with two levels "no child" versus "≥ 1 child") does not influence claim propensities but the degree of risk aversion, or prudence behind the wheel (hidden characteristic) does. To simplify the situation, assume that there are only two categories of drivers, those who are risk averse and those who are risk prone, or risk seeker. Now, let us assume that the number of claims N produced by an insured driver is such that

$$P[N \geq 1 | \text{risk prone}, \geq 1\text{child}] = P[N \geq 1 | \text{risk prone, no child}]$$
$$= P[N \geq 1 | \text{risk prone}]$$
$$= 0.15$$

and

$$P[N \geq 1 | \text{risk averse}, \geq 1\text{child}] = P[N \geq 1 | \text{risk averse, no child}]$$
$$= P[N \geq 1 | \text{risk averse}]$$
$$= 0.05.$$

Formally, having children is independent of the event "$N \geq 1$", given the degree of risk aversion. This statement is just about conditional independence: without information about risk aversion, these two quantities might well be correlated, as shown next.

In order to generate the announced correlation, assume that the distribution of risk-averse drivers inside the portfolio differs according to the number of children. Specifically,

$$P[\text{risk prone} | \text{no child}] = 1 - P[\text{risk averse} | \text{no child}]$$
$$= 0.9$$
$$P[\text{risk prone} | \geq 1\text{child}] = 1 - P[\text{risk averse} | \geq 1\text{child}]$$
$$= 0.1.$$

This induces some correlation between the degree of risk aversion and the number of children. In other words, having children now brings some information about the degree of risk aversion. Without knowledge about the degree of risk aversion, the number of children thus indirectly influences the claim propensity. This is clearly seen from

$$P[N \geq 1 | \text{no child}] = P[N \geq 1 | \text{risk prone, no child}]P[\text{risk prone} | \text{no child}]$$
$$+ P[N \geq 1 | \text{risk averse, no child}]P[\text{risk averse} | \text{no child}]$$
$$= 0.15 \times 0.9 + 0.05 \times 0.1$$
$$= 0.14$$

whereas

$$P[N \geq 1 | \geq 1\text{child}] = P[N \geq 1 | \text{risk prone}, \geq 1\text{child}]P[\text{risk prone} | \geq 1\text{child}]$$
$$+ P[N \geq 1 | \text{risk averse}, \geq 1\text{child}]P[\text{risk averse} | \geq 1\text{child}]$$
$$= 0.15 \times 0.1 + 0.05 \times 0.9$$
$$= 0.06.$$

Even if the presence of children does not influence the claim propensity once the degree of risk aversion behind the wheel has been accounted for, unconditionally there is some dependence between these two variables.

Therefore, we cannot conclude from a marginal analysis that having no child causes an increase in the claim propensity. In reality, these two variables are both related to the degree of risk aversion behind the wheel. Of course, the situation is much more complicated in reality because having children may also increase the distance traveled (to drive them to school or to their various activities, picking them up late night at their parties). Not to mention that among children there may be young drivers who occasionally borrow their parents' car, causing an increase in claim frequencies. None of these effects are controlled so that they all become confounded in the study performed by the actuary.

Because of possible confounding effects as illustrated in the previous example, the actuary has always to keep in mind that it is generally not possible to disentangle

– a true effect of a risk factor
– from an apparent effect resulting from correlation with hidden characteristics

on the basis of observational data. Also, the effect estimated from portfolio statistics is the dominant one: different stories may apply to different policyholders whereas they are all averaged in the estimates obtained by the actuary.

Notice that correlation with hidden risk factors may even reverse the influence of an available risk factor on the response. This is the case for instance when the feature is negatively correlated with the response but positively correlated with a hidden characteristic, the latter being positively related to the response. The actuary may then observe a positive relationship between this feature and the response, despite the true correlation is negative. This is the essence of Simpson's paradox, which shows that the sign of the association between a risk factor and a response may change when the actuary corrects for the effect of another risk factor. This phenomenon will be considered in the next chapters, when discussing omitted variable bias effects.

Remark 1.4.2 Even if a correlation exists between the rating factor and the risk covered by the insurer, there may be no causal relationship between that factor and the risk. Requiring that insurance companies establish such a causal relationship to be allowed to use a rating factor may drastically change risk classification schemes. Now that consumers require a high level of transparency and that every form of discrimination tends to banned in the European Union, this may become a serious issue in the future if the regulators start considering that correlation is not enough to allow insurers to implement premium segmentation but that causality must be established to that end.

1.4.5 Format of the Data

The data required to perform analyses carried out in this book generally consist of linked policy and claims information at the individual risk level. The appropriate definition of individual risk level varies according to the line of business and the type of study. For instance, an individual risk generally corresponds to a vehicle in motor insurance or to a building in fire insurance.

The database must contain one record for each period of time during which a policy was exposed to the risk of filling out a claim, and during which all risk factors remained unchanged. A new record must be created each time risk factors change, with the previous exposure curtailed at the point of amendment. The policy number then allows the actuary to track the experience of the individual risks over time. Policy cancellations and new business also result in the exposure period to be curtailed. For each record, the database registers policy characteristics together with the number of claims and the total incurred losses. In addition to this policy file, there is a claim file recording all the information about each claim, separately (the link between the two files being made using the policy number). This second file also contains specific features about each claim, such as the presence of bodily injuries, the number of victims, and so on. This second file is interesting to build predictive models for the cost of claims based on the information about the circumstances of each insured event. This allows the insurer to better assess incurred losses.

The information available to perform risk classification is summarized into a set of features x_{ij}, $j = 1, \ldots, p$, available for each risk i. These features may have different formats:

– categorical (such as gender, with two levels, male and female);
– integer-valued, or discrete (such as the number of vehicles in the household);
– continuous (such as policyholder's age).

Categorical covariates may be ordered (when the levels can be ordered in a meaningful way, such as education level) or not (when the levels cannot be ranked, think for instance to marital status, with levels single, married, cohabiting, divorced, or widow, say).

Notice that continuous features are generally available to a finite precision so that they are actually discrete variables with a large number of numerical values. For example, age last birthday (i.e. age rounded from below) is often recorded in the database and used in ratemaking so that age could be considered as an ordered categorical feature. However, as the number of age categories is substantial and as a smooth progression of losses with age is expected, age last birthday is generally treated as a continuous feature using the methods presented in Chap. 6. Models for continuous data usually ignore this measurement precision.

When the x_{ij} are numeric, integer or continuous, they can directly enter the calculations. However, qualitative data must first be processed before entering calculations. Precisely, categorical features are transformed into one or several dummies, indicator or binary features, as shown in the next example.

Example 1.4.3 Suppose the area of residence of a policyholder is considered to be a potential risk factor in a model for claim frequency. If there are three areas, A, B, and C, say, then two indicator (dummy) variables x_{i1} and x_{i2} are needed to represent the available information concerning policyholder i. Specifically, x_{i1} and x_{i2} are defined as

$$x_{i1} = \begin{cases} 1 \text{ if policyholder } i \text{ resides in area A,} \\ 0 \text{ otherwise,} \end{cases}$$

and
$$x_{i2} = \begin{cases} 1 \text{ if policyholder } i \text{ resides in area B,} \\ 0 \text{ otherwise.} \end{cases}$$

The coding of the categorical feature can be summarized in the following table:

	x_{i1}	x_{i2}
Policyholder i resides in A $\Rightarrow$	1	0
Policyholder i resides in B $\Rightarrow$	0	1
Policyholder i resides in C $\Rightarrow$	0	0

No indicator for C is needed as it corresponds to $x_{i1} = x_{i2} = 0$: the left-out category C is called the base, or reference level. It corresponds to the most populated area (for reasons that will become clear in the next chapters).

This coding is automatically performed by statistical softwares when dealing with categorical features, so that the actuary does not have to manually define the associated dummies. However, it is important to keep in mind that every categorical feature enters the model as a collection of unrelated dummies, in order to properly interpret the results of the statistical analyses.

1.4.6 Data Quality Issues

As in most actuarial textbooks, we assume here that the available data are reliable and accurate. This assumption hides a time-consuming step in every actuarial study, during which data are gathered, checked for consistency, cleaned if needed and sometimes connected to external data bases to increase the volume of information. Setting up the data basis often takes the most time and does not look very rewarding. Data preparation is however of crucial importance because, as the saying goes, "garbage in, garbage out": there is no hope to get a reliable technical price list from a database suffering many limitations.

Once data have been gathered, it is important to spend enough time on exploratory data analysis. This part of the analysis aims at discovering which features seem to influence the response, as well as subsets of strongly correlated features. This traditional, seemingly old-fashioned view may well conflict with the modern data science approach, where practitioners are sometimes tempted to put all the features in a black-box model without taking the time to even know what they mean. But we firmly believe that such a blind strategy can sometimes lead to disastrous conclusions in insurance pricing so that we strongly advise to dedicate enough time to discover the kind of information recorded in the data basis under study.

Data exploration using appropriate graphical displays and tabulations is a first step in model building aimed to discover

(i) relationships between the response and features, suggesting explanatory variables to be included in the analysis, and their likely effects on the response.
(ii) relationships among features, highlighting which features are associated and thus encode the same information.

The understanding of the correlation structure among the available features is essential for model building as strongly related features are included in a model with care.

Exploratory data analysis is generally based on one-way analyses (sometimes supplemented with two-way ones, to detect possible interactions). In a one-way analysis, the effect of a particular feature on claims is studied without taking account of the effect of other features contained in the database. The major flaw with one-way analyses is thus that they can be distorted by correlations: premium differentials reflected by one-way analyses may double-count the effect of some features.

This first step can be automated, creating standardized reports describing the main effect of each feature. It may rise interesting questions about the actual meaning of negative claims (where the third party is liable for the accident, for instance) or zero claims (i.e. claims closed without payment). It is important to check whether such claims have been consistently recorded in the database. Even if these claims appear to be costless in terms of insurance benefits, it is important to model them for allocating operating expenses. Such claims should be excluded from the cost-based analysis.

Another common situation is a disproportionate number of fixed-amount claims. They may result from particular settlement rules, for instance the automatic imputation of a fixed incurred loss amount to claims which have just been notified. They may also be attributable to particular market mechanisms to speed up the settlement of small claims: insurers sometimes agree to pay each other a fixed, average amount in case the claim falls under the conditions of being treated in this simple, effective way. It can also be due to business characteristics. For instance, windscreen claims may all have about the same costs.

Additional features can be added to the data basis to account for some specific conditions, if necessary. For instance, if the actuary combines data from several companies, or from several sales channels, indicators variables can be used to distinguish among these subgroups. Such indicators can also be used to absorb some temporary phenomena that are not expected to repeat in the future. Of course, it is always preferable to use data devoid of such disruptive effects.

Databases are generally based on a one or several calendar-accident year periods. The experience gathered about policies during these accident years is recorded in the data basis. For instance, if the data basis relates to calendar years 2017–2018, this means that all policies with anniversary between January the first of year 2016 and December 31, 2019, having been in force at least one day during calendar years 2017–2018 appear in the data basis. The claims under consideration are those occurred between January the first of 2017 until December 31, 2018. Incurred losses are evaluated as of mid-2019 (typically, the end of June 2019).

Considering calendar year as an additional feature allows the actuary to track possible trends over time when longer observation periods are considered. Notice that the calendar year indicator may also account for some delays in claim reporting: in case some claims are filed only later, this will typically impact the last observation periods to a larger extent and be reflected in the calendar year indicator. It is worth to mention that indicator variables corresponding to quarters or months may involve an element of seasonality. In motor insurance for instance, accidents occur more often during winter months because of weather conditions, that may vary from one year to the next.

Recall that incurred losses sum up paid amounts and a prediction of the future amounts to be paid in relation to the particular claim. This is why it is appropriate to wait some time after the end of the observation period, to allow for claims to be reported and for the case estimates to develop. Incurred losses should be based on the most recent reserve estimates, waiting until the last review of case estimates before building the database. Incurred losses should not be truncated according to any large threshold at this stage, to allow the actuary to perform sensitivity analysis testing of different large loss thresholds.

1.5   Bibliographic Notes and Further Reading

This chapter summarizes the main aspects of insurance data and loss models. We refer the reader to Meyers and Cummings (2009) for a clear exposition of the very aim of ratemaking, which is not to predict the actual losses Y but to create accurate estimates of $\mu(X)$, which is unobserved. In Trufin et al. (2019), we explain how to assess the relative performances of premium candidates $\pi(X)$ with reference to the true pure premium $\mu(X)$.

Besides marginal, or one-way analyses, as well as low-dimensional descriptive statistics (considering couples of features to discover some interaction effects, for example), there are techniques allowing the actuary to discover the features which seem to have the highest impact on the mean response. Appropriate testing procedures can be used (Pechon et al. 2019) as well as unsupervised learning techniques (Hainaut et al. 2019).

Also, pure premium calculations only deal with expected values so that this chapter confines to this particular statistical indicator. For risk management purposes, higher-order moments, quantiles and tail-behavior for instance, are also important. These aspects are not covered in this book but we refer the reader to Kaas et al. (2008) and Denuit et al. (2005) for more details.

References

Denuit M, Dhaene J, Goovaerts MJ, Kaas R (2005) Actuarial theory for dependent risks: measures, orders and models. Wiley, New York
Hainaut D, Trufin J, Denuit M (2019) Effective statistical learning methods for actuaries—neural networks and unsupervised methods. Springer Actuarial Series

Kaas R, Goovaerts MJ, Dhaene J, Denuit M (2008) Modern actuarial risk theory using R. Springer, New York

Meyers G, Cummings AD (2009) "Goodness of Fit" vs. "Goodness of Lift". Actuarial Rev 36:16–17

Pechon F, Denuit M, Trufin J (2019) Preliminary selection of risk factors in P&C ratemaking. Variance, in press

Trufin J, Denuit M, Hainaut D (2019) Effective statistical learning methods for actuaries—tree-based methods. Springer Actuarial Series

Chapter 2
Exponential Dispersion (ED) Distributions

2.1 Introduction

All models in this book aim to analyze the relationship between a variable whose outcome needs to be predicted and one or more potential explanatory variables. The variable of interest is called the response and is denoted as Y. Insurance analysts typically encounter non-Normal responses such as

$$Y = \begin{cases} N \text{ number of claims} \\ I \text{ claim propensity } I[N \geq 1] \\ C_k \text{ cost per claim} \\ \overline{C} \text{ average cost per claim} \\ S \text{ yearly total claim cost} \\ T \text{ time-to-event, such as lifetime} \\ \text{etc.} \end{cases}$$

Claim counts, claim sizes or claim propensities on a single policy do not obey the Normal distribution, even not approximately. This is either because the probability mass is concentrated on a small set of integer values, or because the probability density function is not symmetric but positively skewed. Generally, the variance increases with the mean response in insurance applications. To account for these data particularities, actuaries then often select the distribution of the response from the exponential dispersion (or ED) family.

Before proceeding with the thorough study of the ED family, let us recall some basic probability concepts. Claim numbers are modeled by means of non-negative integer-valued random variables (often called counting random variables). Such random variables are described by their probability mass function: given a counting random variable Y valued in the set $\{0, 1, 2, \ldots\}$ of non-negative integers, its probability mass function p_Y is defined as

$$y \mapsto p_Y(y) = P[Y = y], \ y = 0, 1, 2, \ldots$$

© Springer Nature Switzerland AG 2019
M. Denuit et al., *Effective Statistical Learning Methods for Actuaries I*,
Springer Actuarial, https://doi.org/10.1007/978-3-030-25820-7_2

and we set p_Y to zero otherwise. The support S of Y is defined as the set of all values y such that $p_Y(y) > 0$. Expectation and variance[1] are then respectively given by

$$E[Y] = \sum_{y=0}^{\infty} y p_Y(y) \text{ and } \mathrm{Var}[Y] = \sum_{y=0}^{\infty} \left(y - E[Y]\right)^2 p_Y(y).$$

Claim amounts are modeled by non-negative continuous random variables possessing a probability density function. Precisely, the probability density function f_Y of such a random variable Y is defined as

$$y \mapsto f_Y(y) = \frac{\mathrm{d}}{\mathrm{d}y} P[Y \leq y], \quad y \in (-\infty, \infty).$$

In this case,

$$P[Y \approx y] = P\left[y - \frac{\Delta}{2} \leq Y \leq y + \frac{\Delta}{2}\right] \approx f_Y(y)\Delta$$

for sufficiently small $\Delta > 0$, so that f_Y also indicates the region where Y is most likely to fall. In particular, $f_Y = 0$ where Y cannot assume its values. The support S of Y is then defined as the set of all values y such that $f_Y(y) > 0$. Expectation and variance are then respectively given by

$$E[Y] = \int_{-\infty}^{\infty} y f_Y(y) \mathrm{d}y \text{ and } \mathrm{Var}[Y] = \int_{-\infty}^{\infty} \left(y - E[Y]\right)^2 f_Y(y) \mathrm{d}y.$$

Even if the Normal distribution itself is not a good candidate for modeling responses of interest in insurance studies, it still plays a central role in the analyses performed using ED distributions. This is why we start from this basic probability model to proceed with the definition of the ED family, and the study of its properties.

2.2 From Normal to ED Distributions

2.2.1 Normal Distribution

2.2.1.1 Normal Probability Density Function

The oldest distribution for errors in a regression setting is certainly the Normal distribution, also called Gaussian, or Gauss-Laplace distribution after its inventors.

---

[1] In this book, we assume that all moments of the random variables under consideration exist and are finite when they are written (except in the Log-Gamma, or Pareto case, this is always the case with distributions discussed here).

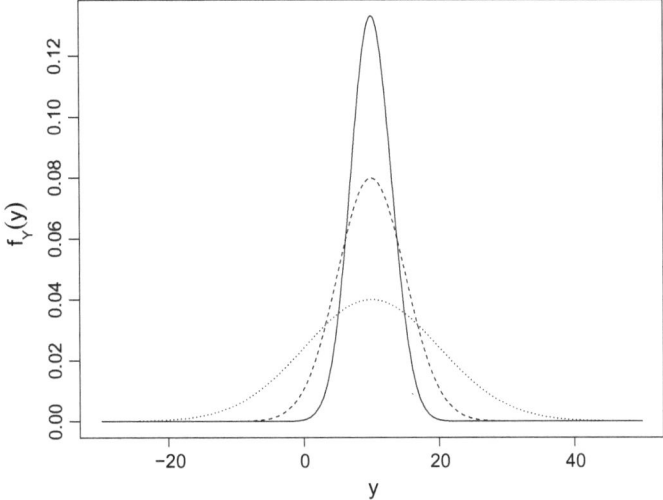

Fig. 2.1 Probability density functions of $Nor(10, 3^2)$ in continuous line, $Nor(10, 5^2)$ in broken line, and $Nor(10, 10^2)$ in dotted line

The family of ED distributions in fact extends the nice structure of this probability law to more general errors.

Recall that a response Y valued in $S = (-\infty, \infty)$ is Normally distributed with parameters $\mu \in (-\infty, \infty)$ and $\sigma^2 > 0$, denoted as $Y \sim Nor(\mu, \sigma^2)$, if its probability density function f_Y is

$$f_Y(y) = \frac{1}{\sigma\sqrt{2\pi}} \exp\left(-\frac{1}{2\sigma^2}(y - \mu)^2\right), \quad y \in (-\infty, \infty). \tag{2.1}$$

Considering (2.1), we see that Normally distributed responses can take any real value, positive or negative as $f_Y > 0$ over the whole real line $(-\infty, \infty)$.

Figure 2.1 displays the probability density function (2.1) for different parameter values. The $Nor(\mu, \sigma^2)$ probability density function appears to be a symmetric bell-shaped curve centered at μ, with σ^2 controlling the spread of the distribution. The probability density function f_Y being symmetric with respect to μ, positive or negative deviations from the mean μ have the same probability to occur. To be effective, any analysis based on the Normal distribution requires that the probability density function of the data has a shape similar to one of those visible in Fig. 2.1, which is rarely the case in insurance applications.

2.2.1.2 Moments

If $Y \sim Nor(\mu, \sigma^2)$ then $E[Y] = \mu$ and $Var[Y] = \sigma^2$. We see that the variance σ^2 is not related to the mean μ. This is a location-scale family as

$$Y \sim Nor(\mu, \sigma^2) \Leftrightarrow Z = \frac{Y - \mu}{\sigma} \sim Nor(0, 1).$$

The whole family of Normal distributions can thus be obtained from the standard Normal distribution $Nor(0, 1)$. The corresponding distribution function is henceforth denoted as Φ, i.e.

$$\Phi(y) = P[Z \le y] = \frac{1}{\sqrt{2\pi}} \int_{-\infty}^{y} \exp\left(-\frac{1}{2}z^2\right) dz, \quad y \in (-\infty, \infty).$$

If $Y \sim Nor(\mu, \sigma^2)$ then

$$P[Y \le y] = P\left[\frac{Y - \mu}{\sigma} \le \frac{y - \mu}{\sigma}\right] = \Phi\left(\frac{y - \mu}{\sigma}\right), \quad y \in (-\infty, \infty).$$

The Normal distribution enjoys the convenient convolution stability property: the sum of independent, Normally distributed random variables remain Normally distributed. This result remains valid as long as both random variables are jointly Normal (as it will be discussed in Chap. 4).

2.2.1.3 Central-Limit Theorem

The Normal distribution naturally arises from the central-limit theorem, as the approximate distribution of a sum of sufficiently many independent and identically distributed random variables with finite variances (and this fundamental result explains why the Normal distribution remains useful in the context of the ED family). Recall that according to the central-limit theorem, having independent and identically distributed random variables $Y_1, Y_2, Y_3, \ldots$ with finite mean μ and variance σ^2, the standardized average

$$\frac{\overline{Y}_n - \mu}{\sigma/\sqrt{n}} = \frac{\sum_{i=1}^{n} Y_i - n\mu}{\sigma\sqrt{n}}$$

approximately obeys the $Nor(0, 1)$ distribution provided n is large enough. Formally,

$$P\left[\frac{\sum_{i=1}^{n} Y_i - n\mu}{\sigma\sqrt{n}} \le z\right] \to \Phi(z) \text{ for all } z \text{ as } n \to \infty.$$

2.2.2 ED Distributions

Let us now rewrite the $\mathcal{N}or(\mu, \sigma^2)$ probability density function to demonstrate how to extend it to a larger class of probability distributions sharing some convenient properties: the ED family. The idea is as follows. The parameter of interest in insurance pricing is the mean μ involved in pure premium calculations. This is why we isolate components of the Normal probability density function where μ appears. This is done by expanding the square appearing inside the exponential function in (2.1), which gives

$$f_Y(y) = \frac{1}{\sigma\sqrt{2\pi}} \exp\left(-\frac{1}{2\sigma^2}(y^2 - 2y\mu + \mu^2)\right)$$

$$= \exp\left(\frac{y\mu - \frac{\mu^2}{2}}{\sigma^2}\right) \frac{\exp\left(-\frac{y^2}{2\sigma^2}\right)}{\sigma\sqrt{2\pi}}. \tag{2.2}$$

The second factor appearing in (2.2) does not involve μ so that the important component is the first one. We see that it has a very simple form, being the exponential (hence the vocable "exponential" in ED) of a ratio with the variance σ^2, i.e. the dispersion parameter, appearing in the denominator. The numerator appears to be the difference between the product of the response y and the canonical Normal mean parameter μ with a function of μ, only. Notice that the derivative of this second term $\frac{\mu^2}{2}$ is just the mean μ. Such a decomposition allows us to define the whole ED class of distributions as follows.

Definition 2.2.1 Consider a response Y valued in a subset S of the real line $(-\infty, \infty)$. Its distribution is said to belong to the ED family if Y obeys a probability mass function p_Y or a probability density function f_Y of the form

$$\left.\begin{array}{c}p_Y(y) \\ f_Y(y)\end{array}\right\} = \exp\left(\frac{y\theta - a(\theta)}{\phi/\nu}\right) c(y, \phi/\nu), \quad y \in S, \tag{2.3}$$

where

θ = real-valued location parameter, called the canonical parameter

ϕ = positive scale parameter, called the dispersion parameter

ν = known positive constant, called the weight

$a(\cdot)$ = monotonic convex function of θ

$c(\cdot)$ = positive normalizing function.

In the majority of actuarial applications, the weight corresponds to some volume measure, hence the notation ν.

The parameters θ and ϕ are essentially location and scale indicators, extending the mean value μ and variance σ^2 to the whole family of ED distributions. Considering (2.2), we see that it is indeed of the form (2.3) with

$$\theta = \mu$$
$$a(\theta) = \frac{\mu^2}{2} = \frac{\theta^2}{2}$$
$$\phi = \sigma^2$$
$$\nu = 1$$
$$c(y, \phi) = \frac{\exp\left(-\frac{y^2}{2\sigma^2}\right)}{\sigma\sqrt{2\pi}}.$$

Remark 2.2.2 Sometimes, (2.3) is replaced with the more general form

$$\exp\left(\frac{y\theta - a(\theta)}{b(\phi, \nu)}\right) c(y, \phi, \nu).$$

The particular case where ϕ and ν are combined into ϕ/ν, i.e.

$$b(\phi, \nu) = \frac{\phi}{\nu} \text{ and } c(y, \phi, \nu) = c\left(y, \frac{\phi}{\nu}\right)$$

appears to be enough for actuarial applications and is kept throughout this book.

Remark 2.2.3 The function $a(\cdot)$ is sometimes called the cumulant function. This terminology can be explained as follows. Consider a continuous, say, random variable Z valued in S with probability density function $f_Z(z) = c(z)$ and moment generating function

$$m_Z(t) = \mathrm{E}[\exp(tZ)] = \int_S \exp(tz)c(z)\mathrm{d}z.$$

Define $a(\theta) = \ln m_Z(\theta)$, the latter function being referred to as the cumulant generating function of Z. Then,

$$\int_S \exp(\theta z)c(z)\mathrm{d}z = \exp\left(a(\theta)\right)$$

or

$$\int_S \exp\left(\theta z - a(\theta)\right)c(z)\mathrm{d}z = 1 \text{ for all } \theta.$$

Hence,

$$z \mapsto \exp\left(\theta z - a(\theta)\right)c(z)$$

defines an ED distribution with canonical parameter θ. Such a construction is often used in applied probability and called an exponential tilting of Z. Notice that for $\theta = 0$ we simply recover f_Z.

2.2.3 Some Distributions Related to the Normal

Before exploring the ED family, let us briefly recall the definition of some standard distributions closely related to the Normal one.

2.2.3.1 LogNormal Distribution

In actuarial applications, the Normal distribution is often used on the log scale, i.e. assuming that $\ln Y \sim Nor(\mu, \sigma^2)$. The response Y is then said to be LogNormally distributed. The corresponding probability density function is given by

$$f_Y(y) = \frac{1}{y\sigma\sqrt{2\pi}} \exp\left(-\frac{1}{2\sigma^2}(\ln y - \mu)^2\right), \quad y > 0.$$

The LogNormal distribution is often used for modeling claim sizes. However, the prior transformation of the data to the logarithmic scale may bias the prediction unless some appropriate corrections are implemented. The main interest in the use of the ED family is to avoid this preliminary transformation and work with the original, untransformed response Y.

If $\ln Y \sim Nor(\mu, \sigma^2)$ then $\exp(\mu)$ is not the mean but the median, as

$$P[Y \leq \exp(\mu)] = P[\ln Y \leq \mu] = \frac{1}{2}.$$

The mean and variance are respectively given by

$$E[Y] = \exp\left(\mu + \frac{\sigma^2}{2}\right),$$

and

$$\text{Var}[Y] = \left(\exp(\sigma^2) - 1\right)\exp\left(2\mu + \sigma^2\right) = \left(\exp(\sigma^2) - 1\right)\left(E[Y]\right)^2.$$

Hence, we see that the variance is a quadratic function of the mean. As $\exp(\sigma^2) > 1$, the variance increases with the mean.

The skewness coefficient γ of a random variable is defined as follows: given a random variable Y with mean μ and variance σ^2,

$$\gamma[Y] = \frac{E[(Y - \mu)^3]}{\sigma^3}.$$

If the probability density function of Y is symmetric (as in the Normal case) then $\gamma[Y] = 0$. The larger $|\gamma[Y]|$, the more asymmetric is the distribution. For the Log-Normal distribution, the skewness coefficient

$$\gamma[Y] = \left(\exp(\sigma^2) + 2\right)\sqrt{\exp(\sigma^2) - 1}$$

is always positive and only depends on σ^2, not on μ.

2.2.3.2 Chi-Square Distribution

The Chi-Square distribution with k degrees of freedom corresponds to the distribution of a sum

$$Y = Z_1^2 + Z_2^2 + \ldots + Z_k^2$$

of independent $\mathcal{N}or(0, 1)$ random variables $Z_1, Z_2, \ldots, Z_k$, squared. Henceforth, we denote $Y \sim \chi_k^2$ in such a case.

Let us give the probability density function of the Chi-Square distribution. Recall that the Gamma function $\Gamma(\cdot)$ is defined for $\alpha \geq 0$ as the integral

$$\Gamma(\alpha) = \int_0^\infty \exp(-z)z^{\alpha-1}dz.$$

Clearly, $\Gamma(0) = 1$ and an integration by parts shows that

$$\Gamma(\alpha) = (\alpha - 1)\Gamma(\alpha - 1) \text{ for any } \alpha > 1.$$

Thus, the Gamma function may be thought of as a continuous version of the factorial function in that when α is a non-negative integer,

$$\Gamma(\alpha + 1) = \alpha! = \prod_{j=0}^{\alpha-1}(\alpha - j).$$

The probability density function of the Chi-Square distribution with k degrees of freedom is given by

$$f_Y(y) = \frac{y^{k/2-1}\exp(-y/2)}{2^{k/2}\Gamma(k/2)}, \quad y > 0.$$

The Chi-Square distribution is used primarily in hypothesis testing and not for direct modeling of responses. It arises in the likelihood-ratio test for nested models, for instance.

2.2.3.3 Student's t-Distribution

Student's t-distribution (or simply t-distribution) is related to the mean of Normally distributed responses in situations where the sample size is small and population standard deviation is unknown. Specifically, let $Y_1, \ldots, Y_n$ be independent random variables obeying the $\textit{Nor}(\mu, \sigma^2)$ distribution and define

$$\overline{Y} = \frac{1}{n} \sum_{i=1}^{n} Y_i \text{ and } S^2 = \frac{1}{n-1} \sum_{i=1}^{n} (Y_i - \overline{Y})^2.$$

These quantities are referred to as the sample mean and the sample variance. Then,

$$\frac{\overline{Y} - \mu}{\sigma/\sqrt{n}} \sim \textit{Nor}(0, 1)$$

whereas the random variable

$$T = \frac{\overline{Y} - \mu}{S/\sqrt{n}}$$

has the Student's t-distribution with $n - 1$ degrees of freedom. The corresponding probability density function is

$$f_T(t) = \frac{\Gamma(n/2)}{\sqrt{(n-1)\pi}\,\Gamma(\frac{n-1}{2})} \left(1 + \frac{t^2}{n-1}\right)^{-n/2}, \quad t \in (-\infty, \infty).$$

The Student's t probability density function is symmetric, and its overall shape resembles the bell shape of a Normally distributed random variable with mean 0 and variance 1, except that it has thicker tails. It becomes closer to the Normal distribution as the number of degrees of freedom increases.

Student's t-distribution with d degrees of freedom can be generally defined as the distribution of the random variable

$$T = \frac{Z}{\sqrt{C/d}}$$

where $Z \sim \textit{Nor}(0, 1)$, $C \sim \chi_d^2$, Z and C being independent.

2.2.3.4 Fisher Distribution

Consider two independent Chi-Square random variables C_1 and C_2 with d_1 and d_2 degrees of freedom, respectively. The F-distribution with parameters d_1 and d_2 corresponds to the ratio

$$\frac{C_1/d_1}{C_2/d_2}.$$

This distribution naturally appears in some testing procedures.

2.3   Some ED Distributions

2.3.1   Gamma Distribution

2.3.1.1   Gamma Probability Density Function

The Gamma distribution is right-skewed, with a sharp peak and a long tail to the right. These characteristics are often visible on empirical distributions of claim amounts. This makes the Gamma distribution a natural candidate for modeling accident benefits paid by the insurer.

Precisely, a random variable Y valued in $S = (0, \infty)$ is distributed according to the Gamma distribution with parameters $\alpha > 0$ and $\tau > 0$, which will henceforth be denoted as $Y \sim \mathcal{G}am(\alpha, \tau)$, if its probability density function is given by

$$f_Y(y) = \frac{y^{\alpha-1}\tau^{\alpha}\exp(-\tau y)}{\Gamma(\alpha)}, \quad y > 0. \tag{2.4}$$

The parameter α is often called the shape of the Gamma distribution whereas τ is referred to as the scale parameter.

2.3.1.2   Moments

The mean and the variance of $Y \sim \mathcal{G}am(\alpha, \tau)$ are respectively given by

$$E[Y] = \frac{\alpha}{\tau} \text{ and } Var[Y] = \frac{\alpha}{\tau^2} = \frac{1}{\alpha}(E[Y])^2. \tag{2.5}$$

We thus see that the variance is a quadratic function of the mean, as in the LogNormal case. The Gamma distribution is useful for modeling a positive, continuous response when the variance grows with the mean but where the coefficient of variation

$$CV[Y] = \frac{\sqrt{Var[Y]}}{E[Y]} = \frac{1}{\sqrt{\alpha}}$$

stays constant. As their names suggest, the scale parameter in the Gamma family influences the spread (and incidentally, the location) but not the shape of the distribution, while the shape parameter controls the skewness of the distribution. For $Y \sim \mathcal{G}am(\alpha, \tau)$, we have

$$\gamma[Y] = \frac{2}{\sqrt{\alpha}}$$

so that the Gamma distribution is positively skewed. As the shape parameter gets larger, the distribution grows more symmetric.

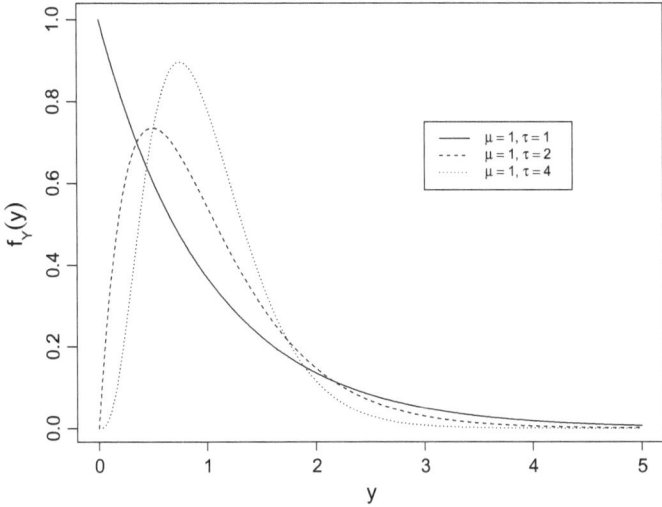

Fig. 2.2 Probability density functions of $\mathcal{G}am(\alpha, \tau)$ with $a = \tau \in \{1, 2, 4\}$

Figure 2.2 displays Gamma probability density functions for different parameter values. Here, we fix the mean $\mu = \frac{\alpha}{\tau}$ to 1 and we take $\tau \in \{1, 2, 4\}$ so that the variance is equal to $1, 0.5$, and 0.25. Unlike the Normal distribution (whose probability density function resembles a bell shape centered at μ whatever the variance σ^2), the shape of the Gamma probability density function changes with the parameter α. For $\alpha \leq 1$, the probability density function has a maximum at the origin whereas for $\alpha > 1$ it is unimodal but skewed. The skewness decreases as α increases.

Gamma distributions enjoy the convenient convolution stability property for fixed scale parameter τ. Specifically,

$$\left. \begin{array}{l} Y_1 \sim \mathcal{G}am(\alpha_1, \tau) \\ Y_2 \sim \mathcal{G}am(\alpha_2, \tau) \\ Y_1 \text{ and } Y_2 \text{ independent} \end{array} \right\} \Rightarrow Y_1 + Y_2 \sim \mathcal{G}am(\alpha_1 + \alpha_2, \tau). \tag{2.6}$$

2.3.1.3 Particular Cases: Negative Exponential, Erlang and Chi-Square

When $\alpha = 1$, the Gamma distribution reduces to the Negative Exponential one with probability density function

$$f_Y(y) = \tau \exp(-\tau y), \quad y > 0.$$

In this case, we write $Y \sim \mathcal{E}xp(\tau)$. This distribution enjoys the remarkable memoryless property:

$$P[Y > s + t | Y > s] = \frac{\exp(-\tau(s + t))}{\exp(-\tau s)}$$
$$= \exp(-\tau t) = P[Y > t].$$

This corresponds to a constant

$$\frac{f_Y(t)}{1 - F_Y(t)} = \tau$$

death, or failure rate: an individual with lifetime $Y \sim \mathcal{E}xp(\tau)$ does not age, being subject to a constant death rate τ throughout his entire lifetime.

When α is a positive integer, the corresponding Gamma distribution function is then given by

$$F(y) = 1 - \sum_{j=0}^{\alpha-1} \exp(-y\tau) \frac{(y\tau)^j}{j!}, \quad y \geq 0.$$

This particular case is referred to as the Erlang distribution. The Erlang distribution corresponds to the distribution of a sum

$$Y = Z_1 + Z_2 + \ldots + Z_\alpha$$

of independent random variables $Z_1, Z_2, \ldots, Z_\alpha$ with common $\mathcal{E}xp(\tau)$ distribution, by virtue of (2.6). Hence, when $\alpha = 1$ the Erlang distribution reduces to the Negative Exponential one.

The Chi-Square distribution with k degrees of freedom appears to be a particular case of the $\mathcal{G}am(\alpha, \tau)$ distribution with $\alpha = k/2$ and $\tau = 1/2$.

2.3.1.4   ED Form

Let us now establish that the Gamma distribution belongs to the ED family. To this end, we let the mean parameter $\mu = \alpha/\tau$ enter the expression of the probability density function and rewrite the $\mathcal{G}am(\alpha, \tau)$ probability density function (2.4) as follows:

$$f_Y(y) = \frac{\tau^\alpha}{\Gamma(\alpha)} y^{\alpha-1} \exp(-\tau y)$$
$$= \frac{\alpha^\alpha}{\Gamma(\alpha)} \mu^{-\alpha} y^{\alpha-1} \exp\left(-y\frac{\alpha}{\mu}\right) \text{ with } \mu = \frac{\alpha}{\tau} \Leftrightarrow \tau = \frac{\alpha}{\mu}$$
$$= \exp\left(\alpha\left(-\frac{y}{\mu} - \ln\mu\right)\right) \frac{\alpha^\alpha}{\Gamma(\alpha)} y^{\alpha-1}$$

which is well of the form (2.3) with

$$\theta = -\frac{1}{\mu}$$
$$a(\theta) = \ln \mu = -\ln(-\theta)$$
$$\phi = \frac{1}{\alpha}$$
$$c(y, \phi) = \frac{\alpha^{\alpha}}{\Gamma(\alpha)} y^{\alpha-1}.$$

The Gamma distribution thus belongs to the ED family.

2.3.1.5 Power-Gamma, or Weibull Distributions

The Weibull distribution is often used in reliability studies as it corresponds to the distribution of the minimum over a large number of independent random variables, each of them corresponding to the time until a different type of failure. Such an item thus fails as soon as one of its components does. The Weibull distribution has also been applied in life insurance studies, considering the human body as such a series system.

Consider $Y = Z^{1/\alpha}$ for some $\alpha > 0$ where $Z \sim \mathcal{E}xp(\tau)$. Then, Y obeys the Weibull distribution with probability density function

$$f_Y(y) = \alpha \tau y^{\alpha-1} \exp(-\tau y^{\alpha}), \quad y > 0.$$

The mean of the Weibull distribution is

$$E[Y] = \beta^{-1/\alpha} \Gamma\left(1 + \frac{1}{\alpha}\right)$$

and the variance is

$$\text{Var}[Y] = \beta^{-2/\alpha} \left(\Gamma\left(1 + \frac{2}{\alpha}\right) - \left(\Gamma\left(1 + \frac{1}{\alpha}\right)\right)^2\right).$$

The Weibull distribution is closely related to accelerated failure time modeling. Notice also that if Y is Weibull then $\ln Y = \frac{\ln Z}{\alpha}$ obeys the Extreme Value distribution (that will be studied in Chap. 9).

2.3.1.6 Log-Gamma, or Pareto Distributions

Log-Gamma distributions are useful to model heavy-tailed responses. Considering $\ln Y \sim \mathcal{E}xp(\xi)$, we have

$$P[Y \le y] = \begin{cases} 0, & \text{if } y \le 1, \\ 1 - y^{-\xi}, & \text{if } y > 1. \end{cases}$$

This model differs from the preceding ones in that the probability density functions decreases polynomially to 0, and not exponentially. This slower decrease to zero means that more probability mass can be located in the upper tail of the distribution. Such a response may therefore deviate to a much larger extent from its mean (and the latter may even become infinite when $\xi \leq 1$).

To get the Pareto distribution, it suffices to consider

$$Y = \tau\big(\exp(Z) - 1\big) \text{ with } Z \sim \mathcal{E}xp(\xi)$$

so that

$$
\begin{aligned}
P[Y > y] &= P\left[Z > \ln\left(1 + \frac{y}{\tau}\right)\right] \\
&= \left(1 + \frac{y}{\tau}\right)^{-\xi} \\
&= \left(\frac{\tau}{y + \tau}\right)^{\xi}.
\end{aligned}
$$

Henceforth, we denote as $\mathcal{P}ar(\alpha, \tau)$ the Pareto distribution with positive parameters α and τ. As mentioned before for the LogNormal case, the prior transformation to the logarithmic scale is not optimal from a statistical point of view.

The mean of the Pareto distribution is infinite if $\alpha \leq 1$. For $\alpha > 1$, the mean is equal to $\frac{\theta}{\alpha-1}$. The variance is defined for $\alpha > 2$, being equal to $\frac{\alpha\theta^2}{(\alpha-2)(\alpha-1)^2}$ in that case.

2.3.2 Inverse Gaussian Distribution

2.3.2.1 Inverse Gaussian Probability Density Function

The Inverse Gaussian distribution sometimes appears in the actuarial literature, as another good candidate for modeling positive, right-skewed data. Its properties resemble those of the Gamma and LogNormal distributions. Its name derives from the inverse relationship that exists between the cumulant generating function of the Inverse Gaussian distribution and the one of the Normal (or Gaussian) distribution.

Recall that a random variable Y valued in $\mathcal{S} = (0, \infty)$ is distributed according to the Inverse Gaussian distribution with parameters $\mu > 0$ and $\alpha > 0$, which will be henceforth denoted as $Y \sim \mathbb{I}\mathcal{G}au(\mu, \alpha)$, if its probability density function is given by

$$f_Y(y) = \sqrt{\frac{\alpha}{2\pi y^3}} \exp\left(-\frac{\alpha(y - \mu)^2}{2y\mu^2}\right), \quad y > 0. \tag{2.7}$$

The Inverse Gaussian distribution with $\mu = 1$ is known as the Wald distribution.

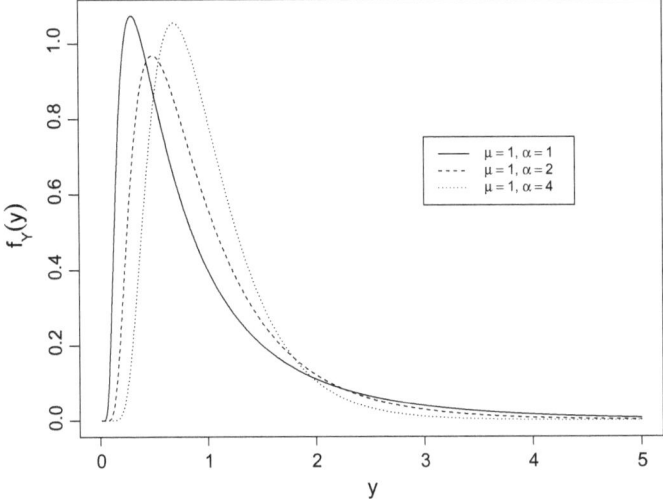

Fig. 2.3 Probability density functions of $I\!Gau(\mu, \alpha)$ with $\mu = 1$ and $\alpha \in \{1, 2, 4\}$

2.3.2.2 Moments

If $Y \sim I\!Gau(\mu, \alpha)$ then the moments are given by

$$E[Y] = \mu \text{ and } \mathrm{Var}[Y] = \frac{\mu^3}{\alpha} = \frac{1}{\alpha} \left(E[Y]\right)^3. \tag{2.8}$$

Like the Gamma distribution, the variance of the Inverse-Gaussian distribution increases with its mean, but at a faster rate (cubic instead of quadratic).

The Inverse Gaussian distribution has gained attention in describing and analyzing right-skewed data. The main appeal of Inverse Gaussian models lies in the fact that they can accommodate a variety of shapes, from highly skewed to almost Normal. Moreover, they share many elegant and convenient properties with Gaussian models. In applied probability, the Inverse Gaussian distribution arises as the distribution of the first passage time to an absorbing barrier located at a unit distance from the origin in a Wiener process (or Brownian motion).

Figure 2.3 displays Inverse Gaussian probability density functions for different parameter values. Here, we take the same mean and variances as in the Gamma case above, i.e. $\mu = 1$ and $\alpha = 1, 2, 4$.

2.3.2.3 ED Form

In order to show that the Inverse Gaussian distribution belongs to the ED family, let us rewrite the $I\!Gau(\mu, \alpha)$ probability density function (2.7) as

$$f_Y(y) = \sqrt{\frac{\alpha}{2\pi y^3}} \exp\left(-\frac{\alpha(y-\mu)^2}{2y\mu^2}\right)$$

$$= \exp\left(-\frac{\alpha(y^2 - 2y\mu + \mu^2)}{2y\mu^2}\right)\sqrt{\frac{\alpha}{2\pi y^3}}$$

$$= \exp\left(\frac{y\left(-\frac{1}{2\mu^2}\right) - \left(-\frac{1}{\mu}\right)}{1/\alpha}\right)\exp\left(-\frac{\alpha}{2y}\right)\sqrt{\frac{\alpha}{2\pi y^3}}$$

which is well of the form (2.3) with

$$\theta = -\frac{1}{2\mu^2}$$

$$a(\theta) = -\frac{1}{\mu} = -\sqrt{-2\theta}$$

$$\phi = \frac{1}{\alpha}$$

$$c(y, \phi) = \exp\left(-\frac{\alpha}{2y}\right)\sqrt{\frac{\alpha}{2\pi y^3}}.$$

The Inverse Gaussian distribution thus belongs to the ED family.

2.3.2.4 Comparison of LogNormal, Gamma, and Inverse Gaussian Distributions

It is natural to wish to compare the right-skewed probability density functions corresponding to the LogNormal, Gamma and Inverse Gaussian distributions. To this end, actuaries perform the comparison for identical mean and variance. Specifically, consider random variables X obeying the Gamma distribution, Y obeying the Inverse Gaussian distribution and Z obeying the LogNormal distribution, such that

$$E[X] = E[Y] = E[Z] \text{ and } \text{Var}[X] = \text{Var}[Y] = \text{Var}[Z].$$

Then, Kaas and Hesselager (1995, Theorem 3.1) established that the inequalities

$$E[(X - t)_+^2] \leq E[(Y - t)_+^2]$$
$$E[(X - t)_+^2] \leq E[(Z - t)_+^2]$$
$$E[(Y - t)_+^3] \leq E[(Z - t)_+^3]$$

hold for all $t \geq 0$. Thus, the inequalities

$$E[(X - t)_+^3] \leq E[(Y - t)_+^3] \leq E[(Z - t)_+^3] \text{ hold for all } t \geq 0,$$

showing that the tails of the LogNormal distribution are heavier compared to those of
the Inverse Gaussian distribution, which are themselves heavier compared to those
of the Gamma distribution.

As an example, Fig. 2.4 displays the three probability density functions with the
same mean 1 and variance 0.5. We can see there that the three probability density
functions indeed agree with this ranking for sufficiently large values of their argument.

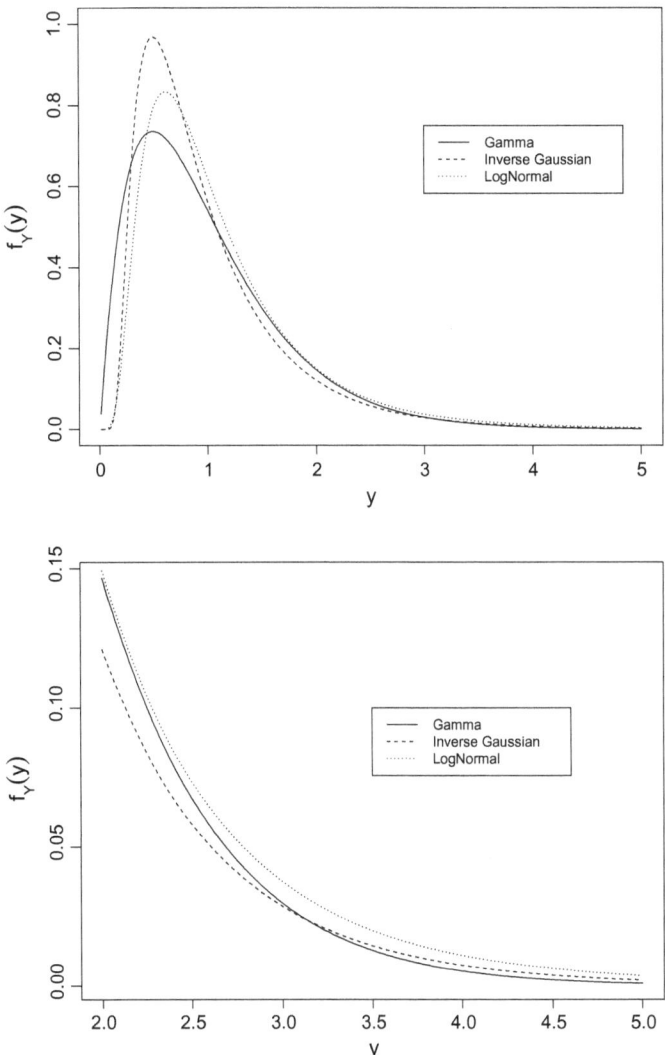

Fig. 2.4 Probability density functions of the LogNormal, the Gamma and the Inverse Gaussian
distributions with the same mean 1 and variance 0.5 (top panel) and zoom on the region [2, 5]
(bottom panel)

2.3.3   Binomial Distribution

2.3.3.1   Bernoulli Trial

Many counting distributions are defined by means of Bernoulli trials. In a Bernoulli
trial, a given experiment leading either to success or to failure is repeated under
identical conditions (so that previous outcomes do not influence subsequent ones).
Success means that some specific event E occurs. The outcome Y of a Bernoulli
trial is thus dichotomous, e.g., success or failure, alive or dead, presence or absence,
0 or 1, yes or no. The variable of interest is related to the number of successes. In
Bernoulli trials, the analyst is interested in the probability q that the event will occur.
This is in essence a propensity, or incidence model.

2.3.3.2   Bernoulli Distribution

The response Y valued in $\mathcal{S} = \{0, 1\}$ is Bernoulli distributed with success probability
q, which is denoted as $Y \sim \mathcal{B}er(q)$. The corresponding probability mass function is
given by

$$p_Y(y) = \begin{cases} 1 - q \text{ if } y = 0 \\ q \text{ if } y = 1 \\ 0 \text{ otherwise.} \end{cases}$$

There is thus just one parameter: the success probability q. The mean and variance
of $Y \sim \mathcal{B}er(q)$ are given by

$$\text{E}[Y] = q \text{ and } \text{Var}[Y] = q(1 - q). \tag{2.9}$$

Such a response Y is an indicator variable: for some event E of probability q,

$$Y = \text{I}[E] = \begin{cases} 1 \text{ if } E \text{ occurs} \\ 0 \text{ otherwise.} \end{cases}$$

More generally, we denote as $\text{I}[\cdot]$ the indicator function equal to 1 if the condition
appearing within the brackets is fulfilled, and to 0 otherwise.

To check whether the Bernoulli distribution belongs to the ED family, we have
to show that the corresponding probability mass function p_Y is of the form (2.3). To
this end, let us write

$$p_Y(y) = q^y(1 - q)^{1-y}$$
$$= \exp\left(y \ln \frac{q}{1 - q} + \ln(1 - q)\right),$$

which corresponds to the ED probability mass function (2.3) with

$$\theta = \ln \frac{q}{1-q}$$
$$a(\theta) = -\ln(1-q) = \ln(1+\exp(\theta))$$
$$\phi = 1$$
$$v = 1$$
$$c(y, \phi) = 1.$$

This shows that the Bernoulli distribution indeed belongs to the ED family.

2.3.3.3 Binomial Distribution

The Binomial distribution corresponds to the number of successes recorded from a sequence of m independent Bernoulli trials, each with the same probability q of success. Denoting as Y the number of successes valued in $S = \{0, 1, \ldots, m\}$, the probability that success is the result obtained in exactly y of the trials is

$$p_Y(y) = \binom{m}{y} q^y (1-q)^{m-y}, \quad y = 0, 1, \ldots, m, \tag{2.10}$$

and 0 otherwise, where the Binomial coefficient defined as

$$\binom{m}{y} = \frac{m!}{y!(m-y)!}$$

is the number of different ways to select the y successes out of the m trials. Formula (2.10) defines the Binomial distribution. There are now two parameters: the number of trials m (also called the exponent, or size) and the success probability q. Often, m is known to the actuary whereas q is the parameter of interest. Henceforth, we write $Y \sim Bin(m, q)$ to indicate that Y is Binomially distributed, with size m and success probability q. Notice that the definition of success with respect to failure is arbitrary in that

$$Y \sim Bin(m, q) \Leftrightarrow m - Y \sim Bin(m, 1-q).$$

The shape of the Binomial probability mass function is displayed in the graphs of Fig. 2.5 for $m = 100$ and increasing values of q. Precisely, Fig. 2.5 displays bar charts, where the height is equal to the probability mass located on the corresponding integer. For small q, we see that the probability mass function is highly asymmetric, concentrating the probability mass on the smaller outcomes. When the success probability gets larger, the Binomial probability mass function becomes more symmetric and its shape ultimately appears rather similar to (a discrete version of) the Normal density. This is a consequence of the central-limit theorem since the Binomial distribution corresponds to a sum of independent Bernoulli trials. The convergence to

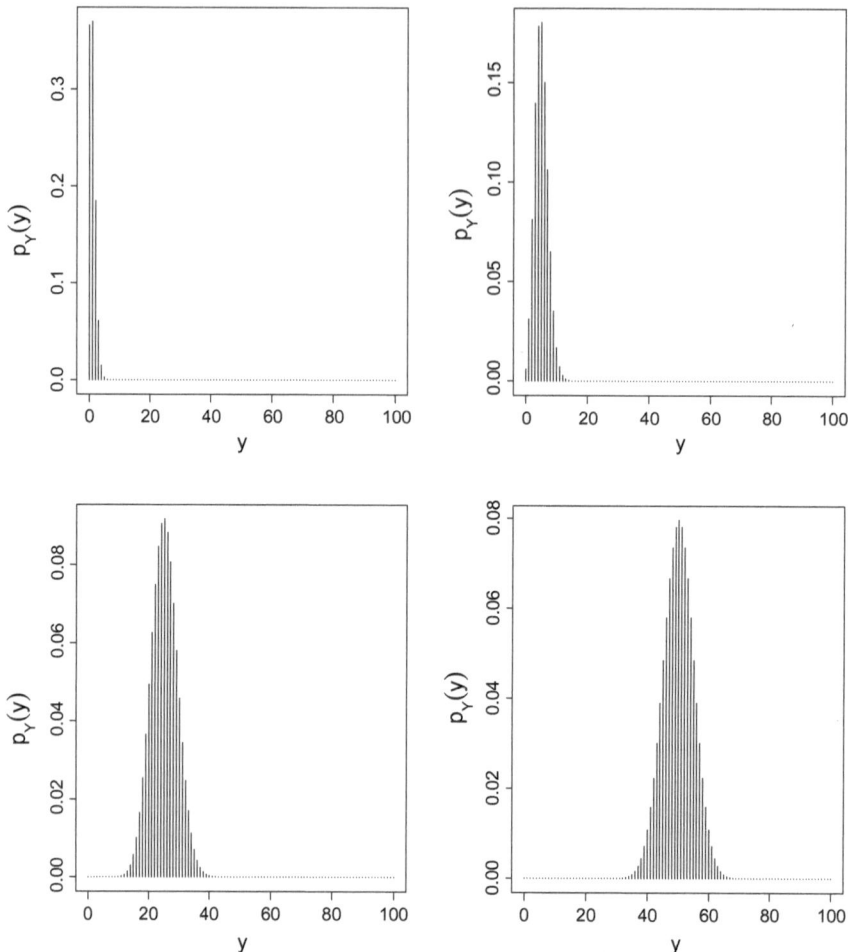

Fig. 2.5 Probability mass functions of $\mathcal{B}in(m, q)$ with $m = 100$ and $q = 0.01, 0.05, 0.25, 0.5$ (from upper left to lower right)

the limiting Normal distribution is faster when the Bernoulli outcomes become more symmetric, that is when q approaches 0.5.

Clearly,

$$\left.\begin{array}{r} Y_1 \sim \mathcal{B}in(m_1, q) \\ Y_2 \sim \mathcal{B}in(m_2, q) \\ Y_1 \text{ and } Y_2 \text{ independent} \end{array}\right\} \Rightarrow Y_1 + Y_2 \sim \mathcal{B}in(m_1 + m_2, q).$$

From (2.9), we then easily deduce that for $Y \sim \mathcal{B}in(m, q)$,

$$E[Y] = mq \text{ and } \mathrm{Var}[Y] = mq(1 - q). \qquad (2.11)$$

It is easy to see that the Binomial variance is maximum when $q = \frac{1}{2}$. We immediately observe that the Binomial distribution is underdispersed, i.e. its variance is smaller than its mean:

$$\mathrm{Var}[Y] = mq(1 - q) \leq E[Y] = mq.$$

The third central moment, measuring skewness, is given by $mq(1 - q)(1 - 2q)$. Hence, the Binomial distribution is symmetric when $q = \frac{1}{2}$. This can be seen from the last panel in Fig. 2.5. For all other values of q, the Binomial probability mass function is skewed (positively skewed for q smaller than 0.5, negatively skewed for q larger than 0.5).

Remark 2.3.1 Consider independent random variables $Y_i \sim \mathcal{B}er(q_i)$, $i = 1, 2, \ldots,$ m, and define $Y_\bullet = \sum_{i=1}^{m} Y_i$. The random variable $Y_\bullet$ does not obey the Binomial distribution because of unequal success probabilities across trials. The first moments of $Y_\bullet$ can be obtained as follows:

$$E[Y_\bullet] = \sum_{i=1}^{m} q_i = m\overline{q} \text{ with } \overline{q} = \frac{1}{m} \sum_{i=1}^{m} q_i$$

and

$$\begin{aligned}
\mathrm{Var}[Y_\bullet] &= \sum_{i=1}^{m} q_i(1 - q_i) \\
&= m\overline{q}(1 - \overline{q}) - m\sigma_q^2
\end{aligned}$$

where

$$\sigma_q^2 = \frac{1}{m} \sum_{i=1}^{m} q_i^2 - (\overline{q})^2$$

is the variance of the q_is. This means that the variance of $Y_\bullet$ is smaller than the variance $m\overline{q}(1 - \overline{q})$ of the Binomial distribution with size m and success probability $\overline{q}$ (having the same mean as $Y_\bullet$). In words, allowing individual success probability to vary produces less-than-standard Binomial variance. Stated differently, the Binomial distribution maximizes the variance of the sum of independent Bernoulli random variables holding the mean constant.

In order to show that the Binomial distribution $\mathcal{B}in(m, q)$ belongs to the ED family, we have to rewrite its probability mass function as follows:

$$p_Y(y) = \binom{m}{y} q^y (1-q)^{m-y}$$

$$= \exp\left(y \ln \frac{q}{1-q} + m \ln(1-q)\right) \binom{m}{y}$$

where we recognize the ED probability mass function (2.3) with

$$\theta = \ln \frac{q}{1-q}$$
$$a(\theta) = -m \ln(1-q) = m \ln(1 + \exp(\theta))$$
$$\phi = 1$$
$$\nu = 1$$
$$c(y, \phi) = \binom{m}{y}.$$

2.3.3.4 Geometric and Pascal Distributions

This distribution arises from Bernoulli trials by considering the number Y of failures before getting m successes, for some fixed positive integer value m. Now, the response Y is no more bounded from above, i.e. $S = \{0, 1, 2, \ldots\}$. Its probability mass function is

$$p_Y(y) = \binom{m+y-1}{y} q^m (1-q)^y, \quad y = 0, 1, 2, \ldots \tag{2.12}$$

This distribution is henceforth referred to as the Pascal distribution, named after its inventor.[2] This is denoted as $Y \sim \mathcal{P}as(m, q)$. In insurance applications, the Pascal distribution may be used to describe the number of claim files under investigation before discovering a given number m of fraudulent ones.

For $m = 1$ we get the Geometric distribution $\mathcal{G}eo(q)$ for which

$$p_Y(y) = q(1-q)^y, \quad y = 0, 1, 2, \ldots \tag{2.13}$$

The Geometric distribution can be seen as the result of the discretization of the Exponential one. Precisely, consider $Z \sim \mathcal{E}xp(\tau)$ and define its integer part as

$$Y = \lfloor Z \rfloor = \max\{\text{integer } k \text{ such that } k \leq Y\}.$$

Then,

$$P[Y = 0] = P[Z \leq 1] = 1 - \exp(-\tau)$$

---

[2]Sometimes, the Pascal distribution is called the Negative Binomial one, where "negative" refers to the counted failures. In this book, we reserve this name for the general case where the parameter m is allowed to assume positive, non-integer values. This extension will be discussed in Chap. 5.

and for $y \in \{1, 2, 3, \ldots\}$,

$$
\begin{aligned}
P[Y = y] &= P[Z < y + 1] - P[Z \leq y] \\
&= \exp(-y\tau) - \exp\big(-(y+1)\tau\big) \\
&= \exp(-y\tau)\big(1 - \exp(-\tau)\big)
\end{aligned}
$$

where we recognize the $\mathcal{G}eo(q)$ distribution with

$$
q = 1 - \exp(-\tau).
$$

As its continuous counterpart, the Geometric distribution enjoys the memoryless property. This can be seen from the discrete failure rate

$$
P[Y = y | Y \geq y] = 1 - q \text{ whatever } y \in \{0, 1, 2, \ldots\},
$$

that appears to be constant.

The mean and variance of $Y \sim \mathcal{P}as(m, q)$ are given by

$$
E[Y] = \frac{m(1-q)}{q} \text{ and } Var[Y] = \frac{m(1-q)}{q^2}.
$$

The variance is thus always larger than the mean. Furthermore, the third central moment is $(2 - q)/\sqrt{m(1-q)}$ so that the Pascal distribution is always positively skewed but becomes more symmetric as m increases. The shape of the Pascal probability mass function is displayed in the graphs of Fig. 2.6. We can see there that the $\mathcal{P}as(m, q)$ probability mass function is indeed always positively skewed but it becomes more symmetric for large values of m.

By definition, it is easily seen that Pascal distributions enjoy the convenient convolution stability property for fixed value of q, i.e.

$$
\left.\begin{aligned}
Y_1 &\sim \mathcal{P}as(m_1, q) \\
Y_2 &\sim \mathcal{P}as(m_2, q) \\
Y_1 \text{ and } Y_2 &\text{ independent}
\end{aligned}\right\} \Rightarrow Y_1 + Y_2 \sim \mathcal{P}as(m_1 + m_2, q). \tag{2.14}
$$

Every response Y obeying the $\mathcal{P}as(m, q)$ distribution can thus be seen as the sum of m independent $\mathcal{G}eo(q)$ distributed terms.

In order to show that the Pascal distribution belongs to the ED family, let us rewrite the probability mass function (2.12) as

$$
p_Y(y) = \exp\big(y \ln(1 - q) + m \ln(q)\big) \binom{m + y - 1}{y}
$$

which is of the form (2.3) with

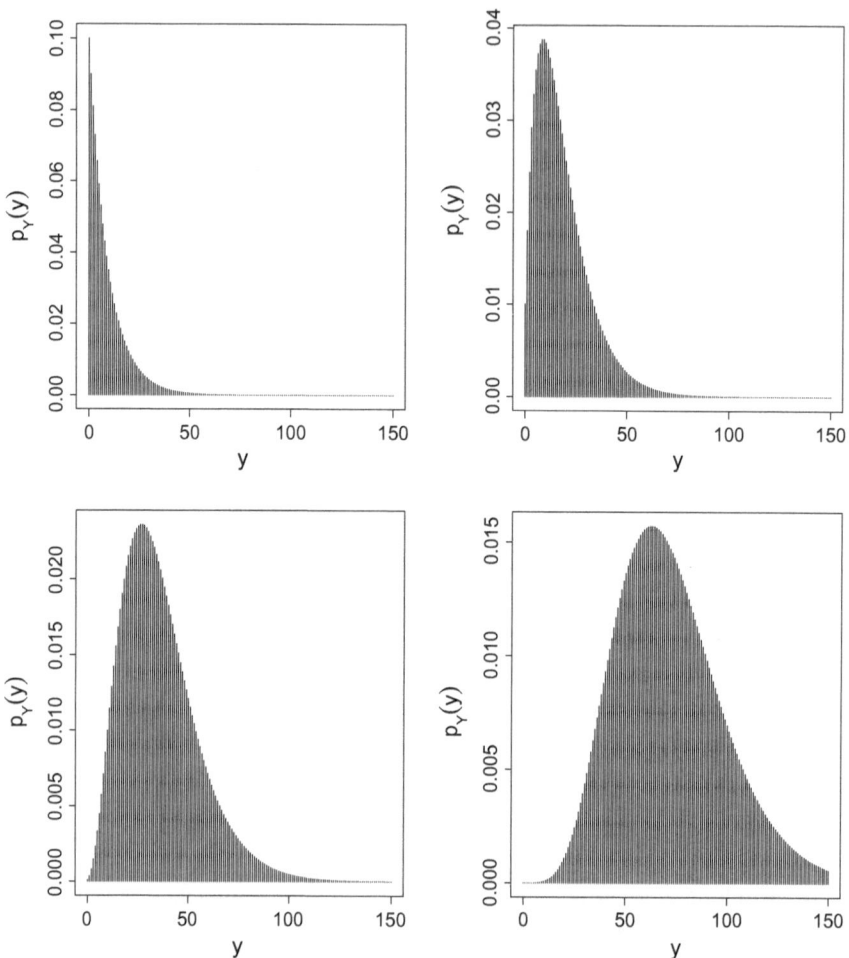

Fig. 2.6 Probability mass functions of the Pascal distributions with parameters $q = 0.1$ and $m = 1, 2, 4, 8$ (from upper left to lower right)

$$\theta = \ln(1 - q)$$
$$a(\theta) = -m \ln(q) = -m \ln(1 - \exp(\theta))$$
$$\phi = 1$$
$$\nu = 1$$
$$c(y, \phi) = \binom{m + y - 1}{y}.$$

Thus, the Pascal distribution belongs to the ED family (recall that m is not a parameter to be estimated, but a known positive integer).

2.3.4 Poisson Distribution

2.3.4.1 Limiting Forms of Binomial Distributions

Two important distributions arise as approximations of Binomial distributions. If m is large enough and the skewness of the distribution is not too great (that is, q is not too close to 0 or 1), then the Binomial distribution is well approximated by the Normal distribution. This is a direct consequence of the central-limit theorem and is clearly visible on Fig. 2.5.

When the number of trials m is large, and the success probability q is small, the corresponding Binomial distribution is well approximated by the Poisson distribution with mean $\lambda = mq$. In order to see why this is the case, consider $Y_m \sim \mathcal{B}in(m, \frac{\lambda}{m})$ for some $\lambda > 0$. Then,

$$P[Y_m = 0] = \left(1 - \frac{\lambda}{m}\right)^m \to \exp(-\lambda) \text{ as } m \to \infty$$

and

$$\frac{P[Y_m = y + 1]}{P[Y_m = y]} = \frac{\frac{m-y}{y+1}\frac{\lambda}{m}}{1 - \frac{\lambda}{m}} \to \frac{\lambda}{y+1} \text{ as } m \to \infty. \tag{2.15}$$

Considering (2.15) as a recurrence relation between successive probabilities, starting from $\exp(-\lambda)$, we thus obtain

$$\lim_{m \to \infty} P[Y_m = y] = \exp(-\lambda)\frac{\lambda^y}{y!}$$

which defines the Poisson probabilities. Hence, the $\mathcal{B}in(m, \frac{\lambda}{m})$ distribution is approximately Poisson with parameter λ for m large enough. More formally, if $m \to \infty$ and $\lambda_m \to 0$ in such a way that $m\lambda_m \to \lambda$ then the $\mathcal{B}in(m, \lambda_m)$ distribution converges to the Poisson distribution with mean λ. The Poisson distribution is thus sometimes called the law of small numbers because it is the probability distribution of the number of occurrences of an event that happens rarely but has many opportunities to happen.

Typically, a Poisson random variable counts the number of events occurring in a certain time interval or spatial area. For example, the number of cars passing a fixed point in a five-minute interval, or the number of claims reported to an insurance company by an insured driver in a given period. Although the Poisson distribution is often called the law of small numbers, there is no need for $\lambda = mq$ to be small. It is the largeness of m and the smallness of $q = \frac{\lambda}{m}$ that are important.

2.3.4.2 Poisson Distribution

A Poisson-distributed response Y takes its values in $S = \{0, 1, 2, \ldots\}$ and has probability mass function

$$p_Y(y) = \exp(-\lambda)\frac{\lambda^y}{y!}, \; y = 0, 1, 2, \ldots. \tag{2.16}$$

Having a counting random variable Y, we denote as $Y \sim \mathcal{P}oi(\lambda)$ the fact that Y is Poisson distributed with parameter λ. The parameter λ is often called the rate, in relation to the Poisson process (see below).

If $Y \sim \mathcal{P}oi(\lambda)$, then the limiting argument given before shows that

$$\mathrm{E}[Y] = \lambda \text{ and } \mathrm{Var}[Y] = \lambda. \tag{2.17}$$

Considering (2.17), we see that both the mean and variance of the Poisson distribution are equal to λ, a phenomenon termed as equidispersion. The skewness coefficient of the Poisson distribution is

$$\gamma[Y] = \frac{1}{\sqrt{\lambda}}.$$

As λ increases, the Poisson distribution thus becomes more symmetric and is eventually well approximated by a Normal distribution, the approximation turning out to be quite good for $\lambda > 20$. But if $Y \sim \mathcal{P}oi(\lambda)$ then $\sqrt{Y}$ converges much faster to the $\mathcal{N}or(\lambda, \frac{1}{4})$ distribution. Hence, the square root transformation was often recommended as a variance stabilizing transformation for count data at a time classical methods assuming Normality (and constant variance) were employed.

The shape of the Poisson probability mass function is displayed in the graphs of Fig. 2.7. For small values of λ, we see that the $\mathcal{P}oi(\lambda)$ probability mass function is highly asymmetric. When λ increases, it becomes more symmetric and ultimately looks like the Normal bell curve.

The Poisson distribution enjoys the convenient convolution stability property, i.e.

$$\left.\begin{array}{r} Y_1 \sim \mathcal{P}oi(\lambda_1) \\ Y_2 \sim \mathcal{P}oi(\lambda_2) \\ Y_1 \text{ and } Y_2 \text{ independent} \end{array}\right\} \Rightarrow Y_1 + Y_2 \sim \mathcal{P}oi(\lambda_1 + \lambda_2). \tag{2.18}$$

This property is useful because sometimes the actuary has only access to aggregated data. Assuming that individual data is Poisson distributed, then so is the summed count and Poisson modeling still applies.

In order to establish that the Poisson distribution belongs to the ED family, let us write the $\mathcal{P}oi(\lambda)$ probability mass function (2.16) as follows:

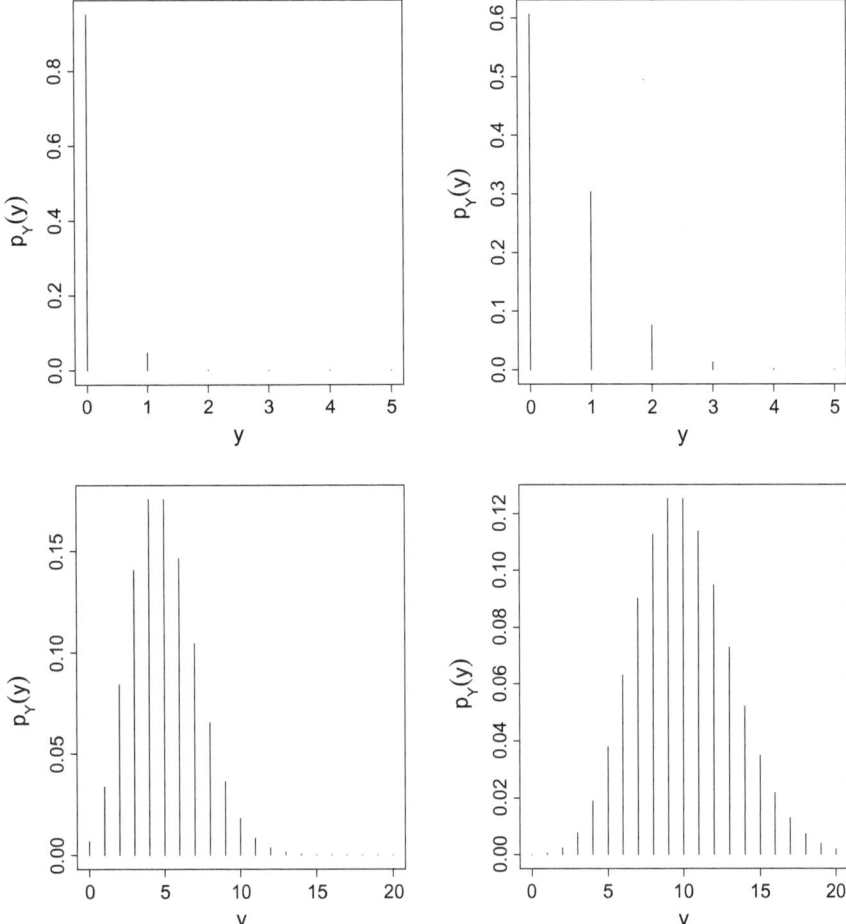

Fig. 2.7 Probability mass functions of $\mathcal{P}oi(\lambda)$ with $\lambda = 0.05, 0.5, 5, 10$ (from upper left to lower right)

$$p_Y(y) = \exp(-\lambda)\frac{\lambda^y}{y!}$$

$$= \exp\left(y \ln \lambda - \lambda\right)\frac{1}{y!}$$

where we recognize the probability mass function (2.3) with

$$\theta = \ln \lambda$$
$$a(\theta) = \lambda = \exp(\theta)$$
$$\phi = 1$$
$$\nu = 1$$
$$c(y, \phi) = \frac{1}{y!}.$$

Thus, the Poisson distribution indeed belongs to the ED family.

2.3.4.3   Exposure-to-Risk

The number of observed events generally depends on a size variable that determines the number of opportunities for the event to occur. This size variable is often the time as the number of claims obviously depend on the length of the coverage period. However, some other choices are possible, such as distance traveled in motor insurance, for instance (as it will be seen in Chap. 5).

The Poisson process setting is useful when the actuary wants to analyze claims experience from policyholders who have been observed during periods of unequal lengths. Assume that the claims occur according to a Poisson process with rate λ. In this setting, claims occur randomly and independently in time. Denoting as $T_1, T_2, \ldots$ the times between two consecutive events, this means that these random variables are independent and obey the $\mathcal{E}xp(\lambda)$ distribution, the only one enjoying the memoryless property. Hence, the kth claim occurs at time

$$\sum_{j=1}^{k} T_j \sim \mathcal{G}am(k, \lambda)$$

where we recognize the Erlang distribution.

Consider a policyholder covered by the company for a period of length e, that is, the policyholder exposes the insurer to the risk of recording a claim during e time units. Then, the number Y of reported claims is such that

$$P[Y \geq k] = P\left[\sum_{j=1}^{k} T_j \leq e\right] = 1 - \sum_{j=0}^{k-1} \exp(-\lambda e)\frac{(\lambda e)^j}{j!}$$

so that it has probability mass function

$$P[Y = k] = P[Y \geq k] - P[Y \geq k + 1] = \exp(-\lambda e)\frac{(\lambda e)^k}{k!}, \quad k = 0, 1, \ldots,$$

that is, $Y \sim \mathcal{P}oi(\lambda e)$.

In actuarial studies, the length e of the observation period is generally referred to as the exposure-to-risk (hence the letter e). It allows the analyst to account for the fact that some policies are observed for the whole period whereas others just entered the portfolio or left it soon after the beginning of the observation period. We see that the exposure-to-risk e simply multiplies the annual expected claim frequency λ in the Poisson process model.

2.3.4.4 Truncated Poisson Distribution

Consider a Poisson response $Y^\star$ and assume that 0 values are not observed. This is typically the case when the actuary considers the number of claims reported by those policyholders who filed at least one claim. This means that the actuary works with the observed response Y defined as $Y^\star$ given that $Y^\star \geq 1$, with probability mass function

$$p_Y(y) = P[Y^\star = y | Y^\star \geq 1] = \frac{\exp(-\lambda)\lambda^y}{y!(1 - \exp(-\lambda))}, \quad y = 1, 2, \ldots$$

Rewriting p_Y as

$$p_Y(y) = \exp\left(y \ln \lambda - \lambda - \ln(1 - \exp(-\lambda))\right)\frac{1}{y!}$$

shows that the truncated Poisson distribution has a probability mass function of the form (2.3) with

$$\theta = \ln \lambda$$
$$a(\theta) = \lambda + \ln(1 - \exp(-\lambda))$$
$$= \exp(\theta) + \ln(1 - \exp(-\exp(\theta)))$$
$$\phi = 1$$
$$\nu = 1$$
$$c(y, \phi) = \frac{1}{y!}$$

so that it belongs to the ED class. This property is interesting when the actuary considers using Hurdle Poisson models (as studied in Chap. 7).

2.3.4.5 Multinomial Distribution

A Multinomial scheme extends a Bernoulli trial in that several outcomes are possible for each experiment (and not only success/failure). Specifically, assume that there are

b types of outcomes, with respective probabilities $q_1, \ldots, q_b$ such that $\sum_{i=1}^{b} q_i = 1$. For Bernoulli trials, $b = 2$ and $q_2 = 1 - q_1$.

Let us denote as Y_i the number of outcomes of type i among m repetitions of this experiment. The random variables $Y_1, \ldots, Y_b$ are negatively correlated, as increasing the number of outcomes of a given type may only decrease the numbers of outcomes of other types since the number m of trials is kept fixed. This means that the random vector $Y = (Y_1, \ldots, Y_b)^\top$ must be considered rather than each Y_j in isolation.

The random vector Y is Multinomially distributed, with joint probability mass function p_Y given by

$$p_Y(y_1, \ldots, y_b) = P[Y_1 = y_1, \ldots, Y_b = y_b] = \frac{m!}{y_1! \ldots y_b!} q_1^{y_1} \ldots q_b^{y_b}$$

if $y_1 + \ldots + y_b = m$ and 0 otherwise, where m is the number of trials. This is henceforth denoted as $Y \sim \mathcal{M}ult(m, q_1, \ldots, q_b)$. Clearly,

$$Y \sim \mathcal{M}ult(m, q_1, \ldots, q_b) \Rightarrow Y_i \sim \mathcal{B}in(m, q_i) \text{ for } i = 1, \ldots, b.$$

The interest of the Multinomial distribution in insurance applications comes from the following result.

Property 2.3.2 *Let Y be the total number of claims, such that $Y \sim \mathcal{P}oi(\lambda)$. Assume that the Y claims may be classified into b categories, according to a Multinomial partitioning scheme with probabilities $q_1, \ldots, q_b$. Let Y_i represent the number of claims of type i, $i = 1, \ldots, b$. Then, $Y_1, \ldots, Y_b$ are independent and*

$$Y_i \sim \mathcal{P}oi(\lambda q_i), \ i = 1, \ldots, b.$$

Proof To show that $Y_i \sim \mathcal{P}oi(\lambda q_i)$, let us write

$$P[Y_i = y] = \sum_{z=y}^{\infty} P[Y_i = y | Y = z] P[Y = z]$$

$$= \sum_{z=y}^{\infty} \binom{z}{y} q_i^y (1 - q_i)^{z-y} \exp(-\lambda) \frac{\lambda^z}{z!}$$

$$= \exp(-\lambda) \frac{(\lambda q_i)^y}{y!} \sum_{z=0}^{\infty} \frac{((1 - q_i)\lambda)^z}{z!}$$

$$= \exp(-\lambda q_i) \frac{(\lambda q_i)^y}{y!}.$$

The mutual independence of $Y_1, \ldots, Y_b$ follows from

$$P[Y_1 = y_1, \ldots, Y_b = y_b] = P[Y_1 = y_1, \ldots, Y_b = y_b | Y = y_1 + \cdots + y_b] \exp(-\lambda) \frac{\lambda^{y_1 + \cdots + y_b}}{(y_1 + \cdots + y_b)!}$$

$$= \frac{(y_1 + \cdots + y_b)!}{y_1! \cdots y_b!} q_1^{y_1} \cdots q_b^{y_b} \exp(-\lambda) \frac{\lambda^{y_1 + \cdots + y_b}}{(y_1 + \cdots + y_b)!}$$

$$= \prod_{j=1}^{b} \exp(-\lambda q_j) \frac{(\lambda q_j)^{y_j}}{y_j!}$$

$$= \prod_{j=1}^{b} P[Y_j = y_j].$$

This ends the proof of the announced result.

In particular, having independent $Y_i \sim \mathcal{P}oi(\lambda_i)$, we see that the distribution of any Y_i conditional on the sum $\sum_{i=1}^{n} Y_i = m$ is $\mathcal{B}in(m, q_i)$ with

$$q_i = \frac{\lambda_i}{\sum_{j=1}^{n} \lambda_j}.$$

2.4 Some Useful Properties

Table 2.1 summarizes the main findings obtained so far. For each distribution under consideration, we list the canonical parameter θ, the cumulant function $a(\cdot)$, and the dispersion parameter ϕ entering the general definition (2.3) of ED probability mass or probability density function. We also give the two first moments. In this section, we establish several properties of ED distributions that appear to be particularly useful in the analysis of insurance data.

Table 2.1 Examples of ED distributions (with $\nu = 1$)

Distribution	θ	$a(\theta)$	ϕ	$\mu = E[Y]$	Var$[Y]$
$\mathcal{B}er(q)$	$\ln \frac{q}{1-q}$	$\ln(1 + \exp(\theta))$	1	q	$\mu(1 - \mu)$
$\mathcal{B}in(m, q)$	$\ln \frac{q}{1-q}$	$m \ln(1 + \exp(\theta))$	1	mq	$\mu \left(1 - \frac{\mu}{m}\right)$
$\mathcal{G}eo(q)$	$\ln(1 - q)$	$-\ln(1 - \exp(\theta))$	1	$\frac{1-q}{q}$	$\mu(1 + \mu)$
$\mathcal{P}as(m, q)$	$\ln(1 - q)$	$-m \ln(1 - \exp(\theta))$	1	$m\frac{1-q}{q}$	$\mu \left(1 + \frac{\mu}{m}\right)$
$\mathcal{P}oi(\mu)$	$\ln \mu$	$\exp(\theta)$	1	μ	μ
$\mathcal{N}or(\mu, \sigma^2)$	μ	$\frac{\theta^2}{2}$	σ^2	μ	ϕ
$\mathcal{E}xp(\mu)$	$-\frac{1}{\mu}$	$-\ln(-\theta)$	1	μ	μ^2
$\mathcal{G}am(\mu, \alpha)$	$-\frac{1}{\mu}$	$-\ln(-\theta)$	$\frac{1}{\alpha}$	μ	$\phi \mu^2$
$\mathcal{IG}au(\mu, \alpha)$	$-\frac{1}{2\mu^2}$	$-\sqrt{-2\theta}$	$\frac{1}{\alpha}$	μ	$\phi \mu^3$

2.4.1 Averages

Averaging independent and identically distributed ED responses does not modify their distribution, just the value of the weight. This remarkable property, which appears to be particularly useful in insurance data analysis is easily established using the moment generating function (that has already been encountered in Remark 2.2.3).

Recall that the moment generating function m_Y of the random variable Y is defined as

$$m_Y(t) = E[\exp(tY)].$$

Moment generating functions play an important role in probability. When it exists, the moment generating function is unique and characterizes the probability distribution (that is, two random variables with the same moment generating function are identically distributed). In addition, when it exists, the kth derivative of m_Y evaluated at the origin is just the kth moment of Y (hence the name moment generating function), i.e.

$$E[Y^k] = \frac{d^k}{dt^k} m_Y(t) \Big|_{t=0}.$$

Moreover, the moment generating function allows the actuary to deal with sums of independent random variables, hence its interest for studying averages. Formally, the moment generating function of the sum of independent random variables Y_1 and Y_2 is obtained as follows:

$$m_{Y_1+Y_2}(t) = E\left[\exp\left(t(Y_1 + Y_2)\right)\right] = E[\exp(tY_1)]E[\exp(tY_2)] = m_{Y_1}(t)m_{Y_2}(t).$$

Thus, we see that adding independent random variables amount to multiplying their respective moment generating functions. As $m_{Y_1+Y_2}$ characterizes the distribution of $Y_1 + Y_2$, this provides the actuary with a convenient method to obtain the distribution of sums of independent responses (whereas their distribution function is obtained from convolution products which are generally more difficult to handle).

The next result establishes the expression of the moment generating function of ED distributions. As expected, the cumulant function a is the key ingredient to this formula.

Property 2.4.1 *The moment generating function m_Y of the response Y whose distribution belongs to the ED family is given by*

$$m_Y(t) = \exp\left(\frac{a(\theta + t\phi/\nu) - a(\theta)}{\phi/\nu}\right).$$

Proof We give the proof for the continuous case, only. The proof for the discrete case easily follows by replacing probability density functions with probability mass functions and integrals with sums. Let us write

$$m_Y(t) = \int_S \exp(yt) f_Y(y) dy$$

$$= \int_S \exp(yt) \exp\left(\frac{y\theta - a(\theta)}{\phi/v}\right) c(y, \phi/v) dy$$

$$= \int_S \exp\left(\frac{y(\theta + t\phi/v) - a(\theta + t\phi/v)}{\phi/v}\right) c(y, \phi/v) dy$$

$$\exp\left(\frac{a(\theta + t\phi/v) - a(\theta)}{\phi/v}\right).$$

The integral appearing in the last formula is just the integral of the ED probability density function with canonical parameter $\theta + t\phi/v$, dispersion parameter ϕ and weight v so that it equals 1. Hence, we obtain the announced result.

Now that we have established the analytical expression for m_Y, we can study the effect of averaging the responses before the analysis starts. If Y_1 and Y_2 are independent with common moment generating function m_Y as in Property 2.4.1 then $\frac{Y_1+Y_2}{2}$ has moment generating function

$$m_{\frac{Y_1+Y_2}{2}}(t) = \exp\left(\frac{t}{2}(Y_1 + Y_2)\right)$$

$$= \left(m_Y\left(\frac{t}{2}\right)\right)^2$$

$$= \exp\left(\frac{a(\theta + t\phi/(2v)) - a(\theta)}{\phi/(2v)}\right).$$

Hence, $\frac{Y_1+Y_2}{2}$ obeys the same ED distribution, except that the weight v becomes $2v$. Arithmetic averaging $\frac{Y_1+Y_2}{2}$ is thus equivalent to doubling the weight v, all the other components of the ED distribution remaining unchanged. More generally, we have the following result.

Property 2.4.2 *If $Y_1, Y_2, \ldots, Y_n$ are independent with common moment generating function m_Y given in Property 2.4.1 then their average*

$$\overline{Y} = \frac{1}{n}\sum_{i=1}^n Y_i$$

has probability density function

$$\exp\left(\frac{y\theta - a(\theta)}{\phi/(nv)}\right) c(y, \phi/(nv)).$$

The distribution for $\overline{Y}$ is thus the same as for each Y_i except that the weight v is replaced with nv.

Averaging observations is accounted for by modifying the weights in the ED family. Notice that, for example, halving ϕ has the same effect as doubling the weight (sample size) has. Weights are needed to model average claim sizes. By not taking the weights into account, one disregards the fact that an average over many observations has been measured with much more precision.

Example 2.4.3 (Binomial proportions) Often, analysts concentrate on proportions rather than counts Y, i.e. they work in relative instead of absolute terms. Considering $Y \sim \mathcal{B}in(m, q)$, this means that the response is no more the number of success Y but the corresponding proportion $\widetilde{Y} = \frac{Y}{m}$. Hence, the number of success becomes $m\widetilde{Y}$ in this setting. Working with proportions, we have $\mathcal{S} = \{0, \frac{1}{m}, \ldots, 1\}$ and $E[\widetilde{Y}] = q$. Let us show that the distribution of $\widetilde{Y}$ still belongs to the ED family. This result can be seen as an application of Property 2.4.2 as $\widetilde{Y}$ is the average of m independent $\mathcal{B}er(q)$ random variables. Here, we also provide an alternative, direct proof. To this end, let us rewrite the probability mass function of $\widetilde{Y}$ as

$$p_{\widetilde{Y}}(\widetilde{y}) = \binom{m}{m\widetilde{y}} q^{m\widetilde{y}} (1-q)^{m(1-\widetilde{y})}$$
$$= \exp\left(\frac{\widetilde{y}\ln\frac{q}{1-q} + \ln(1-q)}{1/m}\right)\binom{m}{m\widetilde{y}}$$

which is well of the form (2.3) with

$$\theta = \ln\frac{q}{1-q}$$
$$a(\theta) = -\ln(1-q) = \ln(1+\exp(\theta))$$
$$\phi = 1$$
$$v = m$$
$$c(\widetilde{y}, \phi/v) = \binom{m}{m\widetilde{y}}.$$

The distribution of the Binomial proportion $\widetilde{Y}$ thus also belongs to the ED family. Compared to Y, we see that the function $a(\cdot)$ does no more depend on the size m, which is now taken as a weight. The support is included in the unit interval $[0, 1]$ whatever the size m.

2.4.2 Mean

We know from Property 2.4.1 that the function a appearing in (2.3) determines the moments of the response Y. In the Normal case, the derivative of $a(\theta) = \frac{\theta^2}{2}$ is $\theta = \mu$ so that we recover the mean response from a'. This turns out to be a property generally valid for all ED distributions. Precisely, the next result shows that the first derivative of a corresponds to the mean response.

Property 2.4.4 *If the response Y has probability density/mass function of the form (2.3) then*

$$E[Y] = a'(\theta).$$

Proof Let us give two different proofs of this result. The first one makes use of the expression for the moment generating function m_Y obtained in Property 2.4.1:

$$E[Y] = \frac{d}{dt} m_Y(t)\Big|_{t=0} = a'(\theta + t\phi/v)m_Y(t)\Big|_{t=0} = a'(\theta).$$

Another way to get the same result consists in starting from

$$\int_S f_Y(y)dy = 1$$

so that, assuming that we are allowed to let the derivative enter the integral (which is the case in the ED family), we get

$$\begin{aligned} 0 &= \frac{d}{d\theta} \int_S f_Y(y)dy \\ &= \int_S \left(\frac{d}{d\theta} f_Y(y) \right) dy \\ &= \int_S \left(\frac{y - a'(\theta)}{\phi/v} \right) f_Y(y)dy \end{aligned}$$

which finally implies that

$$\int_S y f_Y(y)dy = a'(\theta),$$

as announced.

2.4.3 Variance

The mean response Y corresponds to the first derivative of the function $a(\cdot)$ involved in (2.3). The next result shows that the variance is proportional to the second derivative of $a(\cdot)$.

Property 2.4.5 *If the response Y has probability density/mass function of the form (2.3) then*

$$Var[Y] = \frac{\phi}{v}a''(\theta).$$

Proof Again, let us give two different proofs of this result. The first one makes use of the moment generating function:

$$
\begin{aligned}
E[Y^2] &= \left.\frac{d^2}{dt^2} m_Y(t)\right|_{t=0} \\
&= \left.\frac{\phi}{\nu} a''(\theta + t\phi/\nu) m_Y(t)\right|_{t=0} + \left.(a'(\theta + t\phi/\nu))^2 m_Y(t)\right|_{t=0} \\
&= \frac{\phi}{\nu} a''(\theta) + (a'(\theta))^2 \\
&= \frac{\phi}{\nu} a''(\theta) + (E[Y])^2 \text{ by Property 2.4.4,}
\end{aligned}
$$

whence the announced result follows. Another way to get the same result is as follows: assuming that we are allowed to let the derivative enter the integral (which is the case in the ED family), we get

$$
\begin{aligned}
0 &= \frac{d^2}{d\theta^2} \int_S f_Y(y) dy \\
&= \int_S \left(\frac{d^2}{d\theta^2} f_Y(y) \right) dy \\
&= \int_S \left(\left(\frac{y - a'(\theta)}{\phi/\nu} \right)^2 - \frac{a''(\theta)}{\phi/\nu} \right) f_Y(y) dy
\end{aligned}
$$

which gives the announced result as $a'(\theta) = E[Y]$ by Property 2.4.4. This ends the proof.

Because the variance is obviously always positive, Property 2.4.5 implies that the second derivative of the function a used to define the ED distribution must be non-negative, i.e. that the function a must be a convex function (so that $a'' \geq 0$) as assumed in the definition (2.3) of the ED family. Considering the tilting construction presented in Remark 2.2.3, this condition is obviously fulfilled as all derivatives of every moment generating function are non-negative.

Notice also that increasing the weight thus decreases the variance whereas the variance increases linearly in the dispersion parameter ϕ. The impact of θ on the variance is given by the factor

$$
a''(\theta) = \frac{d}{d\theta} \mu(\theta)
$$

expressing how a change in the canonical parameter θ modifies the expected response. In the Normal case, $a''(\theta) = 1$ and the variance is just constantly equal to $\phi/\nu = \sigma^2/\nu$, not depending on θ. In this case, the mean response does not influence its variance. For the other members of the ED family, a'' is not constant and a change in θ modifies the variance.

2.4.4 Variance Function

The variance function $V(\cdot)$ indicates the relationship between the mean and variance of an ED distribution. Notice that an absence of relation between the mean and the variance is only possible for real-valued responses (such as Normally distributed ones, where the variance σ^2 does not depend on the mean μ). Indeed, if Y is non-negative (i.e. $Y \geq 0$) then intuitively the variance of Y tends to zero as the mean of Y tends to zero. That is, the variance is a function of the mean for non-negative responses.

The variance function $V(\cdot)$ is formally defined as

$$V(\mu) = \frac{d^2}{d\theta^2}a(\theta) = \frac{d}{d\theta}\mu(\theta).$$

The variance function thus corresponds to the variation in the mean response $\mu(\theta)$ viewed as a function of the canonical parameter θ. In the Normal case, $\mu(\theta) = \theta$ and $V(\mu) = 1$. The other ED distributions have non-constant variance functions. Again, we see that the cumulant function $a(\cdot)$ determines the distributional properties in the ED family.

The variance of the response can thus be written as

$$\text{Var}[Y] = \frac{\phi}{\nu}V(\mu).$$

It is important to keep in mind that the variance function is not the variance of the response, but the function of the mean entering this variance (to be multiplied by ϕ/ν). The variance function is regarded as a function of the mean μ, even if it appears as a function of θ; this is possible by inverting the relationship between θ and μ as we known from Property 2.4.4 that $\mu = E[Y] = a'(\theta)$. The convexity of $a(\cdot)$ ensures that the mean function a' is increasing so that its inverse is well defined. Hence, we can express the canonical parameter in terms of the mean response μ by the relation

$$\theta = (a')^{-1}(\mu).$$

The variance functions corresponding to the usual ED distributions are listed in Table 2.2. Notice that

$$V(\mu) = \mu^\xi \text{ with } \xi = \begin{cases} 0 \text{ for the Normal distribution} \\ 1 \text{ for the Poisson distribution} \\ 2 \text{ for the Gamma distribution} \\ 3 \text{ for the Inverse Gaussian distribution.} \end{cases}$$

These members of the ED family thus have power variance functions. The whole family of ED distributions with power variance functions is referred to as the Tweedie

Table 2.2 Variance functions for some selected ED distributions

Distribution	Variance function $V(\mu)$
$Ber(q)$	$\mu(1-\mu)$
$Geo(q)$	$\mu(1+\mu)$
$Poi(\mu)$	μ
$Nor(\mu, \sigma^2)$	1
$Gam(\mu, \alpha)$	μ^2
$Gau(\mu, \alpha)$	μ^3

class, which will be studied later on in Sect. 2.5. Not all ED distributions admit a power variance function. In the Bernoulli case for instance,

$$
\begin{aligned}
V(\mu) &= \frac{d^2}{d\theta^2} \ln(1 + \exp(\theta)) \\
&= \frac{\exp(\theta)}{(1 + \exp(\theta))^2} \\
&= \mu(1 - \mu).
\end{aligned}
$$

2.4.5 Weights

Often in actuarial studies, responses appear to be ratios with the aggregate exposure or even premiums in the denominator (in case loss ratios are analyzed). The numerator may correspond to individual data, or to grouped data aggregated over a set of homogeneous policies. This means that the size of the group has to be accounted for as the response ratios will tend to be far more volatile in low-volume cells than in high-volume ones. Actuaries generally consider that a large-volume cell is the result of summing smaller independent cells, leading to response variance proportional to the inverse of the volume measure. This implies that weights vary according to the business volume measure.

We have already considered Binomial proportions in Example 2.4.3. Here, we further consider weighted averages of responses with distribution (2.3).

Property 2.4.6 *Consider independent responses $Y_1, \ldots, Y_n$ obeying ED distributions (2.3) with common mean μ, dispersion parameter ϕ and specific weights v_i. Define the total weight*

$$
v_\bullet = \sum_{i=1}^n v_i.
$$

Then, the weighted average

$$\frac{1}{v_\bullet} \sum_{i=1}^{n} v_i Y_i$$

still follows an ED distribution (2.3) *with mean* μ, *dispersion parameter* ϕ *and weight* $v_\bullet$.

Proof The moment generating function of the weighted average is given by

$$E\left[\exp\left(\frac{t}{v_\bullet} \sum_{i=1}^{n} v_i Y_i\right)\right] = \prod_{i=1}^{n} \exp\left(v_i \frac{a\left(\theta + t\frac{\phi}{v_\bullet}\right) - a(\phi)}{\phi}\right)$$

$$= \exp\left(v_\bullet \frac{a\left(\theta + t\frac{\phi}{v_\bullet}\right) - a(\phi)}{\phi}\right),$$

which ends the proof.

Property 2.4.6 extends Property 2.4.2 to weighted averages of ED responses.

2.5 Tweedie Distributions

2.5.1 Power Variance Function

Tweedie models are defined within the ED family by a power variance function of the form

$$V(\mu) = \mu^\xi$$

where the power parameter ξ controls the shape of the distribution. The Normal ($\xi = 0$), Poisson ($\xi = 1$), Gamma ($\xi = 2$), and Inverse Gaussian ($\xi = 3$) distributions all belong to the Tweedie subclass of the ED family. It can be shown that the ED distribution with such a power variance function always exists except when $0 < \xi < 1$. When $v = 1$, this implies a variance equal to

$$\mathrm{Var}[Y] = \phi\mu^\xi.$$

For a probability density function of the form (2.3) with $v = 1$, it comes

$$\frac{d}{d\mu} \ln f = \frac{d}{d\theta} \ln f \times \frac{d\theta}{d\mu}$$

$$= \frac{d}{d\theta}\left(\frac{y\theta - a(\theta)}{\phi}\right) \times \frac{d}{d\mu}\left((a')^{-1}(\mu)\right)$$

$$= \frac{y - a'(\theta)}{\phi} \frac{1}{a''\left((a')^{-1}(\mu)\right)}$$

$$= \frac{y - a'(\theta)}{\phi a''(\theta)}$$

$$= \frac{y - \mu}{\phi V(\mu)}.$$

With the Tweedie power variance function, we get

$$\ln f(y) = \int \frac{y - m}{\phi m^{\xi}}\, dm$$

$$= \frac{1}{\phi}\int \left(ym^{-\xi} - m^{-\xi+1}\right) dm$$

$$= \frac{1}{\phi}\left(y\frac{\mu^{-\xi+1}}{-\xi+1} - \frac{\mu^{-\xi+2}}{-\xi+2}\right) + \ln c \qquad (2.19)$$

where c is a constant with respect to m. Identifying (2.19) with (2.3) enables us to determine that the Tweedie canonical parameter is given by

$$\theta = \frac{\mu^{1-\xi}}{1 - \xi} \Leftrightarrow \mu = \left((1-\xi)\theta\right)^{\frac{1}{1-\xi}}$$

while

$$a(\theta) = \frac{\mu^{2-\xi}}{2 - \xi} = \frac{\left((1-\xi)\theta\right)^{\frac{2-\xi}{1-\xi}}}{2 - \xi}.$$

2.5.2 Scale Transformations

Although ED distributions are closed under averaging, this family is in general not closed under scale transformations. This means that δY may not obey an ED distribution even if Y does, for a given constant $\delta > 0$. The Tweedie subclass of the ED family is closed under this type of scale transformation, as shown next.

Property 2.5.1 *Consider Y obeying the Tweedie distribution with mean μ, dispersion parameter ϕ (setting $v = 1$), and power parameter ξ. For any constant $\delta > 0$, δY is Tweedie distributed with mean $\delta\mu$, dispersion parameter $\phi(\delta)^{2-\xi}$ and power parameter ξ.*

Proof Let us start from the Tweedie probability density function deduced from (2.19), i.e.

$$f_Y(y) = \exp\left(\frac{1}{\phi}\left(y\frac{\mu^{1-\xi}}{1-\xi} - \frac{\mu^{2-\xi}}{2-\xi}\right)\right)c(y,\phi).$$

Then, δY has probability density function

$$
\begin{aligned}
f_{\delta Y}(y) &= \frac{1}{\delta}f_Y\left(\frac{y}{\delta}\right) \\
&= \exp\left(\frac{1}{\phi}\left(\frac{y}{\delta}\frac{\mu^{1-\xi}}{1-\xi} - \frac{\mu^{2-\xi}}{2-\xi}\right)\right)\tilde{c}(y,\delta,\phi) \\
&= \exp\left(\frac{1}{\phi(\delta)^{2-\xi}}\left(y\frac{(\delta\mu)^{1-\xi}}{1-\xi} - \frac{(\delta\mu)^{2-\xi}}{2-\xi}\right)\right)\tilde{c}(y,\delta,\phi)
\end{aligned}
$$

so that δY is Tweedie with mean $\delta\mu$ and dispersion parameter $\phi(\delta)^{2-\xi}$.

Considering Property 2.5.1, we see that the actuary is thus allowed to work with ratios, dividing the total claim cost Y by some appropriate volume measure (such as the salary mass in workers' compensation insurance, for instance).

Example 2.5.2 (Poisson rates) For Binomial responses, we have seen in Example 2.4.3 that the actuary was allowed to work either with the number of successes or with the proportion of successes, provided the weights are modified accordingly. The same idea applies to the Poisson model where the actuary is free to work with claim rates instead of claim counts.

Precisely, considering the Poisson process setting described above, this means that the actuary works with the response

$$\widetilde{Y} = \frac{Y}{e} \text{ where } Y \sim \mathcal{P}oi(\lambda e)$$

where e is the exposure-to-risk measuring the time during which the individual has been exposed to the risk covered by the insurance company (and was thus susceptible of reporting claims to the company). The parameter λ is then the yearly claim rate, or the expected number of claims reported for a unit exposure (one year, say).

The claim rate $\widetilde{Y}$ is valued in $S = \{0, \frac{1}{e}, \frac{2}{e}, \ldots\}$ and such that $E[\widetilde{Y}] = \lambda$. To show that the distribution of $\widetilde{Y}$ still belongs to the ED family, let us consider $\tilde{y} \in S$ and write its probability mass function as

$$
\begin{aligned}
p_{\widetilde{Y}}(\tilde{y}) &= p_Y(\tilde{y}e) \\
&= \exp(-e\lambda)\frac{(e\lambda)^{\tilde{y}e}}{(\tilde{y}e)!} \\
&= \exp\left(\frac{\tilde{y}\ln\lambda - \lambda}{1/e}\right)\frac{e^{\tilde{y}e}}{(\tilde{y}e)!}.
\end{aligned}
$$

We recognize a probability mass function of the form (2.3) with

$$\theta = \ln \lambda$$
$$a(\theta) = \lambda = \exp(\theta)$$
$$\phi = 1$$
$$\nu = e$$
$$c(y, \phi) = \frac{e^{\widetilde{y}e}}{(\widetilde{y}e)!}.$$

Thus, the Poisson rate distribution indeed belongs to the ED family. This means that we can equivalently work with the observed claim count (or absolute claim frequency) Y or with the observed claim rate (or relative claim frequency) $\widetilde{Y}$ provided the weight $\nu = e$ enters the analysis.

This finding can be seen as an application of Property 2.5.1 because the Poisson distribution belongs to the Tweedie class and $\widetilde{Y} = \delta Y$ with $\delta = 1/e$. Thus, considering

$$\xi = 1 \text{ and } \delta = \frac{1}{e}$$

Property 2.5.1 shows that $\widetilde{Y}$ still belongs to the Tweedie class.

2.6 Bibliographic Notes and Further Reading

We refer the reader to Bahnemann (2015) for a gentle introduction to standard probability distributions and their applications to P&C insurance. The main reference in this domain remains Klugman et al. (2012) where a detailed account of loss distributions can be found. The books devoted to GLMs mentioned in the closing section of Chap. 4 all contain a brief recap about the ED family including Kaas et al. (2008). Jorgensen (1997) provides a detailed study of the dispersion models, including ED distributions as special cases.

References

Bahnemann D (2015) Distributions for actuaries. CAS monograph series, vol 2

Jorgensen B (1997) The theory of dispersion models. CRC Press

Kaas R, Goovaerts MJ, Dhaene J, Denuit M (2008) Modern actuarial risk theory using R. Springer, New York

Kaas R, Hesselager O (1995) Ordering claim size distributions and mixed Poisson probabilities. Insur Math Econ 17:193–201

Klugman SA, Panjer HH, Willmot GE (2012) Loss models: from data to decisions, 4th edn. Wiley

Chapter 3
Maximum Likelihood Estimation

3.1 Introduction

Assuming that the responses are random variables with unknown distribution depending on one (or several) parameter(s) θ, the actuary must draw conclusions about the unknown parameter θ based on available data. Such conclusions are thus subject to sampling errors: another data set from the same population would have inevitably produced different results. To perform premium calculations, the actuary needs to select a value of θ hopefully close to the true parameter value. Such a value is called an estimate (or pointwise estimation) of the unknown parameter θ. The estimate is distinguished from the model parameters by a hat, which means that an estimate of θ is denoted by $\widehat{\theta}$. This distinction is necessary since it is generally impossible to estimate the true parameter θ without error. Thus, $\widehat{\theta} \neq \theta$ in general. The estimator is itself a random variable as it varies from sample to sample (in a repeated sampling setting, that is, drawing random samples from a given population and computing the estimated value, again and again). Formally, an estimator $\widehat{\theta}$ is a function of the observed responses.

It seems reasonable to require that a good estimate of the unknown parameter θ would be the value of the parameter that maximizes the chance of getting the data that have been recorded in the database (in which the actuary trusts). Maximum likelihood estimation is a method to determine the likely values of the parameters having produced the available responses in a given probability model. Broadly speaking, the parameter values are found such that they maximize the chances that the model produced the data that were actually observed.

Maximum-likelihood estimation can be traced back to famous pioneers such as Carl Friedrich Gauss or Pierre-Simon Laplace. Its use has been widespread after the seminal works by Ronald Fisher in the early 1900s. This method for estimating the parameters of a statistical model, given observations, attempts to find the parameter values that maximize the likelihood function. The latter quantifies the plausibility of the data, given the current values of the parameters. Formally, the likelihood

© Springer Nature Switzerland AG 2019   69
M. Denuit et al., *Effective Statistical Learning Methods for Actuaries I*,
Springer Actuarial, https://doi.org/10.1007/978-3-030-25820-7_3

function is the joint probability density function, or joint probability mass function in the discrete case, of the responses, viewed as a function of the parameter θ.

Maximum likelihood estimation provides the actuary with an analytic procedure that applies to a wide variety of problems, including censored data (when claim severities have been ceiled in application of policy limits, for instance or when individuals are still alive at the end of the observation period, so that only a lower bound on their lifetime is known to the actuary). Although maximum-likelihood estimators are not necessarily optimal (in the sense that there may be other estimation algorithms that can achieve better results in a particular situation, especially when limited amounts of data are available), they enjoy attractive properties in large samples, considering a sequence of maximum-likelihood estimators based on an increasing number of observations. For instance, maximum-likelihood estimators become unbiased minimum variance estimators as the sample size increases, and approximately obey Normal distributions. The price to pay to benefit from these nice properties is that calculating maximum-likelihood estimates sometimes requires specialized software for solving complex optimization problems (but statistical packages, including R are extending their capability in that respect).

3.2   Statistical Inference

3.2.1   Estimation

The starting point of an actuarial analysis generally consists in an observational study having produced a set of data, called sample values. The sample values may be realizations y_i of responses Y_i, $i = 1, \ldots, n$, as in this chapter, or a more complex data structure, such as the values of a response accompanied by possible explanatory variables as in the next chapters. The fundamental assumption is that the value of the response is random and thus considered as the realization of a random variable with unknown distribution. In the parametric approach, the responses Y_i obey a given distribution with unknown parameter θ, so that the actuary needs to determine the value of θ to perform premium calculations.

Here, we assume that we have independent responses $Y_1, \ldots, Y_n$ with common distribution (2.3) and unit weights. Our goal is to draw conclusions about the unknown parameter θ based on available data $y_1, \ldots, y_n$, subject to sampling errors. Such errors represent variations in the parameter estimates resulting from different samples drawn from the population of interest. Because of randomness, the parameter estimates computed from these different samples fluctuate and these variations across samples are described by the sampling distribution of the estimator.

The actuary needs to select a value of θ (hopefully close to the true parameter value) to perform calculations. Such a value is called an estimate (or pointwise estimation) of the unknown parameter. The function $\widehat{\theta}(\cdot)$ of the data giving the point estimate is called the estimator. An estimator $\widehat{\theta}$ for θ is thus a function of the observations $Y_1, Y_2, \ldots, Y_n$, that is,

$$\widehat{\theta} = \widehat{\theta}(Y_1, Y_2, \ldots, Y_n).$$

It is important at this stage to realize that an estimator is itself a random variable as it varies from sample to sample. The corresponding estimate is $\widehat{\theta}(y_1, y_2, \ldots, y_n)$, computed from the realizations of the responses for a particular sample of observations $y_1, y_2, \ldots, y_n$.

3.2.2 Properties of Estimators

3.2.2.1 Repeated Sampling

In general, there exist several possible estimators for the parameter θ. This is why actuaries evaluate their relative merits by examining their properties. Properties of estimators are generally inherited from an hypothetical repeated sampling procedure. Assume that we are allowed to draw repeatedly samples from a given population. Averaging the estimates $\widehat{\theta}_1, \widehat{\theta}_2, \ldots$ obtained for each of the samples corresponds to the mathematical expectation as

$$E[\widehat{\theta}] = \lim_{k \nearrow \infty} \frac{1}{k} \sum_{j=1}^{k} \widehat{\theta}_j$$

by the law of large number. In this setting, $\mathrm{Var}[\widehat{\theta}]$ measures the stability of the estimates across these samples.

3.2.2.2 Unbiasedness

An estimator $\widehat{\theta}$ for the parameter θ is unbiased if

$$E[\widehat{\theta}] = \theta.$$

Unbiasedness means that the estimation errors $\widehat{\theta}_k - \theta$ where $\widehat{\theta}_k$ has been obtained from independent samples $k = 1, 2, \ldots$, cancel on average (i.e. over infinitely many such samples).

For instance, the sample mean

$$\overline{Y} = \frac{1}{n} \sum_{i=1}^{n} Y_i$$

is an unbiased estimate $\widehat{\mu}$ of the common mean $\mu = E[Y_i]$ as

$$E[\overline{Y}] = \frac{1}{n} \sum_{i=1}^{n} E[Y_i] = \mu.$$

3.2.2.3   Consistency

Consistency is another asymptotic property obtained by letting the sample size increase to infinity (whereas for unbiasedness, the numbers of samples drawn from the population grows to infinity). Let us denote as $\widehat{\theta}_n$ the estimator obtained with a sample of size n, that is, with n observations $Y_1, Y_2, \ldots, Y_n$. The estimator $\widehat{\theta}_n = \widehat{\theta}(Y_1, Y_2, \ldots, Y_n)$ is said to be consistent for the parameter θ if

$$\lim_{n \nearrow \infty} P[|\widehat{\theta}_n - \theta| < \epsilon] = 1 \text{ for all } \epsilon > 0.$$

This asymptotic property ensures that, as the sample size gets large the estimator is increasingly likely to fall within a small region around the true value of the parameter. This expresses the idea that $\widehat{\theta}_n$ converges to the true value θ of the parameter as the sample size becomes infinitely large. Hence, a consistent estimate provides the actuary with the correct answer provided enough information becomes available. Consistency is a minimal requirement for an estimator.

If $\widehat{\theta}_n$ is an unbiased estimator for the parameter θ, that is, $E[\widehat{\theta}_n] = \theta$, and if

$$\lim_{n \nearrow \infty} \text{Var}[\widehat{\theta}_n] = 0$$

then $\widehat{\theta}_n$ is consistent for θ. For instance, $\widehat{\mu} = \overline{Y}$ is a consistent estimator of $\mu = E[Y_i]$ as

$$\text{Var}[\overline{Y}] = \frac{1}{n^2} \sum_{i=1}^{n} \text{Var}[Y_i] = \frac{\text{Var}[Y_1]}{n} \searrow 0 \text{ as } n \nearrow \infty.$$

Notice that the consistency of $\overline{Y}$ for the mean is a direct consequence of the law of large numbers.

3.2.2.4   Mean-Squared Error

The accuracy of $\widehat{\theta}$ is traditionally measured by its mean-squared error

$$\text{MSE}(\widehat{\theta}) = E[(\widehat{\theta} - \theta)^2].$$

This corresponds to the average squared estimation error $\widehat{\theta} - \theta$ over infinitely many samples in the hypothetical repeated sampling setting.

The variance of an unbiased estimator indicates its precision. This is because

$$\widehat{\theta} \text{ unbiased } \Rightarrow \text{MSE}(\widehat{\theta}) = \text{Var}[\widehat{\theta}].$$

Despite its intuitive contents, unbiasedness is not necessarily a property that has always to be imposed to estimators, as it will be seen in the next chapters. This is because we can decompose the mean-squared error into

$$\text{MSE}(\widehat{\theta}) = \text{Var}[\widehat{\theta}] + \left(\text{E}[\widehat{\theta}] - \theta\right)^2$$

involving the square of the bias $\text{E}[\widehat{\theta}] - \theta$. In order to minimize the MSE, it might appear desirable to sacrifice a bit on the bias provided we can reduce $\text{Var}[\widehat{\theta}]$ to a large extent. This is typically the case with smoothing or regularization, for instance. This bias-variance decomposition of the MSE is also used for justifying the performances of ensemble learning techniques (as it will be seen in Chap. 6 of Hainaut et al. 2019).

3.3 Likelihood-Based Statistical Inference

3.3.1 Maximum-Likelihood Estimator

The parameter value that makes the observed $y_1, \ldots, y_n$ the most probable is called the maximum-likelihood estimate. Formally, the likelihood function $\mathcal{L}(\theta)$ is defined as the joint probability mass/density function of the observations. In our case, for independent observations $Y_1, \ldots, Y_n$ obeying the same ED distribution (2.3) with unit weights, the likelihood function is given by

$$\mathcal{L}(\theta) = \prod_{i=1}^{n} \exp\left(\frac{y_i \theta - a(\theta)}{\phi}\right) c(y_i, \phi).$$

The likelihood function $\mathcal{L}(\theta)$ can be interpreted as the probability or chance of obtaining the actual observations $y_1, \ldots, y_n$ under the parameter θ. It is important to remember that the likelihood is always defined for a given set of observed values $y_1, \ldots, y_n$. In case repeated sampling properties are discussed, the numerical values y_i are replaced with unobserved random variables Y_i.

The maximum-likelihood estimator is the value $\widehat{\theta}$ which maximizes the likelihood function. Equivalently, $\widehat{\theta}$ maximizes the log-likelihood function

$$L(\theta) = \ln \mathcal{L}(\theta)$$

which is given by

$$L(\theta) = \sum_{i=1}^{n} \left(\frac{y_i \theta - a(\theta)}{\phi} + \ln c(y_i, \phi) \right)$$

$$= \frac{n(\bar{y}\theta - a(\theta))}{\phi} + \sum_{i=1}^{n} \ln c(y_i, \phi)$$

where $\bar{y}$ is the sample mean, i.e. the arithmetic average

$$\bar{y} = \frac{1}{n} \sum_{i=1}^{n} y_i$$

of the available observations $y_1, y_2, \ldots, y_n$.

Models often involve several unknown parameters whereas only a subset of them is of interest. The latter are called the parameters of interest whereas the remaining parameters are called the nuisance parameters and are not of interest in themselves. In risk classification, the parameter of interest is the mean μ of the response entering the pure premium calculation. Working with ED distributions, this means that the parameter of interest is θ (that is, the analyst in primarily interested in the mean value of the response), so that ϕ is a nuisance parameter. To ease the exposition, we assume in this section that ϕ is known (the estimation of ϕ will be addressed in the next chapters). A full likelihood inference on both parameters (θ, ϕ) is often difficult beyond the Normal case and the method of moments is generally used in a second stage, to obtain $\widehat{\phi}$.

3.3.2   Derivation of the Maximum-Likelihood Estimate

The likelihood has a unique maximum interior to the parameter space for most models considered in this part of the book. The desired $\widehat{\theta}$ can then easily be obtained by solving the likelihood equation

$$\frac{\mathrm{d}}{\mathrm{d}\theta} L(\theta) = 0.$$

This gives

$$0 = \frac{\mathrm{d}}{\mathrm{d}\theta} L(\theta)$$

$$= \sum_{i=1}^{n} \frac{\mathrm{d}}{\mathrm{d}\theta} \frac{n(\bar{y}\theta - a(\theta))}{\phi}$$

$$= \frac{n(\bar{y} - a'(\theta))}{\phi}$$

so that the maximum-likelihood estimate of θ is the unique root of the equation

$$\bar{y} = a'(\theta) \Leftrightarrow \widehat{\theta} = (a')^{-1}(\bar{y}).$$

The solution indeed corresponds to a maximum as the second derivative

$$\frac{d^2}{d\theta^2} L(\theta) = -\frac{na''(\theta)}{\phi} = -\frac{n}{\phi^2} \text{Var}[Y_1]$$

is always negative. We see that the individual observations $y_1, \ldots, y_n$ are not needed to compute $\widehat{\theta}$ as long as the analyst knows $\bar{y}$ (which thus summarizes all the information contained in the observations about the canonical parameter). Also, notice that the nuisance parameter ϕ does not show up in the estimation of θ.

3.3.3 Properties of the Maximum-Likelihood Estimators

Let us briefly discuss the properties of the maximum-likelihood estimators. In order to establish these properties, two fundamental results are used repeatedly:

Law of large numbers

As recalled in Chap. 1, having independent and identically distributed random variables $Z_1, Z_2, Z_3, \ldots$ with finite mean μ, the average $\bar{Z}_n = \frac{1}{n} \sum_{i=1}^{n} Z_i$ converges to the mean μ with probability 1.

Central-limit theorem

As recalled in Chap. 2, having independent and identically distributed random variables $Z_1, Z_2, Z_3, \ldots$ with finite mean μ and variance σ^2, the standardized average

$$\frac{\bar{Z}_n - \mu}{\sigma/\sqrt{n}} = \frac{\sum_{i=1}^{n} Z_i - n\mu}{\sigma\sqrt{n}}$$

approximately obeys the $\mathcal{N}or(0, 1)$ distribution provided n is large enough. Formally,

$$\text{P}\left[\frac{\sum_{i=1}^{n} Z_i - n\mu}{\sigma\sqrt{n}} \leq z \right] \to \Phi(z) \text{ for all } z \text{ as } n \to \infty,$$

where Φ denotes the $\mathcal{N}or(0, 1)$ distribution function.

Both results extend to the multivariate case, with random variables Z_i possibly obeying different distributions.

We are now ready to list the most relevant properties of the maximum-likelihood estimators.

3.3.3.1 Consistency

In the ED family, consistency of the maximum-likelihood estimator is a direct consequence of the law of large numbers as $\overline{Y}$ is known to converge to μ with probability 1 so that $\widehat{\theta} = (a')^{-1}(\overline{Y})$ converges to $(a')^{-1}(\mu) = \theta$.

3.3.3.2 Invariance

If h is a one-to-one function then $h(\widehat{\theta})$ is the maximum-likelihood estimate for $h(\theta)$, that is,

$$\widehat{h(\theta)} = h(\widehat{\theta})$$

when maximum-likelihood is used for estimation. This ensures that for every distribution in the ED family, the maximum-likelihood estimate fulfills

$$\widehat{\mu} = \overline{y}.$$

For instance, consider a response Y obeying the Poisson distribution with mean λ. If the actuary is interested into the no-claim probability

$$P[Y = 0] = \exp(-\lambda)$$

and wishes to estimate it by maximum likelihood, then

$$\widehat{\exp(-\lambda)} = \exp(-\widehat{\lambda})$$

where $\widehat{\lambda}$ is the maximum-likelihood estimate for the Poisson mean λ. In practice, this means that it is not needed to write another likelihood function, with argument $\exp(-\lambda)$ to be maximized, but that it suffices to plug in the latter function the maximum-likelihood estimate $\widehat{\lambda}$ of the original parameter λ.

3.3.3.3 Asymptotic Unbiasedness

The expectation $E[\widehat{\theta}]$ approaches θ as the sample size increases. Again, this is a consequence of the law of large numbers since $\overline{Y}$ tends to μ with probability 1.

3.3.3.4 Minimum Variance

In the class of all estimators, for large samples, $\widehat{\theta}$ has the minimum variance and is therefore the most accurate estimator possible.

We see that many attractive properties of the maximum-likelihood estimation principle hold in large samples. As actuaries generally deal with massive amounts of data, this makes this estimation procedure particularly attractive to conduct insurance studies.

3.4 Fisher's Score and Information

3.4.1 Limited Taylor Expansion of the Log-Likelihood

Expanding the log-likelihood around $\widehat{\theta}$ with the help of Taylor formula, we get

$$
L(\theta) = L(\widehat{\theta}) + \underbrace{L'(\widehat{\theta})}_{=0}\left(\theta - \widehat{\theta}\right) + \frac{1}{2}L''(\widehat{\theta})\left(\theta - \widehat{\theta}\right)^2 + \cdots
$$

$$
= L(\widehat{\theta}) + \frac{1}{2}L''(\widehat{\theta})\left(\theta - \widehat{\theta}\right)^2 + \cdots
$$

If the log-likelihood is well approximated by a quadratic function, i.e. if the approximation

$$
L(\theta) \approx L(\widehat{\theta}) + \frac{1}{2}L''(\widehat{\theta})\left(\theta - \widehat{\theta}\right)^2
$$

is accurate enough, then the likelihood function (or the underlying model) is said to be regular. When the sample size n becomes large, this is generally the case. Many properties which are exactly true for Normally distributed responses remain approximately true in regular models.

3.4.2 Observed Information

In a regular model, the behavior of $L(\theta)$ near $\widehat{\theta}$ is largely determined by the second derivative $L''(\widehat{\theta})$ which measures the local curvature of the log-likelihood, in the neighborhood of the maximum-likelihood estimate $\widehat{\theta}$. A large curvature is associated to a tight, or strong peak. This intuitively indicates less uncertainty about θ. This leads to define the observed Fisher information as

$$
I(\widehat{\theta}) = -L''(\widehat{\theta}) \geq 0.
$$

A large curvature translates into a large, positive value of $I(\widehat{\theta})$. The log-likelihood exhibits in this case a marked peak at $\widehat{\theta}$ and the available sample values bring a lot of information about θ. On the contrary, if $I(\widehat{\theta})$ is close to zero then the graph of the likelihood is rather flat around $\widehat{\theta}$ and there remains a lot of uncertainty about θ.

The observed information is a quantity that varies from sample to sample. It can be regarded as an indicator of the strength of preference for $\widehat{\theta}$ with respect to the other points of the parameter space for the particular sample under study.

3.4.3 Fisher's Score and Expected Information

The first derivative of the log-likelihood L is called Fisher's score,[1] and is denoted by

$$U(\theta) = L'(\theta) = \frac{n(\overline{Y} - a'(\theta))}{\phi}. \tag{3.1}$$

Then one can find the maximum-likelihood estimator by setting the score to zero, i.e. by solving the likelihood equation

$$\overline{Y} = a'(\theta).$$

We know from Property 2.4.4 that $E[Y_1] = a'(\theta)$ when the observations obey an ED distribution. Then,

$$E[U(\theta)] = \frac{n}{\phi}E[\overline{Y} - E[Y_1]] = 0$$

so that the score is centered. This property is generally valid.

Fisher's (expected) information is defined as the variance of Fisher's score U, i.e.

$$I(\theta) = \text{Var}[U(\theta)] = E\left[U^2(\theta)\right].$$

This terminology refers to the fact that, if the data are strongly in favor of a particular value θ of the parameter then the likelihood will be sharply peaked and $U^2(\theta)$ will be large whereas data containing little information about the parameter result in an almost flat likelihood and small $U^2(\theta)$. Hence, $E\left[U^2(\theta)\right]$ indeed measures the amount of information about the parameter of interest in the repeated sampling setting. Notice that from the definition (3.1) of Fisher's score and from $\text{Var}[\overline{Y}] = \frac{\text{Var}[Y_1]}{n}$, we obtain in the ED case

$$I(\theta) = \frac{n\text{Var}[Y_1]}{\phi^2}. \tag{3.2}$$

---

[1] We comply here with standard statistical terminology, keeping in mind that the score has a very different meaning in actuarial applications, as it will become clear from the next chapters. To make the difference visible, we always speak of Fisher's score to designate the statistical concept.

3.4.4 Alternative Expression

There is a useful alternative expression for the Fisher information

$$
\begin{aligned}
\mathcal{I}(\theta) &= -\mathrm{E}\left[L''(\theta)\right] \\
&= -\mathrm{E}\left[U'(\theta)\right] \\
&= \frac{na''(\theta)}{\phi}
\end{aligned}
\tag{3.3}
$$

which coincides with (3.2). For responses obeying an ED distribution, Fisher's information $\mathcal{I}$ and observed information I both coincide as $\mathcal{I}(\theta) = I(\theta)$. This is a remarkable property of the ED family that remains valid as long as we work with the canonical parameter θ (see e.g. Sect. 8.3 of Pawitan 2001, for more details).

Contrarily to the observed information $I(\widehat{\theta})$ that relates to a given sample with maximum likelihood estimate $\widehat{\theta}$, Fisher's information $\mathcal{I}$ quantifies the expected amount of information available in a sample of size n. As a function of θ, Fisher's information $\mathcal{I}$ tells how hard it is to estimate θ as parameters with greater information can be estimated more easily (requiring a smaller sample size to achieve a required precision level).

Written in the alternative form (3.3), $\mathcal{I}(\theta)$ can be interpreted as follows: the information contained in a typical sample about the canonical parameter θ

- increases linearly in the sample size n;
- decreases with the dispersion parameter ϕ;
- increases with

$$
a''(\theta) = \frac{\mathrm{d}}{\mathrm{d}\theta}\mu(\theta),
$$

that is, increases with the variations of the mean response in the canonical parameter.

Thus,

- a larger sample size provides more information about θ.
- a smaller dispersion parameter ϕ translates into more information about θ. Identity (3.3) shows that halving ϕ has the same effect as doubling the sample size in terms of information.
- a sample provides more information about θ when the mean value is more responsive to a change in θ. Samples corresponding to parameter θ such that $a''(\theta)$ is large brings thus more information.

3.5 Sampling Distribution

A first-order Taylor expansion of Fisher's score gives

$$\underbrace{U(\widehat{\theta})}_{=0} \approx U(\theta) + U'(\theta)\left(\widehat{\theta} - \theta\right) = I(\theta)\left(\theta - \widehat{\theta}\right)$$

where the second equality comes from the identity

$$U'(\theta) = -\frac{na''(\theta)}{\phi} = -I(\theta).$$

The central-limit theorem ensures that

$$Z = \frac{\sum_{i=1}^{n} Y_i - nE[Y_1]}{\sqrt{n\text{Var}[Y_1]}} = \frac{\sum_{i=1}^{n} Y_i - na'(\theta)}{\sqrt{n\phi a''(\theta)}} \approx \mathcal{N}or(0, 1)$$

for n sufficiently large. Hence,

$$U(\theta) = \frac{n\left(\overline{Y} - a'(\theta)\right)}{\phi} = Z\sqrt{\frac{n}{\phi}a''(\theta)} = Z\sqrt{I(\theta)}$$

is approximately Normally distributed provided the sample size n is large enough. Precisely, we have

$$U(\theta) \approx \mathcal{N}or\left(0, I(\theta)\right) \text{ for large sample size } n,$$

so that the estimation error

$$\widehat{\theta} - \theta = \left(I(\theta)\right)^{-1} U(\theta) \approx \mathcal{N}or\left(0, \left(I(\theta)\right)^{-1}\right).$$

In practice, we use the large-sample approximation

$$\widehat{\theta} \approx \mathcal{N}or(\theta, \widehat{\sigma}_{\widehat{\theta}}^2) \text{ with } \widehat{\sigma}_{\widehat{\theta}}^2 = \left(I(\widehat{\theta})\right)^{-1}.$$

3.6 Numerical Illustrations

3.6.1 Death Counts

Assume that we observe 100 "identical" policyholders (same age, gender, socio-economic status, health status, etc.) so that their remaining lifetimes can be considered as independent and identically distributed random variables $T_1, \ldots, T_{100}$. These 100 individuals are observed during a whole year and the number of deaths

$$Y = \sum_{i=1}^{100} I[T_i \leq 1]$$

is recorded, where $I[\cdot]$ is the indicator function. Here, individual observations are

$$Y_i = I[T_i \leq 1], \quad i = 1, \ldots, 100.$$

In this setting, the parameter of interest is the one-year death probability q defined as

$$q = P[T_1 \leq 1].$$

Observations Y_i are independent and obey the $\mathcal{B}er(q)$ distribution. Clearly, $Y \sim \mathcal{B}in(100, q)$. Since q can be considered as the population version of the proportion of deaths recorded in an homogeneous group of individuals, a natural estimator for q is $Y/100$.

Assume that after one year, 30 individuals among the group of size 100 died. Based on this information, we would like to estimate the one-year death probability q. Let us derive the maximum-likelihood estimate for q. The likelihood associated with this observational study is

$$\mathcal{L}(q) = \binom{100}{30} q^{30}(1-q)^{70}$$

which is the probability to observe 30 deaths out of 100 individuals for a one-year death probability equal to q. The corresponding log-likelihood is

$$L(q) = \ln \mathcal{L}(q) = \ln \binom{100}{30} + 30 \ln q + 70 \ln(1-q)$$
$$= 30 \ln q + 70 \ln(1-q) + \text{constant}.$$

Let us rewrite the log-likelihood in function of the canonical parameter

$$\theta = \ln \frac{q}{1-q} \Leftrightarrow q = \frac{\exp(\theta)}{1+\exp(\theta)}$$

of the Binomial distribution. This gives

$$\mathcal{L}(\theta) = \binom{100}{30} \left(\frac{\exp(\theta)}{1 + \exp(\theta)} \right)^{30} \left(1 - \frac{\exp(\theta)}{1 + \exp(\theta)} \right)^{70}$$

$$= \binom{100}{30} \left(\exp(\theta) \right)^{30} \left(1 + \exp(\theta) \right)^{-100}$$

$$L(\theta) = 30\theta - 100 \ln \left(1 + \exp(\theta) \right) + \text{constant}.$$

The log-likelihood L corresponding to this sample is displayed in the left panel of Fig. 3.1. We can see there that there is a unique maximum corresponding to the maximum-likelihood estimate $\widehat{q}$.

Let us now compute the maximum-likelihood estimate analytically. Fisher's score is equal to

$$U(\theta) = \frac{d}{d\theta} L(\theta) = 30 - 100 \frac{\exp(\theta)}{1 + \exp(\theta)}$$

so that the maximum-likelihood estimate of θ is the solution to the likelihood equation $U(\theta) = 0$, that is,

$$0 = 30 - 100 \frac{\exp(\theta)}{1 + \exp(\theta)}.$$

The solution is given by

$$\widehat{\theta} = \ln \frac{3}{7} \Leftrightarrow \widehat{q} = 0.3.$$

The advantage of working with θ instead of q is that θ is allowed to take any real value whereas q is restricted to the unit interval $[0, 1]$. This makes the large-sample Normal approximation to $\widehat{\theta}$ more accurate. The observed information is

$$I(\widehat{\theta}) = -U'(\widehat{\theta}) = 100 \frac{\exp(\widehat{\theta})}{\left(1 + \exp(\widehat{\theta}) \right)^2} = 21.$$

Intuitively speaking, we feel that we would have been more comfortable with our estimate if it were based on a larger group of observed individuals. Assume now that we observed $y = 300$ deaths out of $n = 1,000$ individuals. We end up with the same estimate as before but we now intuitively feel more comfortable with this estimate. This is reflected by the observed information which is now equal to

$$1,000 \times 0.3 \times 0.7 = 210,$$

ten times more than with $n = 100$. To formalize this idea, it is convenient to use the relative likelihood

$$R(\theta) = \frac{\mathcal{L}(\theta)}{\mathcal{L}(\widehat{\theta})}. \tag{3.4}$$

The graph of $\theta \mapsto \ln R(\theta)$ for $n = 100$ (continuous line), $n = 10$ (broken line) and $n = 1,000$ (dotted line) is displayed in the right panel of Fig. 3.1. We clearly see

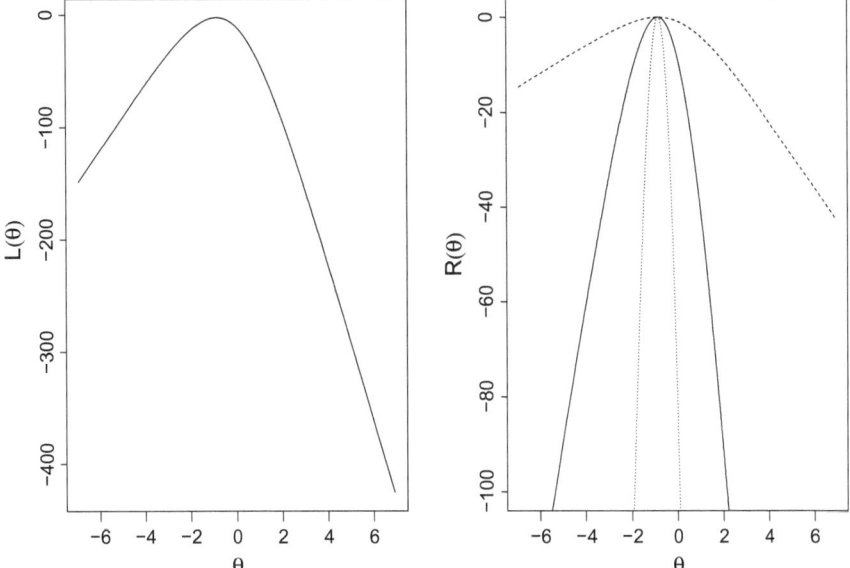

Fig. 3.1 Graph of the Binomial log-likelihood $\theta \mapsto L(\theta)$ in the left panel and of the log-relative likelihood $\theta \mapsto \ln R(\theta)$ by varying the number of individuals in the right panel: the broken line (- - -) corresponds to $n = 10$ and the dotted line ($\cdots$) corresponds to $n = 1,000$

there that the log-likelihood becomes more peaked when n gets larger. Considering the information as a value of q, we see that it is maximum for $\theta = 0 \Leftrightarrow q = 0.5$ and decreases when q departs from its central value. Hence, it appears to be more difficult to estimate smaller values of q, which typically arise in mortality studies.

In actuarial applications, computing $\widehat{\theta}$ is not enough and an assessment of the degree of accuracy of the estimate (error margin) is needed. This amounts to determine the set of plausible values of the parameter, including $\widehat{\theta}$ as one of its elements. Even if $\widehat{\theta}$ is the preferred point estimate because it has the highest likelihood, other parameter values close to $\widehat{\theta}$ are also quite plausible. It is therefore natural to look for an interval which should presumably contain the parameter. The set of the plausible values for θ is called the confidence interval for θ.

Actuaries generally determine confidence intervals on the basis of the large-sample approximation to the law of the maximum-likelihood estimator $\widehat{\theta}$. For $n = 100$, the large-sample approximation to the estimation error distribution is given by

$$\widehat{\theta} - \theta \approx \mathcal{N}or\left(0, \frac{1}{21}\right).$$

This ensures that the estimation error $\widehat{\theta} - \theta$ approximately fulfills

$$0.95 \approx P\left[-1.96\frac{1}{\sqrt{21}} \leq \widehat{\theta} - \theta \leq 1.96\frac{1}{\sqrt{21}}\right]$$

$$= P\left[\widehat{\theta} - 1.96\frac{1}{\sqrt{21}} \leq \theta \leq \widehat{\theta} + 1.96\frac{1}{\sqrt{21}}\right]$$

so that the interval $[\widehat{\theta} \pm 1.96/\sqrt{21}]$ contains θ with approximate probability 95%; the latter is referred to as the confidence level or coverage probability of this interval. The error margin is the half width of the confidence interval, that is, $1.96/\sqrt{21}$ in our example. Often, the value 1.96 is approximated by 2 so that we recover the "2 standard deviations" rule.

If we compare this confidence interval based on the large-sample Normal approximation to the exact confidence interval for the Binomial parameter q obtained from the R function `binomial.test`, we get

$$[0.2124064; 0.3998147] \text{ (exact)}$$

versus

$$[0.2184017; 0.3966147] \text{ (large-sample approximation).}$$

This shows that the Normal approximation is reasonably accurate in this case.

3.6.2   Claim Frequencies

Let us now consider a slightly more elaborate example, where individuals differ in terms of the length of their observation period. Precisely, assume that policyholder i has been observed during a period of length e_i (which thus represents the exposure to risk for policyholder i) and produced y_i claims, $i = 1, 2, \ldots, n$. Assuming that the numbers of claims Y_i are independent and $Poi(\lambda e_i)$ distributed, the likelihood is

$$\mathcal{L}(\lambda) = \prod_{i=1}^{n} P[Y_i = y_i] = \prod_{i=1}^{n}\left(\exp(-\lambda e_i)\frac{(\lambda e_i)^{y_i}}{y_i!}\right).$$

Notice that the formula derived above cannot be applied directly as the canonical parameter here varies from one response to the other as

$$\theta_i = \ln(e_i \lambda) = \ln e_i + \ln \lambda = \ln e_i + \theta.$$

Here, $\theta = \ln \lambda$ is the parameter of interest as it gives the expected number of claims over one period: if $e = 1$ then $E[Y] = \lambda = \exp(\theta)$. The exposures to risk e_i account for observation periods of different lengths among individual policyholders.

Switching from the mean λ to the canonical parameter θ, the corresponding log-likelihood is

$$L(\theta) = -\exp(\theta)\sum_{i=1}^{n} e_i + \sum_{i=1}^{n} y_i\left(\theta + \ln e_i\right) - \sum_{i=1}^{n} \ln y_i!$$

$$= -\exp(\theta)\sum_{i=1}^{n} e_i + \theta\sum_{i=1}^{n} y_i + \text{constant with respect to } \theta.$$

The maximum-likelihood estimate of θ is then obtained from

$$U(\theta) = \frac{d}{d\theta}L(\theta) = 0 \Leftrightarrow -\exp(\theta)\sum_{i=1}^{n} e_i + \sum_{i=1}^{n} y_i = 0$$

$$\Rightarrow \widehat{\theta} = \ln\frac{\sum_{i=1}^{n} y_i}{\sum_{i=1}^{n} e_i}$$

which indeed corresponds to the maximum of L as

$$\frac{d^2}{d\theta^2}L(\theta) = -\exp(\theta)\sum_{i=1}^{n} e_i < 0 \text{ for any } \theta.$$

Consider the observed claim distribution displayed in Table 3.1. We can read there the number n_k of policies having reported k claims, $k = 0, 1, 2, 3, 4$, together with the corresponding total exposure e_k for these n_k policies. The left panel of Fig. 3.2 displays log-likelihood function associated to the data in Table 3.1, that is,

$$L(\theta) = -\exp(\theta)\sum_{k=0}^{4} e_k + \theta\sum_{k=1}^{4} kn_k + \text{constant with respect to } \theta.$$

We can see there that there is a clear maximum to the likelihood function, which corresponds to

Table 3.1 Observed claim frequency distribution

Number of claims k	Number of policies n_k	Total exposure e_k (in years)
0	12 962	10 545.94
1	1 369	1 187.13
2	157	134.66
3	14	11.08
4	3	2.52
Total	14 505	11 881.35

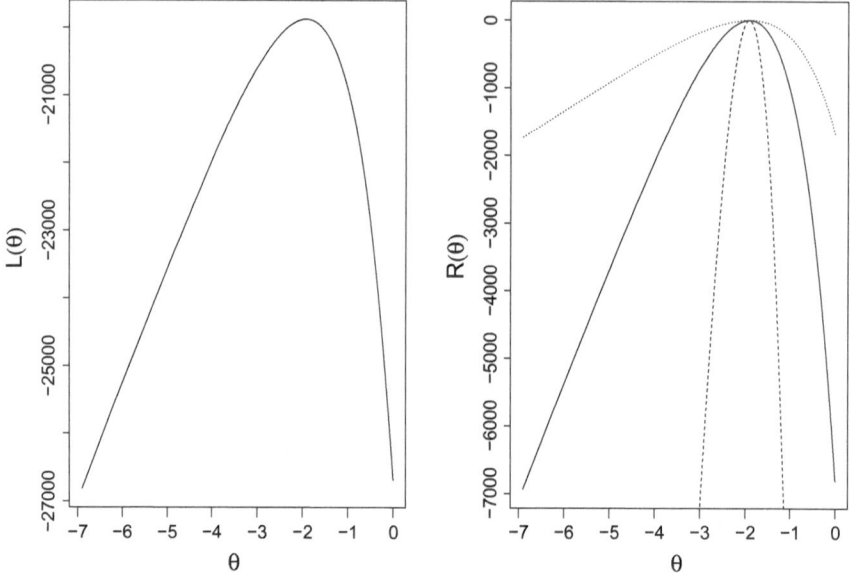

Fig. 3.2 Graph of the Poisson log-likelihood $\theta \mapsto L(\theta)$ in the left panel and of the log-relative likelihood $\theta \mapsto \ln R(\theta)$ by varying the total exposure in the right panel: the dotted line $(\cdots)$ corresponds to the initial exposure divided by 4 while the broken line (- - -) corresponds to 10 times the initial exposure

$$
\begin{aligned}
\widehat{\theta} &= \ln \frac{\sum_{k=1}^{4} k n_k}{\sum_{k=0}^{4} e_k} \\
&= \ln \frac{1369 \times 1 + 157 \times 2 + 14 \times 3 + 3 \times 4}{10545.94 + \ldots + 2.52} \\
&= \ln 0.1462 \\
&= -1.922809.
\end{aligned}
$$

The observed information is equal to

$$
I(\widehat{\theta}) = -L''(\widehat{\theta}) = \exp(\widehat{\theta}) \sum_{i=1}^{n} e_i = \exp(\widehat{\theta}) \sum_{k=0}^{4} e_k = 1{,}737.
$$

Increasing the total exposure $\sum_{i=1}^{n} e_i$ means that more information becomes available. Hence, we intuitively feel that we should become more comfortable with the estimate. Also, the higher the observed frequency, the more information in the sample. Another look at this formula shows that the more claims are recorded, the more information is contained in the sample data as $I(\widehat{\theta}) = \sum_{i=1}^{n} y_i$.

As in the Binomial case, it is again convenient to use the relative likelihood (3.4) to formalize this idea. The graph in the right panel of Fig. 3.2 displays the curve $\theta \mapsto \ln R(\theta)$ when varying $\sum_{i=1}^{n} e_i$. We clearly see that increasing the total exposure, holding $\widehat{\theta}$ constant, the relative likelihood becomes more and more peaked.

Fisher's expected information is equal to

$$I(\theta) = \exp(\theta) \sum_{i=1}^{n} e_i.$$

The information contained in a typical sample thus depends on the total observation period, or total exposure $\sum_{i=1}^{n} e_i$, as well as on the underlying expected claim rate $\lambda = \exp(\theta)$. Increasing the total exposure $\sum_{i=1}^{n} e_i$ increases the information provided by the available sample about θ. Also, a larger value of θ, that is, a larger λ, is easier to estimate. In comparison, a small λ requires a larger exposure to achieve the same level of accuracy. Equivalently, we see that Fisher's expected information is equal to the expected number of claims for a set of observations with total exposure $\sum_{i=1}^{n} e_i$. This is in contrast to the observed information which equals the total number of claims actually observed for that sample. Once estimated, the numerical values of these two quantities just coincide.

The large-sample approximation to the estimation error distribution is given by

$$\widehat{\theta} - \theta \approx \mathcal{N}or\left(0, \frac{\exp(-\widehat{\theta})}{\sum_{i=1}^{n} e_i}\right).$$

Hence, an approximate 95% confidence interval for θ (known as the Wald one) is easily obtained as follows: starting from

$$0.95 \approx P\left[-1.96 \leq \frac{\widehat{\theta} - \theta}{\sqrt{\exp(-\widehat{\theta})/\sum_{i=1}^{n} e_i}} \leq 1.96\right]$$

$$= P\left[\widehat{\theta} - 1.96\sqrt{\exp(-\widehat{\theta})/\sum_{i=1}^{n} e_i} \leq \theta \leq \widehat{\theta} + 1.96\sqrt{\exp(-\widehat{\theta})/\sum_{i=1}^{n} e_i}\right]$$

we conclude that the interval

$$\left[\widehat{\theta} \pm 1.96\sqrt{\exp(-\widehat{\theta})/\sum_{i=1}^{n} e_i}\right]$$

contains the true value of the canonical parameter with approximate probability 95%. In our example, this gives

$$\left[-1.922809 \pm 1.96\sqrt{0.000576}\right] = \left[-1.922809 \pm 0.047028\right].$$

If we compare the large-sample approximate confidence interval to the exact one based on the Poisson distribution obtained from the R function `poisson.test`, we get

$$[0.1394007; 0.1532364] \text{ (exact)}$$

and

$$[0.1394779; 0.1532371] \text{ (large-sample approximation)}.$$

The two intervals are in close agreement.

Remark 3.6.1 Notice that the results depend on the chosen parameterization. Keeping the mean parameter λ instead of the canonical parameter θ, we get

$$L(\lambda) = -\lambda \sum_{i=1}^{n} e_i + \ln \lambda \sum_{i=1}^{n} y_i + \text{constant}$$

$$L'(\lambda) = -\sum_{i=1}^{n} e_i + \frac{\sum_{i=1}^{n} y_i}{\lambda}$$

$$\widehat{\lambda} = \frac{\sum_{i=1}^{n} y_i}{\sum_{i=1}^{n} e_i}$$

$$= \exp(\widehat{\theta}) \text{ as expected from the functional invariance property}$$

$$I(\lambda) = \frac{\sum_{i=1}^{n} y_i}{\lambda^2}$$

$$\mathcal{I}(\lambda) = \frac{\sum_{i=1}^{n} e_i}{\lambda}.$$

Even if $I(\lambda) \neq \mathcal{I}(\lambda)$, we see that their estimations coincide, as

$$I(\widehat{\lambda}) = \mathcal{I}(\widehat{\lambda}).$$

We refer the reader to Sect. 8.3 in Pawitan (2001) for more details about the effect of a change in parametrization inside the ED family.

Remark 3.6.2 As $\widehat{\lambda}$ is an unbiased estimator of λ, the maximum-likelihood estimator $\widehat{\theta}$ is biased in small samples. This is because

$$\mathrm{E}[\widehat{\theta}] = \mathrm{E}[\ln \widehat{\lambda}] \leq \ln \mathrm{E}[\widehat{\lambda}] = \ln \lambda = \theta$$

where the inequality follows from Jensen theorem (the logarithm being a concave function). The bias nevertheless becomes negligible when the total exposure becomes large enough.

3.7 Bootstrap

In some applications, analytical results are hard to obtain and numerical methods must be used. The bootstrap approach recalled in this section appears to be particularly useful in that respect.

3.7.1 Estimation Accuracy

Suppose we have independent random variables $Y_1, Y_2, \ldots, Y_n$ with common distribution function F and that we are interested in using them to estimate some quantity $\theta(F)$ associated with F. An estimator

$$\widehat{\theta} = g(Y_1, Y_2, \ldots, Y_n)$$

is available for $\theta(F)$. Its accuracy can be measured by the mean square error

$$\mathrm{MSE}(F) = \mathrm{E}_F\left[\left(g(Y_1, Y_2, \ldots, Y_n) - \theta(F)\right)^2\right]$$

where the subscript "F" to E_F explicitly indicates that the mathematical expectation is taken with respect to F.

Example 3.7.1 Let us consider the case $\theta(F) = \mathrm{E}_F[Y_1]$. Since F is (at least partially) unknown, it is impossible to compute $\mathrm{MSE}(F)$. In this particular case, using the estimator

$$\widehat{\mathrm{E}_F[Y_1]} = \overline{Y}_n = \frac{1}{n}\sum_{i=1}^{n} Y_i$$

we have

$$\mathrm{MSE}(F) = \mathrm{E}_F\left[\left(\overline{Y}_n - \mathrm{E}_F[Y_1]\right)^2\right] = \mathrm{Var}[\overline{Y}_n] = \frac{\mathrm{Var}[Y_1]}{n}.$$

Hence, there is a natural estimate of the MSE

$$\widehat{\mathrm{MSE}}(F) = \frac{S_n^2}{n} \text{ where } S_n^2 = \frac{1}{n-1}\sum_{i=1}^{n}(Y_i - \overline{Y}_n)^2$$

is the unbiased estimator of the common variance of $Y_1, \ldots, Y_n$

In case of more complicated statistics, there is usually no analytical formula available and the bootstrap may give a viable solution, as explained next.

3.7.2 Plug-In Principle

Define the empirical counterpart to F as

$$\widehat{F}_n(x) = \frac{\#\{Y_i \text{ such that } Y_i \leq x\}}{n} = \frac{1}{n}\sum_{i=1}^{n} I[Y_i \leq x].$$

Thus, the empirical distribution function $\widehat{F}_n$ puts an equal probability $\frac{1}{n}$ on each of the observed data points $Y_1, \ldots, Y_n$. The bootstrap estimate of $\text{MSE}(F)$ is

$$\text{MSE}(\widehat{F}_n) = E_{\widehat{F}_n}\left[\left(g(Y_1^\star, Y_2^\star, \ldots, Y_n^\star) - \theta(\widehat{F}_n)\right)^2\right]$$

where the random variables $Y_1^\star, Y_2^\star, \ldots, Y_n^\star$ are independent with common distribution function $\widehat{F}_n$.

In some cases, it is possible to calculate bootstrap estimates analytically but in general resampling is necessary.

Example 3.7.2 Let us consider again the case $\theta(F) = E_F[Y_1]$. Clearly,

$$\theta(\widehat{F}_n) = \bar{y}_n = \frac{1}{n}\sum_{i=1}^{n} y_i$$

corresponds to the classical estimate for the mean parameter. The bootstrap estimate of $\text{MSE}(F)$ is then equal to

$$\text{MSE}(\widehat{F}_n) = E_{\widehat{F}_n}\left[\left(\frac{1}{n}\sum_{i=1}^{n} Y_i^\star - \bar{y}_n\right)^2\right]$$

where the the random variables $Y_i^\star$ are independent and obey the distribution function $\widehat{F}_n$. We then have

$$E_{\widehat{F}_n}\left[\frac{1}{n}\sum_{i=1}^{n} Y_i^\star\right] = E_{\widehat{F}_n}[Y_1^\star] = \bar{y}_n.$$

Hence,

$$\text{MSE}(\widehat{F}_n) = \text{Var}_{\widehat{F}_n}\left[\frac{1}{n}\sum_{i=1}^{n} Y_i^\star\right] = \frac{\text{Var}_{\widehat{F}_n}[Y_1^\star]}{n}.$$

Since

$$\text{Var}_{\widehat{F}_n}[Y_1^\star] = E_{\widehat{F}_n}[(Y_1^\star - \bar{y}_n)^2] = \frac{1}{n}\sum_{i=1}^{n}(y_i - \bar{y}_n)^2 = \widehat{\sigma}_n^2$$

$$\Rightarrow \text{MSE}(\widehat{F}_n) = \frac{\widehat{\sigma}_n^2}{n} \approx \frac{S_n^2}{n} \text{ in large samples.}$$

So that we essentially reach the same conclusion as before.

3.7.3 Bootstrap Estimate

In general,

$$\text{MSE}(\widehat{F}_n) = \frac{1}{n^n} \sum_{i_1=1}^{n} \sum_{i_2=1}^{n} \cdots \sum_{i_n=1}^{n} \left(g(y_{i_1}, y_{i_2}, \ldots, y_{i_n}) - \theta(\widehat{F}_n) \right)^2.$$

The computation of $\text{MSE}(\widehat{F}_n)$ requires, thus, summing n^n terms, which becomes time-consuming, or even out of reach when n gets large. This is why $\text{MSE}(\widehat{F}_n)$ is approximated by using simulation and averaging over a large number of terms.

The idea behind the non-parametric bootstrap is to simulate sets of independent random variables

$$Y_1^{(b)}, Y_2^{(b)}, \ldots, Y_n^{(b)}$$

obeying the distribution function $\widehat{F}_n$, $b = 1, 2, \ldots, B$. This can be done by simulating $U_i \sim \mathcal{U}ni(0, 1)$ and setting

$$Y_i^{(b)} = y_I \text{ with } I = [nU_i] + 1.$$

Then,

$$\text{MSE}(\widehat{F}_n) \approx \frac{1}{B} \sum_{b=1}^{B} \left(g(Y_1^{(b)}, Y_2^{(b)}, \ldots, Y_n^{(b)}) - \theta(\widehat{F}_n) \right)^2$$

for some large enough B.

3.7.4 Nonparametric Versus Parametric Bootstrap

In the non-parametric set-up, resampling is performed from $\widehat{F}_n$, that is, sampling from the observations $y_1, \ldots, y_n$ with replacement. This leads to a distribution of the statistic of interest, from which quantities such as standard errors can be estimated. In the parametric bootstrap, the underlying distribution is estimated from the data by a parametric model, that is,

$$F = F_\theta.$$

Bootstrap samples are then generated by sampling from $F_{\widehat{\theta}}$ rather than from $\widehat{F}_n$.

3.7.5   Applications

Let us use the bootstrap approach to obtain a 95% confidence interval for the one-year death probability q. Recall that the observations consist in 30 deaths and 70 survivors, gathered in the vector of death indicators $Y_i = I[T_i \leq 1]$ (with a 1 appearing 30 thirty times and a 0 appearing 70 times). Without loss of generality, we can re-order the observations $Y_1, \ldots, Y_{100}$ so that the values 1 appear first in the vector.

We then sample 10,000 times with replacement from this vector using the `sample` function available in the statistical software R. For each boostrap data set, we compute the success probability by averaging the values of the re-sampled vector and we store the results. The interval starting at the 2.5% percentile to the 97.5% percentile is [0.21, 0.39]. The obtained results appear to be very close to those based on the large-sample Normal approximation and the exact Binomial confidence interval reported before. Notice that instead of resampling from the 100 observed values, another possibility consists in generating the number of deaths from the Binomial distribution with parameters 100 and $\widehat{q} = 0.3$ using the R function `rbinom` and divide the resulting 10,000 realizations by the size 100. Figure 3.3 (upper panel) displays the histogram of the 10,000 bootstrap values for the one-year death probability, which appears to be in agreement with the large-sample Normal approximation.

The `boot` function of the R library `boot` allows us to get different versions of the bootstrap approach. The mean of the available sample is $\widehat{q} = 0.3$. The difference between $\widehat{q}$ and the mean $\widehat{q}^b$ of the 10,000 bootstrapped samples is

$$\widehat{q} - \frac{1}{10,000} \sum_{b=1}^{10,000} \widehat{q}^b = -0.000279$$

whereas the corresponding standard error equals to 0.04546756. The histogram of the 10,000 values is displayed in the lower panel of Fig. 3.3. Even if the `boot` function does not provide different results compared to a direct use of the `sample` function, it offers advanced bootstrap capabilities, such as the confidence interval based on the large-sample Normal approximation [0.2112, 0.3894], the confidence interval obtained from the 2.5 and 97.5% quantiles of the 10,000 bootstrapped values [0.21, 0.39], and the bias-corrected accelerated percentile confidence interval [0.21, 0.38] considered to be the most accurate (even if in this particular example, all intervals closely agree).

3.8   Bibliographic Notes and Further Reading

Klugman et al. (2012) offer a detailed treatment of estimation in an actuarial context. The reader interested in a specialized analysis of the likelihood paradigm is referred to Pawitan (2001). All the numerical illustrations contained in this chapter, as well as in the remainer of the book, have been performed with the help of the statistical

Fig. 3.3 Histogram of the 10,000 bootstrapped values for the death probability (top panel) where ms is the vector containing the 10,000 bootstrap averages and those produced by the boot function (bottom panel)

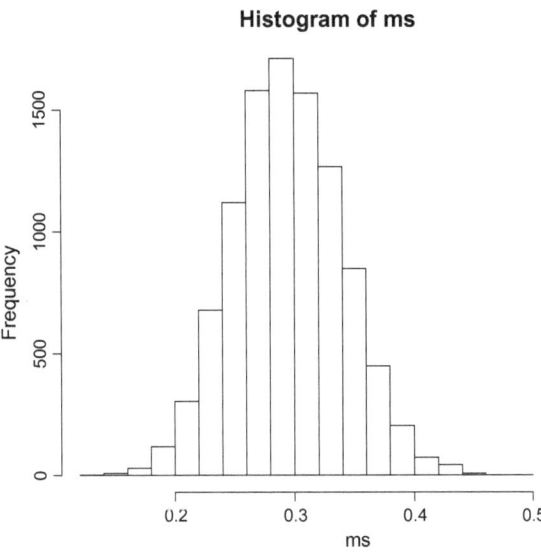

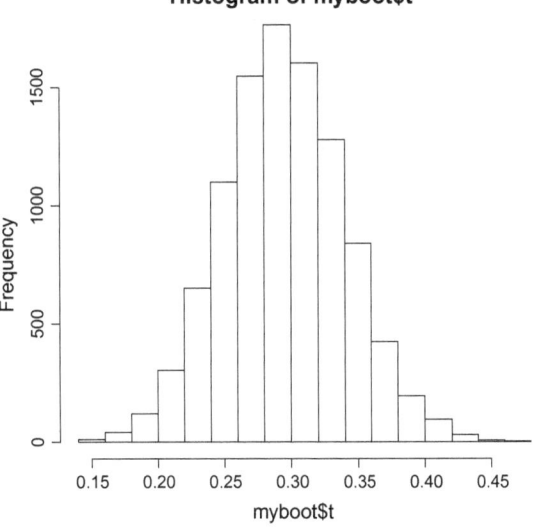

software R. The readers who wish to get an introduction to R are referred to the books included in the Use R! series published by Springer. The R book by Crawley (2007) also appears to be very useful. The bootstrap applications closely follow this book (precisely, the section devoted to the bootstrap in Chap. 8 of Crawley 2007). We also refer the interested reader to the book by Efron and Hastie (2016) for a lucid treatment of the evolution of statistical thinking.

References

Crawley MJ (2007) The R book. Wiley

Efron B, Hastie T (2016) Computer age statistical inference. Cambridge University Press

Klugman SA, Panjer HH, Willmot GE (2012) Loss models: from data to decisions, 4th edn. Wiley

Pawitan Y (2001) In all likelihood: statistical modelling and inference using likelihood. Oxford
   University Press

Part II
Linear Models

Chapter 4
Generalized Linear Models (GLMs)

4.1 Introduction

A GLM is defined by specifying two components:

- the response must obey an ED distribution presented in Chap. 2 and
- a link function describes how the mean response is related to a linear combination of available features (generally called the score in insurance applications, not to be confused with Fisher's score in statistics recalled in Chap. 3).

GLMs were originally introduced in actuarial practice as a method for improving the accuracy of motor insurance pricing. Their use was subsequently rapidly extended to most business lines. Today, GLMs are routinely applied to underwriting, pricing or claims management, for instance. Here are some typical applications of GLMs in insurance:

- Claim numbers modeling.
  The experience relating to a particular insurance contract is generally decomposed into a number of claims together with the respective costs of these claims. The Poisson distribution is the natural candidate to model the number of claims reported by the policyholders. The typical assumption in these circumstances is that the conditional mean claim frequency can be written as an exponential function of a linear score to be estimated from the data.
- Claim amounts modeling.
  Since claim costs are non-negative and positively skewed, the Gamma and Inverse-Gaussian distributions are natural candidates to model them. Here also actuaries generally assume that the conditional mean claim severities can be written as the exponential of a linear score.
- Graduation of mortality and morbidity rates.
  Actuaries are aware that a lot of heterogeneity is present in policyholders' survival. Risk factors affecting mortality generally include age, gender, education, income, occupation, marital status, and health behaviors. In general, individuals

M. Denuit et al., *Effective Statistical Learning Methods for Actuaries I*,
Springer Actuarial, https://doi.org/10.1007/978-3-030-25820-7_4

with higher socio-economic status tend to live longer than those in lower socio-economic groups. The Binomial and Poisson GLMs allow the actuary to perform risk classification in life and health insurance. The approach depends on the kind of data available: open or closed group, and individual data or data grouped into intervals or according to some features (such as the sum insured, for instance).
– Elasticity modeling.
Elasticity models for new and renewal business help companies to predict the impact of various actions (and premium increases) on the market share. The profitability and elasticity models can then be combined to take optimal pricing decisions.
– Loss reserving.
Besides the Overdispersed Poisson GLM corresponding to the classical Chain-Ladder aggregate reserving method, GLMs can also be used in individual claim reserving to predict the time-to-settlement depending on some claim features, as well as the ultimate claim amount. This allows the claim handlers to set up more accurate reserves and to provide early identifications of claims that may be fraudulent or are most likely to end up in a lawsuit.
– Competitive analysis.
GLMs can be used to reconstitute the commercial price list used by competitors based on samples of quotations by these competitors. This is done by reverse-engineering of their price list. This allows the actuaries to extrapolate premiums to other risk profiles beyond those for which quotations are available.

This chapter aims to carefully reviews the GLM machinery, step by step. The more advanced methods presented in the next chapters can be seen as extensions of the basic GLM approach proposed to remedy to some of their drawbacks. This is why a deep understanding of the GLM machinery is needed and justifies the detailed treatment offered in this chapter.

As mentioned above, GLMs are based on a distributional assumption (selecting a member of the ED family to model random departures from the mean), a linear score involving the available features and an assumed link between the expected response and the score. Once the actuary has selected these components in line with the data under consideration, statistical inference is conducted with the help of the maximum likelihood principle reviewed in Chap. 3. Diagnostic measures then allow the actuary to study the relative merits of the scores under consideration, in order to select the optimal one for the insurance data under consideration. All the analysis can be performed with existing software, including the freely available, open-source R used throughout this book.

4.2 Structure of GLMs

The GLM approach is an example of regression modeling, or supervised learning procedure. A regression model aims to explain some characteristics of a response, with the help of features acting as explanatory variables. In insurance applications, actuaries typically wish to explain the mean response value entering the pure premium calculation: expected number of claims and corresponding expected claim amounts, for instance.

GLMs separate systematic features of the data from random variations that must be compensated by insurance. The systematic features are represented by a regression function whereas the random variations are represented by a probability distribution in the ED family. Precisely, a GLM consists of three components:

- the response distribution (random component) selected from the ED family,
- the linear score (systematic component), and
- the link function relating the mean response to the score involving the available features.

The systematic component gives the pure premium corresponding to the observable policyholders' profiles whereas the random component represents the volatility to be compensated by the insurer by virtue of the mutuality, or random solidarity principle. When using a GLM, the actuary implicitly assumes that all the necessary information is available (that is, the random vectors X and X^+ coincide, in the notation of Sect. 1.3.4). If this is not the case, then mixed models must be used instead (as those presented in Chap. 5).

4.2.1 Response Distribution

Consider responses $Y_1, Y_2, \ldots, Y_n$ measured on n individuals. For each response, the actuary has information summarized in a vector $x_i = (1, x_{i1}, \ldots, x_{ip})^\top$ of dimension $p + 1$ containing the corresponding features (supplemented with a 1 in front when an intercept is included in the score, as assumed in the majority of insurance applications). Features typically consist in policy terms or policyholder's characteristics. Think for instance of type of vehicle, age or extent of coverage in motor insurance, or of construction type, building date or amount of insurance in homeowners insurance. Insurers have now access to many possible classification variables that are based on policy conditions or information provided by policyholders, or that are contained in external databases. Sometimes, the actuary is able to supplement insurance with banking or telematics data, expanding a lot the available information. Henceforth, we assume that the features x_{ij} have been recorded without error nor missing values.

Besides the response, we have a target, i.e. a quantity of interest. In ratemaking, the target is the pure premium

$$\mu(x) = \mathrm{E}[Y|X = x],$$

also called regression functions in statistics. Notice that the target is not the response Y itself (as it would be in the case for prediction). This is because the actuary wishes to determine pure premiums, that is, expected losses. For pricing purposes, the actuary is not interested in predicting the actual losses of any policyholder, but well the average losses for an homogeneous group of insured individuals.

Actuaries generally deal with observational data, so that the features are themselves random variables. The analysis is then conducted on a sample (y_i, x_i) of observed values that can be understood as realizations of a random vector joining the response with the vector of features. However, we do not attempt to model the joint distribution of this random vector. Instead, given the information contained in x_i, the responses Y_i are assumed to be independent with a conditional distribution in the ED family thoroughly studied in Chap. 2. It is worth stressing that the responses do not need to be identically distributed. Specifically, Y_i obeys the distribution (2.3) with specific canonical parameter

$$\theta_i = \theta(x_i)$$

expressed in function of the available features x_i, and the same dispersion parameter ϕ. The way θ_i is linked to the information contained in x_i is explained next. The quantity of interest is the policyholder-specific mean response μ_i given by

$$\mu_i = \mu(x_i) = a'(\theta_i) = a'\big(\theta(x_i)\big).$$

Specific weights ν_i can be used to account for varying volumes of information, e.g. when Y_i represents an average of outcomes for a group of homogeneous risks rather than the outcome of individual risks. For instance, the response may be the average loss amount for several claims, all with the same features. The weight ν_i is then taken as the number of losses entering the average, in application of Property 2.4.2. Despite this remarkable property of GLMs, allowing the actuary to perform analyses on grouped data, it is important to realize that it is always preferable not to aggregate data and to keep individual records as much as possible.

In actuarial applications, the weights also sometimes reflect the credibility assigned to a given observation. Prior weights allow the actuary to identify responses carrying more information. Since observations with higher weights have a smaller variance, the model fit will be more influenced by these data points. We will come back to this issue at several instances in this chapter when discussing estimation.

4.2.2 Linear Score

The GLM score for response Y_i is defined as

$$\text{score}_i = x_i^\top \beta = \beta_0 + \sum_{j=1}^{p} \beta_j x_{ij}, \quad i = 1, 2, \ldots, n,$$

where $\boldsymbol{\beta} = (\beta_0, \beta_1, \ldots, \beta_p)^\top$ is a vector of dimension $p + 1$ containing the unknown regression coefficients. Thus, the β_j are parameters to be estimated from the data. As pointed out in Chap. 3, the score refers to the linear combination $\boldsymbol{x}_i^\top \boldsymbol{\beta}$ of the features, and not to Fisher's score.

The quantity β_0 contributes equally to all scores. It is called the intercept because it corresponds to the score for an individual such that $x_{ij} = 0$ for all $j = 1, \ldots, p$. Then, each β_j quantifies the impact on the score of an increase of one unit in the corresponding feature x_{ij}. The presence of an intercept β_0 explains why the first entry of the vector $\boldsymbol{x}_i$ is a 1.

Let us emphasize the meaning of the linearity of the score. The word "linear" in GLM means that the explanatory variables are linearly combined to arrive at the prediction of (a function of) the mean. The linearity meant in GLMs refers to linearity in the β_j coefficients, not in the features. For instance, the scores

$$\text{score}_i = \beta_0 + \beta_1 x_{i1} + \beta_2 x_{i1}^2 \text{ and score}_i = \beta_0 + \beta_1 x_{i1} + \beta_2 x_{i2} + \beta_3 x_{i1} x_{i3}$$

are both "linear" in the GLM sense, despite the square x_{i1}^2 and the product $x_{i1} x_{i3}$. In fact, we are always allowed to transform features x_{ij}, creating new, transformed features like x_{ij}^2 or $\ln x_{ij}$ (although GAMs generally offer a better approach to deal with nonlinear effects on the score scale, as it will be seen in Chap. 6). However, the score

$$\text{score}_i = \beta_0 + \beta_1 x_{i1} + \exp(\beta_2 x_{i2})$$

is not linear (and cannot be considered in the GLM setting, thus). Such nonlinear scores that are not eligible for a GLM analysis will be considered in Chap. 8.

4.2.3 Design Matrix

Let us also introduce the matrix X, often called the design matrix in statistics, obtained by binding all the $\boldsymbol{x}_i$, row-wise. Precisely, the ith row of X is $\boldsymbol{x}_i^\top$ so that

$$X = \begin{pmatrix} \boldsymbol{x}_1^\top \\ \boldsymbol{x}_2^\top \\ \vdots \\ \boldsymbol{x}_n^\top \end{pmatrix} = \begin{pmatrix} 1 & x_{11} & \cdots & x_{1p} \\ 1 & x_{21} & \cdots & x_{2p} \\ \vdots & \vdots & \ddots & \vdots \\ 1 & x_{n1} & \cdots & x_{np} \end{pmatrix}.$$

When the regression model includes an intercept, the design matrix X includes a column of 1. The vector $\boldsymbol{s}$ gathering the scores $s_1, \ldots, s_n$ for all n responses is then obtained from

$$\boldsymbol{s} = (s_1, \ldots, s_n)^\top = X \boldsymbol{\beta}.$$

Throughout this chapter, we always tacitly assume that the design matrix X is of full rank. This implies that the columns of X are linearly independent. This assumption is violated if one of the features is a linear combination of some other features, implying redundancy of information. We will come back to this issue when discussing collinearity.

As an example, consider the data set displayed in Table 4.1 cross-classified according to policyholders' gender (with two levels, male and female) and coverage extent (with three levels, compulsory third-party liability (henceforth abbreviated as TPL) only, limited damage and comprehensive). When the features x_{ij} are numeric, integer or continuous, they can directly enter the score: their contribution is then obtained by multiplying x_{ij} with the corresponding coefficient β_j. However, qualitative, or categorical data must first be processed before entering the score calculation. GLMs efficiently account for the effects of categorical features by "dummying up" for inclusion in the score, as explained in Example 1.4.3.

The two features in Table 4.1 are categorical. They can be coded by means of binary features. Precisely, each policyholder can be represented by a vector $x_i = (1, x_{i1}, x_{i2}, x_{i3})^\top$ with three binary features:

$$x_{i1} = \begin{cases} 1 \text{ if policyholder } i \text{ is a woman} \\ 0 \text{ otherwise} \end{cases}$$

$$x_{i2} = \begin{cases} 1 \text{ if policyholder } i \text{ has only TPL} \\ 0 \text{ otherwise} \end{cases}$$

$$x_{i3} = \begin{cases} 1 \text{ if policyholder } i \text{ has comprehensive} \\ 0 \text{ otherwise.} \end{cases}$$

Here, we have selected the base levels corresponding to the most populated categories, more precisely the category with the largest total exposure here: male for gender ($x_{i1} = 0$) and limited damage for coverage extent ($x_{i2} = x_{i3} = 0$). The reasons for this choice will become clear later on.

The data can then be displayed in the format used for statistical analysis, as shown in Table 4.2. We will show in Sect. 4.3.5 why the GLM analysis can be performed on the data aggregated in tabular form instead of individual records, without loss of information. Therefore, the index i may refer to a single policy or an entire risk class (as it is the case here).

The design matrix for the data displayed in Table 4.1 is

$$X = \begin{pmatrix} 1 & 0 & 1 & 0 \\ 1 & 0 & 0 & 0 \\ 1 & 0 & 0 & 1 \\ 1 & 1 & 1 & 0 \\ 1 & 1 & 0 & 0 \\ 1 & 1 & 0 & 1 \end{pmatrix}.$$

Table 4.1 Observed number of claims together with corresponding exposures-to-risk (in policy-years) appearing between brackets. Motor insurance, hypothetical data

	TPL only	Limited damage	Comprehensive
Males	1,683	3,403	626
	(10,000)	(30,000)	(5,000)
Females	873	2,423	766
	(6,000)	(24,000)	(7,000)

Table 4.2 Data of Table 4.1 arranged in the input format for GLM analysis with the help of a statistical software like R

X_1	X_2	X_3	# claims	Exposure
0	1	0	1,683	10,000
0	0	0	3,403	30,000
0	0	1	626	5,000
1	1	0	873	6,000
1	0	0	2,423	24,000
1	0	1	766	7,000

The corresponding vector of observations is

$$Y = \begin{pmatrix} 1,683 \\ 3,403 \\ 626 \\ 873 \\ 2,423 \\ 766 \end{pmatrix}.$$

4.2.4 Link Function

4.2.4.1 Linking the Mean to the Linear Score

Instead of equating the mean response to the score as in the Normal linear regression model, there is a one-to-one continuous differentiable transformation mapping the mean to the linear score, called the link function in the GLM jargon. Specifically, the mean

$$\mu_i = \mu(\boldsymbol{x}_i)$$

of the response $Y_i, i = 1, 2, \ldots, n$, is linked to the score involving explanatory variables with the help a smooth and invertible linearizing transformation g, that is,

$$g(\mu_i) = \text{score}_i.$$

The function g is monotone, differentiable and is called the link function. Since the link function g is one-to-one, we can invert it to obtain

$$g(\mu_i) = \text{score}_i \Leftrightarrow \mu_i = g^{-1}(\text{score}_i).$$

It is important here to realize that we do not transform the response Y_i but rather its expected value μ_i. From a statistical point of view, a model where $g(Y_i)$ is linear in x_i is not the same as a GLM where $g(\mu_i)$ is linear in x_i. To understand this properly, assume that $\ln Y_i \sim \mathcal{N}or(x_i^\top \beta, \sigma^2)$. Though the model is fitted on the log scale, the actuary is interested in the response on the original scale (in monetary units, for claim amounts). If estimated scores $x_i^\top \widehat{\beta}$ are exponentially transformed to get back to the original scale, $\exp(x_i^\top \widehat{\beta})$ are fitted values for the median response, not for the mean. For the mean, the fitted values are

$$\widehat{\mu}_i = \exp\left(x_i^\top \widehat{\beta} + \frac{\widehat{\sigma}^2}{2}\right)$$

according to the formula for the mathematical expectation associated to the LogNormal distribution.

Notice that a GLM is a conditional model in that it specifies the distribution of the response Y_i given the vector of features $X_i = (X_{i1}, X_{i2}, \ldots, X_{ip})$. Specifically, given $X_i = x_i$, Y_i obeys a distribution in the ED family with

$$\mu(x_i) = g^{-1}\left(\beta_0 + \sum_{j=1}^{p} x_{ij}\right).$$

It is nevertheless important to realize that the random vector $X_i^\top = (X_{i1}, X_{i2}, \ldots, X_{ip})$ may have a complex dependence structure that affects the estimation of β as it will be seen later on. It is also important to keep in mind that unconditionally, Y_i does not obey the same distribution. For instance, if Y_i obeys the $\mathcal{P}oi(\exp(\beta^\top x_i))$ distribution given $X_i = x_i$, unconditionally, the Poisson mean becomes the random variable $\exp(\beta^\top X_i)$ and we are faced with a mixed Poisson random variable (as studied in Chap. 5).

4.2.4.2 Usual Link Functions

In principle, any monotonic, continuous and differentiable function g is eligible for being a link function. But in practice there are some convenient and common choices for the standard GLMs. In a GLM, the link function is specified by the actuary whereas it is estimated in a Single Index Model (SIM). SIMs are also based on an ED distribution for the response and a linear score, but the link function is estimated from the data.

Often, the mean response μ is constrained (being positive or comprised in the unit interval $[0, 1]$, for instance). As the linear score is not constrained (it can take any real value, depending on the sign of the regression coefficients β_j and the range of the features x_{ij}), the link function may account for these conditions so that regression coefficients remain unconstrained (which eases the maximization of the likelihood to estimate the β_js). Thus, one purpose of the link function is to map the linear predictor to the range of the response.

Of course, this is not to say that the choice of the link function is entirely determined by the range of the response variable. The choice of the link function must also be data driven. The SIM approach allows the actuary to estimate g from the data, without specifying it a priori. This nonparametric estimate can serve as a benchmark to select the appropriate functional form. Alternatively, some diagnostic graphs can be used to check the appropriateness of the selected link function once the model has been fitted to the available data. This means that the choice of g is not purely arbitrary and that some data-driven procedures can guide the actuary in that respect.

Table 4.3 lists the usual link functions and their inverses. Some of these link functions impose conditions on the mean μ, and hence on the range of the response Y. For instance, the square root and log-links only apply to non-negative and positive values, respectively. The last four link functions listed in Table 4.3 are for Binomial data, where μ represents the probability of success and thus belongs to the unit interval $[0, 1]$.

In the next section, we now discuss some link functions that are particularly useful to actuaries.

Table 4.3 Some common link functions and their inverses. Here, μ_i is the expected value of the response, s_i is the score, and $\Phi(\cdot)$ is the distribution function of the standard Normal distribution $Nor(0, 1)$

Link function	$s_i = g(\mu_i)$	$\mu_i = g^{-1}(s_i)$
Identity	μ_i	s_i
Log	$\ln \mu_i$	$\exp(s_i)$
Inverse	μ_i^{-1}	s_i^{-1}
Inverse-square	μ_i^{-2}	$s_i^{-1/2}$
Square-root	$\sqrt{\mu_i}$	s_i^2
Logit	$\ln \frac{\mu_i}{1-\mu_i}$	$\frac{\exp(s_i)}{1+\exp(s_i)}$
Probit	$\Phi^{-1}(\mu_i)$	$\Phi(s_i)$
Log-log	$-\ln(-\ln \mu_i)$	$\exp(-\exp(s_i))$
Complementary log-log	$-\ln(-\ln(1-\mu_i))$	$1 - \exp(-\exp(s_i))$

4.2.4.3   Log Link

As explained before, a promising link removes the restriction on the range of expected response. For instance, the response is non-negative in many applications to insurance, where it often represents a claim frequency or a claim severity so that $\mu_i > 0$. A log-link function then removes this constraint as the specification

$$\ln \mu_i = \text{score}_i \Leftrightarrow \mu_i = \exp(\text{score}_i)$$

ensures that this condition is always fulfilled. The log-link function is also often used to build commercial price lists as it provides the actuary with a multiplicative tariff structure:

$$\exp(\text{score}_i) = \exp(\beta_0) \prod_{j=1}^{p} \exp(\beta_j x_{ij}).$$

All fitted values are expressed with respect to the reference $\exp(\beta_0)$ corresponding to individuals with $x_{ij} = 0$ for all $j = 1, \ldots, p$. The factors $\exp(\beta_j x_{ij})$ entering the product can then be seen as corrections to $\exp(\beta_0)$ when $x_{ij} \neq 0$ for some j.

When all the explanatory variables are categorical, each policyholder is represented by a vector x_i with components equal to 0 or 1 (see Example 1.4.3). If $x_{ij} \in \{0, 1\}$ for all $j = 1, \ldots, p$ then the expected response is equal to

$$\exp(\boldsymbol{\beta}^\top x_i) = \exp(\beta_0) \prod_{j | x_{ij}=1} \exp(\beta_j)$$

where

$$\exp(\beta_0) = \text{expected response corresponding to the reference class}$$
$$\text{i.e. to } x_{ij} = 0 \text{ for all } j = 1, 2, \ldots, p$$
$$\exp(\beta_j) = \text{relative effect of the } j \text{ th feature.}$$

Each time the policyholder under consideration differs from the reference class, that is, $x_{ij} = 1$ for some j, $\exp(\beta_0)$ is multiplied by a correcting factor of the form $\exp(\beta_j)$. Hence, the sign of β_j reveals the impact of x_{ij} on the mean response: if

$$\beta_j > 0 \Leftrightarrow \exp(\beta_j) > 1$$

then $x_{ij} = 1$ increases the expected value compared to $x_{ij} = 0$ whereas if

$$\beta_j < 0 \Leftrightarrow \exp(\beta_j) < 1$$

then $x_{ij} = 1$ decreases the expected value compared to $x_{ij} = 0$. Features x_{ij} with associated coefficient $\beta_j > 0$ thus indicate higher risk profiles compared to the reference one whereas features x_{ij} with $\beta_j < 0$ correspond to lower risk profiles.

The factors $\exp(\beta_j)$, $j = 1, \ldots, p$, are called the relativities because they correspond to ratios of expected responses for two policyholders differing only in the jth feature, being equal to 1 for the former (appearing in the numerator) and to 0 for the latter (appearing in the denominator). Formally, considering two policyholders i_1 and i_2 such that

$$x_{i_1 j} = 1, \quad x_{i_2 j} = 0, \quad x_{i_1 k} = x_{i_2 k} \text{ for all } k \neq j,$$

we see that the ratio of the mean responses for policyholders i_1 and i_2 writes

$$\frac{\exp(x_{i_1}^\top \beta)}{\exp(x_{i_2}^\top \beta)} = \exp(\beta_j).$$

The typical output of a GLM analysis with a log-link function is as follows. A base premium $\exp(\beta_0) = 100$ say is determined for a reference profile in the portfolio: say a male driver, middle-aged, living in a big city, and so on. Then, the premiums for the other profiles are obtained from

base premium 100×0.9 for a female driver

$\times 1.3$ for a young, unexperienced driver

$\times 0.95$ for a senior driver

$\times 0.9$ for a driver living in the suburbs

$\times 0.85$ for a driver living in the countryside

and so on.

The factors 0.9, 1.3, 0.95, 0.9, and 0.85 entering the latter formula correspond to estimated relativities $\exp(\beta_j)$. Expressed in this way, the output of a GLM analysis is fully transparent and easy to communicate inside the company, to intermediaries as well as to policyholders. This is why most commercial price lists have a multiplicative structure.

4.2.4.4 Logit, Probit and Log-Log Links

For Binomial proportions, the mean response corresponds to the success probability so that μ_i is confined to the unit interval $[0, 1]$. Distribution functions are therefore natural candidates to transform the real-valued score into the mean response. Logistic and Normal distributions have often been used in this setting, giving the so-called logit and probit link functions. The complementary log-log link function is inherited from the Extreme Value distribution.

The logit, probit and complementary log-log link functions map the unit interval $[0, 1]$ to the entire real line, removing the constraint on the mean response. The logit link

$$\ln \frac{\mu_i}{1 - \mu_i} = \text{score}_i \Leftrightarrow \mu_i = \frac{\exp(\text{score}_i)}{1 + \exp(\text{score}_i)} = \frac{1}{1 + \exp(-\text{score}_i)}$$

is often used for probabilities μ_i in actuarial applications. The typical GLM for modeling retention and new business conversion is a Binomial GLM with a logit link function (also known as a logistic regression model). If the success probability is close to 0 (so that the response is generally 0) then it may also be possible to use a multiplicative Poisson GLM as an approximation, given that the output from a multiplicative GLM is easier to communicate to a non-technical audience. This is because the probability mass is almost entirely concentrated on $\{0, 1\}$ when the Poisson parameter is small enough.

When $x_{ij} \in \{0, 1\}$ for all $j = 1, \ldots, p$, we see that

$$\frac{\mu_i}{1 - \mu_i} = \exp(\beta_0) \prod_{j | x_{ij} = 1} \exp(\beta_j).$$

The features x_{ij} thus enter the so-called odd $\mu_i / (1 - \mu_i)$ in a multiplicative way. If μ_i is small, so that $1 - \mu_i \approx 1$, we see that

$$\mu_i \approx \exp(\beta_0) \prod_{j | x_{ij} = 1} \exp(\beta_j)$$

so that the Poisson and logistic regression model produce similar results in this case.

The probit model uses the distribution function Φ of the $\mathcal{N}or(0, 1)$ distribution to map the score on the unit interval $[0, 1]$. The complementary log-log model uses the Extreme (minimum) Value distribution function to that end. This link function exactly connects the Poisson and Binomial GLMs, as shown next. Consider $Y \sim \mathcal{P}oi(\mu)$ and define the truncated response

$$\tilde{Y} = \min\{1, Y\}.$$

If the log-link function has been used for the Poisson count Y, i.e. $\ln \mu = \text{score}$ then

$$\begin{aligned} E[\tilde{Y}] &= 1 - P[\tilde{Y} = 0] \\ &= 1 - \exp(-\mu) \\ &= 1 - \exp\left(-\exp(\text{score})\right) \end{aligned}$$

where we recognize the complementary log-log link function. Hence, the complementary log-log link function bridges Poisson and Bernoulli GLMs. This constructions also allows the actuary to account for different exposures to risk in a regression model for binary features, by specifying

$$E[\widetilde{Y}] = 1 - \exp\big(-\exp(\ln e + \text{score})\big)$$

where the exposure to risk e becomes apparent. Precisely, the logarithm of the exposure is added to the score as an offset. This concept plays a central role in many applications to insurance. It is studied in details in Sect. 4.3.7.

4.2.4.5 Canonical Links

When the link function makes the score the same as the canonical parameter, it is referred to as a canonical link. Formally, the canonical link function is such that the identity

$$\theta_i = \text{score}_i$$

holds true. As $g(\mu_i) = \text{score}_i$ and $\mu_i = a'(\theta_i)$, the canonical link is defined as

$$\theta_i = \text{score}_i \Leftrightarrow g\big(a'(\theta_i)\big) = \theta_i \Leftrightarrow g^{-1} = a'.$$

Thus, the cumulant function $a(\cdot)$ determines the canonical link function so that each ED distribution possesses its own specific canonical link. For instance, in the Poisson case we know that $a(\theta) = \exp(\theta)$. Hence, the canonical link function is the inverse of $a'(\theta) = \exp(\theta)$, resulting in the log-link function. Table 4.4 summarizes the canonical link functions for the main ED distributions.

The canonical parameter greatly simplifies the analysis based on a GLM, but other link functions may be used as well. This is precisely a great strength of the GLM approach. Often in the past, actuaries transformed the response prior to a Normal linear regression analysis (such as LogNormal modeling). However, the same transformation then has to simultaneously make the transformed response approximately

Table 4.4 Examples of canonical link functions. Notice that the canonical link is $\frac{\mu^\xi}{\xi}$ for Gamma and Inverse-Gaussian distributions but the minus sign is generally removed, as well as the division by 2 in the Inverse-Gaussian case

Link function	$g(\mu)$	Canonical link for
Identity	μ	Normal
Log	$\ln \mu$	Poisson
Power	μ^ξ	Gamma ($\xi = -1$)
		Inverse Gaussian ($\xi = -2$)
Logit	$\ln \frac{\mu}{1-\mu}$	Binomial

Normally distributed and the mean of the transformed response linear in the explanatory variables, i.e.

$$E[g(Y_i)] = \beta_0 + \sum_{j=1}^{p} \beta_j x_{ij}.$$

In a GLM, the transformed expectation of the response is modeled, i.e.

$$g(E[Y_i]) = \beta_0 + \sum_{j=1}^{p} \beta_j x_{ij}$$

so that the sole role of g is to make the expected response linear in the features x_{ij} on that particular scale (called the score scale), not to render the response Y_i approximately Normally distributed on a modified scale. In a GLM, the distribution of the response is selected from the ED family according to the characteristics of the observed data. There is no prior transformation of the response so that it obeys a given distribution. As soon as the link function g is not linear, these two approaches differ (although similar in spirit).

4.2.5 Interactions

Interaction arises when the effect of a particular risk factor is reliant on the value of another. An example in motor insurance is given by driver's age and gender: often, young female drivers cause on average less claims compared to young male ones whereas this gender difference disappears (and sometimes even reverses) at older ages. Hence, the effect of age depends on gender. Contrarily to other regression techniques (considered in the next volumes, such as tree-based methods), GLMs do not automatically account for interactions and such effects need to be manually entered in the score.

Interactions are included in the score with the help of products of features. To understand this properly, think of Taylor expansion of the the mean function on the score scale:

$$g(E[Y|X = x]) \approx \beta_0 + \sum_{j=1}^{p} \beta_j x_{ij}$$

when $\frac{\partial^2}{\partial x_j \partial x_k} g(E[Y|X = x]) = 0$. In this setting,

$$\beta_0 = g(E[Y|X = 0])$$

and

$$\beta_j = \frac{\partial}{\partial x_j} g(\mathrm{E}[Y|X = x])\Big|_{x=0}.$$

This means that the score only contains the main effects of the features and is therefore additive. Now, if the features interact, the second mixed partial derivatives are no more zero and a more accurate approximation to the true regression function is given by

$$g(\mathrm{E}[Y|X = x]) \approx \beta_0 + \sum_{j=1}^{p} \beta_j x_{ij} + \sum_{j=1}^{p} \sum_{k \geq j+1} \gamma_{jk} x_{ij} x_{ik}$$

where

$$\gamma_{jk} = \frac{\partial^2}{\partial x_j \partial x_k} g(\mathrm{E}[Y|X = x])\Big|_{x=0}.$$

Notice that in this approach, an interaction term can only exist if the main terms are also included in the model. This is why it is generally recommended to include interactions only if main effects also enter the score.

Consider a model with two continuous features and a score of the form

$$\mathrm{score}_i = \beta_0 + \beta_1 x_{i1} + \beta_2 x_{i2} + \beta_3 x_{i1} x_{i2}.$$

The linear score involves a third feature $x_{i3} = x_{i1} x_{i2}$, called an interaction term, built from the two first ones. The inclusion of the product $x_{i1} x_{i2}$ in the score indeed induces interactions as a unit increase in x_{i1} increases the linear score by $\beta_1 + \beta_3 x_{i2}$, and hence the effect of the first feature depends on the second one. Similarly, the effect of a unit increase in x_{i2} is $\beta_2 + \beta_3 x_{i1}$ which depends on x_{i1}. Thus, the two features interact. The variables x_{i1} and x_{i2} are called the main effects and are distinguished from the interaction effect $x_{i3} = x_{i1} x_{i2}$.

The product $x_{i1} x_{i2}$ appearing into the score comes from a limited Taylor expansion of a more general contribution $h(x_{i1}, x_{i2})$ on the score scale. In this sense, GAMs introduced in Chap. 6 allow the actuary to include more flexible interactions in the model by estimating the smooth functions $h(\cdot, \cdot)$. Interactions are also effectively captured with the help of tree-based methods (presented in Trufin et al. 2019) and neural networks (presented in Hainaut et al. 2019).

Interacting an ordinary quantitative feature with a binary feature is also often useful in insurance applications. Assume x_{i1} is continuous and x_{i2} is an indicator taking on the values 0 and 1. Then the score is given by

$$\mathrm{score}_i = \beta_0 + \beta_1 x_{i1} + \beta_2 x_{i2} + \beta_3 x_{i1} x_{i2}$$
$$= \begin{cases} \beta_0 + \beta_1 x_{i1} \text{ if } x_{i2} = 0 \\ \beta_0 + \beta_2 + (\beta_1 + \beta_3) x_{i1} \text{ if } x_{i2} = 1. \end{cases}$$

The effect of x_{i1} (that is, the intercept and slope) depends on the group indicated by x_{i2}.

4.2.6 Recap' of the Assumptions for Regression with GLMs

To sum up, the GLM approach to insurance data modeling proceeds from the following assumptions:

- Response Y_i obeying the ED distribution with probability mass function or probability density function of the form

$$f_{\theta_i}(y) = \exp\left(\frac{y\theta_i - a(\theta_i)}{\phi/\nu_i}\right) c(y, \phi/\nu_i)$$

  with a common dispersion parameter ϕ and specific weights ν_i.
- Linear score

$$\text{score}_i = \text{score}(\boldsymbol{x}_i) = \boldsymbol{x}_i^\top \boldsymbol{\beta} = \beta_0 + \sum_{j=1}^{p} \beta_j x_{ij}.$$

- Mean modeling: expected response $\mu_i = \mu(\boldsymbol{x}_i)$ of the form

$$a'(\theta_i) = \mu_i$$
$$g(\mu_i) = \text{score}_i \Leftrightarrow \mu_i = g^{-1}(\text{score}_i)$$
$$g \in \{\ln, \text{logit}, \ldots\}.$$

- The monotonic link function g is specified by the actuary, not estimated from the data.

Notice that the choice of the ED distribution amounts to specify the cumulant function $a(\cdot)$, so that it is equivalent to select a variance function $V(\cdot) = a''(\cdot)$ reflecting the mean-variance relationship supported by the data under consideration.

4.3 Likelihood Equations in GLMs

4.3.1 General Form

Once the GLM has been specified in terms of link function and ED distribution, or equivalently once a variance function has been selected by the actuary, estimated parameters are obtained by the method of maximum likelihood (recalled in Chap. 3). This provides the actuary not only with estimates of the regression coefficients β_j but also with estimated large-sample standard errors. In essence, this approach seeks to estimate the GLM that has the highest probability to produce the actual data.

Some of the ED distributions on which GLMs are based include an unknown dispersion parameter ϕ. Although this parameter could also be estimated by maximum-likelihood, a moment estimator is generally used instead, in a second step once $\boldsymbol{\beta}$

has been estimated. In fact, it will be seen that the knowledge of ϕ is not needed to estimate $\boldsymbol{\beta}$ so that the two sets of parameters can be dealt with separately. This is why we view here the likelihood as a function of $\boldsymbol{\beta}$, only.

The likelihood function is the product of probabilities of observing the value of each response, with the probability density function used in place of the probability for continuous responses. As explained in Chap. 3, it is usual to switch to the logarithm of the likelihood because it involves a sum over the observations rather than a product (and any parameter value maximizing the log-likelihood also maximizes the likelihood itself). Thus, the maximum-likelihood estimates of the parameters $\beta_0, \beta_1, \ldots, \beta_p$ maximize

$$L(\boldsymbol{\beta}) = \sum_{i=1}^{n} \ln f_{\theta_i}(y_i) = \sum_{i=1}^{n} \frac{y_i \theta_i - a(\theta_i)}{\phi/v_i} + \text{constant with respect to } \boldsymbol{\beta}$$

where θ_i involves the linear score $s_i = \boldsymbol{x}_i^\top \boldsymbol{\beta}$ since $\theta_i = g^{-1}(\boldsymbol{x}_i^\top \boldsymbol{\beta})$.

To find the maximum-likelihood estimate for $\boldsymbol{\beta}$, L is differentiated with respect to β_j and equated to 0. Using the chain rule, this gives

$$\frac{\partial}{\partial \beta_j} L(\boldsymbol{\beta}) = \sum_{i=1}^{n} \frac{\partial}{\partial \beta_j} \left(\frac{y_i \theta_i - a(\theta_i)}{\phi/v_i} \right)$$

$$= \sum_{i=1}^{n} \frac{\partial}{\partial \theta_i} \left(\frac{y_i \theta_i - a(\theta_i)}{\phi/v_i} \right) \frac{\partial \theta_i}{\partial \mu_i} \frac{\partial \mu_i}{\partial \beta_j}$$

$$= \sum_{i=1}^{n} \left(\frac{y_i - \mu_i}{\phi/v_i} \right) \frac{\partial \theta_i}{\partial \mu_i} \frac{\partial \mu_i}{\partial \beta_j}.$$

Now, as

$$\frac{\partial \mu_i}{\partial \theta_i} = a''(\theta_i) = \frac{v_i \operatorname{Var}[Y_i]}{\phi}$$

$$\frac{\partial \mu_i}{\partial \beta_j} = \frac{\partial \mu_i}{\partial s_i} \frac{\partial s_i}{\partial \beta_j} = \frac{\partial \mu_i}{\partial s_i} x_{ij}$$

we have to solve

$$\sum_{i=1}^{n} \frac{(y_i - \mu_i) x_{ij}}{\operatorname{Var}[Y_i]} \frac{\partial \mu_i}{\partial s_i} = 0 \Leftrightarrow \sum_{i=1}^{n} \frac{(y_i - \mu_i) x_{ij}}{V(\mu_i)/v_i} \frac{\partial \mu_i}{\partial s_i} = 0. \qquad (4.1)$$

The system (4.1) does not possess en explicit solution and must be solved numerically, using an iterative algorithm. The noticeable exception is the Normal linear regression model with identity link where $\widehat{\boldsymbol{\beta}}$ possesses an analytical expression. The resolution of (4.1) will be explained in Sect. 4.4.

4.3.2 Fisher Information

To obtain Fisher information, let us write element (j, k) of the Fisher information matrix $\boldsymbol{I}(\boldsymbol{\beta})$:

$$
\begin{aligned}
\boldsymbol{I}_{jk}(\boldsymbol{\beta}) &= \mathrm{E}\left[\frac{\partial}{\partial \beta_j} L(\boldsymbol{\beta}) \frac{\partial}{\partial \beta_k} L(\boldsymbol{\beta})\right] \\
&= \mathrm{E}\left[\sum_{i=1}^{n} \frac{(Y_i - \mu_i) x_{ij}}{\mathrm{Var}[Y_i]} \frac{\partial \mu_i}{\partial s_i} \sum_{l=1}^{n} \frac{(Y_l - \mu_l) x_{lk}}{\mathrm{Var}[Y_l]} \frac{\partial \mu_l}{\partial s_l}\right] \\
&= \sum_{i=1}^{n}\sum_{l=1}^{n} \mathrm{Cov}[Y_i, Y_l] \frac{x_{ij} x_{lk}}{\mathrm{Var}[Y_i]\mathrm{Var}[Y_l]} \frac{\partial \mu_i}{\partial s_i} \frac{\partial \mu_l}{\partial s_l}.
\end{aligned}
$$

As the responses are independent, the covariance terms vanish, that is, $\mathrm{Cov}[Y_i, Y_l] =$ for $i \neq l$. Thus, only the terms corresponding to $i = l$ remain. We then get

$$
\boldsymbol{I}_{jk}(\boldsymbol{\beta}) = \sum_{i=1}^{n} \frac{x_{ij} x_{ik}}{\mathrm{Var}[Y_i]} \left(\frac{\partial \mu_i}{\partial s_i}\right)^2.
$$

Let us define the auxiliary weight

$$
\tilde{v}_i = \frac{1}{\mathrm{Var}[Y_i]}\left(\frac{\partial \mu_i}{\partial s_i}\right)^2 = \frac{v_i}{\phi V(\mu_i)}\left(\frac{\partial \mu_i}{\partial s_i}\right)^2 \tag{4.2}
$$

and the the $n \times n$ diagonal matrix $\widetilde{\boldsymbol{W}}$ with entries $\tilde{v}_1, \tilde{v}_2, \ldots, \tilde{v}_n$. In matrix form, we then have

$$
\boldsymbol{I}(\boldsymbol{\beta}) = \boldsymbol{X}^\top \widetilde{\boldsymbol{W}} \boldsymbol{X} = \sum_{i=1}^{n} \tilde{v}_i \boldsymbol{x}_i \boldsymbol{x}_i^\top. \tag{4.3}
$$

4.3.3 Impact of the Variance Function

Considering the GLM likelihood Eq. (4.1), it is important to realize how the choice of the variance function $V(\cdot)$ affects the model fit. We see there that the larger $V(\mu_i)$, the less observation i contributes to the sum appearing in the likelihood equations. Other things being equal, this means that larger differences $y_i - \mu_i$ are tolerated when $V(\mu_i)$ is large.

We know from Chap. 2 that typical variance functions are of the Tweedie form $V(\mu) = \mu^\xi$. Within the Tweedie family,

– For $\xi = 0$ (Normal case), each observation has the same fixed variance regardless of its mean and fitted values are equally attracted to observed data points.

- For $\xi = 1$ (Poisson case), the variance is proportional to the expected value of each observation: observations with a smaller expected value have a smaller assumed variance, which results in greater weights (or credibility) when estimating the parameters. The model fit is thus more influenced by observations with smaller expected value than higher expected value (and hence higher assumed variance).
- When $\xi = 2$ (Gamma case) and $\xi = 3$ (Inverse Gaussian case), the GLM fit is even more impacted by the observations with smaller expected values, tolerating more errors for observations with larger expected values.

Prior weights ν_i also act on the variance and result in data points having more or less impact on the model fit. Indeed, increasing the weight reduces the variance and the corresponding observation plays a more important role in the model fit.

The choice of the variance function is therefore of primary importance to get an accurate model fit. As it will be seen from quasi-likelihood techniques, GLM-based techniques appear to be rather robust with respect to errors in distribution as long as the variance function has been correctly specified.

4.3.4 Likelihood Equations and Fisher's Information with Canonical Link

With canonical link function, the canonical parameter is equal to the score: $\theta_i = s_i$. In this case, we have

$$\frac{\partial \mu_i}{\partial s_i} = \frac{\partial \mu_i}{\partial \theta_i} = \frac{\partial a'(\theta_i)}{\partial \theta_i} = a''(\theta_i).$$

Hence, (4.1) further simplifies and we have to solve

$$\frac{\partial}{\partial \beta_j} L(\boldsymbol{\beta}) = 0 \Leftrightarrow \sum_{i=1}^{n} \left(\frac{y_i - \mu_i}{\phi / \nu_i} \right) x_{ij} = 0.$$

If $\nu_i = 1$ for all i then the likelihood equations ensure that the fitted mean values

$$\widehat{\mu}_i = \mu_i(\widehat{\boldsymbol{\beta}}) = g^{-1}(x_i^\top \widehat{\boldsymbol{\beta}})$$

fulfill the condition

$$\sum_{i=1}^{n}(y_i - \widehat{\mu}_i)x_{ij} = 0 \Leftrightarrow \sum_{i=1}^{n} y_i x_{ij} = \sum_{i=1}^{n} \widehat{\mu}_i x_{ij} \text{ for } j = 0, 1, \ldots, p. \qquad (4.4)$$

This can be rewritten as $\boldsymbol{X}^\top (\boldsymbol{y} - \widehat{\boldsymbol{\mu}}) = \boldsymbol{0}$ in matrix form. In general, define $\boldsymbol{W}$ to be the diagonal matrix with ith element ν_i, that is,

$$
W = \begin{pmatrix} v_1 & 0 & \cdots & 0 \\ 0 & v_2 & \cdots & 0 \\ \vdots & \vdots & \ddots & \vdots \\ 0 & 0 & \cdots & v_n \end{pmatrix}.
$$

With the canonical link function, the likelihood equations are

$$
X^\top W \left(y - \mu(\widehat{\beta}) \right) = 0. \tag{4.5}
$$

The maximization of the log-likelihood is well posed with canonical link functions. This is because with canonical link function, the log-likelihood is given by

$$
L(\beta) = \frac{1}{\phi} \sum_{i=1}^{n} v_i \left(y_i s_i - a(s_i) \right) + \text{constant with respect to } \beta
$$

which is concave in β (since a is convex and the score s_i is linear in β).

Another interesting property of canonical link functions concerns the Fisher information matrix. When the canonical link function has been selected, we obtain

$$
\begin{aligned}
\frac{\partial^2}{\partial \beta_j \partial \beta_k} L(\beta) &= - \sum_{i=1}^{n} \frac{v_i x_{ij}}{\phi} \frac{\partial \mu_i}{\partial \beta_k} \\
&= - \sum_{i=1}^{n} \frac{v_i x_{ij}}{\phi} \frac{\partial \mu_i}{\partial s_i} x_{ik} \\
&= - \sum_{i=1}^{n} \frac{v_i x_{ij} x_{ik}}{\phi} a''(\theta_i).
\end{aligned}
$$

Therefore, the entry (j, k) of the observed information matrix is given by

$$
H_{jk}(\beta) = \sum_{i=1}^{n} \frac{x_{ij} x_{ik}}{\text{Var}[Y_i]} \left(a''(\theta_i) \right)^2
$$

which does not depend on the responses. Provided $\theta_i = s_i$, we thus have

$$
\frac{\partial^2}{\partial \beta_j \partial \beta_k} L(\beta) = E\left[\frac{\partial^2}{\partial \beta_j \partial \beta_k} L(\beta) \right]
$$

so that expected and observed information matrices coincide.

4.3.5 Individual or Grouped Data

Let us consider the data displayed in Table 4.5. The portfolio of interest has been grouped in 4 classes according to the two binary features encoding the information available about each policyholder (gender and annual distance traveled above of below 20,000 kms per year). The response Y_i is the number of claims reported by policyholder i. We assume that $Y_1, \ldots, Y_n$ are independent and Poisson distributed, with respective means $\mu_1, \ldots, \mu_n$. The expected number of claims for policy i in the portfolio is expressed as

$$\mu_i = e_i \exp(\beta_0 + \beta_1 x_{i1} + \beta_2 x_{i2})$$

where e_i is the risk exposure for policyholder i, x_{i1} records annual mileage, and x_{i2} records gender. Precisely,

$$x_{i1} = \begin{cases} 1 \text{ if policyholder } i \text{ drives less than 20,000 kms a year,} \\ 0 \text{ otherwise,} \end{cases}$$

and

$$x_{i2} = \begin{cases} 1 \text{ if policyholder } i \text{ is female,} \\ 0 \text{ otherwise.} \end{cases}$$

The reference levels correspond to the most populated categories in the data set: more than 20,000 kms for distance ($x_{i1} = 0$) and male for gender ($x_{i2} = 0$).

Table 4.5 summarizes the experience of the whole portfolio in just four figures: one for each risk class (that is, aggregating all policies that are identical with respect to the two features x_{i1} and x_{i2}). Let us now show that such a grouping is allowed when working with Poisson distributed responses (and more generally, with any ED distribution). This can easily be understood as follows. Let us start from individual records y_i for policyholder i. Here, y_i denotes the observed number of claims for policyholder i for an exposure e_i. The corresponding likelihood function is

Table 4.5 Number of claims and corresponding risk exposures (in policy-years, appearing within brackets) for an hypothetical motor insurance portfolio observed during one calendar year

Number of (exposure claims to risk)		Annual distance traveled	
		<20,000 kms	≥20,000 kms
Gender	Male	143	1,967
		(2,000)	(18,000)
	Female	278	354
		(6,000)	(4,000)

$$\mathcal{L}_{\text{ind}}(\beta_0, \beta_1, \beta_2) = \prod_{i=1}^{n} \exp(-\mu_i)\frac{\mu_i^{y_i}}{y_i!}.$$

The factors appearing in the product can be re-ordered according to the 4 categories in Table 4.5. Precisely, $\mathcal{L}_{\text{ind}}(\beta_0, \beta_1, \beta_2)$ can be rewritten as

$$\mathcal{L}_{\text{ind}} = \prod_{i|x_{i1}=x_{i2}=0} \exp(-\mu_i)\frac{\mu_i^{y_i}}{y_i!} \prod_{i|x_{i1}=0, x_{i2}=1} \exp(-\mu_i)\frac{\mu_i^{y_i}}{y_i!}$$

$$\prod_{i|x_{i1}=1, x_{i2}=0} \exp(-\mu_i)\frac{\mu_i^{y_i}}{y_i!} \prod_{i|x_{i1}=x_{i2}=1} \exp(-\mu_i)\frac{\mu_i^{y_i}}{y_i!}$$

$$\propto \exp\left(-\exp(\beta_0) \sum_{i|x_{i1}=x_{i2}=0} e_i\right) \left(\exp(\beta_0)\right)^{\sum_{i|x_{i1}=x_{i2}=0} k_i}$$

$$\exp\left(-\exp(\beta_0 + \beta_2) \sum_{i|x_{i1}=0, x_{i2}=1} e_i\right) \left(\exp(\beta_0 + \beta_2)\right)^{\sum_{i|x_{i1}=0, x_{i2}=1} k_i}$$

$$\exp\left(-\exp(\beta_0 + \beta_1) \sum_{i|x_{i1}=1, x_{i2}=0} e_i\right) \left(\exp(\beta_0 + \beta_1)\right)^{\sum_{i|x_{i1}=1, x_{i2}=0} k_i}$$

$$\exp\left(-\exp(\beta_0 + \beta_1 + \beta_2) \sum_{i|x_{i1}=x_{i2}=1} e_i\right) \left(\exp(\beta_0 + \beta_1 + \beta_2)\right)^{\sum_{i|x_{i1}=x_{i2}=1} k_i},$$

where "$\propto$" indicates that the two expressions are proportional up to constant factors (not involving any regression coefficient β_j to be estimated). This shows that $\mathcal{L}_{\text{ind}}(\beta_0, \beta_1, \beta_2)$ is proportional to the likelihood based on the data aggregated into 4 risk classes as it only involves the total exposure and the total claim numbers for each risk class (i.e., the data displayed in Table 4.5).

Thus, working with individual data $Y_1, \ldots, Y_n$ or with the four aggregated counts

$$Y_{00} = \sum_{i|x_{i1}=x_{i2}=0} Y_i$$

$$Y_{01} = \sum_{i|x_{i1}=0, x_{i2}=1} Y_i$$

$$Y_{10} = \sum_{i|x_{i1}-1, x_{i2}=0} Y_i$$

$$Y_{11} = \sum_{i|x_{i1}=x_{i2}=1} Y_i$$

results in likelihood functions that are proportional so that the maximum-likelihood procedure produces exactly the same estimates. With the Poisson distribution, we know that the four totals Y_{00}, Y_{01}, Y_{10} and Y_{11} are still Poisson distributed. Precisely,

$$Y_{00} \sim \mathcal{P}oi\,(\lambda_{00}) \text{ with } \lambda_{00} = e_{00}\exp(\beta_0) \text{ where } e_{00} = \sum_{i|x_{i1}=x_{i2}=0} e_i$$

$$Y_{01} \sim \mathcal{P}oi\,(\lambda_{01}) \text{ with } \lambda_{01} = e_{01}\exp(\beta_0 + \beta_2) \text{ where } e_{01} = \sum_{i|x_{i1}=0,x_{i2}=1} e_i$$

$$Y_{10} \sim \mathcal{P}oi\,(\lambda_{10}) \text{ with } \lambda_{10} = e_{10}\exp(\beta_0 + \beta_1) \text{ where } e_{10} = \sum_{i|x_{i1}=1,x_{i2}=0} e_i$$

$$Y_{11} \sim \mathcal{P}oi\,(\lambda_{11}) \text{ with } \lambda_{11} = e_1\exp(\beta_0 + \beta_1 + \beta_2) \text{ where } e_{11} = \sum_{i|x_{i1}=x_{i2}=1} e_i.$$

Hence, the likelihood function $\mathcal{L}_{\text{group}}(\beta_0, \beta_1, \beta_2)$ associated to grouped data is the product of four Poisson probabilities, that is,

$$\mathcal{L}_{\text{group}} = \exp\left(-e_{00}\exp(\beta_0)\right)\frac{\left(e_{00}\exp(\beta_0)\right)^{y_{00}}}{y_{00}!}$$
$$\exp\left(-e_{01}\exp(\beta_0 + \beta_2)\right)\frac{\left(e_{01}\exp(\beta_0 + \beta_2)\right)^{y_{01}}}{y_{01}!}$$
$$\exp\left(-e_{10}\exp(\beta_0 + \beta_1)\right)\frac{\left(e_{10}\exp(\beta_0 + \beta_1)\right)^{y_{10}}}{y_{10}!}$$
$$\exp\left(-e_{11}\exp(\beta_0 + \beta_1 + \beta_2)\right)\frac{\left(e_{11}\exp(\beta_0 + \beta_1 + \beta_2)\right)^{y_{11}}}{y_{11}!}.$$

Clearly,

$$\mathcal{L}_{\text{ind}}(\beta_0, \beta_1, \beta_2) \propto \mathcal{L}_{\text{group}}(\beta_0, \beta_1, \beta_2)$$

so that we can work with aggregated data for estimation purposes, that is, aggregating the experience of the whole portfolio at the level of the 4 homogeneous risk classes does not result in any loss of information.

However, let us stress that this property is only valid under the ED assumption for the responses. It can be seen as an application of Property 2.4.6 ensuring that weighted averages of observations obeying an ED distribution still obeys the same distribution with updated dispersion parameter and associated weight. In general, with distributions outside the ED family, aggregating the data may result in a loss of information so that the actuary should always refrain from proceeding in this way, and keep data bases recording individual, policyholder-specific experience whenever possible.

4.3.6 Marginal Totals Under Canonical Link

Let us now provide an intuitive explanation for the likelihood equations with canonical link (4.4). The first likelihood equation (corresponding to $j = 0$) ensures that the global balance

$$\underbrace{\sum_{i=1}^{n} \widehat{\mu}_i}_{\text{fitted total}} = \underbrace{\sum_{i=1}^{n} y_i}_{\text{observed total}} \tag{4.6}$$

must hold. Therefore, provided that an intercept β_0 is included in the scores, the sum of fitted values obtained from the regression model equals its observed counterpart. The equality holds for the observation period and not necessarily for the future, but estimations are reliable as long as future experience is similar to past one (stationary assumption), possibly subject to corrections for inflation or other previsible effects. The inclusion of an intercept β_0 is therefore the rule in actuarial applications as long as the global balance (4.6) is meaningful.

To figure out the actual meaning of the remaining GLM likelihood equations with canonical link, consider again the data displayed in Table 4.5, with x_{i1} coding the distance traveled (equal to 1 if less than 20,000 kms per year) and x_{i2} coding gender (equal to 1 if policyholder i is a woman). Then (4.4) shows that $\widehat{\boldsymbol{\beta}}$ is such that

$$\sum_{<20,000 \text{ kms}} y_i = \sum_{<20,000 \text{ kms}} \widehat{\mu}_i$$

and

$$\sum_{\text{women}} y_i = \sum_{\text{women}} \widehat{\mu}_i$$

both hold. This mean that the model fits exactly the total claims reported by those policyholders driving less than 20,000 kms per year, on the one hand, and by female policyholders, on the other hand. Thus, combined with the global balance (4.6) guaranteed by the inclusion of an intercept into the score, the likelihood equations ensure that there is no cross-subzidizing between those policyholders driving less than 20,000 kms per year and those driving more, nor between men and women.

This is equivalent to the actuarial method of marginal totals (MMT) approach. The idea is that an acceptable set of relativities should reproduce the experience for each level of the risk factors and also the overall experience, i.e. be balanced for each level and in total. This approach predates GLMs in actuarial practice. It has been proposed in the 1960s by North-American actuaries.

Remark 4.3.1 (From technical to commercial price list). Likelihood Eq. (4.4) impose natural constraints to switch from technical to commercial price list, avoiding premium transfers between categories in the portfolio. All the desirable commercial positioning can be imposed by appropriately constraining some of the regression coefficients. The response variable is the technical premium $\mu(\boldsymbol{X})$ and explanatory variables are the rating factors, only. As multiplicative premiums are often desired in practice, a log-link is used and the corresponding response distribution is therefore Poisson. This explains why the Poisson assumption is sometimes made for responses beyond counts, not because the Poisson assumption is reasonable but as a convenient way to implement the MMT constraints (4.4).

For any GLM with canonical link, there is thus an exact balance between fitted and observed aggregated responses over the whole data set as soon as the score contains an intercept. This is also true for any level of the categorical features. This balance holds for Poisson regression with log-link or for Binomial regression with logit-link, for instance, not to mention the classical Normal linear regression with identity link. This sometimes gives the wrong impression that after fitting any GLM, the fitted values always match the observations in aggregate. However, these balance equations are only valid when canonical links are used. For instance, that balance is not necessarily preserved in case of a Gamma GLM with log-link, which is a very common actuarial model. The reason is that the log-link is not the canonical link function for the Gamma distribution.

As likelihood equations express global balances, this makes the GLM fit very stable from one data set to another as long as claim figures remain stable over time. This is an attractive property for insurance applications.

4.3.7 Offsets

4.3.7.1 Definition

Offsets are used in the vast majority of actuarial studies. Not to be confused with weights, offsets enter the score in an additive way, without being multiplied with a regression coefficient to be estimated from the data. Offsets thus represent known contributions and a priori information to the additive score. Contrarily to the intercept β_0 entering all the scores, offsets usually vary from one individual to another.

Specifically, an offset is a quantity added to the score being multiplied by a known, or pre-specified coefficient not to be estimated from the data. Thus, an offset can be seen as an additional feature whose coefficient is constrained, say to be 1. The score then becomes
$$\text{score}_i = \text{offset}_i + x_i^\top \beta$$

where offset_i denotes the offset value specific to the ith individual.

In the Normal linear regression model, we have
$$Y_i = \text{offset}_i + x_i^\top \beta + Z_i \text{ where } Z_i \sim \mathcal{N}or(0, \sigma^2).$$

Because the offset is a known quantity, this amounts to subtracting offset_i to the response Y_i before running the regression on $Y_i - \text{offset}_i$. The latter quantity can be regarded as a residual effect, to be explained by x_i. This is why offsets are not discussed in Normal linear regression with identity link. Sometimes, offset$_i$ corresponds to the fitted values $\widehat{\mu}_i$ for the mean of Y_i obtained from a previous regression model. Then, offsetting is equivalent to using the residuals from the preliminary regression of the response (which forms the basis of boosting, as it will be seen in Chap. 6).

4.3.7.2   Poisson Counts and Poisson Rates

A typical use of the offset is to incorporate a measure of exposure when modeling rates. For example, if some records in motor insurance correspond to policies in force for 6 months whereas other records correspond to policies in force for the whole year then it is appropriate with log-link to use the log of the coverage period as an offset, as explained next. In the Poisson regression model, the exposure is generally the number of vehicle-years insured and the response is the number Y_i of reported claims. Considering the Poisson process setting of Chap. 2, we know that this exposure multiplies the annual claim rate. As the offset must be on the score scale, the logarithm of the exposure is used as offset: denoting the exposure as e_i,

$$\ln \mu(x_i) = \ln e_i + x_i^\top \beta \Rightarrow \text{offset}_i = \ln e_i. \tag{4.7}$$

In the Poisson case, this is equivalent to replace the claim count Y_i with the observed claim rate $\widetilde{Y}_i = \frac{Y_i}{e_i}$ using exposures e_i as weights v_i and dispensing with the offset so that

$$\ln \widetilde{\mu}_i(x_i) = x_i^\top \beta$$

which is obviously equivalent to (4.7). This is a direct application of the results established in Example 2.5.2.

4.3.7.3   Some Other Applications of Offsets

Another useful application of the offset is constraining some rating factors to take on pre-specified values when switching from the technical to the commercial price list. Such constraints are often motivated by competition and legal or marketing considerations. This is done by constraining the value of some regression coefficients, allowing the coefficients of the remaining rating factors to optimally adapt to these constraints.

Offsetting may even help to revise or refine an existing price list. The latter is then put in offset and the estimated regression coefficients then indicate the corrections needed to improve the current price list.

In life insurance, portfolio exposures are sometimes limited and a market (or another reference) life table is often available. This suggests to work in relative terms, with respect to a meaningful reference life table. This is done by entering in the Poisson regression score the logarithm of the expected number of deaths according to this life table to explain the portfolio-specific force of mortality. The same approach applies to morbidity rates in health insurance.

Offsets are also particularly useful when the score is estimated using the backfitting algorithm designed for Generalized Additive Models (or GAMs, see Chap. 6) or using a forward stagewise approach (or boosting, see Sect. 6.8 in Chap. 6). By forward stagewise we mean building a score by accounting for variations not already explained by the current model. The current model thus serves as an offset when refining the score.

4.3.7.4 Offset in the Tweedie Case with Log-Link

Let us now comment on the subtle difference between weights and offsets. As long as the log-link is used, that is, the mean response is the exponential transform of the score, weights and offsets can be used interchangeably in the Tweedie case provided the responses are modified accordingly. This is precisely shown next.

Consider the Tweedie GLM for the response Y_i, with log-link function. In the likelihood Eq. (4.1), we then use

$$\frac{\partial \mu_i}{\partial s_i} = \mu_i \text{ as } \mu_i = \exp(s_i)$$

and $V(\mu_i) = \phi \mu_i^\xi$ with $1 < \xi < 2$. The likelihood equations are thus given in this case by

$$\sum_{i=1}^{n} v_i \frac{y_i - \mu_i}{\mu_i^{\xi-1}} x_{ij} = 0, \qquad j = 0, 1, \ldots, p. \tag{4.8}$$

Assume that we now introduce an offset so that $\mu_i = z_i \tilde{\mu}_i$, the offset being $\text{offset}_i = \ln z_i$. The likelihood Eq. (4.8) become

$$\sum_{i=1}^{n} v_i \frac{y_i - z_i \tilde{\mu}_i}{(z_i \tilde{\mu}_i)^{\xi-1}} x_{ij} = 0, \qquad j = 0, 1, \ldots, p, \tag{4.9}$$

$$\Leftrightarrow \sum_{i=1}^{n} v_i z_i^{2-\xi} \frac{\frac{y_i}{z_i} - \tilde{\mu}_i}{\tilde{\mu}_i^{\xi-1}} x_{ij} = 0, \qquad j = 0, 1, \ldots, p. \tag{4.10}$$

With the Tweedie GLM, we thus see that the inclusion of an offset z_i is equivalent to use a GLM with the transformed responses $\frac{y_i}{z_i}$ and weights $v_i z_i^{2-\xi}$. The results derived here by means of the Tweedie likelihood Eq. (4.8) can be seen as particular cases obtained with $\delta = \frac{1}{z_i}$ and log-link (so that the logarithm of the factor z_i can enter the score) using Property 2.5.1.

Let us now consider two particular cases:

Poisson case ($\xi = 1$) Including an offset $\ln z_i$, i.e. multiplying the expected response with z_i, is equivalent to run a GLM with transformed responses $\frac{y_i}{z_i}$ and weights $v_i z_i$.

In the particular case of a Poisson GLM with canonical log-link, it is thus equivalent to model claim counts with an offset term equal to the logarithm of the exposures (and unit weights) or to model claim frequencies with no offset term but with weights equal to the exposure of each observation. This coincides with the findings of Sect. 4.3.7.2.

Gamma case ($\xi = 2$) Including an offset $\ln z_i$, i.e. multiplying the expected response with z_i, is equivalent to run a GLM with transformed responses $\frac{y_i}{z_i}$ and weights v_i. In the Gamma case with log-link, the weights thus remain unaffected by this transformation of the response.

This approach can also be used to show that in some very special cases, estimations obtained from different GLMs may in fact coincide. For instance, due to the Poisson process setting, analyzing count data with Poisson regression or the times between events using Gamma regression is equivalent. In life insurance for instance, this means that the actuary is allowed to fit the model by Poisson regression on the numbers of deaths or by Gamma regression on the exposures to risk, or total survival times.

4.4 Solving the Likelihood Equations in GLMs

4.4.1 Back to the Normal Linear Regression Model

Even if a symmetric Normally distributed random variable with a fixed variance does not adequately describe claim counts nor claim amounts, the classical Normal linear regression model remains nevertheless very useful to fit GLMs. The iterative algorithm used to estimate regression coefficients involved in GLMs in fact uses a Normal linear regression model at each step, applying it to working responses derived from the actual ones as it will be seen in the next section. This is why we start by recalling the estimation of the regression coefficients in the multiple linear regression model

$$Y_i \sim \mathcal{N}or\left(\text{score}_i, \frac{\sigma^2}{v_i}\right), \quad i = 1, 2, \ldots, n, \text{ where score}_i = x_i^\top \beta.$$

This is the Normal GLM with identity link. Thus, we work under canonical link and likelihood equations write

$$X^\top W\left(y - X\widehat{\beta}\right) = 0,$$

where W is a diagonal matrix with the weights $v_1, \ldots, v_n$ appearing along the diagonal. This is just a particular case of (4.5). Provided $X^\top W X$ is invertible, this gives

$$\widehat{\beta} = (X^\top W X)^{-1} X^\top W Y. \tag{4.11}$$

We thus have an explicit expression for $\widehat{\beta}$ in this case.

Defining the hat matrix H as

$$H = W^{1/2} X (X^\top W X)^{-1} X^\top W^{1/2} \tag{4.12}$$

we see that the estimator $\widehat{Y} = X\widehat{\beta}$ of E[Y] is $\widehat{Y} = \widehat{E[Y]} = HY$. The fitted values $\widehat{Y}$ are thus obtained in one step, by multiplying the observations Y with the hat matrix H. The latter thus maps, or project Y onto $\widehat{Y}$.

4.4.2 IRLS Algorithm

The likelihood Eq. (4.1) do not admit explicit solutions (except in the Normal case, as shown above) and must therefore be solved iteratively. One of the reasons that contributed to the success of GLMs is the possibility of using a single algorithm for solving the likelihood Eq. (4.1) in all cases. Starting from an appropriate initial value for $\widehat{\beta}$, the Newton-Raphson algorithm can be used to obtain the solutions. The Newton-Raphson algorithm uses the Hessian matrix, that is, the second derivatives of the objective function. When maximizing the log-likelihood function, the Hessian matrix corresponds to the observed Fisher information, up to a change of sign (as recalled in Chap. 3). Compared to the Newton-Raphson algorithm, Fisher scoring algorithm replaces the observed information matrix with its expected counterpart, that is, with the Fisher information matrix. When both information matrices coincide, Fisher scoring algorithm corresponds to the Newton-Raphson method. The expected and observed information matrices are equal when canonical link functions are used (remember that the log-likelihood function is always concave in this case so that the maximum-likelihood estimator is always unique, if it exists). Usually, just a few steps are needed to get $\widehat{\beta}$ when canonical links are used. Each iteration of the Newton-Raphson algorithm can be interpreted as the fit of a Normal linear regression model on working responses, giving rise to the iterative re-weighted least-squares (or IRLS) approach explained next. IRLS thus operates by solving a sequence of least-squares problems, for which well-behaved numerical techniques are available.

In a GLM, the mean response is related to the linear score by the link function. To estimate the parameters, we switch to the old, alternative approach consisting in transforming the responses by the link function. But instead of using $g(y_i)$ we define the working response z_i as local linear approximation to $g(y_i)$. Specifically, a limited Taylor expansion

$$g(y_i) \approx g(\mu_i) + (y_i - \mu_i)g'(\mu_i)$$
$$= s_i + (y_i - \mu_i)\frac{\partial s_i}{\partial \mu_i} \text{ as } g(\mu_i) = s_i$$
$$= z_i.$$

Solving the likelihood equations with GLMs is thus equivalent to fitting the working responses z_i by weighted least-squares in an iterative way. As in any weighted least-squares fit, the weights to be used are inversely proportional to the variance of the working response, given by

$$\text{Var}[Z_i] = \text{Var}[Y_i]\left(\frac{\partial s_i}{\partial \mu_i}\right)^2 = \frac{\phi}{v_i}V(\mu_i)\left(\frac{\partial s_i}{\partial \mu_i}\right)^2. \tag{4.13}$$

Inverting (4.13) leads to the auxiliary weights $\tilde{v}_i$ defined in (4.2). Since the weights $\tilde{v}_i$ depend on $\boldsymbol{\beta}$, an iterative procedure is needed. Precisely, the process is initiated with some appropriate starting values such as

$$\begin{cases} \widehat{\beta}_0^{(0)} = g^{-1}(\bar{y}) \text{ and} \\ \widehat{\beta}_j^{(0)} = 0 \text{ for } j = 1, \ldots, p. \end{cases}$$

This means that the fitting process starts from the homogeneous case in which all the scores are equal to the intercept β_0. Alternatively, $\widehat{\boldsymbol{\beta}}^{(0)}$ can be taken as the regression parameters in the linear model $(g(y_i), \boldsymbol{x}_i), i = 1, \ldots, n$, modifying the definition for $g(y_i)$ in case y_i is such that g is not defined (for instance, $y_i = 0$ in case of Poisson regression with log link).

Define the working response at step r as

$$z_i^{(r)} = \widehat{s}_i^{(r)} + (y_i - \widehat{\mu}_i^{(r)})\frac{\partial s_i}{\partial \mu_i}\Big|_{\boldsymbol{\beta}=\widehat{\boldsymbol{\beta}}^{(r)}}$$

where

$$\widehat{s}_i^{(r)} = \sum_{j=0}^{p} \widehat{\beta}_j^{(r)}x_{ij} \text{ and } \widehat{\mu}_i^{(r)} = g^{-1}(\widehat{s}_i^{(r)}).$$

Recall that

$$\frac{\partial s_i}{\partial \mu_i} = g'(\mu_i).$$

The corresponding working weights entering the IRLS algorithm are given by

$$\tilde{v}_i^{(r)} = \frac{v_i}{V(\widehat{\mu}_i^{(r)})}\left(\frac{\partial \mu_i}{\partial s_i}\Big|_{\boldsymbol{\beta}=\widehat{\boldsymbol{\beta}}^{(r)}}\right)^2 \text{ where } \frac{\partial \mu_i}{\partial s_i} = \frac{1}{g'(\mu_i)}.$$

The maximum-likelihood estimate $\widehat{\boldsymbol{\beta}}$ for a GLM is then obtained with the help of a sequence of weighted least-squares fits of the working response $z_i^{(r)}$ on the features x_i, with weights $\widetilde{v}_i^{(r)}$, that is,

$$\widehat{\boldsymbol{\beta}}^{(r+1)} = \left(\boldsymbol{X}^\top \widetilde{\boldsymbol{W}}^{(r)} \boldsymbol{X}\right)^{-1} \boldsymbol{X}^\top \widetilde{\boldsymbol{W}}^{(r)} \boldsymbol{z}^{(r)} \tag{4.14}$$

which coincides with the formula (4.11) giving the estimated regression parameters by weighted least-squares, with $z_i^{(r)}$ as working responses at iteration $r+1$ and weights $\widetilde{v}_i^{(r)}$ filling the diagonal matrix $\widetilde{\boldsymbol{W}}^{(r)}$. Fitting a GLM is, thus, equivalent to fitting several Normal linear regression models in a row. IRLS implements the so-called Fisher method of scoring to fit GLMs. Despite its apparent simplicity, formula (4.14) may sometimes become numerically unstable, especially in high dimension, that is, when the number p of features becomes large. This is because of matrix inversion. Gradient descent methods studied in Hainaut et al. (2019) offer a robust alternative in this case.

Notice that the working weights $\widetilde{v}_i^{(r)}$ are inversely proportional to the variance of the working response $z_i^{(r)}$ given in (4.13). The dispersion parameter ϕ simplifies so that it disappears from the iterative formula (4.14) for $\widehat{\boldsymbol{\beta}}^{(r+1)}$. This explains why it does not show up in the definition of the working weights $\widetilde{v}_i^{(r)}$ given above, compared to (4.2).

The stopping rule for the IRLS algorithm generally consists in keeping the current iteration when

$$\left\|\widehat{\boldsymbol{\beta}}^{(r)} - \widehat{\boldsymbol{\beta}}^{(r+1)}\right\| \quad \text{or} \quad \frac{L(\widehat{\boldsymbol{\beta}}^{(r+1)}) - L(\widehat{\boldsymbol{\beta}}^{(r)})}{L(\widehat{\boldsymbol{\beta}}^{(r)})} \quad \text{becomes small enough.}$$

Let r_{stop} be the number of steps before the IRLS algorithm stops. The final iteration $\widehat{\boldsymbol{\beta}}^{(r_{\text{stop}})}$ is taken as the maximum likelihood estimate $\widehat{\boldsymbol{\beta}}$. The working weights $\widetilde{v}_i^{(r_{\text{stop}})}$ at final iteration are stored in the diagonal matrix

$$\widetilde{\boldsymbol{W}}^{(r_{\text{stop}})} = \begin{pmatrix} \widetilde{v}_1^{(r_{\text{stop}})} & 0 & \cdots & 0 \\ 0 & \widetilde{v}_2^{(r_{\text{stop}})} & \cdots & 0 \\ \vdots & \vdots & \cdots & \vdots \\ 0 & 0 & \cdots & \widetilde{v}_n^{(r_{\text{stop}})} \end{pmatrix}.$$

Notice that the fitting procedure for a GLM only uses the link function $s = g(\mu)$ and the mean-variance relationship $V(\cdot)$, but requires no further knowledge of the response distribution.

In most cases, the convergence of the IRLS algorithm is very rapid. However, difficulties sometimes arise. When convergence fails to occur, it can be due to a problem with the data under consideration. For instance, when the two groups are linearly separable in the feature space with binary responses, the data can be fitted perfectly and unstable parameter estimates with high standard errors are typically obtained. This results in convergence problems in the IRLS algorithm.

As an example, consider again the motor insurance data displayed in Table 4.1. The responses $Y_1, \ldots, Y_n$ represent the number of claims reported by the n policyholders, assumed to be independent and to obey the Poisson distribution. There are two explanatory variables:

– Gender with two levels, male and female, and
– Coverage extent with three levels, TPL only, limited damage and comprehensive.

We know that the GLM analysis can be performed on the data aggregated in tabular form instead of individual records, without loss of information. Therefore, the index i now refers to an entire risk class.

The reference class for the portfolio is such that $x_{ij} = 0$ for all j. Here, this corresponds to male policyholders having subscribed coverage for limited damage in addition to the compulsory TPL. The corresponding annual expected claim number is $\exp(\beta_0)$. All results are interpreted with respect to the reference class (for which $x_{i1} = x_{i2} = x_{i3} = 0$). The portfolio is partitioned into 6 risk classes according to policyholders' gender and coverage extent. For instance, the sequence $(0, 1, 0)$ represents a male policyholder with TPL only.

The scores for each cell are given by the linear combination $\boldsymbol{\beta}^\top \boldsymbol{x}_i = \beta_0 + \beta_1 x_{i1} + \beta_2 x_{i2} + \beta_3 x_{i3}$. The annual expected claim frequency for each cell are given by

$$\exp(\boldsymbol{\beta}^\top \boldsymbol{x}_i) = \exp(\beta_0 + \beta_1 x_{i1} + \beta_2 x_{i2} + \beta_3 x_{i3})$$

$$= \begin{cases} \exp(\beta_0) \text{ for a male policyholder with limited damage} \\ \exp(\beta_0 + \beta_2) \text{ for a male policyholder with TPL only} \\ \exp(\beta_0 + \beta_3) \text{ for a male policyholder with comprehensive} \\ \exp(\beta_0 + \beta_1) \text{ for a female policyholder with limited damage} \\ \exp(\beta_0 + \beta_1 + \beta_2) \text{ for a female policyholder with TPL only} \\ \exp(\beta_0 + \beta_1 + \beta_3) \text{ for a female policyholder with comprehensive} \end{cases}$$

Let us now apply the IRLS algorithm to estimate the regression coefficients β_j. Here, we work with the claim rate Y_i/e_i as response so that we use weights $\nu_i = e_i$ corresponding to the total exposure-to-risk for each risk class. In the Poisson case with log-link, we have

$$s_i = \ln \mu_i \Rightarrow \frac{\partial s_i}{\partial \mu_i} = \frac{1}{\mu_i}$$

so that the working responses are given by

$$z_i = s_i + \frac{y_i - \mu_i}{\mu_i}.$$

The working weights are given by $\widetilde{\nu}_i = \nu_i \mu_i$.

We start the IRLS algorithm with the homogeneous case for which all fitted values are equal to $\exp(\beta_0)$. Here, β_0 replicates the observed average claim frequency within the portfolio, that is,

$$\exp\left(\widehat{\beta}_0^{(0)}\right) = \frac{1,683 + 3,403 + 626 + 873 + 2,423 + 766}{10,000 + 30,000 + 5,000 + 6,000 + 24,000 + 7,000}$$

which gives $\widehat{\beta}_0^{(0)} = -2.126993$. Starting from

$$\widehat{\beta}^{(0)} = \begin{pmatrix} -2.126993 \\ 0.000000 \\ 0.000000 \\ 0.000000 \end{pmatrix},$$

we get after three steps

$$\widehat{\beta}^{(1)} = \begin{pmatrix} -2.16632152 \\ -0.12493520 \\ 0.42641819 \\ 0.08540113 \end{pmatrix}, \quad \widehat{\beta}^{(2)} = \begin{pmatrix} -2.17244198 \\ -0.12633430 \\ 0.38501342 \\ 0.09002269 \end{pmatrix}, \quad \text{and } \widehat{\beta}^{(3)} = \begin{pmatrix} -2.17245491 \\ -0.12635812 \\ 0.38384364 \\ 0.09004595 \end{pmatrix}$$

which is exactly the output of the `glm` function of R implementing the IRLS algorithm.

4.4.3 Choice of the Base Levels for Categorical Features

GLM softwares automatically replace every categorical feature with a set of indicator variables, one for each level other than the base level. The latter must be selected with care as all the analysis is performed in relative terms, with respect to the chosen base level.

Consider for instance Example 1.4.3, where the area of residence of a policyholder is considered to be a potential explanatory variable in a model for claim frequency Y. There are three areas, A, B, and C, coded by means of two indicator (dummy) variables x_{i1} (equal to 1 if policyholder resides in area A, and 0 otherwise) and x_{i2} (equal to 1 if policyholder resides in area B, and 0 otherwise). No indicator is needed for the reference level C. The intercept β_0 accounts for the effect of the reference level, taken here as area C. Then, β_1 quantifies the difference between areas A and C, while β_2 quantifies the difference between areas B and C. Notice that the actuary is not limited to make comparisons relative to the base level: differences between areas A and B are quantified by $\beta_1 - \beta_2$.

The choice of the base level is up to the actuary, but it should not be sparse. Assume for instance that there are no policyholders in area C. Then for each policyholder either x_{i1} or x_{i2} is equal to 1, the other being equal to 0. This in turn implies that

$x_{i1} + x_{i2} = 1$ for all i and hence the design matrix X is singular: the sum of the last two columns is equal to the first one (column of 1s, corresponding to the intercept). This makes the estimation of β impossible as the IRLS algorithm needs to invert the matrix $X^\top X$. Intuitively speaking, one of the two binary features in redundant in such a case and the model cannot disentangle their respective regression coefficients β_1 and β_2: every modification in β_1 can be compensated by a corresponding variation in β_2, and vice versa.

More realistically, if the base level has very few cases then the columns of X are nearly dependent which makes the computation of $\widehat{\beta}$ numerically unstable. This is not always taken into account with automated coding implemented in softwares so that the default choice must sometimes be modified by the actuary.

Notice that it is not sensible to define a feature

$$\text{area}_i = \begin{cases} 2 \text{ if policyholder } i \text{ resides in area A,} \\ 1 \text{ if policyholder } i \text{ resides in area B,} \\ 0 \text{ if policyholder } i \text{ resides in area C.} \end{cases}$$

The corresponding GLM score is then equal to

$$\beta_0 + \beta_{\text{area}}\text{area}_i = \begin{cases} \beta_0 + 2\beta_{\text{area}} \text{ if policyholder } i \text{ resides in area A,} \\ \beta_0 + \beta_{\text{area}} \text{ if policyholder } i \text{ resides in area B,} \\ \beta_0 \text{ if policyholder } i \text{ resides in area C,} \end{cases}$$

so that this specification implies an equal "spacing" between A, B, and C: the score $\beta_0 + \beta_{\text{area}}\text{area}_i$ gives a difference of β_{area} between areas A and B, and the same difference β_{area} between areas B and C. This is not necessarily in accordance with the data and may bias the analysis.

Even if the GLM analysis is performed conditionally on the values x_{ij} of the features X_{ij}, the correlation structure of the random vectors $(X_{i1}, \ldots, X_{ip})$ may impact on the estimation of the regression coefficients. The correlation among the random variables X_{ij} means that some features may serve as substitutes for the remaining ones in the calculation of the score.

4.4.4 Correlated Features, Collinearity

Most of the time, the features X_{ij} are correlated in observational studies. When these correlations are moderate, GLMs are able to isolate the effect of each risk factor on the mean response. This is one of the great strength of GLMs compared to univariate, or marginal analyses which are often distorted because the same effect is counted multiple times. When the correlation between some features becomes large, however, GLMs may run into trouble. Such high correlations mean that the same information is encoded in several features, leading to numerical problems as explained next.

Aliasing occurs when there is a linear dependency among the features. For instance, one feature may be identical to some linear combination of some other ones. Consider the following explanatory variables:

$$x_{i1} = \text{age at which driving commenced}$$
$$x_{i2} = \text{seniority of the driving license}$$
$$x_{i3} = \text{current age for policyholder } i.$$

Obviously, the identity

$$x_{i3} = x_{i1} + x_{i2}$$

holds true for all i, so that x_{i3} is redundant in explaining the response. The score can be written as a function of two features, only. For instance, we can write

$$\beta_0 + \beta_1 x_{i1} + \beta_2 x_{i2} + \beta_3 x_{i3} = \beta_0 + (\beta_1 + \beta_3)x_{i1} + (\beta_2 + \beta_3)x_{i2}.$$

Hence, the three variables x_{i1}, x_{i2}, and x_{i3} explain the same as any two of the variables. Thus, the individual effects of each of the three variables cannot be assessed.

Such variables are perfectly collinear, or aliased. Aliasing thus corresponds to a linear dependence among the columns of the design matrix X. When two features are perfectly correlated, they are said to be aliased and the parameter estimates are not unique. In these cases, the design matrix X has an exact linear dependence between its columns and hence $X^\top X$ is singular which invalidates the IRLS approach.

Apart from exact collinearity illustrated above, it is more usual to encounter situations of near collinearity (or near aliasing). For instance, in a data set where policyholder's area of residence, income level and education level are strongly correlated, the inclusion of the three explanatory variables makes $X^\top X$ near singular and results in difficulties for computing the inverse $(X^\top X)^{-1}$. Tikhonov regularization and ridge regression (thoroughly discussed in Hainaut et al. 2019) may be helpful in that respect.

If the inverse is computable then it will typically have large diagonal entries, implying large standard errors on one or more regression coefficients (for reasons explained in the next section). Collinearity may also mislead the analyst in the choice of explanatory variables: if 2 explanatory variables are highly correlated and both predictive of the response then each one, in the presence of the other, will not contribute much additional explanation. Hypothesis tests for each variable, assuming that the other one is in the model, will show non-significance. The solution is to include only one of these two explanatory variables in the model. This explains why a preliminary examination of the correlation structure of the features is useful before starting the GLM analysis.

Remark 4.4.1 (Collinearity versus Interaction). Collinearity comes from correlation between risk factors, several of them encoding the same information. Interaction has nothing to do with collinearity/correlation. For instance, consider a motor insurance portfolio with the same age structure for males and females (so that the risk factors age and gender are mutually independent). If young male drivers are more dangerous compared to young female drivers whereas this ranking disappears or reverses at older ages, then age and gender interact despite being independent.

4.4.5   Cramer's V

The strength of the dependence for two continuous variables can be assessed by Pearson's linear correlation coefficient (i.e., the covariance divided by the product of the standard deviations) or, preferably, by rank correlation coefficients such as Spearman's rho of Kendall's tau. However, these classical dependence measures are not appropriate for assessing the correlation between non-continuous features. Given the different formats of the features entering GLMs (binary, categorical, discrete or continuous), there is a need for a correlation measure applicable to all these cases. This is why Cramer's V is often used to measure the strength of association between features appearing in insurance studies. This approach is based on data displayed in tabular form so that it effectively deals with discrete, or categorical features. Continuous features must be discretized (or banded) by partitioning their domain into disjoint intervals and hence be treated as discrete ones.

Consider for instance policyholder's age and gender. Age is first categorized in a few classes: for instance, the "young driver" category gathering policyholders aged below 25, the "middle-aged driver" category gathering policyholders aged from 26 to 50 and the "senior driver" category with policyholders aged 51 and older. The portfolio can then be split into 6 classes, crossing the 3 aforementioned age categories with the 2 genders. In each cell, the actuary records the observed responses and the associated exposure. However, such a preliminary banding of a continuous variable (like age in our example) is subjective (the choice of cut-off points is generally made somewhat arbitrary by the analyst) and may lead to a possible loss of information. No optimal cut-off points are available in general.

Cramer's V is based on a contingency table cross-classifying the data according to the pair of features under consideration. Recall that the Chi-Square statistics for testing independence between the two features appearing in the contingency table, denoted as χ^2, is obtained by summing the squared differences between the observed frequency and the expected frequency in each cell of the table, divided by the expected frequency. The expected frequency is the one corresponding to independence between the two features under consideration. It is obtained by multiplying the row total with the column total divided by the grand total. Notice that exposures must be taken into account in the calculation.

Now, given two features X_{ij_1} and X_{ij_2} assuming k_1 and k_2 discrete values, respectively, Cramer's V is defined as

$$
V = \sqrt{\frac{\chi^2/n}{\min\{k_1 - 1, k_2 - 1\}}} \in [0, 1]
$$

where χ^2 is the Pearson's Chi-Square test statistic for independence of features X_{ij_1} and X_{ij_2}. Dividing χ^2 by the number n of observations makes the statistic independent of the number of observations, i.e. multiplying each cell of the contingency table with a positive integer does not alter the value of Cramer's V. Indeed, Cramer's V assumes its values in the unit interval $[0, 1]$, with 0 and 1 corresponding to independence and perfect dependence, respectively. Being based on data displayed in tabular form, Cramer's V is widely applicable.

Moreover, the maximum of χ^2/n is attained when there is total dependence, i.e. each row (or column, depending on the size of the table) exhibits only one strictly positive integer. This means that the value of X_{ij_1} determines the values of X_{ij_2}. The maximum of χ^2/n is equal to the smallest dimension minus 1. Thus, dividing χ^2/n by $\min\{k_1 - 1, k_2 - 1\}$ ensures that V assumes its value in the unit interval $[0, 1]$. Taking square-root guarantees that Cramer's V and Pearson's linear correlation coefficient coincide on 2×2 tables (i.e. for two binary features, with $k_1 = k_2 = 2$).

Example 4.4.2 Let us consider again the data displayed in Table 4.5. This hypothetical data set displayed is cross-classified according to gender and annual mileage (with two categories defined with respect to the threshold 20,000 kms). Cramer's V is equal to 0.533 indicating a rather strong positive association between gender and distance traveled, male policyholders tending to drive longer distances. For the data in Table 4.1, we find that Cramer's V is equal to 0.123 so that the association is weaker.

4.4.6 Omitted Variable Bias

We know from Chap. 2 that the actuary cannot access important information about policyholders' riskiness. If hidden risk factors are correlated with both the response and one or more available features, they will bias the estimate of the corresponding regression coefficient. This phenomenon is commonly known as the omitted variable bias and can be explained as follows.

Consider the number of claims Y_i filed by policyholder i with exposure e_i and features x_i. Assume that

$$E[Y_i] = e_i \exp\left(\beta_0 + \sum_{j=1}^{p} \beta_j x_{ij} + \beta^+ x_i^+\right).$$

If the feature x_i^+ is omitted and x_i^+ is correlated to the remaining features $x_{i1}, \ldots, x_{ip}$ in the sense that the approximation

$$x_i^+ \approx \beta_0^+ + \sum_{j=1}^{p} \beta_j^+ x_{ij}$$

is relatively accurate, then

$$E[Y_i] \approx e_i \exp\left(\beta_0 + \beta^+\beta_0^+ + \sum_{j=1}^{p}(\beta_j + \beta^+\beta_j^+)x_{ij}\right).$$

In such a case, the maximum likelihood estimate $\widehat{\beta}_j$ produced by the IRLS algorithm in fact estimates $\beta_j + \beta^+\beta_j^+$ and not β_j. The true effect β_j of x_{ij} on the score scale cannot be isolated, only the biased effect

$$\beta_j + \beta^+\beta_j^+$$

can be estimated mixing

- the true effect β_j of x_{ij}
- with the effect $\beta^+\beta_j^+$ of the omitted feature x_i^+.

In such a case, we might even have $\widehat{\beta}_j < 0$ whereas $\beta_j > 0$.

Thus, any omitted feature correlated to x_{ij} biases the estimation of β_j obtained from the IRLS algorithm. This is why standard industry models cannot be used to establish causal relationship, just to detect correlations.

Let us now illustrate the effect of omitted variable bias on a simple example. Consider again the hypothetical data set displayed in Table 4.5 cross-classified according to gender and annual mileage (with cut-off 20,000 kms). As both features are categorical with two levels, they can be coded by means of two binary features. Precisely, x_{i1} records annual mileage, with

$$x_{i1} = \begin{cases} 0 \text{ if policyholder } i \text{ drives more than 20,000 kms per year} \\ 1 \text{ otherwise.} \end{cases}$$

Similarly, x_{i2} records gender, with

$$x_{i2} = \begin{cases} 0 \text{ if policyholder } i \text{ is male} \\ 1 \text{ otherwise.} \end{cases}$$

Here, we have selected the base levels corresponding to the most populated categories: male for gender and more than 20,000 kms for the distance traveled per year.

Including both binary features encoding the information in relation to this policy (x_{i1} for gender and x_{i2} for annual mileage), IRLS algorithm produces the following results after three iterations:

$$\widehat{\beta}_0 = -2.20597$$
$$\widehat{\beta}_1 = -0.54753$$
$$\widehat{\beta}_2 = -0.26383.$$

Now, if gender is taken out of the score, keeping only the distance traveled, we get after three iterations:

$$\widehat{\beta}_0 = -2.24904$$
$$\widehat{\beta}_1 = -0.69552.$$

Thus, we can see that the estimated regression coefficients are modified compared to the estimates obtained with gender in the model. In particular, the estimated regression coefficient $\widehat{\beta}_1$ associated to distance traveled becomes more negative, changing from -0.55 to -0.70. This can be explained by the correlation existing between the two features. We know that the association is positive, as indicated by the relatively high value of Cramer's V obtained in Example 4.4.2. In words, male policyholders tend to drive more than 20,000 kms per year whereas female policholders tend to drive less. Hence, if we know that the policyholder drives less than 20,000 kms per year, it is likely that this is a female. Formally,

$$P[\text{female}| < 20,000 \text{ kms}] = \frac{6,000}{6,000 + 2,000} = \frac{3}{4}.$$

Hence, $\widehat{\beta}_1$ becomes more negative when gender is left out of the score because it takes part of the negative effect that being a female driver has on the score.

The preliminary examination of the association existing between the available features indicates that the inclusion or exclusion or offsetting of one of these will have an impact on the others. Features that are highly correlated with the omitted variables will change the most.

4.4.7 Numerical Illustrations

4.4.7.1 Graduation of One-Year Death Probabilities

Consider a closed group of individuals, observed during one year. At each age x from birth to age 100, the response is the number of deaths D_x recorded among l_x individuals aged x on January, the 1st. Figure 4.1 displays the available data D_x and l_x. The corresponding crude one-year death probabilities

$$\widehat{q}_x = \frac{D_x}{l_x}$$

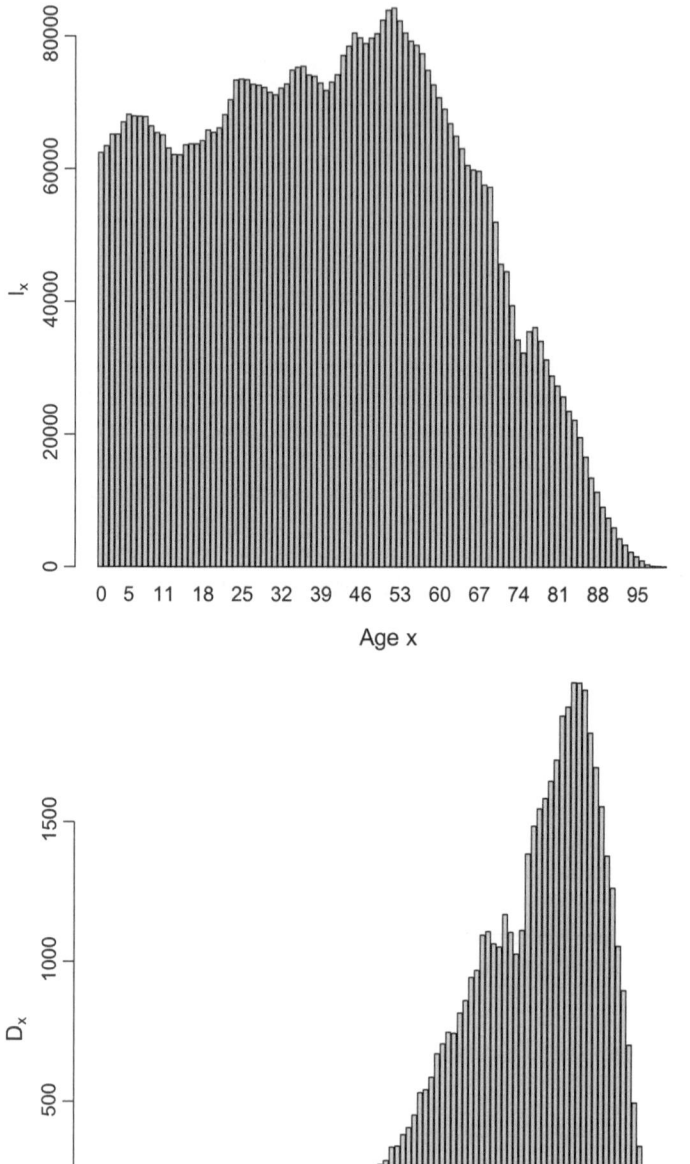

Fig. 4.1 Graph of observed responses D_x (upper panel) and Binomial sizes l_x (lower panel)

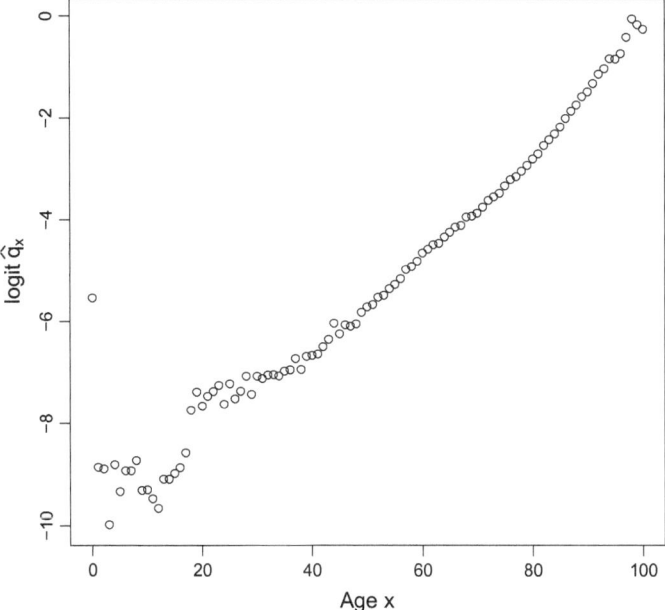

Fig. 4.2 Crude one-year death probabilities $\widehat{q}_x = D_x/l_x$ corresponding to the mortality data displayed in Fig. 4.1, on the logit scale

are the maximum-likelihood estimate of the parameter q_x under the model $D_x \sim \mathcal{B}in(l_x, q_x)$, as shown in Sect. 3.6.1. They are displayed in Fig. 4.2. They can also be obtained with a Binomial GLM treating age x as a categorical feature, that is, using the binary features

$$x_{ij} = \begin{cases} 1 \text{ if age} = j, \\ \\ 0 \text{ otherwise,} \end{cases}$$

for $j = 1, 2, \ldots, 100$ (selecting age 0 as the reference one). Dealing with Binomial responses, we use the canonical parameter $\ln \frac{q_x}{1-q_x}$, that is, the logit of q_x. The corresponding maximum-likelihood estimates are then given by

$$\widehat{\beta}_j = \ln \frac{\widehat{q}_j}{1 - \widehat{q}_j} - \ln \frac{\widehat{q}_0}{1 - \widehat{q}_0}, \quad j = 1, \ldots, 100,$$

and

$$\widehat{\beta}_0 = \ln \frac{\widehat{q}_0}{1 - \widehat{q}_0}.$$

Figure 4.2 shows that the logit of $\widehat{q}_x$ appears to be reasonably linear in age x except at young ages (where the high mortality around birth, the leveling off around

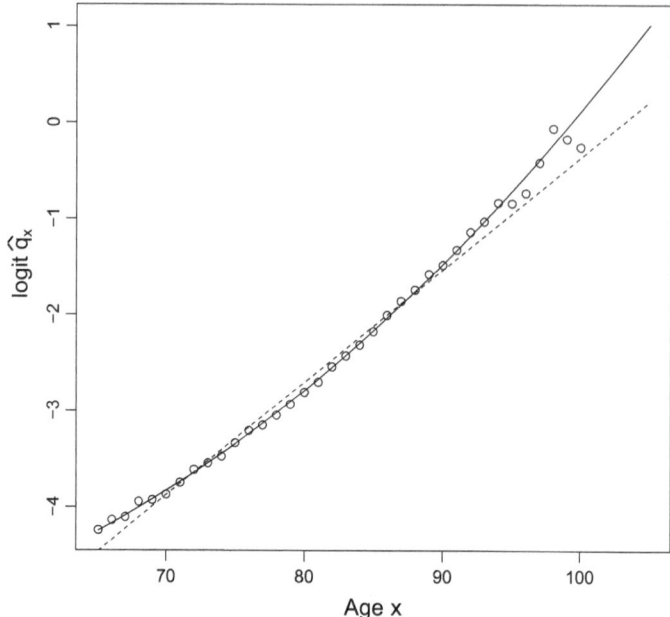

Fig. 4.3 Graph of the fitted values obtained from the quadratic Perks formula (4.15) in continuous line, from the corresponding linear specification ($\beta_2 = 0$) in broken line, together with the crude one-year death probabilities $\widehat{q}_x = D_x / l_x$ on the logit scale, ages 65 and over

age 30 and the accident hump around age 20 are more complicated to capture), with a slight curvature after retirement age. This gave rise to the so-called (quadratic) Perks formula according to which

$$\ln \frac{q_x}{1 - q_x} = \beta_0 + \beta_1 x + \beta_2 x^2 \Leftrightarrow q_x = \frac{\exp(\beta_0 + \beta_1 x + \beta_2 x^2)}{1 + \exp(\beta_0 + \beta_1 x + \beta_2 x^2)}. \qquad (4.15)$$

Age x is treated as a continuous feature and supplemented with its square, used as an additional feature, and the model is restricted to age 65 and over.

Formula (4.15) provides a parsimonious representation of one-year death probabilities at older ages. It can be fitted using the Binomial GLM to the data (D_x, l_x) displayed in Fig. 4.1. The data file comprises the response D_x together with one feature (Age) and Binomial sizes l_x. Using the R function glm, we get $\widehat{\beta}_0 = -3.34$, $\widehat{\beta}_1 = -0.1038$, and $\widehat{\beta}_2 = 0.001384$. The IRLS algorithm converged in only three iterations. The resulting fitted values are displayed in Fig. 4.3. We also represent there the result of a linear fit, which clearly misses the curvature present in the data.

In this example, Fig. 4.3 clearly shows that the quadratic Perks formula outperforms its linear counterpart on the data set under consideration. However, it must be stressed that this conclusion is drawn on the basis of visual inspection, which may become impossible for insurance studies involving many features. This is why we also need formal testing procedures to compare two model specifications, as those presented in the next sections.

4.4.7.2 Loss Reserving

In Property and Casualty (P&C) insurance, claims are sometimes settled over several years. This can be due to long legal procedures associated with liability insurance claims (the litigation required to establish liability in disputed cases may be a long process) or just because the claim may be filed only later. In product liability for instance, it can last long before the damage is discovered and reported, and even longer before the extent of the damage is evaluated.

A typical claim history is depicted in Fig. 4.4. The claim under consideration originates from an event (an accident, for instance) that occurred at time t_1. It has been reported to the insurer at time t_2. The settlement process for this particular claim developed from time t_2 to time t_6, that is, from notification to final settlement, or closure. The claim is said to be open from time t_2 to time t_6. The two levels visible in Fig. 4.4 represent the policyholder's point of view (upper part), with

– occurrence,
– notification and
– payments to the insured or third parties)

and the insurer's point of view (lower part), where

– the insurer is unaware of the claim before its notification,
– builds a reserve, starting from an initial case estimate at time t_2,
– adjusts the reserve over time to reflect the best estimate of the final claim cost, varying according to the information received by the insurer (expert evaluation or judgment, for instance).

Outstanding claims correspond to all claims that are not yet settled. They are classified into different categories:

• IBNR claims (Incurred But Not Reported): claims that have occurred but have not been filed (the claim is classified as IBNR from time t_1 to time t_2 in Fig. 4.4).
• open or RBNS claims (Reported But Not Settled): claims which have been reported to the insurer but not (completely) paid (the claim is classified as RBNS from time t_2 to time t_6 in Fig. 4.4).

At the accounting date, the actuary has a set of open claims and computes the corresponding RBNS reserve, plus a set of hidden claims for which an IBNR reserve has to be determined. The outstanding claims reserve means the total of RBNS plus IBNR.

Clearly, the pattern of reporting and of claims settlement varies according to business lines. In general, property damage leads to a short tail in the run-off (meaning

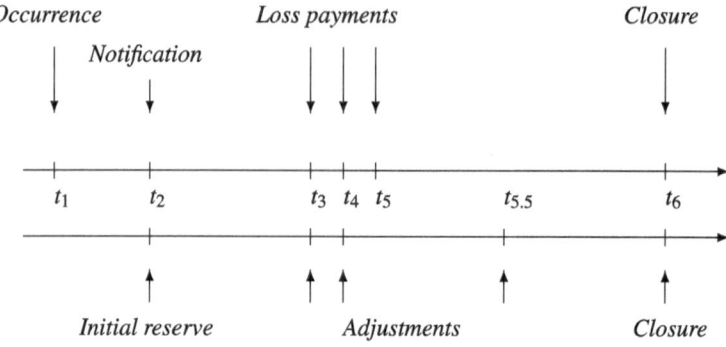

Fig. 4.4 Typical claim history, from occurrence to final settlement, or closure

that claim are rapidly reported and settled) and liability to a medium or long tail. Until all claims have been settled, insurers have to build reserves representing their estimate of outstanding liabilities for claims that occurred on or before the valuation date. For a general insurance company, the largest balance sheet item is often the reserve for outstanding claims. In practice, these reserves need to be assessed by actuarial methods

– to avoid subjective judgments and for reporting to the regulatory authorities.
– because tax authorities require that amounts of reserves are computed from well-established methods/generally accepted principles.
– to account for hidden, IBNR claims.

Reserving calculation has traditionally been performed on the basis of aggregated data summarized in run-off triangles with rows corresponding to accident years and columns corresponding to development years. Such data exhibit three dimensions: for each accident (or occurrence, or underwriting) year AY and development period (considered here as one year, even if it can be semester or a quarter) $DY = 1, 2, \ldots$, we read in cell (AY, DY) inside the triangle the total amount paid by the insurer in calendar year $CY = AY + DY - 1$ for claims originating in year AY, the number of payments made or the average amount per payment. On the basis of the claim figures in the run-off triangle, the actuary wants to make predictions about payments to be made in future calendar years. The goal of the actuarial reserving techniques is to predict these figures, so as to complete the triangle into a rectangle.

Often, reserving calculations are carried on a single triangle gathering yearly total payments by AY. This approach mixes the frequency and severity dynamics, and this often obscures the analysis. Conducting separate analyses for frequencies and severities allows the actuary to isolate each effect, which eases model interpretation and predictions.

Assume for instance that the tables used in court cases to calculate future losses in personal injury and fatal accident cases are revised in calendar year $c^\star$ (such tables are available in most countries; they are called Ogden tables[1] in the UK). This may be due to a change in the discount rate, for instance. As a result, all the amounts paid for the claims yet to be settled will be increased, sometimes to a considerable extent.

Considering the triangle with (incremental) yearly total payments, this is expected to produce an increase along the c^* and next CY diagonals. However, this effect only impacts on costs and does not modify the frequency component. This becomes clear when the actuary works with two separate triangles, that is,

Triangle 1:   number of payments per AY and DY, and
Triangle 2:   average amount of payment per AY and DY.

The changes in the reference tables manifest in triangle 2, causing an increase in diagonals c^* and next. In regression models, additional features such that $I[CY \geq c^*]$ can be introduced in the analysis of triangle 2 to account for this effect. Many problems can be avoided if separate count and severity triangles are analyzed.

Let us consider claim data from the motor third party liability insurance portfolio of an insurance company operating in the European Union. The available information covers 11 accident years. Table 4.6 displays descriptive statistics for the payments per accident year and development period. We can see there the number of payments, the average amounts per payments as well as the corresponding standard deviation. Data are displayed per accident year and development. The response considered here is the average paid amount. Table 4.6 shows the typical increase of mean payments with development, together with the corresponding decrease in the number of payments.

The amounts appearing Table 4.6 are expressed in euros. No prior correction for inflation has been performed. Insurance payments are generally corrected for inflation before starting the GLM analysis. Amounts of benefits corresponding to bodily injuries can be corrected using a wage index whereas consumer price index is used for material damage, for instance. But there often remains some inflation effect even after this correction has been made. This is because insurance benefits are subject to super-inflation, that is, they depart from macroeconomic indices (and generally increase at a faster rate). Regression models like GLMs can be used to estimate inflation effects. This is done by including CY in the features entering the score. Estimated regression coefficients corresponding to CY then reveal the inflation applying to the insurance product under consideration. If amounts corrected for inflation enter the GLM analysis then these regression coefficients correct the inflation indices used in the pre-processing of claim severity data.

The three effects AY, DY and CY present in run-off triangles capture different aspects of insurance loss dynamics:

AY   the accident year AY (row effect), also called year of occurrence or year of origin, represent variations in the size of the portfolio.

---

[1]The full, and official, name of the tables is Actuarial Tables with explanatory notes for use in Personal Injury and Fatal Accident Cases but the unofficial name has now become common parlance, after Sir Michael Ogden having been the chairman of the Working Party for the first four editions.

Table 4.6 Descriptive statistics for payments per accident year and development period, namely the number of payments, the average paid amounts as well as the standard deviation

DY	1	2	3	4	5	6	7	8	9	10	11
AY 1											
Number of payments	2,848	1,459	236	124	68	39	18	18	12	8	7
Mean	1,133	1,877	2,713	4,349	4,446	9,894	16,765	4,422	18,072	12,314	21,263
Standard deviation	2,378	5,317	4,861	9,405	7,918	26,576	27,037	9,768	24,203	14,436	50,490
AY 2											
Number of payments	3,001	1,492	207	97	53	42	24	21	11	11	
Mean	1,112	1,659	3,168	5,455	5,132	14,882	25,781	8,997	4,230	1,347	
Standard deviation	1,847	2,932	6,081	18,278	10,270	41,070	77,046	19,947	2,817	883	
AY 3											
Number of payments	3,007	1,659	268	117	61	41	21	10	10		
Mean	1,164	1,624	5,799	4,494	7,287	6,055	6,141	4,688	12,205		
Standard deviation	2,972	2,932	49,737	7,632	22,190	12,682	9,173	4,594	26,907		
AY 4											
Number of payments	3,246	1,893	322	170	79	48	24	16			
Mean	1,159	1,905	2,679	3,500	7,401	8,243	12,140	13,148			
Standard deviation	2,258	4,984	6,159	5,831	13,989	14,717	20,625	22,292			
AY 5											
Number of payments	3,574	1,816	304	125	71	37	22				
Mean	1,104	1,720	2,189	4,203	4,611	7,775	6,310				
Standard deviation	1,837	3,644	3,524	8,791	9,908	12,249	7,275				
AY 6											
Number of payments	3,545	1,877	300	131	90	51					
Mean	1,142	1,919	3,981	4,379	6,896	9,129					
Standard deviation	1,926	5,710	19,797	11,584	17,446	18,474					

(continued)

Table 4.6 (continued)

DY	1	2	3	4	5	6	7	8	9	10	11
AY 7											
Number of payments	2,874	2,072	338	161	75						
Mean	1,663	1,984	3,637	5,147	14,935						
Standard deviation	4,012	5,832	11,419	13,420	60,912						
AY 8											
Number of payments	2,777	1,930	327	119							
Mean	1,601	1,982	2,441	5,171							
Standard deviation	2,333	4,004	4,119	14,476							
AY 9											
Number of payments	2,860	1,749	282								
Mean	1,716	2,328	4,390								
Standard deviation	4,587	10,085	31,803								
AY 10											
Number of payments	2,924	1,844									
Mean	1,637	2,230									
Standard deviation	4,120	11,414									
AY 11											
Number of payments	2,723										
Mean	1,662										
Standard deviation	2,360										

DY   the development year DY (column effect), also called reporting delay or set-
        tlement delay, represents delays in reporting by policyholders and claim handling
        procedure by the company.
CY   the calendar year CY (diagonal effect), also called payment year, represents
        inflation and jurisprudence effect.

Let us now switch from the classical triangle representation to the data displayed
for a GLM analysis. Here, we have three features: AY, DY, and CY, linked by the
relationship $CY = AY + DY - 1$. In the first step of the analysis, these features
are treated as categorical ones to maximize flexibility. This means that there is one
parameter associated to each period for explaining the average paid amount. Each
observation i corresponds to one cell of the run-off triangle. Henceforth, we denote as
$AY(i)$, $DY(i)$, and $CY(i)$ the respective accident year, development year and calendar
year for observation i.

The total payment Y_i in calendar year $CY(i)$ for claims originating in accident
year $AY(i)$ is decomposed into the compound sum

$$Y_i = \sum_{k=1}^{N_i} P_{ik} = N_i \overline{P}_i,$$

where

$$N_i = \text{number of payments made in calendar year } CY(i)$$
$$\text{for claims originating in accident year } AY(i)$$
$$\text{still open at development } DY(i);$$
$$P_{ik} = \text{corresponding amounts;}$$
$$\overline{P}_i = \frac{Y_i}{N_i}$$
$$= \text{average payment made in calendar year } CY(i)$$
$$\text{for the claims originating in accident year } AY(i)$$
$$\text{still open at development } DY(i).$$

All these random variables are assumed to be mutually independent. For given i, the
random variables P_{i1}, P_{i2}, ... are assumed to be identically distributed. Notice that
here, payments related to individual policies are not tracked, only payments for the
collective are modeled, and all payments related to the same claim are aggregated in
the single, yearly amount P_{ik}.

All random variables $\overline{P}_i$ are assumed to be mutually independent, given the AY,
DY and CY effects included in the score. Notice that the knowledge of the average amount of payments is enough to estimate the parameters in the GLM setting
(there is no loss of information due to the aggregation of all individual payments
P_{i1}, P_{i2}, ..., P_{iN_i} into the single value $\overline{P}_i$ in this case as explained before).

The data file comprises the features AY and DY, together with the number of payments, the average amount per payment and the corresponding standard deviation. Henceforth, we assume that the data basis covers m accident years. We denote as ω the maximum number of years needed to settle all the claims originating in a given calendar year. Clearly, $\omega \geq m$, but here, we assume $\omega = m$ for simplicity.

Considering the number of payments, let us adopt a modeling in line with the classical Chain-Ladder approach to loss reserving. To this end, we assume that the mean number of payments in cell i can be written as

$$E[N_i] = \alpha_{AY(i)}\delta_{DY(i)}.$$

Sometimes, a volume measure is used as an offset (such as total premium income, for instance). To avoid over-parametrization, actuaries generally impose that the identity

$$\sum_{j=1}^{\omega} \delta_j = 1$$

holds true. Then,

$$\alpha_k = \alpha_k \sum_{j=1}^{\omega} \delta_j = \sum_{i|AY(i)=k} E[N_i]$$

so that α_k represents the expected number of payments needed to settle all the claims originating in accident year k. With this constraint, δ_j is the proportion of each α_k corresponding to development j. The intuitive idea behind this model is that a stable percentage δ_j of the total number of payments needed to settle all the claims is made at each accident year, that is, the columns of the run-off triangle are roughly proportional (because the corresponding expected values are assumed to be proportional).

The parameters α_k and δ_j are estimated by maximum likelihood in the model

$$N_i \sim \mathcal{P}oi\left(\alpha_{AY(i)}\delta_{DY(i)}\right) \text{ independent.}$$

This means that we are using a Poisson GLM with a score of the form

$$\text{score}_i = \sum_{k=1}^{\omega} \beta_k I[AY(i) = k] + \sum_{l=1}^{\omega} \beta_{\omega+l} I[DY(i) = l]$$

where the binary features $I[AY(i) = k]$ and $I[DY(i) = l]$ code the categorical features AY and DY, and $\beta_k = \ln \alpha_k$ while $\beta_{\omega+k} = \ln \delta_k$, $k = 1, 2, \ldots, \omega$.

The likelihood equations can easily be solved directly, without using the IRLS algorithm. The idea (known as Verbeek's algorithm in the actuarial community) is to proceed as follows. We know that since we used the canonical link function for the Poisson GLM, the likelihood equations consist in equating the sum all the the numbers of payments in each row and in each column to their corresponding expected value. Computing the sum of all values along the first row of the run-off triangle (corresponding to the first accident year, assumed to be fully settled as $m = \omega$) and equating the result to the sum of expected values, we get

$$\sum_{i|AY_i=1} N_i = \sum_{j=1}^{\omega} \alpha_1 \beta_j \Rightarrow \widehat{\alpha}_1 = \sum_{i|AY_i=1} N_i.$$

Now, consider the last column of the run-off triangle, for which there is a single cell with an observed value. The likelihood equation then gives

$$N_{i|AY_i=1 \text{ and } DY_i=\omega} = \alpha_1 \beta_\omega \Rightarrow \widehat{\beta}_\omega = \frac{N_{i|AY_i=1 \text{ and } DY_i=\omega}}{\widehat{\alpha}_1}.$$

Let us now turn to the second row of the triangle, recording the observations for the second accident year. All the values appearing in the second row have been observed except for the last cell (at development ω). Summing all the values appearing in this row and equating the result to the corresponding expected values, we get

$$\sum_{i|AY_i=2} N_i = \sum_{j=1}^{\omega-1} \alpha_2 \beta_j = \alpha_2(1 - \beta_\omega) \Rightarrow \widehat{\alpha}_2 = \frac{\sum_{i|AY_i=2} N_i}{1 - \widehat{\beta}_\omega}.$$

We then consider column $\omega - 1$, row 3, column $\omega - 2$, and so on and so forth until all parameters will have been estimated. Of course, making the link with GLMs supplement the classical Chain-Ladder output with standard errors and diagnostic measures that appear to be useful to assess the appropriateness of this approach on a particular run-off triangle.

Let us now consider the number of payments N_i displayed in Table 4.6. If we use the multiplicative specification corresponding to the Chain-Ladder model, factoring $E[N_i]$ into a product of a row and a column effect, this means that a score of the form

$$\text{score}_i = \sum_{k=1}^{11} \beta_k I[AY(i) = k] + \sum_{l=1}^{11} \beta_{11+l} I[DY(i) = l]$$

is used in the Poisson GLM. The column effect is now subject to the identifiability constraint

$$\sum_{j=12}^{22} \exp(\beta_j) = 1.$$

Hence, $\exp(\beta_j)$, $j = 11, \ldots, 22$, can be interpreted as the proportion of the expected total number of payments $\exp(\beta_l)$ needed to settle all the claims originating in accident year $l \in \{1, \ldots, 11\}$ made at development $j - 11$. Figure 4.5 displays the estimated α_j and δ_j that have been obtained by Poisson maximum likelihood.

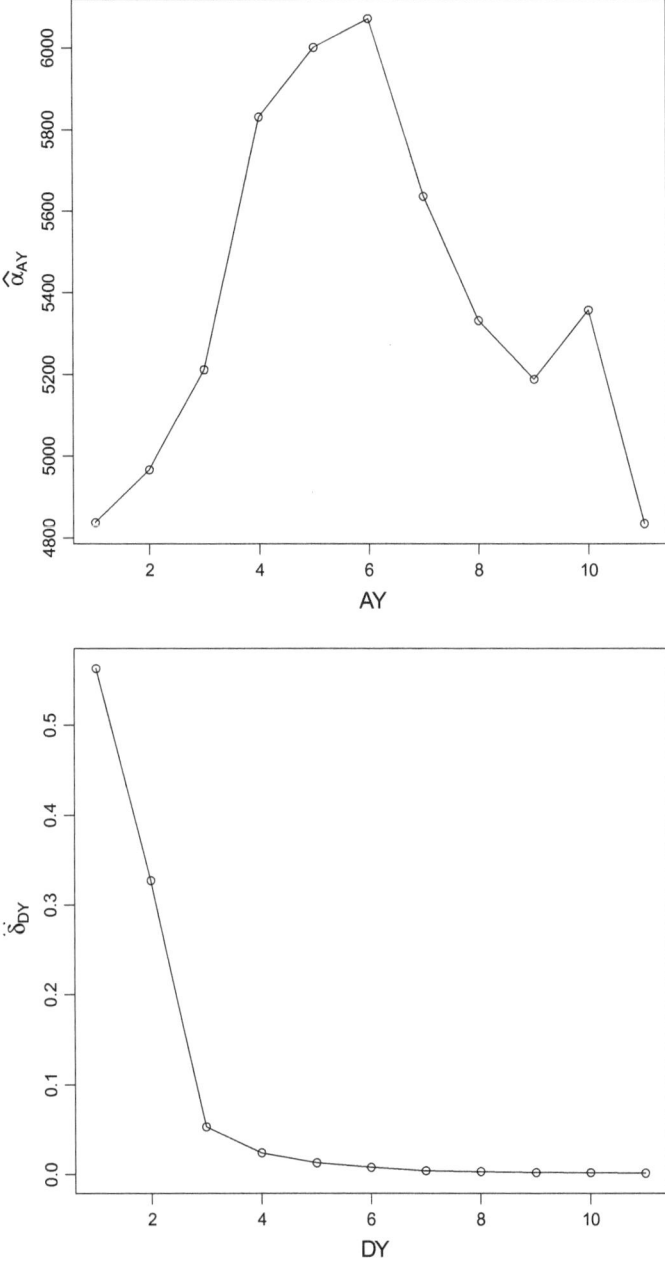

Fig. 4.5 Estimated parameters α_j related to AY j, $j = 1, \ldots, 11$, for the numbers of payments (upper panel) and estimated parameters δ_j related to DY j, $j = 1, \ldots, 11$ (lower panel)

The Chain-Ladder algorithm applied to the same data would have produced the same fitted values (but in another way, as Chain-Ladder operates on cumulative values and only estimates link factors, allowing the actuary to switch from one development to the next in the runoff triangle). We can see that the estimated expected numbers of payments, that is, the estimated parameters α_j related to AY, $j = 1, \ldots, 11$, displayed in Fig. 4.5, first increase, reaching a maximum before decreasing. This reflects the particular change in volume for the portfolio under consideration. Intuitively, we would have expected that the estimated α_j would remain roughly stable over time, suggesting that the volume of business stays unchanged, or moderately increases being compensated by the progressive reduction in claim frequencies in motor insurance. In this case, the bell shape can be explained by circumstances specific to this insurance company, including the progressive integration of business previously sold through bank.

The shape of the estimated parameters δ_j related to DY, $j = 1, \ldots, 11$, displayed in Fig. 4.5, is quite standard. We can see that the majority of payments are made during the accident year (about 55%) and the year after (about 35%), later developments accounting for the remaining claims. In motor third-party liability insurance, experience generally shows that the vast majority of claims are settled at developments 1–2.

Let us now turn to the average payments. Clearly, the Poisson assumption underlying the Chain-Ladder approach is not very appealing for claim amounts. As explained in Sect. 4.11, the important modeling aspect is the mean-variance relationship. The Poisson assumption corresponds to the case where the variance of the paid amount is equal to its expected value. This is the crucial assumption hidden behind the Chain-Ladder approach: in case the mean paid amount is equal to 1 million, the variance is also equal to this amount and the standard deviation is one thousand. Given that the Poisson distribution with such a high mean is very close to the Normal distribution, this means that we have a very high concentration around the mean value, with a dispersion ranging in the few thousands, only. Such a small dispersion may be overly optimistic and the mean-variance relationship must be challenged and not trusted blindly, as explained next.

Considering the average amounts of payment $\overline{P}_i$, we could use a Gamma or an Inverse Gaussian specification. We know that the two models differ by their variance function: $V(\mu) = \mu^2$ for the Gamma distribution whereas $V(\mu) = \mu^3$ for the Inverse Gaussian distribution. To select the more appropriate response distribution, we fit a cubic and quadratic regression model with response the variance of the response (without intercept). Adjusted R-squared is 73.66% (100% representing a perfect fit) for the cubic link function, while it reduces to 52.53% for the quadratic link function. This suggests to use the Inverse-Gaussian distribution for the response. Notice that the Poisson variance function $V(\mu) = \mu$ corresponding to the Chain-Ladder model is not a good candidate for the data under consideration as the adjusted R-squared is only 31.41% in that case.

First, we compare a model decomposing the expected amount per payment into a product of an DY effect and a CY effect with a model decomposing it into a product of a AY effect and a DY effect. The fit obtained in the latter case was much better so that we assume henceforth that

$$\overline{P}_i \sim \mathbb{I}\mathcal{G}au(\mu_i, \tau)$$

with the average paid amount decomposed into a product of an AY effect and a DY effect:

$$\mu_i = \alpha_{\mathrm{AY}(i)}\delta_{\mathrm{DY}(i)}.$$

Notice that this specification accounts for a constant inflation rate ι, as the inflation applicable to the payment $\overline{P}_i$ is

$$(1 + \iota)^{\mathrm{AY}(i)+\mathrm{DY}(i)-2}$$

taking the first accident year as the base year for inflation. This factors into

$$(1 + \iota)^{\mathrm{AY}(i)+\mathrm{DY}(i)-2} = (1 + \iota)^{\mathrm{AY}(i)-1}(1 + \iota)^{\mathrm{DY}(i)-1}$$

where each $(1 + \iota)^{\mathrm{AY}(i)-1}$ and $(1 + \iota)^{\mathrm{DY}(i)-1}$ can be absorbed into the AY effects α and the DY effects δ, respectively. The inflation rate ι is then implicitly contained in the estimated row and column effects.

The corresponding score is

$$\mathrm{score}_i = \sum_{k=1}^{11} \beta_k \mathrm{I}[\mathrm{AY}(i) = k] + \sum_{l=1}^{11} \beta_{11+l}\mathrm{I}[\mathrm{DY}(i) = l]$$

where the binary features $\mathrm{I}[\mathrm{AY}(i) = k]$ and $\mathrm{I}[\mathrm{DY}(i) = l]$ code the categorical features AY and DY. When considering the outputs of the GLM analysis, we have to keep in mind that the fit is perfect in the last cell of the first row ($\mathrm{AY} = 1, \mathrm{DY} = 11$). This is because this is the only observation involving the regression coefficient β_{22} which is thus only determined by this observation. Similarly for the last cell in the first column ($\mathrm{AY} = 11, \mathrm{DY} = 1$) which is the only one involving β_{11}.

The model is fitted using the R function glm, with a log-link and the number of payments in each cell as weight. The IRLS algorithm converged in 7 steps. The results are displayed in Fig. 4.6. We can see there that the estimated regression coefficients corresponding to accident years are first stable over AY 1-6 and then increase over the last years. Considering the estimated regression coefficients associated with development years, we see that they first increase regularly with the development lag, but exhibit a less stable behavior over the last developments.

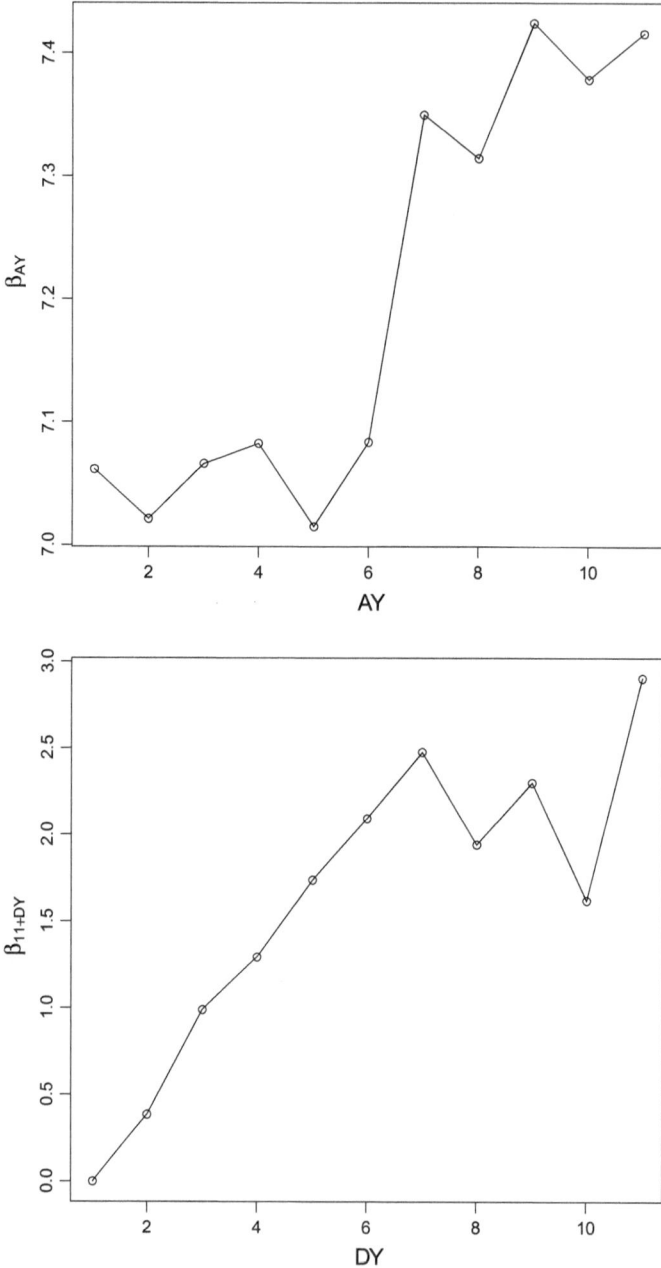

Fig. 4.6 Estimated regression coefficients β_j for the average amounts of payments, related to AY, $j = 1, \ldots, 11$ (upper panel) and to DY, $j = 12, \ldots, 22$ (lower panel)

4.5 Deviance

4.5.1 Null Model

The null model corresponds to the case where the features do not bring any information about the response. In the absence of relation between features and the response,
we fit a common mean μ to all observations. There is thus a single parameter β_0 for
all observations, such that

$$a'(\theta_i) = \bar{y} \Leftrightarrow \widehat{\mu}_i = \bar{y} \text{ for } i = 1, 2, \ldots, n \Leftrightarrow \widehat{\beta}_0 = g^{-1}(\bar{y}).$$

This result directly follows from the properties of maximum-likelihood estimates for
ED distributions established in Chap. 3.

In the null model, the data are represented entirely as random variations around the
common mean μ. If the null model has a good fit then the data are homogeneous and
there is no reason to charge different premium amounts to subgroups of policyholders.

4.5.2 Full, or Saturated Model

For a chosen ED distribution, the model with the best-possible fit counts as many
parameters as observations. The fit is thus perfect and the model estimate $\widetilde{\mu}_i$ for the
mean response is just the corresponding observation y_i. This model is called the full,
or saturated one. It is used as a reference to assess the goodness of the fit of any GLM
based on the same ED distribution.

The full model ensures complete flexibility in fitting so that the corresponding
maximum-likelihood estimate $\widetilde{\theta}_i$ solves

$$a'(\widetilde{\theta}_i) = y_i \Leftrightarrow \widetilde{\mu}_i = y_i \text{ for } i = 1, 2, \ldots, n,$$

as established in Chap. 2. The solution is denoted as

$$\widetilde{\theta}_i = (a')^{-1}(y_i), \quad i = 1, 2, \ldots, n.$$

Thus, each fitted value is equal to the observation and the full model fits perfectly.
However, this model does not extract any structure from the data, but merely repeats
the available observations without condensing them.

4.5.3 Deviance

4.5.3.1 Likelihood Ratio Statistic

A statistical model describes how the actuary partitions the variability present in the data into a systematic structure (as reflected in the GLM score) and random departures from the expected values (variations induced by the assumed ED distribution, or mean-variance relationship). The null model represents one extreme where the data are purely random whereas the saturated or full model represents the data as being entirely systematic. The full model provides the actuary with a measure of how well any model based on the considered ED distribution can fit the data. As the full model gives the highest attainable log-likelihood with the ED distribution under consideration the difference between the log-likelihood L_{full} of the full model and the log-likelihood $L(\widehat{\boldsymbol{\beta}})$ of the GLM under interest, measures the goodness of the fit obtained with this GLM. This leads to the likelihood ratio statistic corresponding to twice this difference. In the ED family, this likelihood ratio statistic is given by

$$\text{LR} = 2\left(L_{\text{full}} - L(\widehat{\boldsymbol{\beta}})\right) = \frac{2}{\phi} \sum_{i=1}^{n} v_i\left(y_i\left(\widetilde{\theta}_i - \widehat{\theta}_i\right) - a\left(\widetilde{\theta}_i\right) + a\left(\widehat{\theta}_i\right)\right)$$

where $\widetilde{\theta}_i$ are the estimates under the full model and $\widehat{\theta}_i$ are the estimates under the model of interest. A too large value of LR indicates that the model under consideration does not satisfactorily fit the actual data whereas a too small value of LR may be the signal that the considered model has low explanatory power. Here, "too small" refers to the quantile of the Chi-Square distribution which appears to provide a reasonable approximation to the distribution of LR under mild regularity conditions.

4.5.3.2 Deviance

When working with GLMs, it is useful to have a quantity that can be interpreted as the sum of squared errors (or residuals) in the Normal linear regression model. This quantity is called the deviance, or residual deviance of the model and is defined as

$$
\begin{aligned}
D(\boldsymbol{y}, \widehat{\boldsymbol{\mu}}) &= \phi \text{LR} \\
&= 2\phi\left(L_{\text{full}} - L(\widehat{\boldsymbol{\beta}})\right) \\
&= 2\sum_{i=1}^{n} v_i\left(y_i\left(\widetilde{\theta}_i - \widehat{\theta}_i\right) - a\left(\widetilde{\theta}_i\right) + a\left(\widehat{\theta}_i\right)\right).
\end{aligned}
$$

It is a measure of distance between a particular model and the observed data defined by means of the saturated model. Similarly to the sum of squared errors, it quantifies the variations in the data that are not explained by the model under consideration.

Table 4.7 Deviance associated to GLMs based on some members of the ED family of distributions

Distribution	Deviance
Binomial	$2 \sum_{i=1}^{n} \left(y_i \ln \frac{y_i}{\widehat{\mu}_i} + (n_i - y_i) \ln \frac{n_i - y_i}{n_i - \widehat{\mu}_i} \right)$ where $\widehat{\mu}_i = n_i \widehat{q}_i$
Poisson	$2 \sum_{i=1}^{n} \left(y_i \ln \frac{y_i}{\widehat{\mu}_i} - (y_i - \widehat{\mu}_i) \right)$ where $y \ln y = 0$ if $y = 0$
Normal	$\sum_{i=1}^{n} \left(y_i - \widehat{\mu}_i \right)^2$
Gamma	$2 \sum_{i=1}^{n} \left(-\ln \frac{y_i}{\widehat{\mu}_i} + \frac{y_i - \widehat{\mu}_i}{\widehat{\mu}_i} \right)$
Inverse Gaussian	$\sum_{i=1}^{n} \frac{\left(y_i - \widehat{\mu}_i \right)^2}{\widehat{\mu}_i^2 y_i}$

Because the saturated model must fit the data at least as well as any other model, the residual deviance is never negative. The larger the deviance, the larger the differences between the actual data and the fitted values. The deviance of the null model is called the null deviance.

Table 4.7 displays the deviance associated to GLMs based on some members of the ED family of distributions. The second term $\sum_{i=1}^{n}(y_i - \widehat{\mu}_i)$ in the Poisson deviance is usually 0 when an intercept is included in the score so that the Poisson deviance reduces to $2 \sum_{i=1}^{n} y_i \ln \frac{y_i}{\widehat{\mu}_i}$. In the Poisson case, the deviance is also called the G-statistic. Notice that in that table, $y \ln y$ is taken to be 0 when $y = 0$ (its limit as $y \to 0$).

4.5.3.3 Scaled Deviance

In families with a known dispersion parameter ϕ, such as the Binomial or Poisson for which $\phi = 1$, the deviance provides a basis for testing lack of fit of a model or for comparing nested models. If ϕ must be estimated from the data then the scaled deviance must be used instead. Precisely, the scaled deviance is given by

$$\widetilde{D}(y, \widehat{\mu}) = \frac{1}{\phi} D(y, \widehat{\mu}).$$

Notice that the scaled deviance does depend on the dispersion parameter ϕ. When $\phi = 1$ (like in the Binomial and Poisson cases), the deviance and scaled deviance are the same. It is common practice to simply plug in an estimate $\widehat{\phi}$ for computing $\widetilde{D}$.

4.5.3.4 Sampling Distribution of the Scaled Deviance

The concept of deviance extends to all members of the ED family the familiar sum of squared residuals used in Normal linear regression. The number of degrees of freedom associated to the scaled deviance is equal to the number n of observations minus the number $p + 1$ of regression parameters $\beta_0, \beta_1, \ldots, \beta_p$ to be estimated. In large samples, the scaled deviance is approximately Chi-Square with $n - (p + 1)$ degrees of freedom, i.e.

$$\tilde{D} \approx \chi^2_{n-p-1} \text{ provided } n \text{ is large enough.} \tag{4.16}$$

A large deviance indicates a badly fitting model. Precisely, if the GLM fits the data reasonably well then the scaled deviance $\tilde{D}$ should be close to the number of residual degrees of freedom for the model, i.e. to $n - p - 1$.

Notice that the Chi-square approximation in (4.16) must be considered with some care. There are situations where it does not hold at all, for instance with binary responses (as shown in the next section).

4.5.4   Bernoulli Case

Maximizing the log-likelihood or minimizing the deviance is equivalent from a numerical point of view, except in some special cases, as shown next. Assume that $Y_i \sim \mathcal{B}er(q_i)$. Then, the log-likelihood writes

$$L(\boldsymbol{\beta}) = \sum_{i=1}^{n} \left(y_i \ln q_i + (1 - y_i) \ln(1 - q_i) \right).$$

In the saturated model, $\widehat{q}_i = y_i$ so that the log-likelihood is equal to 0 because

$$y_i \ln y_i = (1 - y_i) \ln(1 - y_i) = 0 \text{ when } y_i \in \{0, 1\}.$$

The deviance is then equal to minus twice the log-likelihood of the model under consideration, that is,

$$D = -2L(\widehat{\boldsymbol{\beta}}) = -2 \sum_{i=1}^{n} \left(y_i \ln \widehat{q}_i + (1 - y_i) \ln(1 - \widehat{q}_i) \right)$$

$$= -2 \sum_{i=1}^{n} \left(y_i \ln \frac{\widehat{q}_i}{1 - \widehat{q}_i} + \ln(1 - \widehat{q}_i) \right). \tag{4.17}$$

Differentiating the Bernoulli log-likelihood $L(\boldsymbol{\beta})$ with respect to β_j, we get

$$\frac{\partial}{\partial \beta_j} L(\boldsymbol{\beta}) = \sum_{i=1}^{n} \left(\frac{y_i}{q_i} - \frac{1 - y_i}{1 - q_i} \right) q_i (1 - q_i) x_{ij} = \sum_{i=1}^{n} (y_i - q_i) x_{ij}$$

where we recognize a particular case of (4.4) (because the logit is the canonical link function for the Bernoulli GLM). Hence,

$$\sum_{j=1}^{p} \beta_j \frac{\partial}{\partial \beta_j} L(\boldsymbol{\beta}) = \sum_{i=1}^{n} (y_i - q_i) \sum_{j=1}^{p} \beta_j x_{ij}$$

$$= \sum_{i=1}^{n} (y_i - q_i) \ln \frac{q_i}{1 - q_i}.$$

The left-hand side is zero when evaluated at the maximum likelihood estimate $\widehat{\boldsymbol{\beta}}$ of $\boldsymbol{\beta}$ so that

$$\sum_{i=1}^{n} (y_i - \widehat{q}_i) \text{logit}(\widehat{q}_i) = 0 \Leftrightarrow \sum_{i=1}^{n} y_i \text{logit}(\widehat{q}_i) = \sum_{i=1}^{n} \widehat{q}_i \text{logit}(\widehat{q}_i).$$

Inserting this result into (4.17), we get

$$D = -2 \sum_{i=1}^{n} \left(\widehat{q}_i \text{logit}(\widehat{q}_i) + \ln(1 - \widehat{q}_i) \right).$$

We see that the deviance D only depends on the fitted values $\widehat{q}_i$ of q_i, and no more on the actual observations y_i. For a deviance to measure the goodness of fit, it has to compare the fitted values $\widehat{q}_i$ to the data y_i, but here we have only a function of $\widehat{q}_i$. This shows that the deviance does not bring any information about the goodness of fit when $Y_i \sim \mathcal{B}er(q_i)$ and the logit link function is used. Hence, it cannot be used to measure model adequacy in such a case.

4.5.5 Covariance Penalties and AIC

4.5.5.1 Measuring Goodness-of-Fit

Nested models can be compared using deviances. However, when models are not nested, or postulate different distributions for the response, a direct comparison becomes problematic. In such a case, information criteria allow the actuary to compare different models. The classical criteria include Akaike Information Criteria

$$AIC = -2L + 2\# \text{ parameters},$$

and the Bayesian Information Criteria

$$BIC = -2L + \ln(n)\# \text{ parameters}.$$

Notice that the number of parameters counts all parameters included in the model (the dispersion parameter ϕ is thus also counted as a parameter) Both criteria are based on the log-likelihood L and account for a measure of model complexity in the penalty. Smaller values of AIC or BIC are preferred.

It is common practice to discard parts of the likelihood that are not functions of parameters. This has no consequence when models assuming the same distribution for the response are compared, since the parts discarded are equal in such a case. For responses with different distributions, however, it is essential that all parts of the likelihoods be retained.

A bias corrected version of AIC (denoted as AICC) has been developed for use in small samples or when the number of parameters $p + 1$ is a moderate to large fraction of the sample size n, which is typically the case in loss reserving studies based on data aggregated in runoff triangles. It is formally defined as

$$AICC = AIC + \frac{2(p+2)(p+3)}{n-p-1}.$$

As a rule of thumb, AICC should be used instead of AIC unless

$$n > 40 \times \text{\# of parameters.}$$

4.5.5.2   Back to the Normal Linear Regression Model: Covariance Penalties

The predictive ability, or generalization performance (which is a terminology from the machine learning community) of a model describes its prediction capacity on new test data. The predictive power of a model can thus only be assessed on new, out-of-sample data. Having an estimate of this predictive power is very important for comparing different models or for selecting optimal values of parameters. Cross validation is a general method for estimating such generalization, or out-of-sample performances. For structured models like GLMs, some analytical results are nevertheless available, making the computationally intensive cross validation not needed. Prediction error in a regression framework can be assessed using covariance penalties.

Consider the classical Normal linear regression model for responses

$$Y_i \sim Nor(\mu_i, \sigma^2) \text{ with } \mu_i = \boldsymbol{\beta}^\top \boldsymbol{x}_i.$$

The parameter $\boldsymbol{\beta}$ is estimated from the available data $(y_i, \boldsymbol{x}_i)$, $i = 1, \ldots, n$, also called the training sample since it is used to train the estimator. Evaluating model performances with the help of the apparent error given by the mean sum of squared residuals

$$\widehat{\text{MSSR}} = \frac{1}{n} \sum_{i=1}^{n} (y_i - \widehat{\mu}_i)^2$$

is too optimistic because the model has been fitted to make the predictions $\widehat{\mu}_i$ for the data y_i. If we use the training sample again to measure the predictive power of the model under consideration then the results will be overly optimistic and will always favor the more complex models (i.e. models with more degrees of freedom) because MSSR can only decrease when model complexity increases. This is why information criteria like AIC are so helpful to compare model performances.

Alternatively, we could evaluate the model performances on new test data. Remember that the quantity $\widehat{\text{MSSR}}$ is the sample analog to

$$\text{MSSR} = \frac{1}{n} \sum_{i=1}^{n} \text{E}[(Y_i - \widehat{\mu}_i)^2]$$

based on the sample $(Y_i, \boldsymbol{x}_i)$, $i = 1, \ldots, n$. Here, the responses Y_i are those used to train the model, that is, to obtain $\widehat{\mu}_i$. To remove the bias this cases when model performances are evaluated, we switch to the mean squared error of prediction defined as

$$\text{MSEP} = \frac{1}{n} \sum_{i=1}^{n} \text{E}[(Y_i^{\text{new}} - \widehat{\mu}_i)^2]$$

where the expectation is taken with respect to the new observation Y_i^{new} with feature $\boldsymbol{x}_i$, distributed as Y_i included in the observed sample, and independent of $Y_1, Y_2, \ldots, Y_n$. So, Y_i^{new} has not been used to build $\widehat{\mu}_i$, which depends on the actual observations $Y_1, Y_2, \ldots, Y_n$.

Let us now relate MSEP to MSSR. To this end, we write

$$\begin{aligned}
\text{E}[(Y_i - \widehat{\mu}_i)^2] &= \text{E}[(Y_i - \mu_i)^2] - 2\text{E}[(Y_i - \mu_i)(\widehat{\mu}_i - \mu_i)] + \text{E}[(\widehat{\mu}_i - \mu_i)^2] \\
&= \text{Var}[Y_i] - 2\text{Cov}[Y_i, \widehat{\mu}_i] + \text{E}[(\widehat{\mu}_i - \mu_i)^2].
\end{aligned}$$

Similarly,

$$\begin{aligned}
\text{E}[(Y_i^{\text{new}} - \widehat{\mu}_i)^2] &= \text{E}[(Y_i^{\text{new}} - \mu_i)^2] - 2\text{E}[(Y_i^{\text{new}} - \mu_i)(\widehat{\mu}_i - \mu_i)] + \text{E}[(\widehat{\mu}_i - \mu_i)^2] \\
&= \text{Var}[Y_i^{\text{new}}] - 2\text{Cov}[Y_i^{\text{new}}, \widehat{\mu}_i] + \text{E}[(\widehat{\mu}_i - \mu_i)^2] \\
&= \text{Var}[Y_i] + \text{E}[(\widehat{\mu}_i - \mu_i)^2],
\end{aligned}$$

where the last equality follows from the independence of Y_i^{new} and $\widehat{\mu}_i$ and the fact that Y_i^{new} is distributed as Y_i. We finally get

$$\text{E}[(Y_i^{\text{new}} - \widehat{\mu}_i)^2] = \text{E}[(Y_i - \widehat{\mu}_i)^2] + 2\text{Cov}[Y_i, \widehat{\mu}_i].$$

On average, the apparent error thus underestimates the true prediction error by the covariance penalty $2\text{Cov}[Y_i, \widehat{\mu}_i]$. Intuitively speaking, $\text{Cov}[Y_i, \widehat{\mu}_i]$ measures how much Y_i influences its own prediction $\widehat{\mu}_i$. This provides an effective way to evaluate

MSEP on the basis of the available data (that were used to train the model), using the identity

$$\widehat{\text{MSEP}} = \widehat{\text{MSSR}} + \frac{2}{n} \sum_{i=1}^{n} \widehat{\text{Cov}}[Y_i, \widehat{\mu}_i].$$

If $Y_i \sim Nor(\mu_i, \sigma^2)$ and the fitted values $\widehat{\mu}_i$ are linear, we know that

$$\widehat{\boldsymbol{\mu}} = \boldsymbol{H} \boldsymbol{Y}$$

where $\boldsymbol{H}$ is the known matrix (4.12) mapping the observations to the fitted values. The covariance matrix between $\widehat{\boldsymbol{\mu}}$ and $\boldsymbol{Y}$ is then $\sigma^2 \boldsymbol{H}$ giving

$$\text{Cov}[Y_i, \widehat{\mu}_i] = \sigma^2 h_{ii}$$

where h_{ii} is the ith diagonal element of $\boldsymbol{H}$ measuring leverage. In the Normal case, the covariance penalty approach coincides with Mallow's C_p estimate of the prediction error:

$$\widehat{\text{MSEP}} = \frac{1}{n} \sum_{i=1}^{n} (y_i - \widehat{\mu}_i)^2 + \frac{2\sigma^2}{n} \text{trace}(\boldsymbol{H}).$$

Remark 4.5.1 For the linear regression model, it can be deduced from (4.12) that

$$\text{trace}(\boldsymbol{H}) = \sum_{i=1}^{n} h_{ii} = \dim(\boldsymbol{\beta}) = p + 1. \tag{4.18}$$

As $p + 1$ is also related to the number of degrees of freedom of the regression model, this leads to a general definition of degrees of freedom for a regression model built for Normal responses Y_i. This provides the actuary with an appropriate measure of model complexity. If the equivalent number of degrees of freedom is larger then the model is more flexible and may become too close to the actual data so that it requires bigger covariance penalties for fair comparison.

4.5.5.3 Covariance Penalties in GLMs

Let us now consider responses Y_i obeying an ED distribution. The apparent error is then measured by the deviance. Let

$$d(y, \widehat{\theta}) = 2\left(y\theta_y - a(\theta_y) - \left(y\widehat{\theta} - a(\widehat{\theta})\right)\right) \tag{4.19}$$

denote the contribution of the response y to the deviance, where θ_y is the value of the canonical parameter such that the identity

$$a'(\theta_y) = y$$

holds true (as it must be under the full model). Let us now proceed as in the Normal case. Given a new observation Y_i^{new}, distributed as Y_i and independent of the observations $Y_1, \ldots, Y_n$ having produced $\widehat{\theta}_1, \ldots, \widehat{\theta}_n$, we then get

$$
\begin{aligned}
\mathrm{E}\big[d(Y_i^{\text{new}}, \widehat{\theta}_i)\big] - \mathrm{E}\big[d(Y_i, \widehat{\theta}_i)\big] &= 2\Big(\mathrm{E}\big[Y_i^{\text{new}}\theta_{Y_i^{\text{new}}} - a(\theta_{Y_i^{\text{new}}})\big] - \mathrm{E}\big[Y_i^{\text{new}}\widehat{\theta}_i - a(\widehat{\theta}_i)\big] \\
&\qquad - \mathrm{E}\big[Y_i\theta_{Y_i} - a(\theta_{Y_i})\big] + \mathrm{E}\big[Y_i\widehat{\theta}_i - a(\widehat{\theta}_i)\big]\Big) \\
&= 2\Big(\mathrm{E}\big[Y_i\widehat{\theta}_i\big] - \mathrm{E}\big[Y_i^{\text{new}}\widehat{\theta}_i\big]\Big) \\
&= 2\Big(\mathrm{E}\big[Y_i\widehat{\theta}_i\big] - \mathrm{E}\big[Y_i\big]\mathrm{E}\big[\widehat{\theta}_i\big]\Big) \\
&= 2\mathrm{Cov}\big[Y_i, \widehat{\theta}_i\big].
\end{aligned}
$$

In GLMs, the covariance penalty is $2\mathrm{Cov}[Y_i, \widehat{\theta}_i]$, where $\widehat{\theta}_i = \widehat{\theta}(x_i)$ is the estimated canonical parameter for observation i. The mean predictive deviance, that is, the deviance computed on new, out-of-sample observations can then be estimated from the sum

$$\frac{1}{n}D(y, \widehat{\mu}) + \frac{2}{n}\sum_{i=1}^{n}\mathrm{Cov}\big[Y_i, \widehat{\theta}_i\big].$$

Assume that the actuary specifies the canonical link function. In insurance studies, this essentially means that Poisson or Binomial response distributions are considered (so that we can safely assume that $\phi = 1$). Now, if the estimates have been obtained by maximum likelihood and $\dim(\boldsymbol{\beta}) = p + 1$ then the following approximation is accurate enough for practical purposes:

$$\frac{2}{n}\sum_{i=1}^{n}\mathrm{Cov}[Y_i, \widehat{\theta}_i] = \frac{2}{n}\sum_{i=1}^{n}\mathrm{E}[(Y_i - \mu_i)(x_i^\top\widehat{\boldsymbol{\beta}} - x_i^\top\boldsymbol{\beta})] \approx \frac{2(p+1)}{n}.$$

This approximation is obtained from large-sample properties of the maximum-likelihood estimator $\widehat{\boldsymbol{\beta}}$ in the ED family. We refer the reader to Efron (1986) for more details. The mean predictive deviance is then approximated by

$$\frac{1}{n}D(y, \widehat{\mu}) + \frac{2(p+1)}{n},$$

provided n is large enough.

4.5.5.4 Covariance Penalties and AIC

Consider again Poisson or Binomial responses with canonical link and denote as $p_\theta(\cdot)$ the corresponding probability mass function associated to the canonical parameter θ. The prediction error can then be computed as

$$\frac{1}{n} D(\boldsymbol{y}, \widehat{\boldsymbol{\mu}}) + \frac{2}{n} \sum_{i=1}^{n} \text{Cov}[Y_i, \widehat{\theta}_i] \approx \frac{1}{n} D(\boldsymbol{y}, \widehat{\boldsymbol{\mu}}) + \frac{2(p+1)}{n}$$

$$= -\frac{2}{n} \left(\sum_{i=1}^{n} \ln p_{\widehat{\theta}_i}(y_i) - (p+1) \right) + \text{constant}$$

$$= \frac{\text{AIC}}{n} + \text{constant}$$

involving the Akaike information criterion AIC based on the penalized maximum likelihood (notice that we have assumed $\phi = 1$ so that $p + 1$ counts all the parameters involved in the regression model). Comparing GLMs under consideration on the basis of AIC values thus amounts to account for the optimism in the deviance computed on the training set.

4.6   Estimation of the Dispersion Parameter

Some distributions from the ED family include an unknown dispersion parameter ϕ. This is the case for instance with the Gamma or Inverse-Gaussian distributions. The likelihood equations for $\boldsymbol{\beta}$ do not involve the dispersion parameter ϕ so that it is not necessary to estimate ϕ to get $\widehat{\boldsymbol{\beta}}$. However, the estimated dispersion parameter is needed to conduct inference on the regression coefficients. Although ϕ can in principle be estimated by maximum likelihood, it is more common to use an alternative estimator derived from the method of moments which is defined as follows.

Once the IRLS algorithm has produced $\widehat{\boldsymbol{\beta}}$, the dispersion parameter ϕ can be estimated by dividing the Pearson's goodness-of-fit statistic

$$X^2 = \sum_{i=1}^{n} \frac{(y_i - \widehat{\mu}_i)^2}{V(\widehat{\mu}_i)/v_i}$$

by the residual degrees of freedom for the model, that is,

$$\widehat{\phi} = \frac{1}{n-p-1} \sum_{i=1}^{n} \frac{v_i (y_i - \widehat{\mu}_i)^2}{V(\widehat{\mu}_i)} = \frac{X^2}{n-p-1}.$$

This estimator uses the fact that Pearson's X^2 is asymptotically Chi-Square distributed with $n - p - 1$ degrees of freedom. It is much simpler to compute and, in some cases, offers greater numerical stability than the maximum-likelihood estimate.

Deviance provides an alternative estimator for ϕ. Considering (4.16), we see that the mean of D is approximately equal to $n - p - 1$. This suggests to use the large-sample estimate

$$\widehat{\phi} = \frac{D}{n-p-1}.$$

4.7 Sampling Distribution of Estimated Regression Coefficients

4.7.1 Sampling Distribution

The sampling distribution of model parameters must be understood as follows. Estimates are the results of a random experiment: collecting the data and analyzing them with the help of the GLM. If a different set of data were used, with the same size and all the same underlying characteristics but with different outcomes, the resulting estimated coefficients would be different. The extent of these differences depends on the variability of the observed phenomenon. The sampling distribution describes the behavior of the estimates obtained from different samples of observations, in a repeated sampling setting.

The sampling distribution may be approached by bootstrapping (parametrically or not), as explained in Chap. 3. Nevertheless, this method is computationally intensive. For GLMs, there exists an alternative based on asymptotic properties of maximum-likelihood estimators. Indeed, we know from Chap. 3 that the maximum-likelihood method produces estimators that are approximately Normally distributed when the sample size is large enough. This property is still valid for the vector $\widehat{\boldsymbol{\beta}}$ of estimated regression coefficient. Given that several parameters $\beta_0, \beta_1, \ldots, \beta_p$ are involved in the GLM fit, we have to recall the definition of the multivariate extension of the Normal distribution.

4.7.2 Back to the Normal Linear Regression Model

Let us recall the definition of the multivariate version of the Normal distribution. Let $\boldsymbol{\Sigma}$ be a $n \times n$ positive definite matrix and $\boldsymbol{\mu}$ be a real vector. Then, the random vector $\mathbf{Z} = (Z_1, Z_2, \ldots, Z_n)^\top$ is said to obey the multivariate Normal distribution with parameters $(\boldsymbol{\mu}, \boldsymbol{\Sigma})$, which is henceforth denoted as $\mathbf{Z} \sim Nor(\boldsymbol{\mu}, \boldsymbol{\Sigma})$, if its joint probability density function is of the form

$$f_{\mathbf{Z}}(z) = \frac{1}{\sqrt{(2\pi)^n |\boldsymbol{\Sigma}|}} \exp\left(-\frac{1}{2}(z - \boldsymbol{\mu})^\top \boldsymbol{\Sigma}^{-1}(z - \boldsymbol{\mu})\right), \quad z \in (-\infty, \infty)^n, \quad (4.20)$$

where $|\boldsymbol{\Sigma}|$ denotes the determinant of the matrix $\boldsymbol{\Sigma}$. Here, $\boldsymbol{\mu}$ and $\boldsymbol{\Sigma}$ are the mean vector and the covariance matrix of the multivariate Normal distribution $Nor(\boldsymbol{\mu}, \boldsymbol{\Sigma})$.

We know from Chap. 2 that the sum of n independent, squared $Nor(0, 1)$ random variables obeys the Chi-Square χ_n^2 distribution. This result extends to non-independent summands in that when $\mathbf{Z} \sim Nor(\boldsymbol{\mu}, \boldsymbol{\Sigma})$ where $\boldsymbol{\Sigma}$ has an inverse $\boldsymbol{\Sigma}^{-1}$

$$(\mathbf{Z} - \boldsymbol{\mu})^\top \boldsymbol{\Sigma}^{-1}(\mathbf{Z} - \boldsymbol{\mu}) \sim \chi_n^2.$$

A convenient characterization of the multivariate Normal distribution is as follows: $Z \sim Nor(\mu, \Sigma)$ if, and only if, any random variable of the form $\sum_{i=1}^{n} \alpha_i Z_i$ for real constants $\alpha_1, \ldots, \alpha_n$, obeys the univariate Normal distribution. More generally, the multivariate Normal distribution also has the following useful stability property: Let C be a given $n \times n$ matrix with real entries and let b be a n-dimensional real vector. Then,

$$Z \sim Nor(\mu, \Sigma) \Rightarrow CZ + b \sim Nor(C\mu + b, C\Sigma C^\top).$$

Applying this property to $\widehat{\beta}$ given in (4.11), with

$$
\begin{aligned}
C &= (X^\top W X)^{-1} X^\top W \\
b &= 0 \\
Z^\top &= (Y_1, Y_2, \ldots, Y_n) \\
\mu &= X\beta \\
\Sigma &= \sigma^2 W^{-1}
\end{aligned}
$$

we get

$$C\mu + b = \beta \text{ and } C\Sigma C^\top = \sigma^2 (X^\top W X)^{-1}$$

so that

$$\widehat{\beta} \sim Nor\left(\beta, \sigma^2 (X^\top W X)^{-1}\right).$$

This gives the exact sampling distribution of $\widehat{\beta}$ in the Normal linear regression model.

4.7.3   Sampling Distribution in GLMs

Let us now consider the other GLMs. As $\widehat{\beta}$ has been obtained by maximum likelihood, we know that it is approximately Normally distributed provided n is sufficiently large. Precisely, the estimation error fulfills

$$\widehat{\beta} - \beta \approx Nor\left(0, \widehat{\Sigma}(\widehat{\beta})\right).$$

Because the IRLS algorithm used to git GLMs (as explained in Sect. 4.4) repeatedly fits a Normal linear regression model to working responses, with varying weights, the variance-covariance matrix of the resulting $\widehat{\beta}$ can be obtained from the last step of this procedure, using the formulas valid in the Normal case. Precisely, the estimated large-sample variance-covariance matrix of $\widehat{\beta}$ is given by

$$\widehat{\Sigma}(\widehat{\beta}) = \widehat{\phi}\left(X^\top \widetilde{W}^{(r_{\text{stop}})} X\right)^{-1}$$

where $\widetilde{W}^{(r_{\text{stop}})}$ is taken from the last step of the IRLS algorithm producing $\widehat{\boldsymbol{\beta}}$ (so that it does not include the dispersion parameter and a factor ϕ must be specified to recover formula (4.3)). The standard errors of the parameter estimates are the square root of the diagonal elements of the variance-covariance matrix.

4.7.4 Wald Confidence Intervals

Wald's approach is based on the Normal approximation for the distribution of $\widehat{\boldsymbol{\beta}}$. The standard error is the square root of the estimated $\text{Var}[\widehat{\beta}_j]$. This quantity must be understood as follows. If new samples are drawn repeatedly, the standard deviation of the resulting estimates of β_j would be approximately the estimated standard error. Therefore, a small standard error indicates that $\widehat{\beta}_j$ should be close to the true β_j whereas a large standard error suggests that a wide range of estimates could be achieved through randomness.

The confidence interval can be considered as a reasonable range of estimates for the coefficient of interest. Wald's confidence intervals are based on the large-sample distribution of $\widehat{\beta}_j$, that is,

$$\widehat{\beta}_j - \beta_j \approx \textit{Nor}\left(0, \widehat{\sigma}_{jj}^2\right)$$

where $\widehat{\sigma}_{jj}$ is the element (j, j) of $\widehat{\boldsymbol{\Sigma}}(\widehat{\boldsymbol{\beta}})$, that is, the estimated variance of $\widehat{\beta}_j$. The approximate confidence interval for β_j at level $1 - \alpha$ is then obtained as follows:

$$1 - \alpha \approx P\left[-z_{\alpha/2}\sqrt{\widehat{\sigma}_{jj}} \leq \widehat{\beta}_j - \beta_j \leq z_{\alpha/2}\sqrt{\widehat{\sigma}_{jj}}\right]$$
$$= P\left[\widehat{\beta}_j - z_{\alpha/2}\sqrt{\widehat{\sigma}_{jj}} \leq \beta_j \leq \widehat{\beta}_j + z_{\alpha/2}\sqrt{\widehat{\sigma}_{jj}}\right]$$

so that the interval

$$\text{IC}_{1-\alpha}(\beta_j) = \left[\widehat{\beta}_j \pm z_{\alpha/2}\sqrt{\widehat{\sigma}_{jj}}\right]$$

is expected to contain the true value of the parameter with a probability approximately equal to $1 - \alpha$. Here, the probability $1 - \alpha$ is the (approximate) confidence level, or coverage probability of the corresponding confidence interval.

Since $\widehat{\boldsymbol{\beta}}$ is approximately Normal, the estimated score $\widehat{s}_i = \boldsymbol{x}_i^T\widehat{\boldsymbol{\beta}}$ also is. The bias is approximately 0 and the variance of the estimated score $\widehat{s}_i$ is given by

$$\text{Var}[\widehat{s}_i] = \boldsymbol{x}_i^\top \widehat{\boldsymbol{\Sigma}}(\widehat{\boldsymbol{\beta}})\boldsymbol{x}_i.$$

A confidence interval at approximate confidence level $1 - \alpha$ for the true score s_i is then given by

$$\left[\boldsymbol{x}_i^\top\widehat{\boldsymbol{\beta}} \pm z_{\alpha/2}\sqrt{\boldsymbol{x}_i^\top \widehat{\boldsymbol{\Sigma}}(\widehat{\boldsymbol{\beta}})\boldsymbol{x}_i}\ \right].$$

4.7.5   *Variance Inflation Factor*

Multicollinearity refers to the situation where two or more features are strongly predictive of a third one. However, the latter may not be highly correlated with either of the former ones so that this phenomenon does not show up in a correlation matrix made of values for Cramer's V in case features have different formats. This is because such a matrix only considers pairs of features, not larger subsets of the available features. As a consequence, multicollinearity is much more difficult to detect.

When there are strong linear relationships among the features in a regression analysis, the estimation accuracy for the β_j decreases compared to the case where features are independent. The main reason is that the matrix $X^T W X$ involved in each step of the IRLS algorithm is badly conditioned and therefore hardly invertible. A useful statistic for detecting multicollinearity is the variance inflation factor (VIF). The VIF for any feature measures the increase in the squared standard error for the regression parameter due to the presence of collinearity in the data.

Let ρ_j denote the coefficient of determination from the regression of the jth feature on the remaining $p - 1$ other features. It is determined by running a linear model for each of the features using all other features as inputs, and measuring its predictive accuracy. This means that for each feature x_{ij} in turn, we try to identify the linear combination

$$\gamma_0^{(j)} + \sum_{k \neq j} \gamma_k^{(j)} x_{ik}$$

that best predict x_{ij} from $x_{ik}, k \neq j$. Thus,

$$\widehat{\gamma}_0^{(j)} + \sum_{k \neq j} \widehat{\gamma}_k^{(j)} x_{ik}$$

can best replace x_{ij} in the linear score. In the Normal linear regression model, it can then be shown that

$$\text{Var}[\widehat{\beta}_j] = \frac{\sigma^2}{(1 - \rho_j^2) \sum_{i=1}^{n} (x_{ij} - \overline{x}_j)^2}.$$

This gives

$$\widehat{\text{Var}}[\widehat{\beta}_j] = \frac{\widehat{\sigma}^2}{(n - 1)s_j^2} \times \frac{1}{1 - \widehat{\rho}_j^2}$$

where $\widehat{\sigma}^2$ is the estimated variance σ^2, s_j^2 is the sample variance of $x_{1j}, x_{2j}, \ldots, x_{nj}$ and $\frac{1}{1 - \widehat{\rho}_j^2}$ is the variance inflation factor for the estimator $\widehat{\beta}_j$. This formula shows which components drive the accuracy of the estimation for the regression coefficients:

- the smaller the model variance, the smaller the variance of $\widehat{\beta}_j$ and thus the more accurate the estimation.
- the smaller the linear dependence between the jth feature and the remaining $p - 1$ ones (measured through ρ_j^2), the smaller the variance of $\widehat{\beta}_j$.

  The variance of $\widehat{\beta}_j$ is minimized for $\rho_j^2 = 0$, that is, when the features are uncorrelated. When some of the features are strongly correlated, the estimators can become very imprecise. In the extreme situation where $\rho_j^2 \to 1$, the variance of $\widehat{\beta}_j$ explodes towards infinity (reflecting multicollinearity).
- the larger the variability of the jth feature around its average, the smaller the variance of $\widehat{\beta}_j$.

  The variance inflation factor defined as

$$\text{VIF} = \frac{1}{1 - \widehat{\rho}_j^2}$$

then quantifies the increase in the variance of $\widehat{\beta}_j$ due to linear dependence of the jth feature on the remaining $p - 1$ ones. The VIF is the simplest and most direct measure of the harm produced by collinearity. Specifically, the square root of the VIF indicates how much the confidence interval for β_j is expanded relative to similar uncorrelated data (such data may be purely conceptual). The strong linear relationships correspond to the features having a large VIF. As a benchmark, a serious collinearity problem exists when the VIF is larger than 10. The VIF does not apply to sets of regression coefficients (corresponding to the different levels of categorical features, for instance) but there are Generalized VIF, or GVIF available in this case.

Despite their obvious interest in insurance studies, the calculation of VIF measures appears to be computer-intensive in GLM analyses involving many features, given the computational power currently available. This somewhat limits the practical use of this concept.

4.7.6 Testing Hypothesis on the Parameters

4.7.6.1 Hypotheses on a Single Regression Coefficient

To test whether the assumption $\beta_j = b$ is reasonable, for some constant b, we can compute the Wald statistic

$$z = \frac{\widehat{\beta}_j - b}{\sqrt{\widehat{\text{Var}}[\widehat{\beta}_j]}}.$$

We can then compare z to the standard Normal quantiles to decide whether to reject the assumption $\beta_j = b$ or not. Precisely, the assumption $\beta_j = b$ is rejected when $\widehat{\beta}_j$

is too far from its assumed value b, that is, when $|z|$ is large. Working at confidence level α, this means that the assumption $\beta_j = b$ is rejected if $|z|$ exceeds the $Nor(0, 1)$ quantile at probability level $\alpha/2$, equal to $\Phi^{-1}(1 - \alpha/2)$.

Adopting this procedure means that the selected α quantifies the risk of rejecting the assumption $\beta_j = b$ whereas it is in reality true (called type 1 error in statistics). This is easily understood as follows. Under the assumption $\beta_j = b$, the test statistics z is approximately $Nor(0, 1)$ distributed, so that there is a probability α that $|z|$ exceeds $\Phi^{-1}(1 - \alpha/2)$, leading to rejection. Such cases correspond to situations where the assumption should not have been rejected. Hence, the selected α represents the level of safety selected by the actuary when rejecting the assumption under consideration.

The level $\alpha = 5\%$, resulting in $\Phi^{-1}(97.5\%) = 1.96 \approx 2$ is routinely used in practice. This leads to 1 erroneous conclusion out of every 20 tests leading to rejection, on average (provided the tests are run independently). Depending on the importance of the conclusion drawn from the analysis, the actuary could decrease the risk of error (to 1%, for instance), or behave less conservatively (increasing α up to 10%, say).

Statistical softwares, including R generally report the result of testing procedures in the form of p-values, to avoid asking users about the confidence level α they wish to apply. Based on the calculated p-value, the actuary then rejects the assumption when it falls below the selected α, and retains it as a working assumption otherwise. Formally, the actuary can never accept the stated assumption (hence the subtle difference between acceptance and non-rejection). This is because the other source of error, called type 2 error in statistics consisting in accepting the assumption whereas it is actually false, is generally not controlled by the testing procedure.

Notice that standard tests for the null hypothesis $H_0 : \beta_j = 0$ are automatically produced by statistical softwares. When categorical features are involved, $\beta_j = 0$ means that the corresponding level is identical to the base, or reference level for that feature (which effect is included in the intercept β_0). Failing to reject H_0 suggests that the level should be merged with the base one. But other groupings may be meaningful and can be detected by testing null hypotheses of the form

$$H_0 : \beta_j = \beta_k \Leftrightarrow \beta_j - \beta_k = 0.$$

The corresponding test statistics is based on $\widehat{\beta}_j - \widehat{\beta}_k$, suitably standardized, obeying the $Nor(0, 1)$ distribution under the null assumption H_0.

When an intercept is included in the score, the GLM analysis proceeds in relative terms, positioning each risk class with respect to the reference one (whose effect corresponds to the intercept). Switching from one reference level to another does not modify model predictions (despite different estimated regression coefficients are obtained because the levels are put in relation to a different base level). However, the p-values change as they question the difference of a given level compared to the base level: modifying the base level thus modifies the test performed. It is therefore important to check that the base level has a considerable exposure (and not simply take the default base level assigned by the software). Notice that this is precisely the choice made in all examples so far in this book.

Remark 4.7.1 The importance of each variable is expressed by the magnitude of its contribution to the score. In GLMs, this is essentially reflected by the associated $\widehat{\beta}_j$ (once all the predictors have been normalized). The relevance of the jth feature can then be tested with the help of the null assumption H_0: $\beta_j = 0$.

Another way to proceed, that extends to more sophisticated statistical learning approaches, is as follows. The idea is to randomly select a fraction π of the original data and use it to fit the GLM. Then, the actuary uses the fitted model to predict the response in the remaining $1 - \pi$ of the database and compute the accuracy of this prediction. The values of the jth feature are then randomly permuted and the response is re-predicted on that basis. The idea is to compare the prediction accuracy with and without permutation: the more important the feature, the more differences in these two measures, the permutation of the jth feature causing a major drop in prediction accuracy.

4.7.6.2 Hypotheses Involving Several Regression Coefficients

Besides simple hypotheses about a single parameter, more elaborate ones can also be tested. Let us consider the base model

$$\mathcal{M}_0 : \boldsymbol{\beta} = \boldsymbol{\beta}_0 = (\beta_1, \beta_2, \ldots, \beta_q)^\top$$

and the alternative model

$$\mathcal{M}_1 : \boldsymbol{\beta} = \boldsymbol{\beta}_1 = (\beta_1, \beta_2, \ldots, \beta_q, \beta_{q+1}, \ldots, \beta_p)^\top$$

including additional explanatory variables. Under $\mathcal{M}_0$, the features $x_{i,q+1}, \ldots, x_{i,p}$ do not impact on expected responses. This means that $\mathcal{M}_0$ corresponds to the null assumption

$$H_0 : \beta_{q+1} = \ldots = \beta_p = 0$$

whereas the alternative hypothesis is that at least one regression coefficient among $\beta_{q+1}, \ldots, \beta_p$ is nonzero.

When a GLM uses a subset of the features of a larger model, the smaller model is said to be nested in the larger one. In our case, $\mathcal{M}_0$ is nested in $\mathcal{M}_1$. For GLMs without dispersion parameter (i.e. $\phi = 1$ like in the Binomial and Poisson cases), the likelihood ratio test statistic is simply the difference in the deviances for nested models $\mathcal{M}_0$ and $\mathcal{M}_1$. Denoting as D_0 and D_1 the respective deviances associated to $\mathcal{M}_0$ and $\mathcal{M}_1$, the test statistic is

$$\begin{aligned} \Delta &= D_0 - D_1 \\ &= 2 \left(L(\widehat{\boldsymbol{\beta}}_1) - L(\widehat{\boldsymbol{\beta}}_0) \right) \end{aligned}$$

where $\widehat{\beta}_0$ and $\widehat{\beta}_1$ are the maximum-likelihood estimators for β under $\mathcal{M}_0$ and $\mathcal{M}_1$, respectively. If the restrictions on model $\mathcal{M}_1$ represented by model $\mathcal{M}_0$ are correct, that is, if H_0 holds true, then the test statistic Δ is approximately Chi-Square distributed, that is, $\Delta \approx \chi^2_{p-q}$, provided the sample size is large enough.

However, we have to keep in mind that adding features in a model always reduces the deviance (even if the additional feature is not related to the response), simply because more parameters are involved in the larger model, which can only fit the data better). Hence, $D_0 - D_1$ can only be positive. The meaningful question is therefore whether the additional features significantly reduce that deviance, i.e. more than if they were not related to the response. The approximate Chi-Square distribution χ^2_{p-q} governing $D_0 - D_1$ defines the threshold beyond which a difference in the observed deviances is large enough to justify the inclusion of the additional features. Formally, if model $\mathcal{M}_1$ is the true model then it will tend to have a substantially higher likelihood compared to model $\mathcal{M}_0$ so that twice the difference in log-likelihoods would be too large. Thus, H_0 is rejected if the observed value Δ_{obs} of the test statistic Δ is "too large". In large samples, H_0 is then rejected if $\Delta_{\text{obs}} > \chi^2_{p-q;1-\alpha}$ where $\chi^2_{p-q;1-\alpha}$ denotes the quantile at probability level $1 - \alpha$ of the χ^2_{p-q} distribution.

For GLMs in which there is a dispersion parameter to estimate (like in the Normal, Gamma and Inverse-Gaussian cases), we can compare nested models by an F-test, based on the test statistic

$$F = \frac{\frac{D_0 - D_1}{p - q}}{\widehat{\phi}}$$

involving the respective scaled deviances $\frac{D_0}{\phi}$ and $\frac{D_1}{\phi}$. Here, the estimated dispersion parameter $\widehat{\phi}$ is taken from the largest model fit to the data (which is not necessarily model $\mathcal{M}_1$). If the largest model has $k + 1$ regression coefficients then, under the assumption that the restrictions on model $\mathcal{M}_1$ represented by model $\mathcal{M}_0$ are correct, F approximately obeys the Fisher distribution with $p - q$ and $n - k - 1$ degrees of freedom.

4.7.7 Numerical Illustrations

4.7.7.1 Graduation of One-Year Death Probabilities

Let us come back to the graduation of one-year death probabilities using formula (4.15). Recall that the response is the number of deaths D_x recorded among the l_X individuals aged x on January, the 1st. The parameter of interest is the one-year death probability q_x. The Binomial GLM with canonical link is used, with a score involving age x and its squared, that is,

$$\text{logit}(q_x) = \beta_0 + \beta_1 x + \beta_2 x^2.$$

Table 4.8 Output of the `glm` function of R for the fit of the quadratic Perks formula to mortality data displayed in Fig. 4.1. Number of Fisher Scoring iterations: 3

Coefficient	Estimate	Std. Error	z value	$\Pr(> |z|)$	
Intercept	-3.340	4.808×10^{-1}	-6.947	3.73×10^{-12}	$\star\star\star$
Age x	-0.1038	1.213×10^{-2}	-8.558	$<2 \times 10^{-16}$	$\star\star\star$
Squared age x^2	1.384×10^{-3}	7.597×10^{-5}	18.214	$<2 \times 10^{-16}$	$\star\star\star$

Alternatively, the model can be fitted to the observed crude death probabilities $\widehat{q}_x = D_x / l_x$ provided the number of individuals l_x is used as weight. Point estimates $\widehat{\boldsymbol{\beta}}$ obtained from the IRLS algorithm have been given earlier in this chapter. Now, we can supplement $\widehat{\boldsymbol{\beta}}$ with standard errors and p-values for the null hypotheses $H_0 : \beta_j = 0$. The detailed output of the `glm` function of R is summarized in Table 4.8.

We can see in the second column of Table 4.8 (headed "Estimate") the point estimates obtained with the IRLS algorithm. The corresponding standard errors, resulting from the large-sample properties of maximum-likelihood estimation, are reported in the next column. The z value is the Wald statistic for testing $\beta_j = 0$, obtained from the ratio of the values appearing in the two preceding columns. The corresponding p-values are displayed next. The last column in Table 4.8 illustrates the conclusion drawn from the statistical test with the help of a simple coding. Recall that the smaller the p-value, the larger the test statistics and the less doubtful the rejection of $\beta_j = 0$. It is then likely that the corresponding feature should be kept in the model (as its contribution to the GLM score is nonzero). This is represented by stars: three stars ($\star\star\star$) for p-values less than 0.1%, two stars ($\star\star$) for p-values between 0.1 and 1%, a single star ($\star$) for p-values between 1 and 5%. A dot is used for p-values in the gray zone 5–10%, where the actuary is dubious about the significance of the corresponding feature. Nothing appears in the last column for p-values larger than 10%, which generally do not lead to rejection.

For proper interpretation, it is important to realize that these stars relate to the statistical degree of significance, not to the actual impact of the feature on the expected response. We might well have a larger $\widehat{\beta}_j$ associated with a larger p-value (despite remaining significant). The jth feature then induces more premium differentials even if it awarded less stars in the R output. Ranking the features based on the p-values is therefore not meaningful for insurance applications.

We can see in Table 4.8 that all p-values for the test of the null hypothesis $H_0 : \beta_j = 0$ are smaller than 10^{-6} so that there is strong evidence against $H_0 : \beta_j = 0$ for every parameter (which is represented by the three stars appearing in the last column, as explained above). The null deviance is 32,760.878 on 35 degrees of freedom whereas the residual deviance is 42.942 on 33 degrees of freedom. The model thus explains a large part of the deviance. The AIC is 354.33.

Graphically, the fit obtained with the quadratic Perks formula was clearly superior to the one obtained with a linear term, only. This is already apparent from the small p-value obtained for the regression coefficient associated to the squared age component reported in Table 4.8. A formal test can be performed using the function anova that returns an analysis-of-deviance table. The difference in the respective deviances of the linear and quadratic Perks formulas is 325.62 with 1 degree of freedom. The resulting p-value is smaller than 10^{-6} so that the null hypothesis that the linear specification of the score is equivalent to the quadratic one is clearly rejected. Thus, we see that the quadratic formula (4.15) significantly outperforms its linear counterpart (obtained by putting $\beta_2 = 0$).

The 3×3 variance-covariance matrix $\widehat{\boldsymbol{\Sigma}}_{\widehat{\boldsymbol{\beta}}}$ of $\widehat{\boldsymbol{\beta}}$ can be obtained with the R function vcov giving

$$\widehat{\boldsymbol{\Sigma}}_{\widehat{\boldsymbol{\beta}}} = \begin{pmatrix} 2.311966 \times 10^{-1} & -5.823715 \times 10^{-3} & 3.631568 \times 10^{-5} \\ -5.823715 \times 10^{-3} & 1.471464 \times 10^{-4} & -9.202314 \times 10^{-7} \\ 3.631568 \times 10^{-5} & -9.202314 \times 10^{-7} & 5.771165 \times 10^{-9} \end{pmatrix}.$$

The score $\widehat{\beta}_0 + \widehat{\beta}_1 x + \widehat{\beta}_2 x^2$ is approximately Normally distributed, with mean

$$\mathrm{E}\left[\widehat{\beta}_0 + \widehat{\beta}_1 x + \widehat{\beta}_2 x^2\right] = \beta_0 + \beta_1 x + \beta_2 x^2$$

and variance

$$\mathrm{Var}\left[\widehat{\beta}_0 + \widehat{\beta}_1 x + \widehat{\beta}_2 x^2\right] = \mathrm{Var}\left[(1, x, x^2)\widehat{\boldsymbol{\beta}}\right] = (1, x, x^2)\widehat{\boldsymbol{\Sigma}}_{\widehat{\boldsymbol{\beta}}}(1, x, x^2)^{\top}.$$

This allows us to get confidence intervals on the score, and hence on the death probabilities q_x.

4.7.7.2 Loss Reserving

Let us come back to the modeling of average payments according to the accident year AY and development year DY, considered earlier in this chapter. The estimated regression coefficients displayed in Fig. 4.6 suggest to structure the AY and DY effects in order to reduce the number of parameters and increase model stability. Considering the regression coefficients associated with DY, we see that they appear to be markedly linear up to DY $= 7$, and then break, exhibiting a similar linear trend beyond, i.e. for DY≥ 8. This suggests to represent the effect of development year with the two features DY and I[DY > 7] entering the score in a linear way. Considering this specification, the difference in the deviance is 0.030928 with 8 degrees of freedom. The corresponding F-value is 0.8603, giving a p-value of 55.63%. There is thus no significant difference between the models with and without structured DY effects.

The estimated regression parameters β_j associated to AY remain mostly unaffected once the DY effects are structured, as it can be seen from Fig. 4.7. This is why we continue with structuring the regression coefficients β_j associated with the 11 accident years. We see that up to AY $= 6$, the estimated β_j are almost constant while they exhibit a linear trend beyond. This suggests to replace the unstructured accident

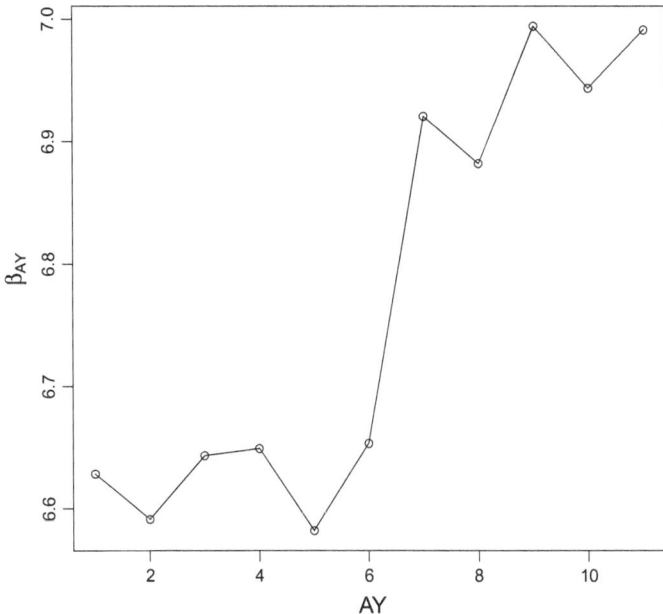

Fig. 4.7 Estimated regression coefficients β_j for the average amounts of payments, related to AY ($j = 1, \ldots, 11$) obtained once the regression coefficients related to DY have been structured

year effect with a new feature AY $\times$ I[AY > 6] entering the score in a linear way, allowing for an intercept. This further simplification does not significantly deteriorate the fit. Precisely, compared to the initial model, the difference in the deviances is equal to 0.068996 with 17 degrees of freedom. The F-value is 0.9032 and the corresponding p-value is 57.85% so that no significant difference is detected. Performing the comparison with the preceding model, we get a difference in deviances of 0.038068, 9 degrees of freedom, a F-value of 0.8155 and a p-value 60.43% which confirms the previous results. Since the structured model tested against the initial model using an analysis of deviance table and against the intermediate one does not show any significant difference (as detected by the F test), the simple structured model is preferred.

The model fit is summarized in Table 4.9. The final model involves the transformed features AY $\times$ I[AY > 6], DY, and I[DY > 7] resulting from the structuring of the effects obtained by treating AY and DY as categorical features. The dispersion parameter for Inverse-Gaussian family is estimated to $\widehat{\phi} = 0.0051574$. The null deviance is 5.48273 on 65 degrees of freedom whereas the residual deviance is equal to 0.25771 on 62 degrees of freedom. The corresponding AIC is 712,569.

Table 4.9 Output of the `glm` function of R for the fit of the loss reserving data displayed in Table 4.6 with structured AY and DY effects. Number of Fisher scoring iterations: 7

Coefficient	Estimate	Std. error	z value	Pr(> |z|)	
Intercept	6.622377	0.030958	213.912	$<2 \times 10^{-16}$	⋆⋆⋆
AYI[AY > 6]	0.035638	0.002832	12.586	$<2 \times 10^{-16}$	⋆⋆⋆
DY	0.427926	0.020153	21.234	$<2 \times 10^{-16}$	⋆⋆⋆
I[DY > 7]	−1.126561	0.648119	−1.738	0.0871	.

4.8 Influence Measures

4.8.1 General Definition

An influential observation is one that, if changed to a small amount or omitted, will modify substantially the parameter estimates of the model. Influence measures aim to evaluate the impact of a particular observation on estimated parameters or fitted values. Leverage is one indication of how much influence an observation has. The general definition of the leverage of the jth observation is the magnitude of the derivative of the ith fitted value with respect to the jth response value.

4.8.2 Diagnostic Measures Based on the Hat Matrix

In the Normal case, we know from (4.11) to (4.12) that the fitted values are obtained by multiplying the observed responses with the hat matrix H. Thus, the diagonal element h_{ii} of H measures the influence of Y_i on $\widehat{Y}_i$ as

$$\widehat{Y}_i = \sum_{j=1}^{n} h_{ij} Y_j = h_{ii} Y_i + \sum_{j \neq i}^{n} h_{ij} Y_j, \quad i = 1, 2, \ldots, n.$$

This is in line with the general definition of leverage recalled previously. Also, h_{ii} measures the impact of Y_i on every fitted value $\widehat{Y}_j$ as

$$h_{ii} = \sum_{j=1}^{n} h_{ij}^2 \text{ and } h_{ij} = h_{ji}.$$

If $h_{ii} = 1$, then $\widehat{Y}_i$ is determined by Y_i, only. In loss reserving applications, this is typically the case for the last cell in the first row of the run-off triangle as well as in the last cell of the first column, for which a model treating AY and DY as categorical features gives a perfect fit. If $h_{ii} = 0$ then y_i has no influence on $\widehat{y}_i$.

Considering the trace of the hat matrix H given in (4.18), the average of the h_{ii} is equal to $\bar{h} = (p+1)/n$. Influential observations have $h_{ii} > 2\bar{h}$; in small samples, the larger threshold $3\bar{h}$ is often used.

In GLMs, leverage is quantified by the hat matrix H coming from the last iteration of the IRLS algorithm. The quantity

$$h_{ii} = x_i^\top (X^\top \tilde{W} X)^{-1} x_i, \quad i = 1, 2, ..., n.$$

is called a leverage or hat value. In Normal linear regression models, it is non-random (i.e. it only depends on the features $x_1, \ldots, x_n$), constrained to belong to the interval $[0, 1]$, depending on the location of x_i relative to the other x_k. A large value of h_{ii} corresponds to an observation with relatively unusual x_i whereas a small h_{ii} corresponds to an observation close to the center of the data. In GLMs however, responses enter the calculations of h_{ii} (through the working weights), making their interpretation more cumbersome beyond the impact on the calculation of the ith fitted value.

4.8.3 Leave-One-Out Estimators and Cook's Distance

Another approach to measure the influence is to examine the estimated regression coefficients obtained by omitting each observation in turn. The subscript "$(-i)$" is used to indicate that the corresponding quantity has been computed using the reduced data set obtained by deleting observation (y_i, x_i) from the training data set. So, $\hat{\beta}_{(-i)}$ gives the estimated regression coefficients based on $(y_1, x_1), \ldots, (y_{i-1}, x_{i-1})$, $(y_{i+1}, x_{i+1}), \ldots, (y_n, x_n)$. It is called a leave-one-out estimate.

In the Normal case, The parameter estimate $\hat{\beta}_{(-i)}$ obtained by omitting the ith observations is easily obtained from

$$\hat{\beta}_{(-i)} = \hat{\beta} - (X^\top W X)^{-1} x_i \frac{y_i - \hat{y}_i}{1 - h_{ii}}.$$

Hence, there is no need to estimate β from the reduced data set since $\hat{\beta}_{(-i)}$ is directly obtained from $\hat{\beta}$ using this identity. Similarly for the estimated variance,

$$(n - p - 2)\hat{\sigma}_{(-i)}^2 = (n - p - 1)\hat{\sigma}^2 - \frac{(y_i - \hat{y}_i)^2}{1 - h_{ii}}.$$

The random variables Y_i et $\hat{Y}_{(-i)}$ are uncorrelated so that

$$\mathrm{Var}[Y_i - \hat{Y}_{(-i)}] = \mathrm{Var}[Y_i] + \mathrm{Var}[\hat{Y}_{(-i)}]$$
$$= \sigma^2 \left(1 + x_i^\top (X_{(-i)}^\top X_{(-i)})^{-1} x_i\right).$$

This shows that in the Normal case, leave-one-out estimation does not require refitting the model n times, without the ith observation. This is unfortunately no more the case for other GLMs but approximations have been developed to reduce the computation time for non-Normal GLMs.

Cook's distance measures the distance between $\widehat{\boldsymbol{\beta}}$ and $\widehat{\boldsymbol{\beta}}_{(-i)}$, large values indicating observations that impact on the estimated regression coefficients. Cook's distance is used to examine how each observation affects the whole vector $\widehat{\boldsymbol{\beta}}$ of parameter estimates. It measures the difference $\widehat{\boldsymbol{\beta}} - \widehat{\boldsymbol{\beta}}_{(-i)}$ in all $p + 1$ regression coefficients. Another influence measure is based on the difference of the fitted values $\widehat{\mu}_i$ using all the data and $\widehat{\mu}_{(-i)}$ omitting observation i.

When n is large (as it is generally the case in insurance applications), omitting a single observation i does not really impact on the estimated regression coefficients, making the leave-one-out approach unattractive. When performed on grouped data or on small-scale problems (such as run-off triangles in loss reserving), this approach is nevertheless informative.

4.9   Raw, Standardized and Studentized Residuals

This sections reviews the different kinds of residuals that are used to identify model deficiencies. Numerical illustrations of the different concepts are gathered in Sect. 4.9.4.

4.9.1   Back to the Normal Linear Regression Model

Residuals represent the difference between the data and fitted values produced by the model under consideration. They are thus important to check the adequacy of the GLM assumption. In the Normal linear regression model, the raw residuals R_i are defined as the difference between the response Y_i and the fitted values $x_i^\top \widehat{\boldsymbol{\beta}}$, i.e.

$$R_i = Y_i - x_i^\top \widehat{\boldsymbol{\beta}}.$$

We then have

$$E[R_i] = 0, \quad \text{Var}[R_i] = \sigma^2(1 - h_{ii}) \text{ and } \text{Cov}[R_i, R_j] = -\sigma^2 h_{ij}$$

where h_{ij} is the (i, j)-element of the hat matrix $\boldsymbol{H}$ defined in (4.12). The diagonal elements h_{ii} are such that the inequalities $1/n \leq h_{ii} \leq 1$ hold for all i and $\sum_{i=1}^{n} h_{ii} = p + 1$, as stated in (4.18). We see that, contrarily to the observations, the raw residuals are correlated. This correlation is easily understood since they are obtained by subtracting the fitted values from the observations, and the fitted values have been produced using the whole data set. The variance of the residual is smaller when h_{ii} is large; it can even be zero if $h_{ii} = 1$. But $\widehat{Y}$ and R remain uncorrelated.

Raw residuals for observations with high leverage (i.e. large h_{ii}) have smaller variances. To correct for the non-constant variance, we can divide R_i by an estimate of its standard deviation. The standardized residuals are defined as

$$T_i = \frac{y_i - \widehat{y}_i}{\widehat{\sigma}\sqrt{1 - h_{ii}}} = \frac{R_i}{\widehat{\sigma}\sqrt{1 - h_{ii}}}.$$

Standardized residuals measure by how many standard deviations any observation is away from the fitted model. However, the numerator and denominator in T_i are not independent, preventing it to follow a Student t-distribution. They have a unit variance but their sample distribution is unknown as the numerator and denominator are correlated (the ith observation entering both of them). This is why Studentized residuals are often used instead.

To bypass the problem of dependence, we resort to leave-one-out estimators $\widehat{\beta}_{(-i)}$ and $\widehat{\sigma}_{(-i)}$ which are based on all observations expecting the ith one. Studentized residuals are then given by

$$T_i^\star = \frac{R_i}{\widehat{\sigma}_{(-i)}\sqrt{1 - h_{ii}}}, \quad i = 1, 2, \ldots, n.$$

It can be shown that

$$T_i^\star = T_i \sqrt{\frac{n - p - 1}{n - p - T_i^2}}$$

so that $T_i^\star$ is easily obtained from T_i, avoiding again to re-fit the model n times in the Normal GLM with canonical link. The residuals $T_i^\star$ obey the Student distribution with $n - p - 2$ degrees of freedom. Often, $T_i^\star$ is preferred over T_i because they identify problematic observations more efficiently. This is because Y_i does not enter $\widehat{\sigma}_{(-i)}$ which is thus robust with respect to the ith observation.

For large n, it is usually the case that

$$T_i \approx T_i^\star \approx \frac{R_i}{\widehat{\sigma}} \approx \mathcal{N}or(0, 1).$$

However, for smaller sample sizes, these approximations do not necessarily hold and the choice of the residual becomes important. In loss reserving applications based on triangles, because of the relatively small number of observations, the actuary must be cautious when selecting the appropriate residuals.

Residual plots may then reveal problems with model fit. There are a variety of ways to display residuals and related quantities. Plotting residuals against the fitted values and versus each feature in turn are the most classical diagnostic graphs. Any systematic variation or isolated point discovered in the plot generally indicates a failure of one or more assumptions underlying the model under consideration.

4.9.2   Residuals in GLMs

In non-Normal GLMs, several types of residuals can be defined in analogy to the
Normal linear regression model. They often refer to the Normal linear regression
model used at the last step of the IRLS algorithm. The raw residual r_i is simply
the difference $y_i - \widehat{\mu}_i$ between the observed response and the estimated mean $\widehat{\mu}_i =
g^{-1}(x_i^\top \widehat{\beta})$. These are called response residuals for GLMs.

Often, actuaries allow for larger departures between observations y_i and fitted
values $\widehat{\mu}_i$ as the mean increases, because the variance is generally an increasing
function of the mean. This makes the plot of raw residuals less informative. To
circumvent this problem, the raw residuals are normalized by the square root of the
variance of the response. This leads to Pearson's residuals defined as

$$r_i^P = \frac{y_i - \widehat{\mu}_i}{\sqrt{V(\widehat{\mu}_i)/v_i}}.$$

They represent the square root of the contribution of each observation to Pearson's
X^2 statistic as

$$\sum_{i=1}^{n} (r_i^P)^2 = X^2.$$

This explains why these residuals are called Pearson ones.

Pearson's residuals are inherited from linear models and do not account for the
shape of the distribution. As the deviance calculation corrects for the skewness of
the response, another residual has been proposed for use in GLMs. The deviance
residuals r_i^D are defined as the signed square root of the contribution of the ith
observation to the deviance. This helps to detect the observations which increase the
deviance to a large extent, and which thus invalidate the model. Specifically, let us
decompose the deviance into

$$D(y, \widehat{\mu}) = 2 \sum_{i=1}^{n} v_i \left(y_i(\widetilde{\theta}_i - \widehat{\theta}_i) - a(\widetilde{\theta}_i) + a(\widehat{\theta}_i) \right) = \sum_{i=1}^{n} d_i$$

where $d_i = d(y_i, \widehat{\theta}_i)$ has been introduced in (4.19). Deviance residuals are then
defined as
$$r_i^D = \text{sign}(y_i - \widehat{\mu}_i)\sqrt{d_i}.$$

Here, the notation "sign" stands for the sign of the quantity within brackets:

$$\text{sign}(y_i - \widehat{\mu}_i) = \begin{cases} 1 \text{ if } y_i > \widehat{\mu}_i \\ -1 \text{ otherwise.} \end{cases}$$

Clearly, the sum of the squares of the deviance residuals gives the deviance, that is

$$\sum_{i=1}^{n}(r_i^D)^2 = D.$$

Deviance residuals are more appropriate in a GLM setting and are often the preferred form of residuals. In the Normal case, Pearson and deviance residuals are identical. However, it should be noted that the analyst is not looking for Normality from the deviance residuals as their sample distribution remains unknown. It is all about lack of fit and looking for patterns visible in the residuals: if the actuary detects some pattern in the residuals plotted against

- fitted values,
- features included in the score or
- features not in the model

then there is something wrong and some action needs to be undertaken. For instance, if the actuary detects some structure in the residuals plotted against an omitted feature then this feature should be integrated in the score. If there are patterns in the graph showing residuals against each feature included in the score then the corresponding feature is not properly integrated into the score (it should be transformed, or the actuary should abandon GLM for an additive model). Sometimes, it is interesting to plot also the residuals against time or spatial coordinates, for instance. Patterns in spread (detected by plotting residuals against fitted values) may indicate overdispersion or more generally the use of a wrong mean-variance relationship.

Standardized residuals account for the different leverages. Consider the hat matrix (4.12) coming from the last iteration of the IRLS algorithm. Hat values h_{ii} measure leverage, or the potential of an observation to affect the fitted values. A large value of the leverage h_{ii} indicates that the fit is sensitive to the response y_i. Large leverages typically mean that the corresponding features are unusual in some way. Standardized residuals, also called scaled residuals, are then given by

$$t_i^P = \frac{r_i^P}{\sqrt{\hat{\phi}(1 - h_{ii})}} \text{ and } t_i^D = \frac{r_i^D}{\sqrt{\hat{\phi}(1 - h_{ii})}}.$$

The Studentized residuals (also called jacknife or deletion residuals) correspond to standardized residuals resulting from omitting each observation in turn. As no simple relationship between $\widehat{\boldsymbol{\beta}}$ and $\widehat{\boldsymbol{\beta}}_{(-i)}$ generally exist for GLMs, calculating them exactly would require refitting the model n times, which is out of reach for large sample size. An approximation is given by the likelihood residuals defined as a weighted average of the standardized deviance and Pearson residuals, so that we take

$$t_i^\star = \text{sign}(y_i - \widehat{\mu}_i)\sqrt{(1 - h_{ii})(t_i^D)^2 + h_{ii}(t_i^P)^2}.$$

Contrarily to the Normal linear regression model, where standardized residuals have a known distribution, the distribution of residuals in GLMs is generally unknown. Therefore, the range of residuals is hard to assess but they can be inspected for structure and dispersion by plotting them against fitted values or some features. Leverage only measures the potential to affect the fit whereas influence measures more directly assess the effect of each observation on the fit. This is typically done by examining the change in the fit from omitting this observation. In a GLM, this can be done by looking at the changes in regression coefficients. The Cook statistics measures the distance between $\widehat{\boldsymbol{\beta}}$ and $\widehat{\boldsymbol{\beta}}_{(-i)}$ obtained by omitting the i data point $(y_i, \boldsymbol{x}_i)$ from the training set.

4.9.3 Individual or Grouped Residuals

The calculation of the contribution of each observation to the deviance can adjust for the shape but not for the discreteness of observations. Often, individual residuals for discrete responses are hard to interpret as they inherit the structure of the response values. For instance, for binary responses, residuals often cluster around two curves, one for "0" observations and the other one for "1" observations. For counts, the same phenomenon occurs and individual residuals concentrate around curves corresponding to "0" observations, "1" observations, "2" observations, etc.

By aggregating individual residuals into large groups of similar risks, the actuary avoids the apparent structure produced by the integer-valuedness of the response. This is because we know from Chap. 2 that responses obeying the Poisson distribution with a large mean or the Binomial distribution with a large size are approximately Normal. The deviance residuals calculated on the aggregated data give much better indications about the appropriateness of the model for the response.

Also, recall that GLMs are used to compute pure premiums. Hence, the analyst is interested in the estimation of expected values, not in predicting any particular outcome. This is why data can be grouped before model performances are evaluated.

4.9.4 Numerical Illustrations

4.9.4.1 Graduation

Figure 4.8 displays the residuals for the quadratic formula (4.15). We can see there that the model is less accurate for the very last ages which is somewhat expected because the exposures are limited at advanced ages. More importantly, the residuals do not display any visible structure when plotted against fitted values or again the features entering the score (age and its square).

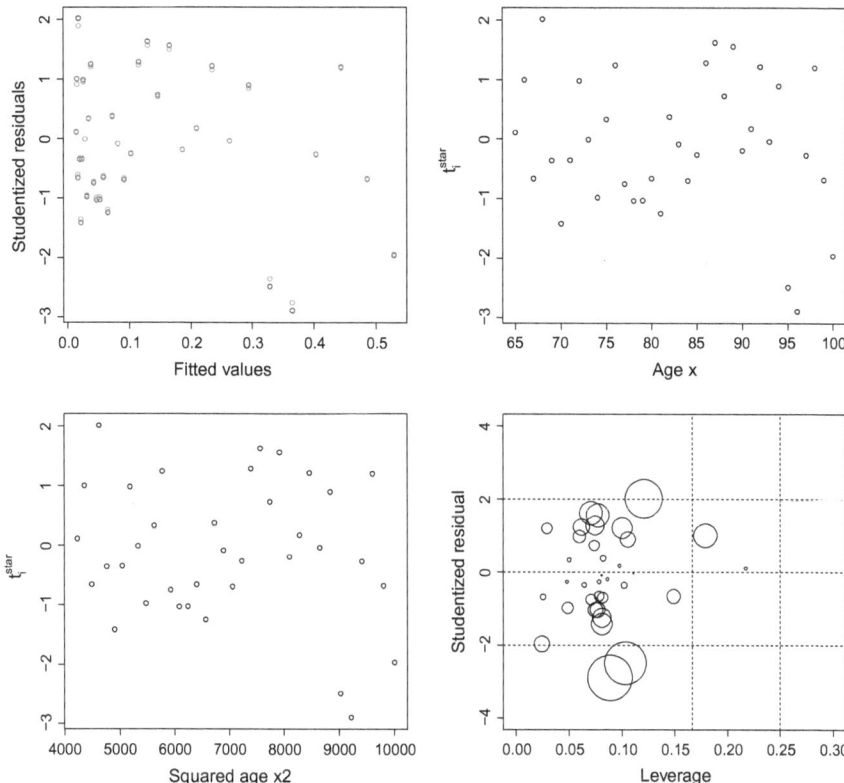

Fig. 4.8 Studentized residuals t_i^* against fitted values (top left panel), against age x (top right panel), against squared age x^2 (bottom left panel), and against leverage with Cook's distances (bottom right panel)

The last panel of Fig. 4.8 combines residuals, hat values and Cook's distances. Precisely, the areas of the circles are proportional to the Cook's distance for the observation. Horizontal lines are drawn at $-2, 0$, and 2 on the Studentized residuals scale, vertical lines at twice and three times the average hat value. The critical region corresponds to the upper and lower right parts of the graph (combining a large hat value and a large absolute Studentized residual value), where no data point falls in the present case. Large Cook's distances occur at older ages (lower left part of the graph), as expected because of the limited exposure near the end of the life table.

4.9.4.2 Loss Reserving

Figure 4.9 displays the residuals for average amounts of payments against fitted values and explanatory variables. Except that more observations are available for smaller values of AY and DY, and for larger values of CY, no structure seems to emerge from the four residual plots in Fig. 4.9.

Figure 4.10 combines residuals, hat values and Cook's distances. It can be interpreted as the last panel of Fig. 4.8. Precisely, Fig. 4.10 shows that no data point falls in the critical regions corresponding to the upper and lower right quadrants of the graph. Some cells of the triangle nevertheless appear to have large Cook's distances, associated to low or moderate leverage.

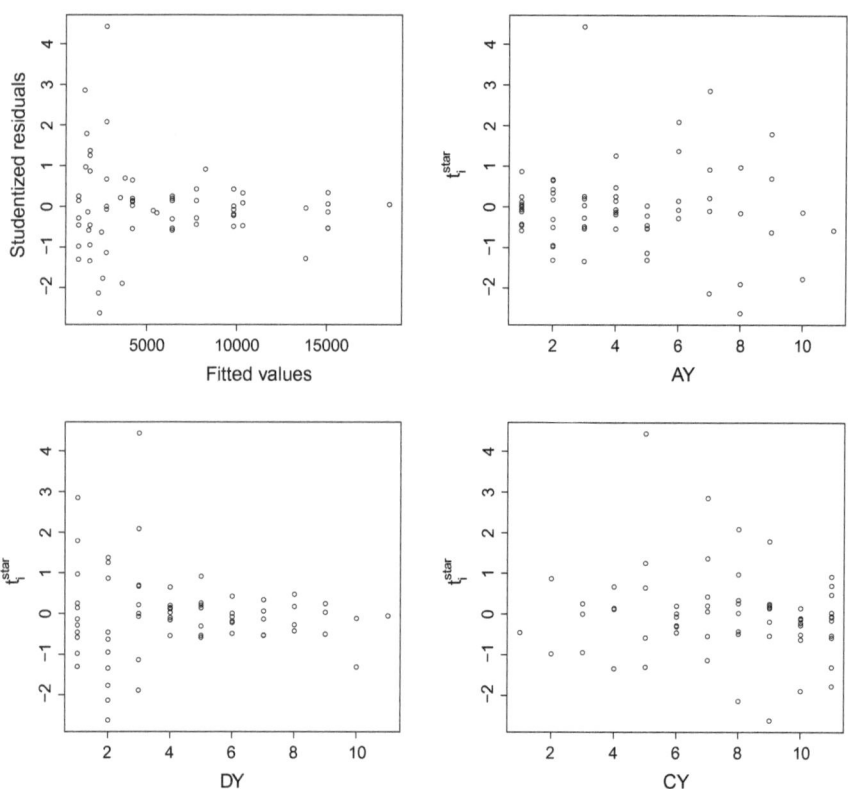

Fig. 4.9 Studentized residuals t_i^* against fitted value (top left panel), against AY (top right panel), against DY (bottom left panel) and against CY (bottom right panel)

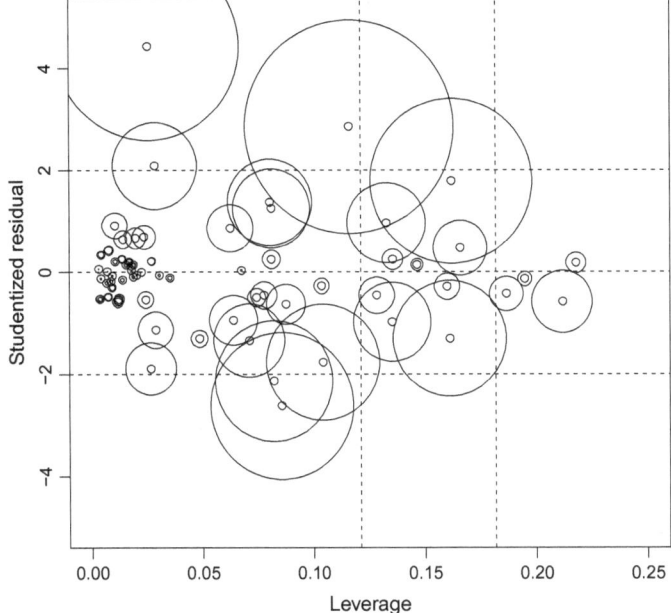

Fig. 4.10 Studentized residuals t_i^* against leverage, with Cook's distances

4.10 Risk Classification in Motor Insurance

In order to recap' all the notions presented so far, let us now study a larger data set related to motor insurance. The response is here the number of claims filed by the insured driver in third-party liability insurance.

4.10.1 Description of the Data Set

4.10.1.1 Available Features and Responses

Let us briefly present the data used to illustrate the techniques described in this chapter. The data set relates to a Belgian motor third-party liability insurance portfolio observed in the late 1990s. It comprises 14,504 contracts with a total exposure of 11,183.68 policy-years. These are vehicle-years, as every policy covers a single vehicle but some contracts are in force for only a part of the year.

The variables contained in the file are as follows:

AgePh   policyholder's age last birthday (on January 1), in 5 groups G1 to G5 of increasing seniority.

Gender   Policyholder's gender, male or female.

`PowerCat`   Power of the car, with four categories from C1 to C4 of increasing
power.
`ExpoR`   Number of days the policy was in force during the observation period.
`City`   Size of the city where the policyholder live, with three categories, large,
midsize or small.

In addition to these features, the number of claims `Nclaim` filed by each policyholder
during the observation period has been recorded. This is the response investigated in
this section.

4.10.1.2   Composition of the Portfolio with Respect to Features

The upper left panel in Fig. 4.11 gives of the distribution of the exposure-to-risk in
the portfolio. About 60% of the policies have been observed during a whole year.
Considering the distribution of the exposure-to-risk, we see that policy renewals and
lapses are randomly spread over the year. Notice that insurance companies often
create a new record each time there is a change in the features (for instance, the
policyholder buys a new car or moves to another location). This explains the relatively
large number of contracts with exposures strictly less than unity. The age structure
of the portfolio is also described in Fig. 4.11 (see the upper right panel there), as well
as the composition of the data set with respect to gender, size of the city where the
policyholder resides and power of the car. We can see that the majority of insured
drivers are in age class G2, about two thirds of contracts cover a male driver, the
distribution of power exhibits a U shape, and the contracts are evenly balanced in the
three types of cities.

The lower right panel in Fig. 4.11 displays the histogram of the observed claim
numbers per policy. We can see there that the majority of the policies (12,458 precisely) did not produce any claim. About 10% of the portfolio produced one claim
(1,820 contracts). Then, 206 contracts produced 2 claims, 17 contracts produced 3
claims, 2 contracts 4 claims and there was a policy with 5 claims. This gives a global
claim frequency equal to 20.53% of at the portfolio level (this is quite large compared
to nowadays experience, with claim frequencies of about 6–7% at market level in
Belgium).

4.10.1.3   Association Between Features

Let us use Cramer's V to detect the features that are related to each other. Figure 4.12
displays the values of Cramer's V between the features, and between the response and
the features. We can see there that the associations are rather weak except between
`Power` and `Gender`. But it remains relatively moderate in every case. Also, the
response appears to be weakly associated with the individual features. This is typically the case with insurance count data because of the heterogeneity present in the
data and the discreteness of the response.

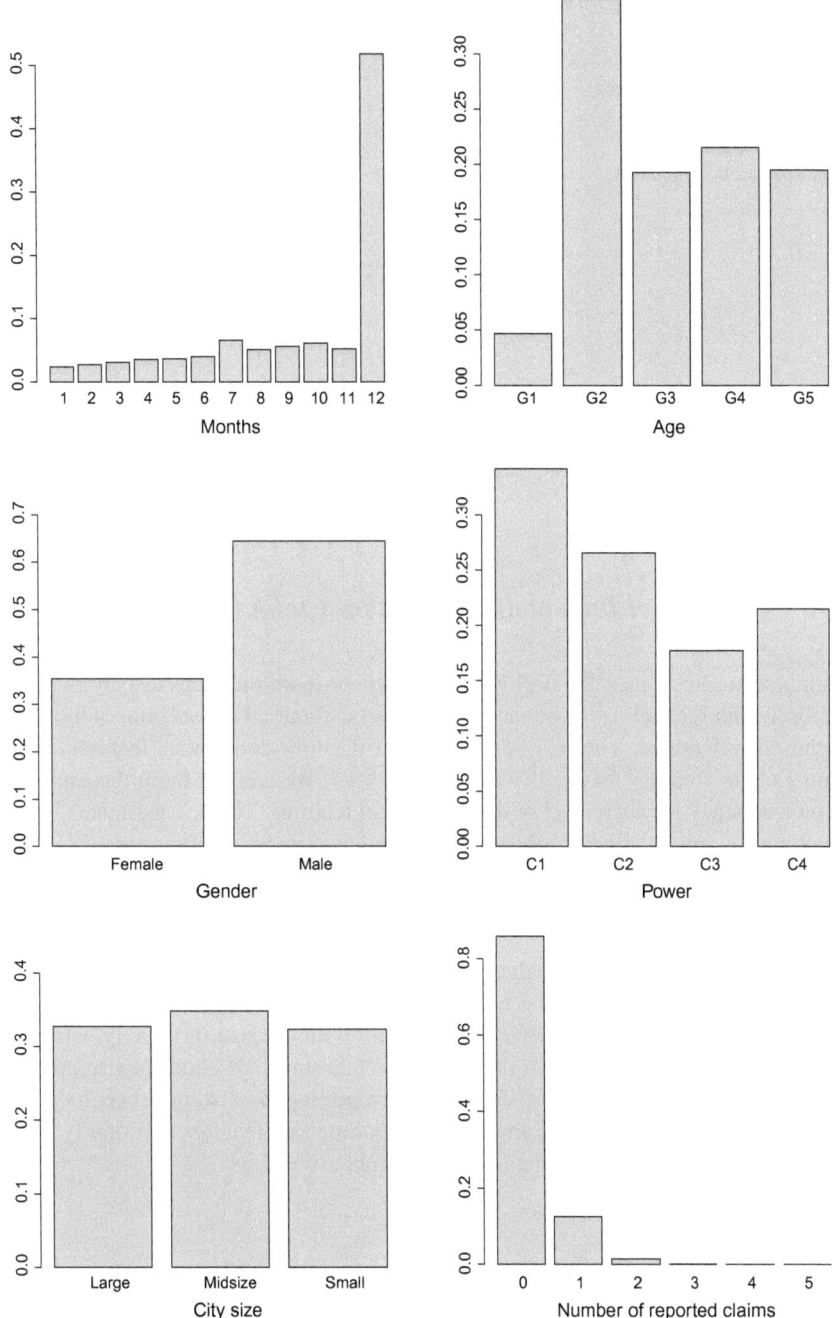

Fig. 4.11 Histograms of risk exposures (in months), policyholder's age and gender, power of the car, size of the city where the policyholder resides, and observed number of claims

Fig. 4.12 Correlation plot for Cramer's V between policyholder's age and gender, power of the car, size of the city where the policyholder resides, and observed number of claims

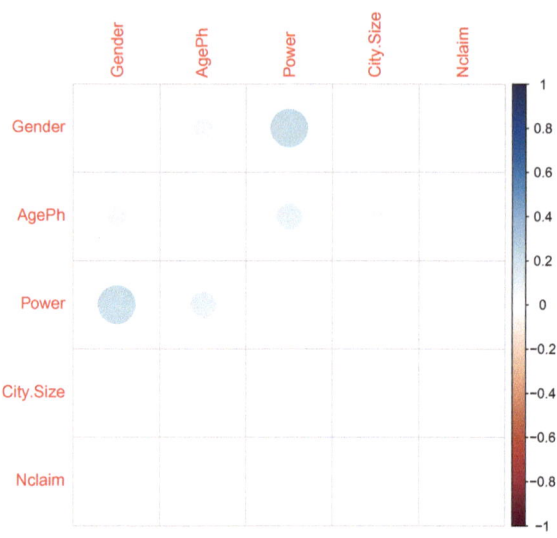

4.10.2 Marginal Impact of Features on Claim Counts

Insurance studies generally start with one-way, or marginal analyses summarizing response data for each value of each feature, but without taking account of the effect of the other features. The marginal effects of the four continuous features on the annual claim frequencies are depicted in Fig. 4.13. We can see there the estimated claim frequency for each level of the categorical features. The best estimates (corresponding to circles) are surrounded with confidence intervals. These results can be obtained with a Poisson GLM analysis including a single feature, so that confidence intervals are easily derived from the sample distribution of the estimated regression coefficients.

Despite the relatively low values of Cramer's V, we see there that the claim frequencies seem to depend on the features in the usual way, decreasing with policyholder's age, larger for male drivers, increasing with the size of the city, while the effect of power is more difficult to interpret at this stage. Of course, these are only marginal effects which can be distorted by the correlation existing between the features so that we need the GLM analysis incorporating all features to properly assess the impact of each feature on the expected number of claims.

4.10.3 Poisson GLM Analysis for Claim Counts

The Poisson distribution is the natural candidate to model the number of claims reported by the policyholders. The typical assumption in these circumstances is that

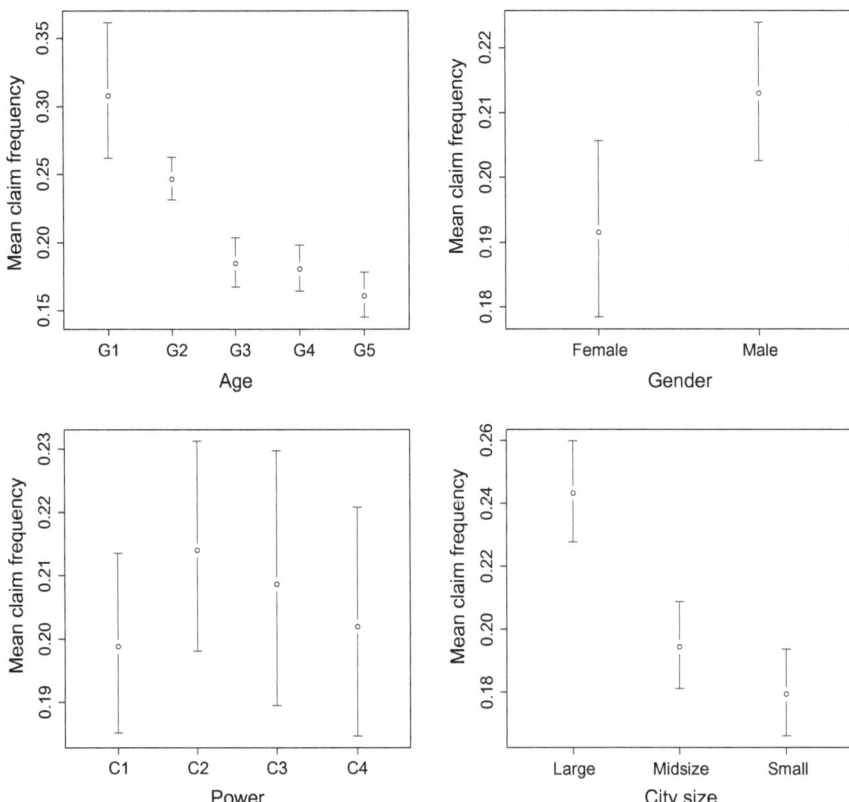

Fig. 4.13 Marginal impacts of the features on the average claim frequencies

the conditional mean claim frequency can be written as an exponential function of a linear score with coefficients to be estimated from the data.

We start with a model incorporating all features. The reference levels for the binary features are the most populated ones, in accordance with the guidelines explained above. We use

$$\texttt{offset}_i = \ln(\texttt{ExpoR}_i/365).$$

The results obtained with the help of the $\texttt{glm}$ function of the software R are displayed in Table 4.10.

We can see there that the effect of age, gender and city size is in line with the preliminary marginal analysis. Power does not seem to significantly impact on the expected claim frequency. However, the results reported in Table 4.10 are not enough for us to conclude that power is not needed in the GLM. The reason is that this categorical feature has been coded with the help of three dummies so that three tests are performed in Table 4.10, one for each level different from the reference category C1. As we do not control the total error inherent to using the conclusions of these

Table 4.10 Output of the `glm` function of R for the fit of the motor TPL data. Number of Fisher Scoring iterations: 6

Coefficient	Estimate	Std. Error	z value	$Pr(> |z|)$	
Intercept	-1.46605	0.05625	-26.063	$< 2 \times 10^{-16}$	$\star\star\star$
Female	-0.10068	0.04557	-2.209	0.02714	$\star$
Age G1	0.25232	0.08865	2.846	0.00442	$\star\star$
Age G3	-0.30633	0.05999	-5.106	3.29×10^{-7}	$\star\star\star$
Age G4	-0.32670	0.05854	-5.581	2.40×10^{-8}	$\star\star\star$
Age G5	-0.46584	0.06215	-7.495	6.62×10^{-14}	$\star\star\star$
Power C2	0.08483	0.05430	1.562	0.11819	
Power C3	0.07052	0.06211	1.135	0.25622	
Power C4	0.07930	0.06082	1.304	0.19230	
CtySize Large	0.24548	0.04955	4.954	7.25×10^{-7}	$\star\star\star$
CitySize Small	-0.09414	0.05340	-1.763	0.07790	

three tests simultaneously, the p-values displayed in Table 4.10 do not provide the actuary with the appropriate tool to decide whether power can be deleted from the score or not.

The null deviance is 9,508.6 on 14,503 degrees of freedom and the residual deviance is 9,363.5 on 14,493 degrees of freedom. This shows than the Poisson regression model only explains a small part of the deviance. The AIC value is 13,625.

As discussed previously, Table 4.10 suggests that power does not seem to be a risk factor. To formally test whether power can be excluded from the GLM score, we compare the model with and without this feature by means of an analysis-of-deviance table. The difference in the deviances of the models with and without `Power` is 3.0654 with 3 degrees of freedom. The corresponding p-value is 38.17%, which confirms that power is not relevant for explaining the number of claims.

Furthermore, it seems that the two last levels of City Size can be grouped together. We test whether this model simplification significantly deteriorates the fit with the help of an analysis of deviance table. The difference in the deviances of the initial model incorporating all features and the model without `Power` and with the two last levels of `CitySize` grouped together is 6.2137 with 4 degrees of freedom. The corresponding p-value 18.37% indicates that the grouping does not significantly deteriorate the fit. Re-fitting the model, we discover that the estimated regression coefficients for age categories G3 and G4 are respectively equal to -0.30054 and -0.31933, with standard errors respectively equal to 0.05984 and 0.05801. This suggests that levels G3 and G4 of `Age` could be grouped together. This is confirmed by an analysis-of-deviance table as the difference in deviances is 6.2867 with 5 degrees of freedom for the initial model compared to the simplified one. The p-value

Table 4.11 Output of the `glm` function of `R` for the fit of the final Poisson regression model to the motor TPL data. Number of Fisher Scoring iterations: 6

Coefficient	Estimate	Std. Error	z value	$Pr(> |z|)$	
Intercept	−1.76608	0.04183	−42.216	$< 2 \times 10^{-16}$	★ ★ ★
Female	−0.11779	0.04435	−2.656	0.00791	★★
Age G1	0.54798	0.08933	6.134	8.55×10^{-10}	★ ★ ★
Age G2	0.31040	0.04756	6.527	6.71×10^{-11}	★ ★ ★
Age G5	−0.14506	0.06276	−2.311	0.02081	★
CitySize Large	0.29026	0.04300	6.751	1.47×10^{-11}	★ ★ ★

27.93% shows that this further grouping is in line with the data. Because the exposure is now larger in the combined G3–G4 age group, the latter becomes the new reference level.

After these steps, we have now obtained the final model summarized in Table 4.11. The null deviance 9,508.6 on 14,503 degrees of freedom is now compared to the residual deviance 9,369.8 on 14,498 degrees of freedom. The latter is higher than the corresponding value for the initial model comprising all features. The simplified model nevertheless offers a better fit as indicated by the AIC value, which has now decreased from 13,625 to 13,621.

Residuals are displayed in Fig. 4.14. We see that residuals based on individual observations inherit the structure of the response, with the lower curve corresponding to policies without claims, the next ones to policies having reported 1 claim, 2 claims, etc. Residuals computed from grouped data avoid this pitfall. Precisely, individual data are grouped according to the features entering the final model (gender, age with 4 levels and city size) and the exposures are added accordingly. This produces a summarized data set with 16 groups. The estimated regression coefficients remain unaffected by the grouping but the residuals can now be re-computed (notice that the deviance residuals are not additive). The resulting grouped residuals are displayed in the lower panel of Fig. 4.14. When plotted against hat values and supplemented with Cook's distances in Fig. 4.15, the obtained residuals seem to support model fit as no risk class simultaneously exhibits large Cook's distances and leverages.

Figure 4.16 plots the actual observations against the expected values. We can see that model predictions are in close agreement to portfolio's experience: risk classes fall very close to the 45-degree line, especially the most populated ones.

4.10.4 Alternative Modeling

Instead of recording the number N_i of claims over the observation period of length e_i (exposure to risk), the actuary could equivalently base the claim frequency analysis

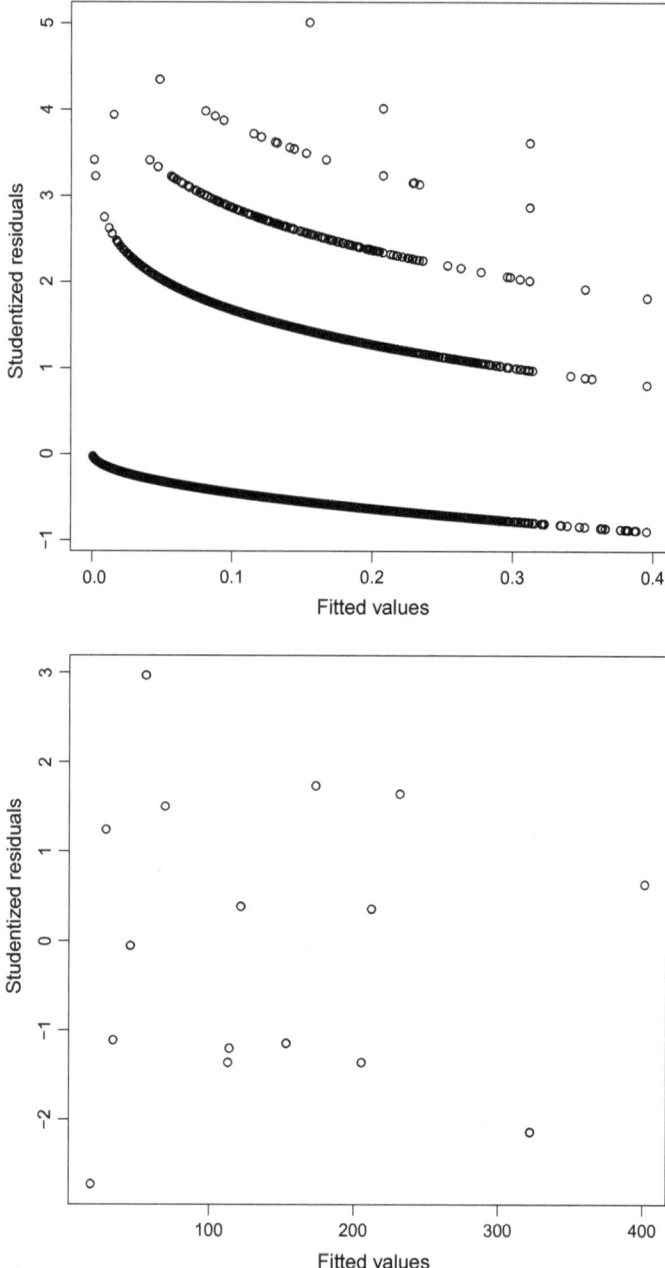

Fig. 4.14 Residuals of the final Poisson GLM based on individual observations (top panel) and on grouped data (bottom panel)

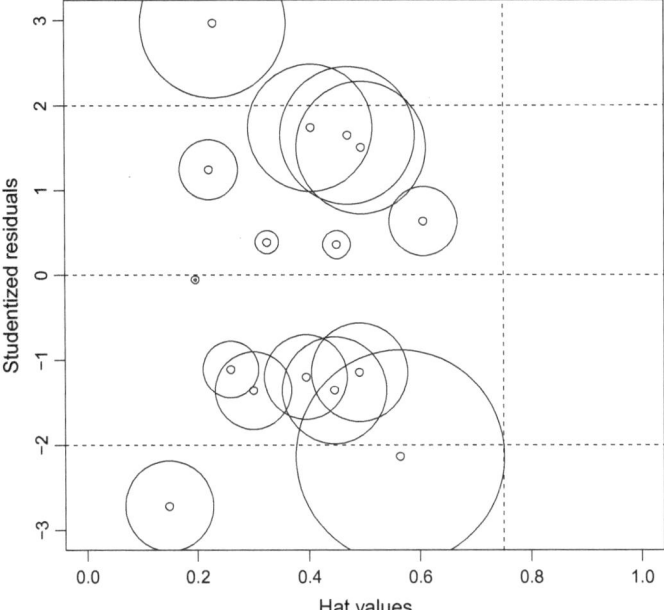

Fig. 4.15 Residuals of the final Poisson GLM based on grouped data together with combined leverage-residual plot with areas of the circles proportional to the Cook's distances

on the times $T_{i1}, \ldots, T_{i,N_i}$ at which these N_i claims occurred. The waiting periods W_{ik} between two consecutive claims filed by policyholder i are defined as

$$W_{ik} = T_{ik} - T_{i,k-1}, \quad k = 1, 2, \ldots$$

where $T_{i0} = 0$, by convention.

Assume that the waiting times W_{ik} are independent and obey the $\mathcal{E}xp(\lambda_i)$ distribution. If policyholder i does not report any claim ($N_i = 0$) then $T_{i1} = W_{i1} > e_i$ and the contribution to the likelihood is

$$\mathrm{P}[W_{i1} > e_i] = \exp(-e_i\lambda_i).$$

If policyholder i reports a single claim ($N_i = 1$) at time t_1 then the exposure to risk can be divided into the time t_1 until the claim and the time $e_i - t_1$ from t_1 until the end of the observation period. The contribution to the likelihood is then given by

$$\lambda_i \exp(-\lambda_i t_1)\mathrm{P}[W_{i2} > e_i - t_1] = \lambda_i \exp\left(-(t_1 + e_i - t_1)\lambda_i\right) = \lambda_i \exp(-e_i\lambda_i).$$

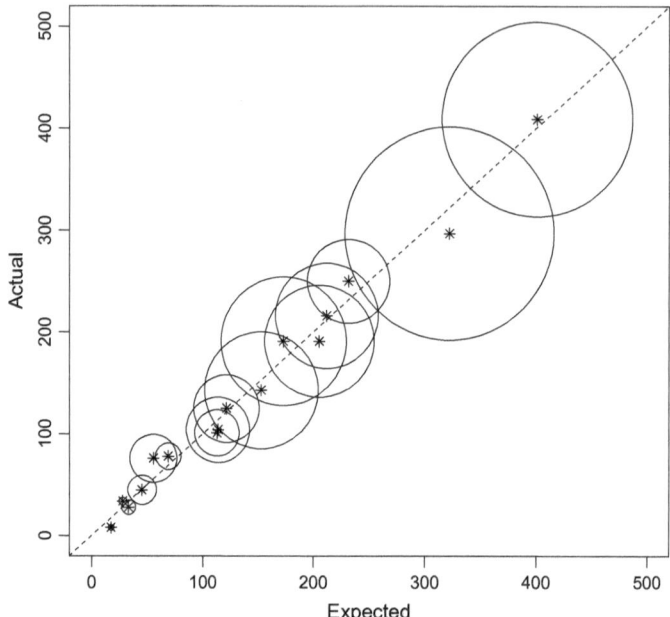

Fig. 4.16 Actual versus expected number of claims, with the area of the circles proportional to the total exposure of the risk classes

If policyholder i reports two claims at times t_1 and t_2 then the contribution to the likelihood is given by

$$\lambda_i \exp(-\lambda_i t_1)\lambda_i \exp\left(-\lambda_i(t_2 - t_1)\right) \exp\left(-\lambda_i(e_i - t_2)\right) = \lambda_i^2 \exp(-e_i\lambda_i).$$

In general, the likelihood associated to the occurrence times of the N_i claims is of the form

$$\mathcal{L} = \prod_{i=1}^{n} \left(\lambda_i^{N_i} \exp(-\lambda_i e_i)\right)$$

where

$$N_i = \max\{k \,|\, T_{i0} + T_{i1} + \ldots + T_{ik} \le e_i\}.$$

Gamma regression on the waiting periods W_{ik} is thus equivalent to Poisson regression on the number of claims.

4.11 Quasi-Likelihood, M-Estimation and Pseudo-Likelihood

4.11.1 Quasi-Likelihood Estimation

We know that selecting a distribution to perform a GLM analysis is tantamount to choosing the variance function that relates the variance of the response to the mean. This is because the IRLS algorithm only involves the conditional mean and variance of the responses Y_i, given the features x_i. Precisely, as long as we can express the transformed mean of the response Y_i as a linear function of the features x_i and can write down a variance function for Y_i (expressing the conditional variance of Y_i as a function of its mean μ_i and a dispersion parameter ϕ), we can apply IRLS algorithm and obtain estimated regression parameters. And this even without fully specifying the distribution for Y_i.

This approach is referred to as quasi-likelihood estimation and retains many good properties of maximum-likelihood estimation provided the sample size is large enough (which is generally the case in insurance applications). Although the quasi-likelihood estimators may not be efficient, they are consistent and asymptotically Normal. They remain asymptotically unbiased, with a variance-covariance matrix that can be estimated from the data. Quasi-likelihood is analogous to least-squares estimation with potentially non-Normal responses: as long as the relationship between the mean response and the explanatory variables is linear, the variance is constant, and the observations are mutually independent, ordinary least-squares estimation applies. The inference is fairly robust to non-Normality as the sample size is large. Considering the GLM framework, the important part of model specification corresponds to the link and variance functions while the conclusions of the regression analysis are less sensitive to the actual distribution of the response in large samples.

There is an advantage of using the quasi-likelihood approach for models with variance functions corresponding to Binomial and Poisson distributions. The regular Binomial and Poisson GLMs assume $\phi = 1$ whereas the corresponding quasi-Binomial and quasi-Poisson GLMs allow for the dispersion ϕ to be a free parameter. This is particularly useful in modeling overdispersion that is generally present in insurance count data, as well as underdispersion resulting from summing Bernoulli indicators with different success probabilities.

Overdispersion refers to a situation where the conditional variance of the response is larger than the variation implied by the ED distribution used to fit the model. This means that $\text{Var}[Y|X = x] > \text{E}[Y|X = x]$ in the Poisson case. There are several reasons why overdispersion manifests in the data, including

- responses Y_{i_1} and Y_{i_2} do not exactly have the same distribution despite sharing the same features $x_{i_1} = x_{i_2}$. This is because some information is missing and there is some residual, unmodeled heterogeneity.
- observations may be correlated or clustered, while the specified covariance structure wrongly assumes independent data.

The quasi-Poisson model is an example of the quasi-likelihood approach, estimating the mean response with a variance equal to $\phi\mu$ (extending the Poisson model for which $\phi = 1$). Overdispersed Poisson, or ODP refers to the particular case where $\phi > 1$. In actuarial applications, the ODP model is often used in loss reserving.

4.11.2   M-Estimation

Maximum-likelihood estimation, as well as method of marginal totals (MMT) and quasi-likelihood can be seen as particular cases of M-estimation. M-estimation offers a method to estimate the regression parameters when there exist $p + 1$ estimating functions $\psi_0, \ldots, \psi_p$ such that the system

$$E[\psi_j(Y, x, \beta)] = 0 \text{ for } j = 0, 1, \ldots, p,$$

uniquely determines β. Then, under certain weak regularity conditions, β can be estimated using an M-estimator $\widehat{\beta}$ implicitly defined as the solution to

$$\frac{1}{n}\sum_{i=1}^{n}\psi_j(y_i, x_i, \beta) = 0 \text{ for } j = 0, 1, \ldots, p.$$

In particular, ψ_j may be the partial derivative of an objective function Ψ, that is,

$$\psi_j(y, x, \beta) = \frac{\partial}{\partial\beta_j}\Psi(y, x, \beta) \text{ for } j = 0, 1, \ldots, p.$$

In such a case, M-estimators maximize (or minimize) Ψ. Under mild technical conditions, an estimator maximizing

$$\Psi_n = \frac{1}{n}\sum_{i=1}^{n}\Psi(y_i, x_i, \beta)$$

is consistent and asymptotically Normal if the limit function $\Psi_\infty = \lim_{n\to+\infty}\Psi_n$ exists and has a unique maximum at the true parameter value. The large-sample variance-covariance matrix can be obtained with the help of so-called sandwich formulas that are available in R.

Examples of M-estimation include

(i) maximum likelihood (where Ψ is the log-likelihood).
(ii) least-squares (where Ψ is the sum of squared residuals).
(iii) quasi-likelihood assuming $E[Y_i] = \mu_i(\beta)$ and $\text{Var}[Y_i] = \phi V(\mu_i(\beta))$ and estimating β as the solution to

$$\sum_{i=1}^{n} \frac{\partial \mu_i}{\partial \boldsymbol{\beta}} \frac{y_i - \mu_i}{V(\mu_i(\boldsymbol{\beta}))} = \mathbf{0}.$$

This means that the ED maximum-likelihood estimators remain consistent for a much wider class of distributions, as long as the mean and variance are correctly specified. This also extends to multivariate outcomes, in which case it is called Generalized Estimating Equations (GEE, see Chap. 5).

4.11.3 Pseudo-Maximum Likelihood

In the pseudo-maximum likelihood (or PML) approach, the only assumption made about the data is that the mean of Y_i belongs to a given family of functions of the features $\boldsymbol{x}_i$. Precisely, there is a function m such that

$$E[Y_i|\boldsymbol{x}_i] = m(\boldsymbol{x}_i, \boldsymbol{\beta}).$$

Often in actuarial applications the function m is specified as $m(\boldsymbol{x}_i, \boldsymbol{\beta}) = \exp(\boldsymbol{\beta}^\top \boldsymbol{x}_i)$. Then, the parameters are estimated by maximizing a likelihood which is not based on the true distribution of the data (hence the name, pseudo-maximum likelihood).

The idea is to find a family $f(y, m)$ of probability density functions indexed by their mean m such that maximizing the pseudo-likelihood function

$$\prod_{i=1}^{n} f(y_i, m(\boldsymbol{x}_i, \boldsymbol{\beta}))$$

provides the actuary with a consistent and asymptotically Normal estimator $\widehat{\boldsymbol{\beta}}$ of $\boldsymbol{\beta}$. Thus, PML is a particular case of M-estimation obtained by considering as objective function a pseudo-likelihood function. Consistency and asymptotic Normality hold if, and only if, the chosen probability density function is of the form

$$f(y, m) = \exp(a(m) + b(y) + c(y)m)$$

for some functions a, b and c, which appears to be very similar to the ED specification (2.3). Precisely, the Normal distribution (with known variance), the Poisson distribution and the Gamma distribution (with known scale parameter) have density functions of this form. The corresponding pseudo-likelihood functions produce consistent and asymptotically Normal estimators, whatever the true underlying distribution.

4.12 Conclusion About GLMs

The GLM approach to data modeling enjoys many convenient properties making it particularly attractive to conduct insurance studies. Let us mention a few of these properties that have been reviewed in this chapter:

- the discrete or skewed nature of the response is taken into account without prior transformation of the claim data;
- the linear score is easily interpretable and communicable, making it ideal for the commercial price list (transparency, marketing constraints, etc.);
- GLMs are easily fitted to insurance data by maximum likelihood, with the help of IRLS algorithm that appears to be generally effective from a numerical point of view;
- GLMs extend the MMT pioneered by US actuaries in the 1960s, which eased their adoption by the actuarial community in the 1990s, providing analysts with relevant interpretations for likelihood equations under canonical link function.

Besides these advantages, GLMs also suffer from several severe limitations:

- the linear structure of the score is too crude for the technical price list where nonlinear, spatial or multilevel risk factors need to be properly integrated on the score scale;
- GLMs do not automatically account for interactions;
- collinearity is an issue;
- as well as "large p" cases, making the manual variable selection procedure illustrated in this chapter out of reach.

Remember also that the whole story is about correlations and that no firm conclusions can be drawn about a possible causal relationship between the features and the response.

The next chapters of this book, and the two subsequent volumes (Trufin et al. 2019; Hainaut et al. 2019) precisely demonstrate how alternative modeling strategies remedy these deficiencies. In this first volume, variants of GLMs are explored, sharing many similarities with the basic GLM approach described in this chapter. In the next two volumes, alternative methods for constructing the score are proposed involving regression trees and neural networks. But in all cases, the analysis still relies on ED distributions for the response. Also, the different techniques can be efficiently combined by the actuary.

4.13 Bibliographic Notes and Further Reading

Due originally to Nelder and Wedderburn (1972), GLMs nicely extend Normal linear regression and Bernoulli logistic regression to the whole ED family. Not to mention the seminal book by Mc Cullagh and Nelder (1989), there are several actuarial textbooks dealing with GLMs, including De Jong and Heller (2008) as well as Ohlsson and Johansson (2010). Several chapters of Denuit et al. (2007), Kaas et al. (2008), Charpentier (2014) and Wuthrich and Buser (2017) are also devoted to GLMs and their applications. The pedagogical note published by the Casualty Actuarial Society (Goldburg et al. 2016) provides a very nice introduction to the topic as well as Chap. 5 in Frees et al. (2014). The 2-volume books by Faraway (2005a, b) and Fox (2016a, b) as well as Dunn and Smyth (2019) all contain very interesting material for those readers interested in implementing GLMs in practice while Fahrmeir et al. (2013) offer a comprehensive account of regression techniques. The reader interested in single-index models is referred to the comprehensive book by Horowitz (2009).

The use of offsets in actuarial studies has been demonstrated by Yan et al. (2009) on which Sect. 4.3.7 is based to a large extent. Mildenhall (1999) formally related GLM with minimum bias and the method of marginal totals, bridging the classical actuarial approach to these modern regression models. Method of marginal totals is due to Bailey (1963) and Jung (1968). We refer the reader to Ajne (1986) for a comparison of three methods for fitting multiplicative models to observed, cross-classified insurance data: the method of Bailey-Simon, the method of marginal totals and maximum-likelihood estimation.

GLMs are conveniently fitted with the help of the `glm` function in R. However, this function may fail when dealing with large data sets, because of limited availability of memory. In such a case, the actuary has the function `bigglm` in the package `biglm` at his or her disposal, as well as the function `speedglm`.

References

Ajne B (1986) Comparison of some methods to fit a multiplicative tariff structure to observed risk data. ASTIN Bull 16:63–68

Bailey RA (1963) Insurance rates with minimum bias. Proc Casualty Actuar Soc 93:4–11

Charpentier A (2014) Computational actuarial science with R. The R Series. Chapman & Hall/CRC

De Jong P, Heller GZ (2008) Generalized linear models for insurance data. Cambridge University Press

Denuit M, Marechal X, Pitrebois S, Walhin J-F (2007) Actuarial modelling of claim counts: risk classification, credibility and bonus-malus systems. Wiley

Dunn PK, Smyth GK (2019) Generalized linear models with examples in R. Springer, New York

Efron B (1986) How biased is the apparent error rate of a prediction rule? J Am Stat Assoc 81:461–470

Fahrmeir L, Kneib T, Lang S, Marx B (2013) Regression models, methods and applications. Springer

Faraway JJ (2005a) Linear models with R. Chapman & Hall/CRC

Faraway JJ (2005b) Extending the linear model with R. Chapman & Hall/CRC

Fox J (2016a) Applied regression analysis and generalized linear models. Sage Publications

Fox J (2016b) A R companion to applied regression analysis and generalized linear models. Sage Publications

Frees EW, Derrig RA, Meyers G (2014) Predictive modeling applications in actuarial science. Volume I: predictive modeling techniques. International Series on Actuarial Science. Cambridge University Press

Goldburg M, Khare A, Tevet D (2016) Generalized linear models for insurance pricing. CAS Monograph Series, Number 5. http://www.casact.org/

Hainaut D, Trufin J, Denuit M (2019) Effective statistical learning methods for actuaries—neural networks and unsupervised methods. Springer Actuarial Series

Horowitz JL (2009) Semiparametric and nonparametric methods in econometrics. Springer, New York

Jung J (1968) On automobile insurance ratemaking. ASTIN Bull 5:41–48

Kaas R, Goovaerts MJ, Dhaene J, Denuit M (2008) Modern actuarial risk theory using R. Springer, Berlin

Mc Cullagh P, Nelder JA (1989) Generalized linear models. Chapman & Hall, New York

Mildenhall SJ (1999) A systematic relationship between minimum bias and generalized linear models. Proc Casualty Actuar Soc 86:393–487

Nelder JA, Wedderburn RWM (1972) Generalized linear models. J R Stat Soc Ser A 135:370–384

Ohlsson E, Johansson B (2010) Non-life insurance pricing with generalized linear models. Springer, Berlin

Trufin J, Denuit M, Hainaut D (2019) Effective statistical learning methods for actuaries—tree-based methods. Springer Actuarial Series

Wuthrich MV, Buser C (2017) Data analytics for non-life insurance pricing. Swiss Finance Institute Research Paper No. 16–68. SSRN: https://ssrn.com/abstract=2870308

Yan J, Guszcza J, Flynn M, Wu CSP (2009) Applications of the offset in property-casualty predictive modeling. In: Casualty actuarial society e-forum, winter 2009, pp 366–385

Chapter 5
Over-Dispersion, Credibility Adjustments, Mixed Models, and Regularization

5.1 Introduction

We know from Chap. 4 that GLMs unify regression methods for a variety of discrete and continuous outcomes typically encountered in insurance studies. The basis of GLMs is the assumption that the data are sampled from an ED distribution and that the score is a linear combination of observed features linked to the mean response by a monotonic transformation.

However, GLMs assign full credibility to the observations in the sense that the score (or, equivalently the mean response) is assumed to be known exactly, up to estimation error. In practice, we know from Chap. 1 that there is a lot of information that the insurer cannot access. This means that a part of the score remains generally unknown to the actuary and this typically causes

(i) overdispersion in cross-sectional data (that is, data observed during one period) and
(ii) serial correlation in longitudinal, or panel data (that is, when individuals are followed during a few periods of time and the response is observed at multiple instances).

Formally, let us come back to the decomposition of the information into available features X_i and hidden ones X_i^+ proposed in Chap. 1. Given $X_i = x_i$ and $X_i^+ = x_i^+$, the true score should be

$$\text{true score}_i = \boldsymbol{\beta}^\top x_i + \boldsymbol{\gamma}^\top x_i^+,$$

with some components of $\boldsymbol{\beta}$ possibly equal to zero (as some observable features may become irrelevant once the hidden information becomes available). This means that GLMs only include the score $\boldsymbol{\beta}^\top x_i$ based on available risk factors $X_i = x_i$ whereas there is a missing part $\boldsymbol{\gamma}^\top x_i^+$. To account for the missing part of the score, we explicitly acknowledge that there is an error when using the score $\boldsymbol{\beta}^\top x_i$ and the regression model is based on the working score defined as

© Springer Nature Switzerland AG 2019   197
M. Denuit et al., *Effective Statistical Learning Methods for Actuaries I*,
Springer Actuarial, https://doi.org/10.1007/978-3-030-25820-7_5

$$\text{working score}_i = \boldsymbol{\beta}^\top \boldsymbol{x}_i + \mathcal{E}_i,$$

where the error term $\mathcal{E}_i$ accounts for the missing part $\boldsymbol{\gamma}^\top \boldsymbol{X}_i^+$. However, and this is important for the proper interpretation of the results, we consider that $\mathcal{E}_i$ is only a residual effect, so that the estimated value of $\boldsymbol{\beta}$ using the working score may differ from its estimated value in the true score, because of omitted variable bias (as explained in Chap. 4). Thus, GLMMs split the score into two parts:

(i) the first one $\boldsymbol{x}_i^\top \boldsymbol{\beta}$ accounts for observable features, as in Chap. 4. Here, $\boldsymbol{x}_i^\top \boldsymbol{\beta}$ represents the fixed effects in the score.

(ii) while the second one, called the random effect $\mathcal{E}_i$ represents the missing component due to partial information.

This results in a mixed effect model, including both fixed and random effects. Thus, two types of components are involved in GLMMs: fixed and random effects. A fixed effect is an unknown constant that we want to estimate from the available data. This is the case for the vector $\boldsymbol{\beta}$ of regression coefficients in the GLMs. In contrast, a random effect is a random variable. Therefore, it does not make sense to estimate a random effect. Instead, we aim to estimate the parameters governing the distribution of the random effect, and to revise its distribution once the responses have been observed (a posteriori corrections yielding the predictive distribution).

Random effects can also capture dependencies over time, space or across different groups created inside the portfolio (sharing similar characteristics). For instance, if the response is measured repeatedly on the same policyholder every year then the random term $\mathcal{E}_i$ added to the linear score accounts for serial dependence among observations related to the same contract, followed over time. The same construction can also be used to model spatial dependence in catastrophe insurance or to deal with multi-level factors with some sparsely populated categories. Examples of categorical features with many levels include car model classification in motor insurance or sector of activity[1] in workers' compensation insurance, with many data for some levels and limited exposure for others. Including such a feature directly in a GLM is generally not possible, so that they are effectively treated with the help of random effects in this chapter. This gives rise to GLMMs, opening the door to dependence modeling and credibility corrections based on past claims history.

It is worth noticing that the models presented in this chapter offer powerful tools to implement a wide range of credibility models. For instance, the score may be obtained from other techniques, such as tree-based ones (Trufin et al. 2019) or neural networks (Hainaut et al. 2019). The mixed model then superposes random effects to these scores, which are treated as known constants and included in the offset. The predictive distribution of the random effects then captures the part of claim experience that cannot be explained by the available features and accounts for this effect when computing future premiums.

---

[1]NACE codes are often used in workers' compensation insurance. NACE (Nomenclature of Economic Activities) is the European statistical classification of economic activities grouping organizations according to their business activities. The NACE code is subdivided in a hierarchical, four-level structure.

As an application of the mixed models presented in this chapter, we consider usage-based insurance pricing. Pay-How-You-Drive (PHYD) or Usage-Based (UB) systems for automobile insurance provide actuaries with behavioral risk factors, such as the time of the day, average speeds and other driving habits. These data are collected while the contract is in force with the help of telematic devices installed in the vehicle. They thus fall in the category of a posteriori information that becomes available after contract initiation. For this reason, they must be included in the actuarial pricing by means of credibility updating mechanisms instead of being incorporated in the score as ordinary a priori observable features. In this chapter, we show that multivariate mixed models can be used to describe the joint dynamics of telematics data and claim frequencies. Future premiums, incorporating past experience can then be determined using the predictive distribution of claim characteristics given past history. This approach allows the actuary to deal with a variety of situations encountered in insurance practice, ranging from new drivers without telematics record to contracts with different seniority and drivers using their vehicle to different extent, generating varied volumes of telematics data.

5.2 Mixed Distributions

5.2.1 Heterogeneity and Overdispersion

Hidden information, as well as serial dependence tend to inflate the variance of claim counts beyond that assumed by the selected ED distribution. Let us consider the following simple example explaining why this is indeed the case. Consider the number of claims Y reported by a policyholder over a given period of time. The Poisson model imposes the strong equidispersion constraint:

$$E[Y] = \text{Var}[Y] = \mu.$$

Equidispersion is often violated in practice because of residual heterogeneity, suggesting that the Poisson assumption is not appropriate and favoring a mixed Poisson model.

To figure out why overdispersion is related to heterogeneity, let us consider for instance two risks classes C_1 and C_2 without overdispersion, i.e. such that the respective means μ_1 and μ_2, and the variances σ_1^2 and σ_2^2 in these two risk classes satisfy

$$\sigma_1^2 = \mu_1 \text{ and } \sigma_2^2 = \mu_2.$$

Now, assume that the actuary is not able to discriminate between those policyholders belonging to C_1 and to C_2. Let w_1 and w_2 denote the respective weights of C_1 and C_2 (e.g., the class exposures divided by total exposure). In the class $C_1 \cup C_2$, the expected claim number equals

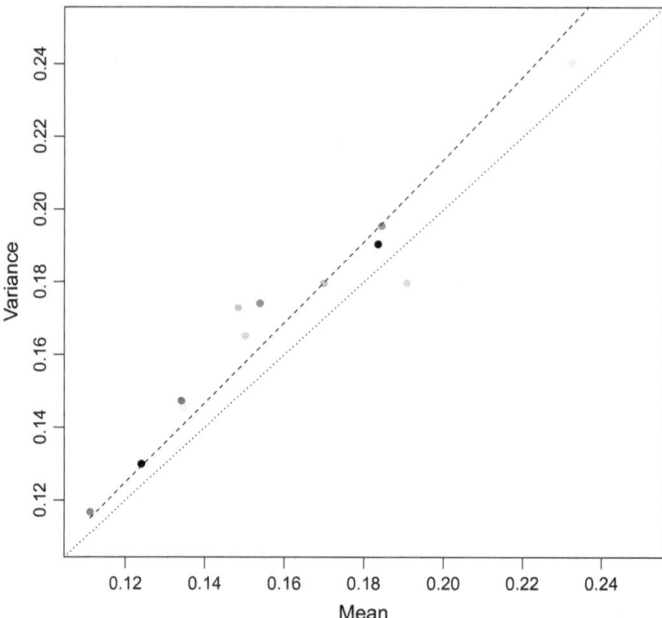

Fig. 5.1 Variance versus mean number of claims, with colors corresponding to the total exposure of the risk classes (from light gray to black)

$$\overline{\mu} = w_1 \mu_1 + w_2 \mu_2.$$

The variance in $C_1 \cup C_2$ is the sum of weighted intra-class and inter-classes variances. Precisely, the variance in $C_1 \cup C_2$ is equal to

$$\underbrace{w_1 \sigma_1^2 + w_2 \sigma_2^2}_{=\overline{\mu}} + w_1 \underbrace{(\mu_1 - \overline{\mu})^2}_{>0} + w_2 \underbrace{(\mu_2 - \overline{\mu})^2}_{>0} > \overline{\mu}.$$

Hence, omitting relevant ratemaking variables inevitably induces overdispersion.

Despite its prevalence as a starting point in the analysis of count data, the Poisson specification is often inappropriate because of unobserved heterogeneity. A convenient way to take this phenomenon into account is to introduce a random effect in this model. This is in line with the classical credibility construction in actuarial science. Figure 5.1 displays the weighted means and variances for the motor insurance data set studied in Sect. 4.10. We clearly see that the mean-variance pairs for the classes with high exposures lie above the diagonal, suggesting that the mean indeed exceeds the variance in the portfolio under consideration.

Properly accounting for overdispersion yields larger standard errors and p-values compared to ordinary Poisson regression. This means that Poisson regression is generally over-optimistic, granting too much importance to the available features.

In practice, this means that features considered as being irrelevant by the Poisson regression can be safely ignored from the subsequent analysis. However, some features selected by Poisson regression may be later discarded when more appropriate inferential tools are applied.

5.2.2 From Homogeneous ED to ED Mixtures

In this section, we enrich the family of distributions that we can use to account for the heterogeneity present in the data, beyond those reviewed in Chap. 2. Consider a group of policyholders, each with a response obeying some given ED distribution, with the same dispersion parameter ϕ but with different canonical parameters θ. The basic idea is to consider a policyholder randomly selected from this group. The canonical parameter for this randomly drawn individual is a random variable Θ, with a distribution reflecting the composition, or the structure of the group under consideration.

This means that the response Y for this policyholder taken at random from the group is ED only if we know the value of Θ. Formally, given $\Theta = \theta$, the response Y obeys the ED distribution with canonical parameter θ and dispersion parameter ϕ, with probability mass or probability density function given in (2.3). Denote as f_Θ the probability density function of Θ. Unconditionally, that is, without knowledge about Θ, the probability density (or probability mass) function of Y writes

$$f_Y(y) = \int \exp\left(\frac{y\theta - a(\theta)}{\phi/\nu}\right) c(y, \phi/\nu) f_\Theta(\theta) d\theta,$$

where the integral runs over the support of Θ, assumed to be a subset of, or to coincide with, the set of admissible values for the canonical parameter of the ED distribution under consideration. Such a response Y obeys a mixture model because it is obtained by mixing several ED distributions, each with its own canonical parameter θ, and to take a random draw out of it. The distribution of Y is then called a mixed distribution, with mixing distribution f_Θ, and mixing parameter, or random effect Θ.

Here, we have assumed that Θ was a continuous random variable. If Θ is a discrete random variable then this construction corresponds to a discrete mixture model for which the same reasoning holds, but replacing integrals with sums. Because the groups under consideration in insurance applications are generally quite large, the discrete mixture models are approximated by continuous ones.

All the calculations can be conducted by conditioning on Θ, and then by applying the results available about the ED distributions. For instance, the mean response is given by

$$E[Y] = E\big[E[Y|\Theta]\big] = E[a'(\Theta)].$$

Similarly, the variance of the response is obtained from

$$\mathrm{Var}[Y] = \mathrm{E}\big[\mathrm{Var}[Y|\Theta]\big] + \mathrm{Var}\big[\mathrm{E}[Y|\Theta]\big]$$
$$= \frac{\phi}{\nu}\mathrm{E}\big[a''(\Theta)\big] + \mathrm{Var}\big[a'(\Theta)\big].$$

In the next section, we study the Poisson mixtures, by far the most commonly used in actuarial studies. The other ED mixtures can be analyzed similarly.

5.2.3 Mixed Poisson Distribution

5.2.3.1 Definition

In words, a mixed Poisson distribution arises when there are different groups of policyholders, each group being characterized by a specific Poisson mean. If all the members of the group share the same claim rate then the Poisson distribution is eligible to model the number of claims. However, if individuals have different levels of claim proneness then the accident rate is heterogeneous across individuals and a mixed Poisson distribution replaces the homogeneous Poisson one when considering the number of claims reported by an individual taken at random from the group.

Let us consider a policyholder randomly drawn from an heterogeneous group. Given $\Theta = \theta$, assume that the number of claims Y reported to the insurer is Poisson distributed with canonical parameter θ. The interpretation we give to this model is that not all policyholders in the portfolio have an identical expected claim frequency $\mu = \exp(\theta)$. Some of them have a higher expected frequency, others have a lower expected frequency. Thus we use a random effect to model this empirical observation, replacing θ with a random variable Θ representing the result of the random draw.

The annual number of claims reported by a policyholder selected at random from the portfolio obeys a mixed Poisson law. In this case, the probability that such a randomly selected policyholder reports y claims to the company is obtained by averaging the conditional Poisson probabilities with respect to Θ. The probability mass function associated with mixed Poisson models is defined as

$$P[Y = y] = \mathrm{E}\left[\exp(-\exp(\Theta))\frac{(\exp(\Theta))^y}{y!}\right]$$
$$= \int_{-\infty}^{\infty} \exp(-\exp(\theta))\frac{(\exp(\theta))^y}{y!}f_\Theta(\theta)\mathrm{d}\theta, \quad y = 0, 1, \dots \quad (5.1)$$

where f_Θ denotes the probability density function of Θ. The mixing distribution described by f_Θ represents the heterogeneity of the portfolio of interest; f_Θ is generally called the structure function in the actuarial literature because it describes the structure of the portfolio with respect to the risk under consideration (here, the number of claims).

The mean response is then given by

$$E[Y] = E\Big[E[Y|\Theta]\Big] = E[\exp(\Theta)]. \tag{5.2}$$

In (5.2), $E[\cdot|\Theta]$ means that we take an expected value considering Θ as a constant. We then average with respect to all the random components, except Θ. Consequently, $E[\cdot|\Theta]$ is a function of Θ. Given Θ, Y is Poisson distributed with mean $\exp(\Theta)$ so that $E[Y|\Theta] = \exp(\Theta)$. The mean of Y is finally obtained by averaging $E[Y|\Theta]$ with respect to Θ.

Every mixed Poisson model (5.1) induces overdispersion because

$$\begin{aligned}
\mathrm{Var}[Y] &= E\Big[\mathrm{Var}[Y|\Theta]\Big] + \mathrm{Var}\Big[E[Y|\Theta]\Big] \\
&= E[\exp(\Theta)] + \mathrm{Var}[\exp(\Theta)] \\
&= E[Y] + \mathrm{Var}[\exp(\Theta)] \\
&> E[Y].
\end{aligned}$$

This theoretical calculation extends to the general case, the particular example with the two risk classes C_1 and C_2 that are merged because of missing information described in the introductory example to this section. Here, Θ replaces the labels 1 and 2 to the risk classes that are mixed together.

5.2.3.2 Interpretation

It is important to figure out the proper interpretation of mixture models. The discrete, or finite mixture model corresponds to (5.1) when Θ has finitely many support points. This corresponds to the random selection from the portfolio under interest, as shown next. Consider independent $Y_i \sim \mathcal{P}oi(\lambda_i)$, $i = 1, 2, \ldots, n$. Then

$$Y_\bullet = \sum_{i=1}^{n} Y_i \sim \mathcal{P}oi(n\bar\lambda) \text{ where } \bar\lambda = \frac{1}{n}\sum_{i=1}^{n}\lambda_i.$$

So, we still have equidispersion as

$$E[Y_\bullet] = \mathrm{Var}[Y_\bullet] = n\bar\lambda.$$

Defining independent random variables Λ_i, $i = 1, 2, \ldots, n$, as

$$\Lambda_i = \begin{cases} \lambda_1 \text{ with probability } \frac{1}{n} \\[1em] \lambda_2 \text{ with probability } \frac{1}{n} \\[1em] \vdots \\[1em] \lambda_n \text{ with probability } \frac{1}{n} \end{cases}$$

and considering Poisson mixtures $\widetilde{Y}_i$, $i = 1, 2, \ldots, n$, such that, given Λ_i, $\widetilde{Y}_i$ is Poisson distributed with mean Λ_i, we have

$$E[\widetilde{Y}_i] = E[\Lambda_i] = \overline{\lambda}$$

whereas

$$\text{Var}[\widetilde{Y}_i] = E[\Lambda_i] + \text{Var}[\Lambda_i] = \overline{\lambda} + \sigma_\lambda^2 > E[\widetilde{Y}_i]$$

where

$$\sigma_\lambda^2 = \frac{1}{n} \sum_{i=1}^n \lambda_i^2 - (\overline{\lambda})^2.$$

However, if $Y_\bullet$ represents the total number of claims in the portfolio,

$$\widetilde{Y}_\bullet = \sum_{i=1}^n \widetilde{Y}_i$$

cannot be interpreted in this way. Rather, $\widetilde{Y}_i$ is the number of claims reported by a policyholder taken at random from the portfolio so that summing the $\widetilde{Y}_i$ makes no concrete sense at the portfolio level.

5.2.3.3 Negative Binomial NB1 Distribution

The NB1 model corresponds to a variance of the form

$$\text{Var}[Y] = \mu + \alpha\mu. \tag{5.3}$$

The NB1 distribution is therefore closely related to the so-called Overdispersed Poisson (or ODP) model. This is because the corresponding variance function is $V(\mu) = (1 + \alpha)\mu$, obtained with dispersion parameter $\phi = 1 + \alpha > 1$.

The NB1 model is obtained by letting the Poisson mean become a random variable obeying some Gamma distribution. Precisely, let us replace $\exp(\Theta)$ with Λ for convenience. Now, assume that given $\Lambda = \lambda$, the response Y is $\mathcal{P}oi(\lambda)$ distributed and that the random variable Λ is such that

$$\Lambda \sim \mathcal{G}am\left(\frac{\mu}{\alpha}, \frac{1}{\alpha}\right),$$

with mean $E[\Lambda] = \mu$ and variance $Var[\Lambda] = \alpha\mu$. Then,

$$E[Y] = E[\Lambda] = \mu$$

and

$$\begin{aligned}
Var[Y] &= E[\Lambda] + Var[\Lambda] \\
&= (1 + \alpha)\mu \\
&> \mu = E[Y]
\end{aligned}$$

so that we indeed recover the variance (5.3).

The resulting mixed Poisson distribution is then given by

$$\begin{aligned}
P[Y = y] &= E\left[\exp(-\Lambda)\frac{(\Lambda)^y}{y!}\right] \\
&= \int_0^\infty \exp(-\lambda)\frac{\lambda^y}{y!}\exp(-\lambda/\alpha)\frac{\lambda^{\frac{\mu}{\alpha}-1}\left(\frac{1}{\alpha}\right)^{\mu/\alpha}}{\Gamma(\mu/\alpha)}d\lambda \\
&= \frac{\left(\frac{1}{\alpha}\right)^{\mu/\alpha}}{y!\Gamma(\mu/\alpha)}\int_0^\infty \exp\left(-\lambda\left(1+\frac{1}{\alpha}\right)\right)\lambda^{y+\frac{\mu}{\alpha}-1}d\lambda \\
&= \frac{\left(\frac{1}{\alpha}\right)^{\mu/\alpha}\Gamma(y+\mu/\alpha)}{y!\Gamma(\mu/\alpha)}\left(1+\frac{1}{\alpha}\right)^{-y-\frac{\mu}{\alpha}} \\
&= \frac{\Gamma(y+\alpha^{-1}\mu)}{y!\Gamma(\alpha^{-1}\mu)}(1+\alpha)^{-\alpha^{-1}\mu}\left(1+\alpha^{-1}\right)^{-y}.
\end{aligned}$$

If $\alpha = 0$ then we obtain the Poisson distribution as a limiting case.

It is worth to mention that the NB1 distribution does not belong to the ED family so that applying the GLM machinery (IRLS algorithm) to fit the NB1 regression model corresponds to a quasi-likelihood approach. Notice that the quasi-Poisson GLM specifies a mean-variance relationship of the form $V(\mu) = \phi\mu$ for some positive dispersion parameter ϕ, that may be larger or smaller than 1. If $\phi > 1$ then this corresponds to overdispersion but we may also have $\phi < 1$ so that the model corresponds to underdispersion, meaning that the variance of the response is smaller compared to a Poisson distribution.

The quasi-Poisson GLM delivers the same point estimates $\widehat{\boldsymbol{\beta}}$ than the ordinary Poisson GLM. The price to pay for introducing a dispersion parameter in the quasi-Poisson GLM is that all the standard errors of the parameters are multiplied with the square root of ϕ. For instance, if $\phi = 4$ then all standard errors are doubled, and the parameters become less significant. This means that some features that seem to significantly contribute to the score under a Poisson GLM may become irrelevant with the quasi-Poisson GLM when ϕ is large. Running a regression analysis based on the Negative Binomial distribution leads to different results when maximum-likelihood is used.

5.2.3.4   Negative Binomial NB2 Distribution

In the NB2 approach, the Poisson parameter is decomposed into the product $\mu\Delta$ where Δ models relative fluctuations around the mean μ, with $E[\Delta] = 1$. Specifically, the NB2 distribution if obtained by assuming that $\Delta \sim Gam(\alpha, \alpha)$ so that $E[\Delta] = 1$ and $Var[\Delta] = 1/\alpha$. The NB2 model has a variance of the form

$$Var[Y] = \mu + \alpha\mu^2. \tag{5.4}$$

Compared to the variance (5.3) of the NB1 model, we see that there is now a quadratic relationship between the mean and the variance (hence the notation NB2).

The probability mass function of Y is then given by

$$
\begin{aligned}
p_Y(y) &= \int_0^{+\infty} \exp(-\mu\delta) \frac{(\mu\delta)^y}{y!} \frac{1}{\Gamma(\alpha)} \alpha^\alpha \delta^{\alpha-1} \exp(-\alpha\delta) d\delta \\
&= \frac{\mu^y}{y!} \frac{\alpha^\alpha}{\Gamma(\alpha)} \int_0^{+\infty} \delta^{\alpha+y-1} \exp(-\delta(\mu+\alpha)) d\delta \\
&= \frac{\mu^y}{y!} \frac{\alpha^\alpha}{\Gamma(\alpha)} \frac{\Gamma(\alpha+y)}{(\mu+\alpha)^{\alpha+y}}
\end{aligned}
$$

so that we finally get

$$
p_Y(y) = \binom{\alpha+y-1}{y} \left(\frac{\mu}{\alpha+\mu}\right)^y \left(\frac{\alpha}{\alpha+\mu}\right)^\alpha, \quad y = 0, 1, \ldots,
$$

where

$$
\binom{\alpha+y-1}{y} = \frac{\Gamma(\alpha+y)}{y!\Gamma(\alpha)}
$$

extends the Binomial coefficient to fractional arguments (this is a particular case of the Beta function).

When α is a positive integer, we recover the Pascal distribution with

$$
q = \frac{\alpha}{\alpha+\mu}.
$$

With the Gamma parameter α fixed to a known value (as in the particular Pascal case), the Negative Binomial distribution belongs to the ED family. This is easily seen from

$$p_Y(y) = \exp\left(y \ln \frac{\mu}{\alpha + \mu} - \alpha \ln \frac{\alpha}{\alpha + \mu}\right) \binom{\alpha + y - 1}{y}$$

where we recognize the probability density function (2.3) with

$$\theta = \ln \frac{\mu}{\alpha + \mu}$$

$$a(\theta) = \alpha \ln \frac{\alpha}{\alpha + \mu} = \alpha \ln(1 - \theta)$$

$$\phi = 1$$

$$c(y, \phi) = \binom{\alpha + y - 1}{y}.$$

Remark 5.2.1 The difference between NB1 and NB2 distributions thus comes from the way individual risks fluctuate around the mean μ. With NB1, $Y \sim \mathcal{P}oi(\Lambda)$ with $\Lambda \sim \mathcal{G}am\left(\frac{\mu}{\alpha}, \frac{1}{\alpha}\right)$ so that fluctuations around the mean response μ are proportional to the mean μ. With NB2, $Y \sim \mathcal{P}oi(\mu\Delta)$ with $\Delta \sim \mathcal{G}am(a, a)$ so that fluctuations are proportional to the square of the mean μ. Working with individual claim numbers, the expected response is usually small, less than 1 in general. Thus, NB1 induces a stronger heterogeneity compared to NB2. For expected counts larger than 1, the opposite holds.

5.2.3.5 Poisson-Inverse Gaussian and Poisson-LogNormal Distributions

Of course, any distribution with support in $(0, \infty)$ can be used instead of the Gamma distribution involved in the NB1 and NB2 distributions. Two common alternative choices to the Gamma for the mixing distribution are the Inverse-Gaussian distribution and the LogNormal distribution.

The Poisson-Inverse Gaussian distribution (also referred to as the Sichel distribution) has greater skewness than the Negative Binomial NB2 distribution, and so may be more suited to modelling heavy-tailed claim frequency distributions. The probability mass function of the Poisson-Inverse Gaussian distribution can be evaluated numerically with the help of modified Bessel functions of the second kind.

Besides the Poisson-Inverse Gaussian distribution, the Poisson LogNormal distribution is often used in statistics. Numerical procedures can be used to calculate the Poisson-LogNormal probability mass function, such as Gaussian quadratures. Routines are now available in standard statistical packages, including R. The Poisson-Log Normal model is interesting because it possesses a natural interpretation. If there are many hidden variables, and if these variables are independent and act multiplicatively on the Poisson mean, then central limit theorems can be invoked in order to establish the LogNormality of the Poisson mean, that is, the Normality of Θ. We will come back to the Poisson LogNormal model in the numerical illustrations proposed in this chapter.

5.2.3.6   Excess Zero Count Models

Often, the number of observed zeros in insurance data sets is much larger than under the Poisson assumption. In motor insurance, for instance, this can be explained by the reluctance of some insured drivers to report their accident. Because of bonus-malus mechanisms or experience rating plans, some claims are not filed to the company because policyholders think it is cheaper for them to defray the third party (or to pay for their own costs in first party coverages) to avoid premium increases. Deductibles also inflate the proportion of zeros, since small claims are not reported by insured drivers.

The Negative Binomial distribution indirectly accounts for this phenomenon as it inflates the probability mass at zero compared to the Poisson law with the same mean. Another, more direct approach consists in using a mixture of two distributions. The zero-inflated Poisson (ZIP) distribution is obtained in this way by combining a degenerated distribution for the zero case with a Poisson distribution. The ZIP probability mass function is then given by

$$
p_Y(y) = \begin{cases} \pi + (1 - \pi)\exp(-\lambda) & \text{for } y = 0 \\[2mm] (1 - \pi)\exp(-\lambda)\frac{\lambda^y}{y!} & \text{for } y = 1, 2, \ldots \end{cases} \tag{5.5}
$$

where λ is the Poisson parameter and π is the additional probability mass at zero.

Zero-inflated models allow for two different sources of zero responses. One source corresponds to individuals who never produce claims (representing a proportion π of the population), the other one to individuals who may report claims but resulted in a zero Poisson response. Even if it produces a better fit to the available data, the underlying data generating process is somewhat questionable in insurance applications as the insurance policy has no value for the proportion π of the population never reporting any claim. Also, when the actuary uses serial data, the mixture collapses to a single Poisson distribution as soon as the policyholder reported at least one claim (making impossible that he or she belongs to the sub-population which never experiences the covered peril). Notice that the ZIP response Y can also be represented as the product JZ where $J \sim \mathcal{B}er(1 - \pi)$ and $Z \sim \mathcal{P}oi(\lambda)$, both random variables being independent.

The two first moments of the ZIP distribution are

$$
\begin{aligned}
\mathrm{E}[Y] &= (1 - \pi)\lambda \\
\mathrm{Var}[Y] &= \mathrm{E}[Y] + \mathrm{E}[Y](\lambda - \mathrm{E}[Y]) \\
&= (1 - \pi)\lambda + \pi(1 - \pi)\lambda^2.
\end{aligned}
$$

We see that both π and λ enter the expected claim number, which differs from the Poisson and Negative Binomial cases examined before (where the mean was one of the parameters). This often makes the interpretation of the results more cumbersome. The variance clearly exceeds the mean so that the ZIP model accounts for the overdispersion generally present in insurance data. Note that the ZIP model can also be seen as a special case of a mixed Poisson distribution obtained when the factor Δ multiplying the Poisson mean is equal either to 0 or to 1 (with respective probabilities π and $1 - \pi$, that is $\Delta \sim \mathcal{B}er(1 - \pi)$) and conditional mean $\lambda \Delta$.

Remark 5.2.2 Notice that ZIP models differ from the hurdle approach, or zero-altered distributions, where one component corresponds to zero-counts and a separate process accounts for the positive counts, once the hurdle at $y = 0$ has been passed. Formally, this leads to a probability mass function of the form

$$
p_Y(y) = \begin{cases} \pi & \text{for } y = 0 \\[2mm] \frac{1-\pi}{1-\exp(-\lambda)} \exp(-\lambda)\frac{\lambda^y}{y!} & \text{for } y = 1, 2, \dots \end{cases} \tag{5.6}
$$

5.3 Structure of GLMMs

GLMMs retain most of the structure of GLMs except that the score now involves a random part. Specifically, given the score, the response Y_i obeys an ED distribution (2.3) with canonical parameter θ_i. The score is now equal to $x_i^\top \beta + \mathcal{E}_i$, where $\mathcal{E}_i$ represents the residual effect of hidden features on the mean response. The random effect $\mathcal{E}_i$ accounts for the missing information, i.e. for the heterogeneity that is present in the portfolio beyond that captured by the available features x_i. In a motor insurance portfolio, for instance, driving abilities or swiftness of reflexes vary from individual to individual beyond what can be explained based on the features x_i. Therefore it is natural to multiply the mean frequency μ_i of the Poisson distribution by a positive random effect $\Delta_i = \exp(\mathcal{E}_i)$. The frequency will vary within the portfolio according to the non-observable random variable Δ_i.

To properly understand which effects are captured by $\mathcal{E}_i$, it is important to remember from Chap. 1 that the available information X_i appears to be correlated to X_i^+ in that the actuary takes advantage of the dependence between some components of X_i^+ with X_i to indirectly account for some risk characteristics in the price list, as shown in the next example.

Example 5.3.1 Consider for instance the use of the vehicle, private or business. This feature is generally comprised in the observables ones X_i. Annual mileage is an important information but harder to measure, so that many insurers do not attempt to record it and distance traveled remains in X_i^+. In such a case, the random effect $\mathcal{E}_i$ does not account for the effect of annual mileage on the response Y_i, but only for the part of annual mileage that is not already explained by the use of the vehicle (cars driven for business tending to travel longer distances).

This means that the estimated regression coefficients are subject to the omitted variable bias. It is crucial to keep this in mind to properly interpret the results from such an analysis, to avoid over-interpreting the estimated parameters: in words, the estimated β_j quantifies not only the impact of the jth feature x_{ij} on the mean response but also of all hidden features x_{ik}^+ correlated to the jth observable one.

It is important to realize here that $g^{-1}(x_i^\top \beta)$ is not the expected response in a GLMM. This is because

$$E[Y_i] = E\Big[E[Y_i|\mathcal{E}_i]\Big] = E\Big[g^{-1}(x_i^\top \beta + \mathcal{E}_i)\Big].$$

Often, $\mathcal{E}_i$ is assumed to be Normally distributed. This can be seen as a consequence of the central-limit theorem because the random effect replaces a sum of missing terms $\gamma_j X_{ij}^+$ and is therefore approximately Normally distributed. If $\mathcal{E}_i \sim Nor(0, \sigma^2)$ and a log-link is specified then

$$E[Y_i] = E\Big[E[Y_i|\mathcal{E}_i]\Big] = \exp\left(x_i^\top \beta + \frac{\sigma^2}{2}\right).$$

In some applications, random effects $\mathcal{E}_i$ are structured in a hierarchical way. When the levels of one factor only vary within the levels of another factor, that factor is said to be nested. This is typically the case when fleets of vehicles are considered, or in workers' compensation insurance, for instance. The random effect $\mathcal{E}_i$ for vehicle i belonging to fleet f is then decomposed into

$$\mathcal{E}_i = \mathcal{E}_f + \mathcal{E}_{fi}$$

where $\mathcal{E}_f$ accounts for unobservable characteristics at the fleet level whereas $\mathcal{E}_{fi}$ accounts for hidden features specific to vehicle i in fleet f. In workers' compensation insurance, NACE codes exhibit a hierarchical structure so that random effects can be structured in a similar way.

Remark 5.3.2 In the statistical literature, GLMMs correspond in reality to a wider class of model, with

$$g(\mu_i) = x_i^\top \beta + z_i^\top B$$

where the random vector B is multivariate Normal with zero mean and variance-covariance matrix Σ_B. This means that the regression coefficients corresponding to the features z_i are no more constant but become random.

Formally, let

$$p_{\text{fixed}} = \# \text{ fixed effects associated with features } x_{ij}$$
$$p_{\text{rand}} = \# \text{ random effects associated with features } z_{ik}.$$

The GLMM score then mixes fixed and random effects

$$\text{score}_i = \beta_0 + \overset{p_{\text{fixed}}}{\underset{j=1}{\sum}} \beta_j x_{ij} + \overset{p_{\text{rand}}}{\underset{k=1}{\sum}} B_k z_{ik} \text{ with } \boldsymbol{B} \sim Nor(\boldsymbol{0}, \boldsymbol{\Sigma}_B).$$

To recover the GLMM with overdispersion, it suffices to consider

$$p_{\text{rand}} = n$$
$$z_{ik} = I[k = i]$$
$$B_i = \mathcal{E}_i$$
$$\boldsymbol{\Sigma}_B = \sigma_{\mathcal{E}}^2 \boldsymbol{I}$$

where n is the number of policies comprised in the portfolio, each of them having its specific random effect $\mathcal{E}_i$.

5.4 Inference in GLMMs for Cross-Sectional Data

5.4.1 Maximum-Likelihood Estimation

Assume that, given $\mathcal{E}_i$, the response Y_i obeys the probability density function or probability mass function

$$\exp\left(\frac{y\theta_i - a(\theta_i)}{\phi/\nu_i}\right) c(y, \phi/\nu_i)$$

with $a'(\theta_i) = \mu_i$ and

$$g(\mu_i) = \text{score}_i = \beta_0 + \sum_{j=1}^{p} \beta_j x_{ij} + \mathcal{E}_i \text{ with } \mathcal{E}_i \sim Nor(0, \sigma_{\mathcal{E}}^2).$$

The expected response μ_i is given by $\mu_i = g^{-1}(\text{score}_i)$ where the link function $g \in \{\ln, \text{logit}, \ldots\}$, is specified by the actuary.

The likelihood can be obtained by integrating out the random effect. Precisely, assuming $\mathcal{E}_i \sim Nor(0, \sigma^2)$, we get

$$\mathcal{L} = \prod_{i=1}^{n} \int_{-\infty}^{\infty} \left(\exp\left(\frac{y_i\theta_i(z) - a(\theta_i(z))}{\phi/\nu_i}\right) c(y_i, \phi/\nu_i)\right) \frac{1}{\sigma_{\mathcal{E}}\sqrt{2\pi}} \exp\left(-\frac{z^2}{2\sigma_{\mathcal{E}}^2}\right) dz$$

where

$$\theta_i(z) = g(\boldsymbol{\beta}^\top \boldsymbol{x}_i + z).$$

In case random effects are structured (in a hierarchical way, for instance), the integral goes over the random effects that are multivariate Normally distributed. Because of the presence of the integrals, maximum likelihood estimation requires numerical procedures. A closed-form expression for the GLMM likelihood is only available with so-called conjugate distributions. An example is the Gamma distribution for $\exp(\mathcal{E}_i)$ combined with a Poisson conditional ED distribution leading to the Negative Binomial distribution, as shown before.

In general, the maximization of the likelihood can be performed for instance

- by approximating the likelihood using the Laplace method. This leads to a quasi-likelihood augmented with a penalty term. This first approach is referred to as the penalized quasi-likelihood (or PQL) for which a version of the Fisher scoring algorithm is available.
- by approximating the integral using numerical integration techniques. Adaptive Gauss-Hermite quadrature formulas can be used to evaluate the integral involved in the likelihood (replacing it with a weighted sum).

Alternatives include the (Monte Carlo) EM algorithm, simulated maximum likelihood or Monte Carlo integration as well as Bayesian implementations of GLMMs.

5.4.2   Numerical Illustrations

Let us consider the motor insurance portfolio used in the introductory example of this chapter (that has been analyzed in Sect. 4.10 with the help of Poisson regression). We know that the data exhibit overdispersion, which rules out the Poisson specification and suggests to move to a mixed Poisson model.

5.4.2.1   Poisson-LogNormal Regression

The Poisson-LogNormal regression model can be fitted using the `glmer` function of the R package `lme4`. To obtain policy-specific random effects $\mathcal{E}_i \sim \textit{Nor}(0, \sigma_{\mathcal{E}}^2)$ with `glmer`, we specify a component `(1|NPOL)` in the score where "1" indicates that the random coefficient is not multiplied with any feature (and thus acts as the intercept, not as a random regression coefficient associated with a given feature) and `|NPOL` indicates that the random effect is nested in policy number, that is, is specific to each policy. The estimated fixed effects are displayed in Table 5.1. Compared to the results contained in Table 4.11, we see that the estimated intercept differs a lot from the Poisson GLM one. Since $\exp(\mathcal{E}_i)$ is LogNormally distributed, the comparison must in fact be based on the intercept adjusted as follows:

$$\widehat{\beta_0} + \frac{\widehat{\sigma}_{\mathcal{E}}^2}{2} = -2.57924 + \frac{1.987}{2} = -1.58774$$

Table 5.1 Output of the `glmer` function of R for the fit of the motor TPL data. Fixed effects

Coefficient	Estimate	Std. error	z value	Pr($>$ $|z|$)	
Intercept	-2.57924	0.06937	-37.182	$<2 \times 10^{-16}$	★★★
Female	-0.12015	0.05484	-2.191	0.0285	★
Age G1	0.58997	0.11560	5.104	3.33×10^{-7}	★★★
Age G2	0.32964	0.05920	5.568	2.57×10^{-8}	★★★
Age G5	-0.15037	0.07585	-1.982	0.0474	★
CitySize large	0.29807	0.05400	5.520	3.39×10^{-8}	★★★

which appears to be closer to its Poisson counterpart. The estimated regression coefficients $\hat{\beta}_j$ are in line with the results obtained from Poisson GLM (this can be explained by the consistency of the Poisson GLM estimates under quasi-likelihood). We see that all the standard errors are larger in Table 5.1, resulting in larger p-values and reduced significance levels. All features remain nevertheless significant at 5% confidence level. The obtained AIC is 13 276.4.

5.4.2.2 NB1 Regression by Quasi-Likelihood Over-Dispersed Poisson GLM

As pointed above, the variance function $V(\mu) = (1 + \alpha)\mu$ of the NB1 distribution corresponding to (5.3) coincides with the Overdispersed Poisson (or ODP) model with dispersion parameter $\phi = 1 + \alpha > 1$. Therefore, even if the NB1 distribution does not belong to the ED family we can still apply the GLM machinery (IRLS algorithm) to fit the NB1 regression model in a quasi-likelihood approach. As every NB1 model induces overdispersion, we must have a dispersion parameter ϕ larger than 1.

Considering the ODP fit summarized in Table 5.2, we see that the quasi-Poisson GLM indeed delivers the same point estimates $\hat{\beta}$ than the ordinary Poisson GLM (displayed in Table 4.11). In the quasi-Poisson GLM, all the standard errors of the

Table 5.2 Output of the `glm` function of R for the fit of the motor TPL data with the NB1 fit using quasi-likelihood

Coefficient	Estimate	Std. error	t value	Pr($>$ $|t|$)	
Intercept	-1.76608	0.05004	-35.295	$<2 \times 10^{-16}$	★★★
Female	-0.11779	0.05305	-2.220	0.0264	★
Age G1	0.54798	0.10685	5.129	2.95×10^{-7}	★★★
Age G2	0.31040	0.05688	5.457	4.92×10^{-8}	★★★
Age G5	-0.14506	0.07506	-1.933	0.0533	
CitySize large	0.29026	0.05143	5.644	1.69×10^{-8}	★★★

parameters are multiplied with the square root of ϕ. Here, the dispersion parameter for quasi-Poisson family is estimated to $\widehat{\phi} = 1.430627$ which is larger than 1, reducing all significance levels compared to ordinary Poisson regression. There is no AIC value as there is no likelihood to be maximized. The number of IRLS iterations is 6.

5.4.2.3   NB2 Regression by Maximum Likelihood

We know that when the Gamma parameter α fixed to a known value, the Negative Binomial distribution belongs to the ED family. A GLM based on that distribution can thus be fitted by IRLS. If, as it is typically the case in insurance studies, α is unknown and must therefore be estimated from the available data, standard methods for GLMs based on ED distributions do not apply. We can however obtain estimates of both the regression coefficients and α by the method of maximum likelihood. It is convenient to use the Poisson maximum-likelihood estimate $\widehat{\boldsymbol{\beta}}$ and a simple moment estimate $\widehat{\alpha}$ as starting values to solve the likelihood equations.

The NB2 likelihood equations possess an interesting interpretation, as shown next. For $\Delta_i \sim \mathcal{G}am(\alpha, \alpha)$, the likelihood writes

$$\mathcal{L}(\boldsymbol{\beta}) = \prod_{i=1}^{n} \frac{\mu_i^{y_i}}{y_i!} \left(\frac{\alpha}{\alpha + \mu_i} \right)^{\alpha} (\alpha + \mu_i)^{-y_i} \frac{\Gamma(\alpha + y_i)}{\Gamma(\alpha)}.$$

The maximum-likelihood estimators for $\boldsymbol{\beta}$ and α then solve

$$\frac{\partial}{\partial \boldsymbol{\beta}} L(\boldsymbol{\beta}) = \sum_{i=1}^{n} \boldsymbol{x}_i \left(y_i - \mu_i \frac{\alpha + y_i}{\alpha + \mu_i} \right) = \boldsymbol{0}.$$

Let us now give an intuitive interpretation for the Negative Binomial likelihood equations. Given $Y_i = y_i$, Δ_i has probability density function $\mathcal{G}am(\alpha + y_i, \alpha + \mu_i)$. This can be established as follows: given $Y_i = y_i$, the probability density function of Δ_i writes

$$\begin{aligned}
f_\Delta(\delta | y_i) &= \frac{P[Y_i = y_i | \Delta_i = \delta] f_\Delta(\delta)}{P[Y_i = y_i]} \\
&= \frac{\exp(-\mu_i \delta) \frac{(\mu_i \delta)^{y_i}}{y_i!} \frac{1}{\Gamma(\alpha)} \alpha^\alpha \delta^{\alpha-1} \exp(-\alpha \delta)}{\int_0^{+\infty} \exp(-\mu_i \xi) \frac{(\mu_i \xi)^{y_i}}{y_i!} \frac{1}{\Gamma(\alpha)} \alpha^\alpha \xi^{\alpha-1} \exp(-\alpha \xi) d\xi} \\
&= \frac{\exp(-\delta_i (\alpha + \mu_i)) \delta_i^{\alpha + y_i - 1}}{\int_0^\infty \exp(-\xi (\alpha + \mu_i)) \xi^{\alpha + y_i - 1} d\xi} \\
&= \exp(-\delta_i (\alpha + \mu_i)) \delta_i^{\alpha + y_i - 1} \frac{(\alpha + \mu_i)^{\alpha + y_i}}{\Gamma(\alpha + y_i)}
\end{aligned}$$

where we recognize the probability density function of the Gamma distribution with updated parameters $\alpha + y_i$ and $\alpha + \mu_i$. This shows that the a posteriori distribution of Δ_i given $Y_i = y_i$ is $\mathcal{G}am\,(\alpha + y_i, \alpha + \mu_i)$. Hence, we have

$$E[\Delta_i | Y_i = y_i] = \frac{\alpha + y_i}{\alpha + \mu_i}.$$

The likelihood equations in the Negative Binomial case are thus similar to the ones obtained in the Poisson case except that μ_i is now replaced with its a posteriori expectation $\mu_i \frac{\alpha + y_i}{\alpha + \mu_i}$, that is, the Negative Binomial maximum-likelihood estimators solve

$$\sum_{i=1}^{n} x_i \left(y_i - \mu_i E[\Delta_i | Y_i = y_i] \right) = \mathbf{0}.$$

Hence, y_i is now compared to its prediction $\mu_i E[\Delta_i | Y_i = y_i]$ based on the information contained in the observed number y_i of claims reported by policyholder i.

Since the Negative Binomial NB2 distribution does not belong to the ED family, the glm function cannot be used in R. But there is a glm.nb function available to perform NB2 regression, which is very similar to glm. Results are displayed in Table 5.3. Again, we see that the point estimates are close to those obtained from the other Poisson and mixed Poisson fits. The standard errors are larger compared to the Poisson GLM ones, but somewhat smaller compared to the other mixed Poisson fits. The obtained AIC is 13,567. This considerably exceeds the corresponding value for the Poisson-LogNormal model, so that the Negative Binomial is not the preferred specification for the data under consideration.

Table 5.3 Output of the glm.nb function of R for the fit of the motor TPL data with the NB2 fit using maximum-likelihood

Coefficient	Estimate	Std. error	z value	$Pr(> |z|)$	
Intercept	−1.75806	0.04381	−40.133	$<2 \times 10^{-16}$	★★★
Female	−0.11777	0.04661	−2.527	0.0115	★
Age G1	0.56146	0.09516	5.900	3.64×10^{-9}	★★★
Age G2	0.31642	0.05008	6.319	2.64×10^{-10}	★★★
Age G5	−0.14398	0.06547	−2.199	0.0279	★
CitySize large	0.29398	0.04540	6.475	9.45×10^{-11}	★★★

5.5 Panel Data

5.5.1 Panel Data

Insurance portfolios constitute open groups of individuals: new entrants correspond to policies sold by the insurers to prospects whereas policyholders canceling their contract leave the portfolio. Policyholders can thus be followed over time as long as they stay insured by the company under consideration. This explains why longitudinal or panel data constitute by far the most commonly encountered grouping structure in insurance studies. In this case, the same individual is followed over time and produces several responses recorded in the database.

Actuaries resort to statistics relating to several years to build their price list in order to

- increase the volume of information
- check the estimated regression coefficients for consistency over time, to establish whether the estimates $\widehat{\beta}_j$ for the different calendar years remain stable from one period to the next. Clearly, a risk factor whose associated estimated regression coefficient $\widehat{\beta}_j$ changes sign over time must be considered with great care, and preferably excluded from the model.
- avoid granting too much importance to a particular calendar year (during which the particular weather conditions could have increased or decreased the number of traffic accidents, for instance).

In such a case, the mutual independence is lost: the observations for the same individual are correlated across the different periods. These data have a panel structure: the same response is observed on a large number of individuals over a relatively short period of time (typically, three to five years). The asymptotic is performed on the portfolio size, not on the length of the observation period (contrarily to time series).

Before embarking in a panel analysis pooling together the observations relating to several years, it is interesting to first work year by year to assess the stability of the effect of each rating factor on the annual expected claim frequency. Specifically, the vector β of the regression coefficients is estimated on the basis of each calendar year and the components are checked for their stability over time. Only stable coefficients are interesting for the purpose of ratemaking. Rating factors with unstable regression coefficients should be excluded from the risk classification scheme unless there is some clear explanation for such variations. To check whether the impact of a feature is consistent over time, it is convenient to allow for an interaction between this feature and calendar time. This produces estimated regression parameters per calendar year that can be checked for stability though time.

In some cases, a time trend is visible for some estimated regression coefficients (this is typically true for the intercept β_0). A time effect can then be incorporated in the model to account for trending coefficients. Including a time-varying intercept may be useful:

– On the one hand, this accounts for inflation or deflation effects. For instance, claim frequencies tend to decrease over time in motor insurance thanks to public road safety campaigns and technological advances whereas claim severities exhibit an opposite trend because of monetary inflation and rising vehicle costs (better equipped vehicles turning out to be more expensive to repair when damaged).
– On the other hand, it can be corrected so that the model exactly reproduces the total costs at portfolio level for each calendar year (a useful property that is known to hold for the Poisson and overdispersed Poisson models, for instance).

These parameters can then be projected to the future in the pricing.

5.5.2 Modeling Panel Data with Static Random Effects

Responses Y_{it} are now doubly indexed, by policy $i = 1, 2, \ldots, n$ and by time $t = 1, \ldots, T_i$, where T_i denotes the number of observation periods for contract i. Given $\mathcal{E}_i$, the T_i observations $Y_{i1}, Y_{i2}, \ldots, Y_{iT_i}$ relating to policyholder i are assumed to be conditionally independent with means that depend on the linear predictor through a specified link function and conditional variances that are specified by a variance function together with a dispersion parameter. These observations are gathered in the random vector $Y_i = (Y_{i1}, Y_{i2}, \ldots, Y_{iT_i})^\top$ of dimension T_i. Even if it is often reasonable to assume independence between the random vectors Y_i, this assumption is generally questionable inside these vectors because of repeated measures on the same individual.

Formally, the basic GLMMs rely on the two following assumptions:

A1 The random vectors $Y_1, Y_2, \ldots, Y_n$ are mutually independent and given $\mu_{it} = a'(\theta_{it})$, the response Y_{it} obeys the ED probability density or probability mass function

$$\exp\left(\frac{y\theta_{it} - a(\theta_{it})}{\phi/v_{it}} \right) c(y, \phi/v_{it}).$$

A2 Given $\mathcal{E}_i$, the responses $Y_{it}, t = 1, \ldots, T_i$, are mutually independent with

$$g(\mu_{it}) = \text{score}_{it} + \mathcal{E}_i$$

$$\text{score}_{it} = \beta_0 + \sum_{j=1}^{p} \beta_j x_{ijt} + \beta_{p+t}$$

$$\text{Var}[Y_{it}] = \frac{\phi}{v_{it}} V(\mu_{it}).$$

Here, the regression coefficient β_{p+t} captures the specificities of observation period t (with time treated as a categorical feature).

A3 The random effects $\mathcal{E}_1, \ldots, \mathcal{E}_n$ are mutually independent with a common distribution.

Often, the random effects $\mathcal{E}_i$ are assumed to conform to the Normal distribution. Since they account for the effect of many hidden risk factors, acting additively on the score scale, the Normality assumption can be justified by the central-limit theorem as explained before for cross-sectional data. In actuarial applications, the co-called conjugate distributions are often used to model the random effects $\mathcal{E}_i$. This is for instance the case with the Gamma distribution for Poisson responses, or the Beta distribution for Binomial responses. As we have seen above, this choice allows the actuary to derive closed-form expressions for a priori and a posteriori distributions.

Static random effects, i.e. $\mathcal{E}_i$ not varying with time t so that the same random effect impacts on all the T_i observations recorded for policyholder i, are often used in insurance applications based on relatively short panels (say $T_i \leq 3$). The direct consequence of such a modeling is that past claims play the same role in the predictive distribution, even if they were reported years ago. To let the importance of claims depend on the time elapsed since their occurrence, the actuary can replace the static random effect model with its dynamic counterpart as explained in the next section.

5.5.3 Modeling Panel Data with Dynamic Random Effects

Sometimes, data show significant changes over time. In such a case, the predictive ability of past experience decreases with the lag between the period of risk prediction and the period of occurrence. Letting the random effects depend on time (i.e., allowing for $\mathcal{E}_{it}$ instead of $\mathcal{E}_i$) accounts for this effect by appropriately discounting past experience.

Formally, the random effects for cell i are now modeled with the help of a random vector

$$(\mathcal{E}_{i1}, \mathcal{E}_{i2}, \ldots, \mathcal{E}_{iT_i})^\top.$$

This amounts to replace assumptions A2-A3 above with

A2' Given $\mathcal{E}_{i1}, \mathcal{E}_{i2}, \ldots, \mathcal{E}_{iT_i}$, the responses $Y_{it}, t = 1, \ldots, T_i$, are mutually independent with

$$g(\mu_{it}) = \text{score}_{it} + \mathcal{E}_{it}$$

$$\text{score}_{it} = \beta_0 + \sum_{j=1}^{p} \beta_j x_{ijt} + \beta_{p+t}$$

$$\text{Var}[Y_{it}] = \frac{\phi}{v_{it}} V(\mu_{it}).$$

A3' The random vectors $(\mathcal{E}_{i1}, \mathcal{E}_{i2}, \ldots, \mathcal{E}_{iT_i})^\top, i = 1, \ldots, n$ are mutually independent with a common distribution (up to different dimensions T_i).

With dynamic random effects, every response Y_{it} is accompanied with its own random effect $\mathcal{E}_{it}$, thus allowing for random effects that develop over time. This is justified since unobservable factors influencing the risk are not constant and policyholders may adjust their efforts for loss prevention according to their experience with past claims, the amount of premium and awareness of future consequences of an accident (due to experience rating schemes). The main technical interest of letting the random effects evolve over time is to take into account the date of claims. This reflects the fact that the predictive ability of a claim depends on its age: a recent claim is a worse sign to the insurer than a very old one.

Another situation where responses are correlated arises when wind or storm perils are studied. Clearly, policyholders in the same area are likely to have positively correlated outcomes, as the losses result from the same storms affecting all insureds in the area. Mixed models are also useful in that respect.

The multivariate Poisson-LogNormal distribution is obtained by assuming that $(\mathcal{E}_{i1}, \mathcal{E}_{i2}, \ldots, \mathcal{E}_{iT_i})$ obeys the multivariate Normal distribution with mean vector $\mathbf{0}$ and variance-covariance matrix $\mathbf{\Sigma}_i$ of dimension $T_i \times T_i$. The variances and covariances are of the form

$$\sigma_{tt} = \text{Var}[\mathcal{E}_{it}] = \sigma^2$$

and

$$\sigma_{s,t} = \text{Cov}[\mathcal{E}_{it}, \mathcal{E}_{is}] = \sigma(|t - s|) \text{ for } s \neq t.$$

We conventionally put $\sigma(0) = \sigma^2$ for convenience. Variances and covariances are defined up to $T_{\max} = \max\{T_1, \ldots, T_n\}$ and stored in the variance-covariance matrix $\mathbf{\Sigma}$ of dimension $T_{\max} \times T_{\max}$. Each $\mathbf{\Sigma}_i$ then corresponds to the upper left part of the matrix $\mathbf{\Sigma}$ with entry $\sigma_{s,t}$.

5.5.4 Generalized Estimating Equations (GEEs)

5.5.4.1 Principle

Observations relating to the same policyholder across time are expected to be correlated. GEEs account for this serial dependence when estimating $\boldsymbol{\beta}$, in a relatively easy way. GEEs are a particular case of quasi-likelihood estimation designed to deal with repeated measures on the same individuals. GEEs provide a practical method with reasonable statistical efficiency to analyze such panel data. GEEs also give initial values for maximum likelihood procedures in models for longitudinal data.

Let $\mathbf{Y}_i$ be the vector of responses $(Y_{i1}, \ldots, Y_{iT_i})$ for policyholder i and let $\text{E}[\mathbf{Y}_i] = \boldsymbol{\mu}_i$ be the vector of expected outcomes. Assume that the components of $\mathbf{Y}_i$ are mutually independent. Then, the variance-covariance matrix $\mathbf{V}_i$ of $\mathbf{Y}_i$ is diagonal with the variances of the responses Y_{it} appearing along the diagonal. The likelihood Eq. (4.1) can be rewritten as

$$\sum_{i=1}^{n}\sum_{t=1}^{T_i} \frac{Y_{it} - \mu_{it}}{\text{Var}[Y_{it}]} \frac{\partial \mu_i}{\partial \beta_j} = 0.$$

In matrix form, we then obtain the system

$$\sum_{i=1}^{n}\left(\frac{\partial \mu_i}{\partial \beta}\right)^{\top} V_i^{-1}(Y_i - \mu_i) = 0 \tag{5.7}$$

of likelihood equations under serial independence. The extension to the case where the components of Y_i are correlated through time is then straightforward, substituting to the diagonal variance-covariance matrix corresponding to serial independence an appropriate matrix V_i accounting for the time dynamics of data.

To account for the serial dependence of the components Y_{it} inside Y_i, GEEs use a variance-covariance matrix V_i depending on a vector α of dependence parameters: here, α represents parameters involved in the modeling of the correlation structure within Y_i. The regression parameters β can then be estimated by solving (5.7) with this more general variance-covariance matrix V_i. Since V_i also depends on the unknown dependence parameters α, we substitute for α in (5.7) a consistent estimate $\widehat{\alpha}$ of it. Notice that no distribution has been specified so far, only the mean and the variance-covariance matrix.

The principle of Generalized Estimating Equation (GEE) is to find a suitable variance-covariance matrix V_i to insert in Eq. (5.7) in replacement of the diagonal one based on serial independence. This matrix should take the overdispersion and the serial dependence into account. A possible specification of this matrix could be

$$V_i = \phi A_i^{1/2} R_i(\alpha) A_i^{1/2}$$

where the "working" correlation matrix $R_i(\alpha)$ takes the serial dependence between the components of Y_i into account and depends on a parameter α while A_i is the variance-covariance matrix under serial independence (that is, a diagonal matrix with the variances of the Y_{it} appearing along the diagonal).

The estimator $\widehat{\beta}$ solving (5.7) is consistent whatever the choice of the matrix $R_i(\alpha)$ but the estimation accuracy will be much better if $R_i(\alpha)$ reflects the true correlation matrix of Y_i. Equation (5.7) is solved thanks to a modified version of the Fisher scoring method for β and a moment estimation for ϕ and α. The iterative procedure is as follows:

1. Compute an initial estimate of β assuming independence.
2. Compute the current "working" correlation matrix based on standardized residuals, current β and the assumed structure of $R_i(\alpha)$.
3. Estimate the covariance matrix V_i.
4. Update β.

Note that GEE is not a likelihood based method of estimation, so that inference based on likelihood is not possible in this case.

5.5.4.2 Modeling Dependence with the "Working" Correlation Matrix

The "working" correlation matrix $R_i(\alpha)$ takes into account the dependence between the observations corresponding to the same policyholder. The form of this matrix must be specified and depends on the parameters vector α.

If $R_i(\alpha) = I$, (5.7) gives exactely the likelihood equations (4.1) under the assumption of independence. Other choices for the "working" correlation matrix include a user-specific correlation matrix (not estimated from the data but specified by the actuary), m-dependence (correlation equal to α_t for lags $t = 1, \ldots, m$, and 0 for higher lags), exchangeable (constant correlation α whatever the lag), unstructured (each correlation coefficient α_{jk} between observations made at times j and k is estimated from the data), and AR1 (autoregressive of order 1, with a correlation coefficient equal to α^t at lag t).

Let us now discuss some particular cases of interest in insurance applications.

Static random effects: Actuaries often consider static random effects $\mathcal{E}_i$ following the same individual through time and inducing serial correlation, as well as overdispersion. In such a case, we adopt a user-specific correlation matrix with all entries equal to 1.

AR1 random effects: AR1 random effects are particularly effective in insurance studies (corresponding to credibility formulas with geometric weights). A particularly simple and efficient dynamic credibility model is obtained by assuming that random effects obey an auto-regressive structure of order one, that is,

$$\mathcal{E}_{it} = \varrho \mathcal{E}_{i,t-1} + \Delta_{it}, \quad t \geq 2,$$

where $\Delta_{it} \sim Nor(0, \sigma^2(1 - \rho^2))$ are independent, $|\varrho| < 1,$, and $\Delta_{i1} \sim Nor(0, \sigma^2)$. In this model, the heterogeneity $\mathcal{E}_{it}$ for period t is influenced by the preceding period $\mathcal{E}_{i,t-1}$ but has also its own characteristics Δ_{it}.

Exchangeable random effects: Exchangeable random effects are also useful. This is the case for instance when there is a static baseline heterogeneity Δ_i for policyholder i which is perturbated by independent and identically distributed yearly effects $\Delta_{i1}, \Delta_{i2}, \ldots$. Specifically, $\mathcal{E}_{it} = \Delta_i + \Delta_{it}$ where Δ_{it} are independent and identically distributed, and independent from Δ_i, all random variables obeying Normal laws. The random effects $\mathcal{E}_{it}$ are then exchangeable.

5.5.4.3 Application to Motor Insurance

Let us now analyze claim frequencies on the basis of panel data. To this end, we consider the motor insurance portfolio studied in Sect. 4.10. In reality, data related to this portfolio were collected during 3 years (from 1997 to 1999). This panel structure has been ignored in the analysis performed in Sect. 4.10 where only the first year has been used.

The same features as for the model where the serial independence was assumed are kept in the final model. This a safe strategy as ignoring serial dependence tends to produce artificially small standard errors. Hence, the features that have been discarded based on serial independence in Sect. 4.10 would also have been discarded when recognizing the panel structure. The analysis is conducted with the features Gender and the simplified versions of AgeP and City2, specifying offset=log(ExpoR/365).

Let us now compare the results produced by the GEE approach using different working correlation structures. First, we consider the general case where the working correlation matrix is left unstructured, with a correlation coefficient specific to each pair of observation years. The analysis is performed with the geeglm function of the R package geepack. The id argument identifies subjects in the input data set. The subject-effect can be a single variable, an interaction effect, a nested effect, or a combination. Each distinct value, or level, of the effect identifies a different subject, or cluster. Responses from different subjects are assumed to be independent, and responses within subjects are assumed to be correlated. Here, as in the majority of actuarial applications, the policy number is typically used as subject-effect. The policy number is used as id to track the experience of the same contract over time, and the general covariance structure is obtained with corstr = "unstructured".

For the sake of comparison, we also fit the same regression model with Overdispersed Poisson GLM, under serial independence with the glm function and with GEE under serial independence, i.e. $R_i(\alpha) = I \Leftrightarrow \mathcal{E}_{i1}, \ldots, \mathcal{E}_{iT_i} \sim \mathcal{N}or(0, \sigma_{\mathcal{E}}^2)$, independent.

The results are displayed in Table 5.4. The dispersion parameter for the quasi-Poisson GLM was estimated to $\widehat{\phi} = 1.343$. We see that ODP and Poisson GEE result in the same point estimates under serial independence, as expected. The IRLS algorithm converged after 6 iterations. The GEE algorithm under serial dependence produced the same estimate $\widehat{\phi} = 1.343$ supplemented with a standard error equal to 0.09502. The estimated regression coefficients remain mostly unaffected by the introduction of a "working" correlation matrix, as it can be seen by comparing the three sets of $\widehat{\beta}$ corresponding to serial independence, unstructured dependence and exchangeability. The unstructured correlation structure is estimated to

	Estimate	Std.err
$\alpha_{1,2}$	0.0525	0.00877
$\alpha_{1,3}$	0.0456	0.01075
$\alpha_{2,3}$	0.0467	0.00973

Correlation parameters thus seem to be equal (about 5%) whatever the lag. This suggests to use exchangeable random effects, i.e. to consider

$$\mathcal{E}_{i1}, \ldots, \mathcal{E}_{iT_i} \sim \mathcal{N}or(0, \sigma_{\mathcal{E}}^2) \text{ and equi-correlated.}$$

This corresponds to the dependence structure known as exchangeability. The corresponding working correlation matrix is

$$R_i(\alpha) = \begin{pmatrix} 1 & \alpha & \cdots & \alpha \\ \alpha & 1 & \cdots & \alpha \\ \vdots & \vdots & \ddots & \vdots \\ \alpha & \alpha & \cdots & 1 \end{pmatrix}.$$

The estimated working correlation parameter obtained with `corstr=`
`"exchangeable"` is

	Estimate	Std.err
α	0.0489	0.00651

Considering the results displayed in Table 5.4, we see that point estimates remain stable whatever the "working" correlation structure. This is because the estimators for β obtained from Poisson maximum-likelihood under serial independence and from GEE are all consistent. The choice of the correlation matrix impacts on the p-value of the test for $\beta_j = 0$. In the example studied here, all features remain significant whatever the "working" correlation matrix adopted for estimation. The GEE approach does not provide the actuary with a fully specified model on which calculations can be performed, just with a way to correct standard errors and assess the relevance of features to explain the responses of interest.

5.6 From Mixed Models to Experience Rating

5.6.1 Predictive Distribution

Let us now connect the mixed models used in statistics to actuarial credibility models for experience rating. Experience rating consists in integrating past claims experience into next year's premium. Formally, let Y_t be the response for a given policyholder during year t. Depending on the situation, Y_t represents the number of claim reported during period t, the average per-claim severity or the total claim amount filed by the policyholder during year t. The predictive distribution of Y_t given past claims history is the distribution of Y_t given $Y_1, \ldots, Y_{t-1}$.

We assume that $Y_1, Y_2, Y_3, \ldots$ are independent and identically distributed given some unknown risk level Θ. The individual, unknown premium is

$$\mathrm{E}[Y_t|\Theta] = \mu(\Theta).$$

The collective premium is

$$\mathrm{E}[Y_t] = \mathrm{E}[\mu(\Theta)] = \mu.$$

Table 5.4 Output of the `glm` and `geeglm` functions of R for the fit of the overdispersed Poisson regression model to the motor TPL data under various serial dependence structures

Overdispersed Poisson GLM, under serial independence

Coefficient	Estimate	Std. error	z value	$\Pr(> |z|)$	
Intercept	−1.7994	0.0268	−67.23	$<2 \times 10^{-16}$	★★★
Female	−0.0722	0.0302	−2.39	0.0170	★
Age G1	0.6350	0.0777	8.18	3.0×10^{-16}	★★★
Age G2	0.2355	0.0332	7.09	1.4×10^{-12}	★★★
Age G5	−0.1444	0.0389	−3.71	0.0002	★★★
CitySize large	0.2191	0.0299	7.32	2.5×10^{-13}	★★★

Poisson GEE, under serial independence

Coefficient	Estimate	Std. error	Wald	$\Pr(> |W|)$	
Intercept	−1.79937	0.02440	5,438.029	$<2 \times 10^{-16}$	★★★
Female	−0.07220	0.02831	6.505	0.0108	★
Age G1	0.63502	0.06938	83.761	$< 2 \times 10^{-16}$	★★★
Age G2	0.23547	0.03054	59.439	1.27×10^{-14}	★★★
Age G5	−0.14436	0.03670	15.477	8.35×10^{-5}	★★★
CitySize large	0.21910	0.02799	61.295	4.88×10^{-15}	★★★

Poisson GEE, unstructured working correlation matrix

Coefficient	Estimate	Std. error	Wald	$\Pr(> |W|)$	
Intercept	−1.7966	0.0244	5,415.55	$<2 \times 10^{-16}$	★★★
Female	−0.0728	0.0283	6.59	0.01023	★
Age G1	0.6264	0.0697	80.78	$< 2 \times 10^{-16}$	★★★
Age G2	0.2359	0.0305	59.79	1.1×10^{-14}	★★★
Age G5	−0.1413	0.0367	14.82	0.00012	★★★
CitySize large	0.2206	0.0280	61.90	3.7×10^{-15}	★★★

Poisson GEE, exchangeable working correlation matrix

Coefficient	Estimate	Std. error	Wald	$\Pr(> |W|)$	
Intercept	−1.7965	0.0244	5,416.09	$<2 \times 10^{-16}$	★★★
Female	−0.0729	0.0283	6.62	0.01008	★
Age G1	0.6260	0.0697	80.66	$< 2 \times 10^{-16}$	★★★
Age G2	0.2360	0.0305	59.82	1.0×10^{-14}	★★★
cre Age G5	−0.1414	0.0367	14.84	0.00012	★★★
CitySize large	0.2206	0.0280	61.91	3.6×10^{-15}	★★★

It is important to realize that even if $Y_1, Y_2, Y_3, \ldots$ are independent given $\Theta = \theta$, independence does no more hold unconditionally. To show it formally, let us define the first two conditional moments

$$\mu(\theta) = \mathrm{E}[Y_t | \Theta = \theta]$$
$$\sigma^2(\theta) = \mathrm{Var}[Y_t | \Theta = \theta].$$

Actuaries routinely used the indicators M^2 and Σ^2 defined next. The indicator M^2 defined as

$$M^2 = \mathrm{Var}[\mu(\Theta)]$$

measures the portfolio heterogeneity. If $M^2 = 0$ then $\mu(\theta) = \mu$ and the portfolio is homogeneous. In general, $M^2 > 0$ indicating that the portfolio is heterogeneous. The indicator Σ^2 defined as

$$\Sigma^2 = \mathrm{E}[\sigma^2(\Theta)]$$

measures pure randomness. The variance σ^2 of the response can then be decomposed into

$$
\begin{aligned}
\mathrm{Var}[Y_t] &= \sigma^2 \\
&= \mathrm{E}\Big[\mathrm{Var}[Y_t|\Theta]\Big] + \mathrm{Var}\Big[\mathrm{E}[Y_t|\Theta]\Big] \\
&= \mathrm{E}[\sigma^2(\Theta)] + \mathrm{Var}[\mu(\Theta)] \\
&= \Sigma^2 + M^2.
\end{aligned}
$$

Past value of the response partly reveals the unknown value of Θ. If past responses Y_t are larger than expected then this suggests that Θ also is. To see that the serial correlation among the responses Y_t is the consequence of unexplained heterogeneity, let us compute for $s \neq t$,

$$
\begin{aligned}
\mathrm{Cov}[Y_t, Y_s] &= \mathrm{E}\Big[\mathrm{Cov}[Y_t, Y_s|\Theta]\Big] + \mathrm{Cov}\Big[\mathrm{E}[Y_t|\Theta_i], \mathrm{E}[Y_s|\Theta]\Big] \\
&= 0 + \mathrm{Var}[\mu(\Theta)] \\
&= M^2.
\end{aligned}
$$

This confirms that serial correlation is induced by portfolio heterogeneity and is thus only apparent.

Now, the actuary may be willing to integrate past claims history $Y_1, Y_2, \ldots, Y_{t-1}$ in the prediction of next year's premium. To this end, the predictive, or a posteriori mean for year t appears to be the natural candidate. It is given by

$$
\begin{aligned}
\mathrm{E}[Y_t|Y_1, \ldots, Y_{t-1}] &= \mathrm{E}\big[\mathrm{E}[Y_t|\Theta, Y_1, \ldots, Y_{t-1}]\big|Y_1, \ldots, Y_{t-1}\big] \\
&= \mathrm{E}\big[\mathrm{E}[Y_t|\Theta]\big|Y_1, \ldots, Y_{t-1}\big] \\
&= \mathrm{E}[\mu(\Theta)|Y_1, \ldots, Y_{t-1}].
\end{aligned}
$$

We will establish in the next section that the predictive mean is optimal in the sense that it minimizes the mean squared error of prediction. Formally, for any other function $\Psi(Y_1, \ldots, Y_{t-1})$ of past claims history, we have that

$$
\mathrm{E}\Big[\big(\Psi(Y_1, \ldots, Y_{t-1}) - \mu(\Theta)\big)^2\Big] > \mathrm{E}\Big[\big(\mathrm{E}[Y_t|Y_1, \ldots, Y_{t-1}] - \mu(\Theta)\big)^2\Big].
$$

Moreover, it is easily seen that

$$E\Big[E\big[Y_t\,|\,Y_1,\ldots,Y_{t-1})\big]\Big] = E[Y_t] = \mu,$$

so that the average total premium income remains stable over time.

5.6.2 ED Family with Conjugate Priors and Credibility Premiums

A credibility premium is a linear combination of the grand mean μ and of the policy-specific historical weighted average. The coefficient multiplying the historical mean is called the credibility coefficient. It turns out that predictive means are credibility premiums provided the actuary selects the mixing distribution appropriately.

Let us consider responses obeying ED mixtures. Given $\Theta = \theta$, $Y_1, Y_2, \ldots$ are independent with probability density, or probability mass function

$$\exp\left(\frac{y\theta - a(\theta)}{\phi/v_t}\right) c(y, \phi/v_t).$$

Assume that the unknown risk level Θ has probability density function

$$f_\Theta(\theta) = \exp\left(\frac{y_0\theta - a(\theta)}{\tau^2}\right) d(y_0, \tau),$$

with parameters y_0 and τ. This probability density function is supposed to tend to 0 when its argument approaches the boundary of the support of Θ, for all y_0. Henceforth, we denote as $\mathrm{supp}(\Theta)$ the support of random variable Θ.

A priori, we see that

$$
\begin{aligned}
E[Y_1] &= E[a'(\Theta)]\\
&= \int_{\theta \in \mathrm{supp}(\Theta)} a'(\theta) \exp\left(\frac{y_0\theta - a(\theta)}{\tau^2}\right) d(y_0, \tau)d\theta\\
&= y_0 - \tau^2 \int_{\theta \in \mathrm{supp}(\Theta)} \frac{y_0 - a'(\theta)}{\tau^2} \exp\left(\frac{y_0\theta - a(\theta)}{\tau^2}\right) d(y_0, \tau)d\theta\\
&= y_0.
\end{aligned}
$$

Assume that the policyholder has been observed during T periods, with observed responses $Y_1 = y_1, \ldots, Y_T = y_T$. Recall that "$\propto$" means "is proportional to" (neglecting factors that disappear from further calculations). As

$$f_\Theta(\theta\,|\,y_1, \ldots, y_T) \propto f_\Theta(\theta) f(y_1, \ldots, y_T\,|\,\theta),$$

the posterior structure function is then obtained from

$$f_\Theta(\theta|y_1, \ldots, y_T) \propto f_\Theta(\theta) \prod_{t=1}^{T} \exp\left(\frac{y_t\theta - a(\theta)}{\phi/v_t}\right) c(y_t, \phi/v_t)$$

$$\propto \exp\left(\frac{y_0\theta - a(\theta)}{\tau^2}\right) \prod_{t=1}^{T} \exp\left(\frac{y_t\theta - a(\theta)}{\phi/v_t}\right)$$

$$= \exp\left(\theta\left(\sum_{t=1}^{T}\frac{y_t v_t}{\phi} + \frac{y_0}{\tau^2}\right) - a(\theta)\left(\sum_{t=1}^{T}\frac{v_t}{\phi} + \frac{1}{\tau^2}\right)\right)$$

$$= \exp\left(\frac{\tilde{y}_T\theta - a(\theta)}{\tilde{\tau}_T^2}\right)$$

with

$$\tilde{\tau}_T^2 = \left(\sum_{t=1}^{T}\frac{v_t}{\phi} + \frac{1}{\tau^2}\right)^{-1} < \tau^2$$

$$\tilde{y}_T = \left(\sum_{t=1}^{T}\frac{v_t}{\phi} + \frac{1}{\tau^2}\right)^{-1} \left(\sum_{t=1}^{T}\frac{y_t v_t}{\phi} + \frac{y_0}{\tau^2}\right).$$

We thus finally get

$$f_\Theta(\theta|y_1, \ldots, y_T) = \exp\left(\frac{\tilde{y}_T\theta - a(\theta)}{\tilde{\tau}_T^2}\right) d(\tilde{y}_T, \tilde{\tau}_T).$$

The exact credibility premium is of the form

$$E[Y_{T+1}|Y_1 = y_1, \ldots, Y_T = y_T]$$
$$= E\left[E[Y_{T+1}|\Theta, Y_1 = y_1, \ldots, Y_T = y_T]\Big|Y_1 = y_1, \ldots, Y_T = y_T\right]$$
$$= E\left[E[Y_{T+1}|\Theta]\Big|Y_1 = y_1, \ldots, Y_T = y_T\right]$$
$$= E\left[a'(\Theta)\Big|Y_1 = y_1, \ldots, Y_T = y_T\right]$$
$$= \int_{\text{supp}(\Theta)} a'(\theta)\exp\left(\frac{\tilde{y}_T\theta - a(\theta)}{\tilde{\tau}_T^2}\right) d(\tilde{y}_T, \tilde{\tau}_T)d\theta$$
$$= \tilde{y}_T - \tilde{\tau}_T^2\int_{\text{supp}(\Theta)}\frac{\tilde{y}_T - a'(\theta)}{\tilde{\tau}_T^2}\exp\left(\frac{\tilde{y}_T\theta - a(\theta)}{\tilde{\tau}_T^2}\right) d(\tilde{y}_T, \tilde{\tau}_T)d\theta$$
$$= \tilde{y}_T.$$

The a posteriori expectation $\widetilde{y}_T$, that is, the exact credibility premium, can be rewritten as

$$
\begin{aligned}
\widetilde{y}_T &= \left(\sum_{t=1}^{T} \frac{v_t}{\phi} + \frac{1}{\tau^2} \right)^{-1} \left(\sum_{t=1}^{T} \frac{y_t v_t}{\phi} + \frac{y_0}{\tau^2} \right) \\
&= \frac{\sum_{t=1}^{T} v_t}{\sum_{t=1}^{T} v_t + \frac{\phi}{\tau^2}} \frac{\sum_{t=1}^{T} v_t y_t}{\sum_{t=1}^{T} v_t} + \left(1 - \frac{\sum_{t=1}^{T} v_t}{\sum_{t=1}^{T} v_t + \frac{\phi}{\tau^2}} \right) y_0 \\
&= \alpha_T \overline{y}^{(T)} + (1 - \alpha_T) y_0
\end{aligned}
$$

with

$$
\alpha_T = \frac{\sum_{t=1}^{T} v_t}{\sum_{t=1}^{T} v_t + \frac{\phi}{\tau^2}}
$$

$$
\overline{y}^{(T)} = \frac{\sum_{t=1}^{T} v_t y_t}{\sum_{t=1}^{T} v_t}.
$$

This shows that the predictive premium is also a credibility premium, where the credibility factor is α_T. In practice, the parameters ϕ and τ entering α_T must be estimated before the actuary can perform credibility re-valuations.

5.6.3   Poisson-Gamma Credibility Model

Let N_{it} be the number of claims reported by policyholder i during period t, $i = 1, 2, \ldots, n, t = 1, 2, \ldots, T_i$. The corresponding exposure to risk (i.e. the duration of the coverage period) for policyholder i during period t is denoted as e_{it}. The available features about policyholder i in period t (like age, gender, power of the car, and so on) summarized in x_{it} is used to predict $\lambda_{it} = E[N_{it}]$.

Let

$$
N_{i\bullet} = \sum_{t=1}^{T_i} N_{it}
$$

be the total number of claims reported by policyholder i during the T_i observation periods. The statistic $N_{i\bullet}$ is a convenient summary of past claims history.

As before, we assume that given $\Delta_i = \delta$, the random variables $N_{it}, t = 1, 2, \ldots$, are independent and obey the $\mathcal{P}oi(\lambda_{it}\delta)$, i.e.

$$
P[N_{it} = k | \Delta_i = \delta] = \exp(-\delta \lambda_{it}) \frac{(\delta \lambda_{it})^k}{k!}, \quad k = 0, 1, 2, \ldots
$$

At the portfolio level, the sequences $(\Delta_i, N_{i1}, N_{i2}, \ldots)$ are assumed to be independent.

In this approach, the dependence between annual claim numbers is a consequence of the heterogeneity of the portfolio. It is only apparent and would disappear if we had a complete knowledge of policy characteristics (then Δ_i would become deterministic). Also, there is no learning effect: an accident may influence the driving behavior, but this is not the dominant effect. The unexplained heterogeneity (which has been modeled through the introduction of the risk parameter Δ_i for policyholder i) is then revealed by the claims history. The predictive distribution is therefore used to included this history in the future premium: past claims history modifies the distribution of Δ_i which in turn modifies the premium.

Let us now assume in addition that the random effects Δ_i are independent and distributed according to the Gamma distribution with unit mean and variance $1/\alpha$. We thus proceed as for the NB2 distribution and consider $\Delta_i \sim Gam(\alpha, \alpha)$ with density

$$f_\Delta(\delta) = \frac{1}{\Gamma(\alpha)} \alpha^\alpha \delta^{\alpha-1} \exp(-\alpha\delta), \quad \delta \geq 0.$$

The joint distribution of the claim counts is then given by

$$P[N_{i1} = k_{i1}, N_{i2} = k_{i2}, \ldots, N_{iT_i} = k_{iT_i}]$$

$$= \int_0^{+\infty} P[N_{i1} = k_{i1}, N_{i2} = k_{i2}, \ldots, N_{iT_i} = k_{iT_i}|\Delta_i = \delta] f_\Delta(\delta) d\delta$$

$$= \int_0^{+\infty} \left(\prod_{t=1}^{T_i} P[N_{it} = k_{it}|\Delta_i = \delta] \right) f_\Delta(\delta) d\delta$$

$$= \int_0^{+\infty} \left(\prod_{t=1}^{T_i} \exp(-\delta\lambda_{it}) \frac{(\delta\lambda_{it})^{k_{it}}}{k_{it}!} \right) f_\Delta(\delta) d\delta$$

$$= \left(\prod_{t=1}^{T_i} \frac{\lambda_{it}^{k_{it}}}{k_{it}!} \right) \left(\frac{\alpha}{\alpha + \sum_{t=1}^{T_i} \lambda_{it}} \right)^\alpha \left(\alpha + \sum_{t=1}^{T_i} \lambda_{it} \right)^{-\sum_{t=1}^{T_i} k_{it}} \frac{\Gamma\left(\alpha + \sum_{t=1}^{T_i} k_{it}\right)}{\Gamma(\alpha)}.$$

In order to derive the predictive distribution of N_{i,T_i+1}, let us first establish the expression of the joint distribution of $(N_{i1}, \ldots, N_{iT_i}, \Delta_i)$. It is obtained from

$$\left(\prod_{t=1}^{T_i} \exp\left(-\delta_i\lambda_{it} \right) \frac{(\delta_i\lambda_{it})^{k_{it}}}{k_{it}!} \right) \frac{1}{\Gamma(\alpha)} \alpha^\alpha \delta_i^{\alpha-1} \exp(-\alpha\delta_i)$$

$$\propto \exp\left(-\delta_i \sum_{t=1}^{T_i} \lambda_{it} \right) \delta_i^{\sum_{t=1}^{T_i} k_{it}+\alpha-1} \exp(-\alpha\delta_i).$$

The density of Δ_i given $N_{it} = k_{it},\quad t = 1, 2, \ldots, T_i$ is deduced from the relation

$$\frac{\exp\left(-\delta_i\left(\alpha + \sum_{t=1}^{T_i} \lambda_{it}\right)\right) \delta_i^{\alpha + \sum_{t=1}^{T_i} k_{it} - 1}}{\int_0^{+\infty} \exp\left(-\xi\left(\alpha + \sum_{t=1}^{T_i} \lambda_{it}\right)\right) \xi^{\alpha + \sum_{t=1}^{T_i} k_{it} - 1} d\xi}$$

$$= \exp\left(-\delta_i\left(\alpha + \sum_{t=1}^{T_i} \lambda_{it}\right)\right) \delta_i^{\alpha + \sum_{t=1}^{T_i} k_{it} - 1} \frac{\left(\alpha + \sum_{t=1}^{T_i} \lambda_{it}\right)^{\alpha + \sum_{t=1}^{T_i} k_{it}}}{\Gamma\left(\alpha + \sum_{t=1}^{T_i} k_{it}\right)}.$$

Given $N_{i1} = k_{i1}, \ldots, N_{iT_i} = k_{iT_i}$, we thus see that

$$\Delta_i \sim \mathcal{G}am\left(\alpha + \sum_{t=1}^{T_i} k_{it}, \alpha + \sum_{t=1}^{T_i} \lambda_{it}\right)$$

so that

$$E[\Delta_i | N_{it} = k_{it}, t = 1, 2, \ldots, T_i] = \frac{\alpha + \sum_{t=1}^{T_i} k_{it}}{\alpha + \sum_{t=1}^{T_i} \lambda_{it}}$$

and

$$E[N_{i,T_i+1} | N_{it} = k_{it}, t = 1, 2, \ldots, T_i] = \lambda_{i,T_i+1} \frac{\alpha + \sum_{t=1}^{T_i} k_{it}}{\alpha + \sum_{t=1}^{T_i} \lambda_{it}}$$

$$= \lambda_{i,T_i+1} \frac{1 + \text{Var}[\Delta_i] \sum_{t=1}^{T_i} k_{it}}{1 + \text{Var}[\Delta_i] \sum_{t=1}^{T_i} \lambda_{it}}.$$

The predictive mean claim frequency can be rewritten as

$$\lambda_{i,T_i+1} \frac{1 + \text{Var}[\Delta_i] N_{i\bullet}}{1 + \text{Var}[\Delta_i] \lambda_{i\bullet}} = \lambda_{i,T_i+1}\left(\frac{1}{1 + \text{Var}[\Delta_i] \lambda_{i\bullet}} \times 1 + \frac{\text{Var}[\Delta_i] \lambda_{i\bullet}}{1 + \text{Var}[\Delta_i] \lambda_{i\bullet}} \times \frac{N_{i\bullet}}{\lambda_{i\bullet}}\right).$$

The predictive mean of Δ_i involved in this expression, that is,

$$\frac{1}{1 + \text{Var}[\Delta_i] \lambda_{i\bullet}} \times 1 + \frac{\text{Var}[\Delta_i] \lambda_{i\bullet}}{1 + \text{Var}[\Delta_i] \lambda_{i\bullet}} \times \frac{N_{i\bullet}}{\lambda_{i\bullet}}$$

appears to be a credibility premium. It is indeed a weighted average between the grand mean $1 = E[\Delta_i]$ and the policy-specific historical mean $\frac{N_{i\bullet}}{\lambda_{i\bullet}}$. The credibility factor is

$$\alpha_{T_i} = \frac{\text{Var}[\Delta_i] \lambda_{i\bullet}}{1 + \text{Var}[\Delta_i] \lambda_{i\bullet}}$$

which appears to be increasing in both $\text{Var}[\Delta_i]$ and $\lambda_{i\bullet}$.

5.7 Usage-Based Motor Insurance Pricing

5.7.1 Context

Technological advances have now supplemented classical risk factors with new ones, reflecting policyholder's actual behavior behind the wheel. Telematics is a branch of information technology that transmits data over long distances. Examples of telematics data include the global position system (GPS) data and the in-vehicle sensor data. The main source for such data is the automotive diagnostic system (or OBD, for On-Board Diagnostics) installed in the vehicle, or the driver's smartphone. Insurance companies also exploit spatial databases about speed limits on existing roads and risks attributed to these roads.

Telematics insurance data offer the opportunity to base actuarial pricing on actual policyholder's behavior. With Pay-How-You-Drive (PHYD) or Usage-Based (UB) motor insurance, premium amounts are based on the total distance traveled, the type of road, the time of the day, average speeds and other driving habits. Thus, premiums are based directly on driver's behavior behind the wheel.

Contrarily to standard risk factors, such as age, gender or place of residence, telematics data evolve over time in parallel to claim experience, progressively revealing the actual behavior of the policyholder behind the wheel. The information contained in past telematics data differs between individuals. For newly licensed drivers, no record is available. For those observed over the past, telematics data are available for the time they were subject to the UB system which may vary among policyholders. Moreover, telematics data are recorded while the policyholders are driving, and some of them regularly use their car (providing a rich information about their driving habits) whereas other ones use their car to a much lesser extent (resulting in limited volume of telematics data). In order to get the multivariate dynamics across insurance periods, past telematics data should not be included in the score like ordinary risk factors but must preferably be modeled jointly with claim experience.

5.7.2 Mixed Poisson Model for Annual Claim Frequencies

Consider an insurance portfolio comprising n policies observed during several periods. Let N_{it} be the number of claims reported by policyholder i, $i = 1, 2, \ldots, n$, during period t, $t = 1, 2, \ldots, T_i$. Compared to classical actuarial studies dealing with annual periods, insurers using telematics data generally work with shorter time periods, like a quarter or a month.

Let x_{it} be the vector of ordinary features for policyholder i, $i = 1, \ldots, n$, during period t, $t = 1, 2, \ldots, T_i$. The exposure-to-risk e_{it} is taken here as the distance driven in kilometers. The score for policyholder i in period t is denoted as $s_{it} = s(x_{it})$ and we define

$$\lambda_{it} = e_{it} \exp(s_{it}) = \exp\left(\ln e_{it} + s_{it}\right).$$

Adding $\ln e_{it}$ to the score s_{it} (i.e. treating this quantity as an offset) means that the insurer's price list is expressed per kilometer, and varies according to traditional risk features included in the vector x_{it}.

In the application considered in this section, we only have three observation periods and four responses considered simultaneously (the number of claims and three signals), so that we restrict ourselves to static random effects. A random effect $\mathcal{E}_i$ is added to the score s_{it} for taking account of residual heterogeneity of the portfolio. Precisely, the model is based on the following assumptions:

A1 given $\mathcal{E}_i = \epsilon$, the random variables $N_{it}, t = 1, 2, \ldots$, are independent and conform to the Poisson distribution with mean

$$\lambda_{it} \exp(\epsilon) = \exp(\ln e_{it} + s_{it} + \epsilon),$$

which is henceforth denoted as $N_{it} \sim \mathcal{P}oi(\lambda_{it} \exp(\epsilon))$. Formally,

$$P[N_{i1} = k_1, \ldots N_{iT_i} = k_{T_i}|\mathcal{E}_i = \epsilon] = \prod_{t=1}^{T_i} P[N_{it} = k_t|\mathcal{E}_i = \epsilon]$$
$$= \prod_{t=1}^{T_i} \left(\exp(-\lambda_{it} \exp(\epsilon)) \frac{(\lambda_{it} \exp(\epsilon))^{k_t}}{k_t!} \right).$$

A2 at the portfolio level, the sequences $(\mathcal{E}_i, N_{i1}, N_{i2}, \ldots)$ are assumed to be independent.
A3 the random effects $\mathcal{E}_i$ are independent, Normally distributed with zero mean and constant variance $\sigma_{\mathcal{E}}^2$.

When the canonical log link function is used in the Poisson regression model, as assumed here, assumption A3 amounts to using a Poisson-LogNormal model for claim counts. Under assumption A3, we then have $E[\exp(\mathcal{E}_i)] = \exp(\sigma_{\mathcal{E}}^2/2)$ according to the formula giving the mathematical expectation for the LogNormal distribution. Therefore, the latter factor has to be included in the calculation of the a priori expected number of claims (with a linear score s_{it}, the intercept of the regression model has thus to be modified accordingly). Formally, the a priori expected number of claims is equal to

$$E[N_{it}] = E[E[N_{it}|\mathcal{E}_i]] = \lambda_{it}E[\exp(\mathcal{E}_i)] = \lambda_{it} \exp(\sigma_{\mathcal{E}}^2/2).$$

Remark 5.7.1 If longer panels are available then the static random effects $\mathcal{E}_i$ can be replaced with dynamic ones $\mathcal{E}_{i1}, \mathcal{E}_{i2}, \ldots$ which discount past observations according to their seniority. This is easily done by replacing $\mathcal{E}_i$ with a random sequence $\mathcal{E}_{i1}, \mathcal{E}_{i2}, \ldots$ obeying a Gaussian process whose covariance structure accounts for the memory effect (AR1, for instance).

5.7.3 Single Behavioral Variable, or Signal

In order to predict the number of claims N_{it} reported by policyholder i during period t, let us assume that the insurer has a signal S_{it} at its disposal about the policyholder's behavior behind the wheel during the same period. Such a signal typically summarizes the individual driving behavior, using information such as the distance traveled, the road used and the time used, the observance of speed limits, the smoothness of acceleration and slowing down.

To refine risk evaluation, we now combine past claims experience with the available signal. Hence, each contract is represented by the sequence

$$(\mathcal{E}_i, \Gamma_i, N_{i1}, S_{i1}, N_{i2}, S_{i2}, N_{i3}, S_{i3}, \ldots)$$

where

$\mathcal{E}_i$ accounts for hidden information influencing claim frequencies N_{it}

Γ_i reflects the quality of driving revealed by the observed signal S_{it}.

It is important to realize here that signals are also influenced by traditional risk factors included in x_{it} so that we need to account for this effect in model design. Here, we assume that the signal consists in recording a number of events, or to round a continuous signal in multiples of a natural unit. This makes the mechanism more transparent, at the cost of a negligible loss of accuracy.

If the signal counts a number of events then we supplement A1-A3 stated above with

A4 Given $\mathcal{E}_i$, the claim counts $N_{i1}, N_{i2}, \ldots$ are independent and independent of $\Gamma_i, S_{i1}, S_{i2}, \ldots$.

A5 Given Γ_i, the signal counts $S_{i1}, S_{i2}, \ldots$ are independent and independent of $\mathcal{E}_i, N_{i1}, N_{i2}, \ldots$, and

$$S_{it} \sim \mathcal{P}oi\left(e_{it} \exp(\eta_{it} + \Gamma_i)\right).$$

where η_{it} is the signal score based on a priori features x_{it} and Γ_i is Normally distributed with zero mean and variance σ_Γ^2.

A6 Given $\mathcal{E}_i$ and Γ_i, all the observable random variables $N_{i1}, S_{i1}, N_{i2}, S_{i2}, \ldots$ are independent.

Assumptions A4–A6 are in line with the traditional actuarial approach to experience rating, in that they postulate that the dependance between signal and claim counts is only apparent and results from missing information. If we had a complete knowledge of policyholder's characteristics, i.e. if we knew $\mathcal{E}_i$, then the signal would not be needed for pricing. Because of limited knowledge about policyholder's driving style, the insurer uses the information contained in the signal that reveals the missing elements in expected claim counts. This is why the signal is separated into three components: the effect η_{it} of the available features x_{it}, the relevant information Γ_i contained in the signal, that may explain expected claim counts beyond the available x_{it}, and the Poisson noise. The correlation $\rho_{\mathcal{E},\Gamma}$ between $\mathcal{E}_i$ and Γ_i can be exploited to improve the estimation of the expected number of claims by combining observed signal values with past claims history.

Notice that claim counts N_{it} and signal values S_{it} are correlated by means of the pair $(\mathcal{E}_i, \Gamma_i)$ of random effects. For a signal consisting in a mixed Poisson count, this is easily seen as follows:

$$\mathrm{Cov}[N_{it}, S_{it}] = \mathrm{Cov}\big[\mathrm{E}[N_{it}|\mathcal{E}_i, \Gamma_i], \mathrm{E}[S_{it}|\mathcal{E}_i, \Gamma_i]\big]$$

because the conditional covariance is zero by virtue of A6. Hence,

$$\mathrm{Cov}[N_{it}, S_{it}] = e_{it}^2 \exp(s_{it} + \eta_{it})\mathrm{Cov}\big[\exp(\mathcal{E}_i), \exp(\Gamma_i)\big].$$

Now, as the pair $(\mathcal{E}_i, \Gamma_i)$ is jointly Normal, with zero mean, variances $\sigma_{\mathcal{E}}^2$ and σ_{Γ}^2, and correlation $\rho_{\mathcal{E},\Gamma}$, we get

$$\mathrm{Cov}\big[\exp(\mathcal{E}_i), \exp(\Gamma_i)\big] = \mathrm{E}\big[\exp(\mathcal{E}_i + \Gamma_i)\big] - \mathrm{E}\big[\exp(\mathcal{E}_i)\big]\mathrm{E}\big[\exp(\Gamma_i)\big]$$
$$= \exp\left(\frac{\sigma_{\mathcal{E}}^2 + \sigma_{\Gamma}^2}{2}\right)\left(\exp\big(\rho_{\mathcal{E},\Gamma}\sigma_{\mathcal{E}}\sigma_{\Gamma}\big) - 1\right)$$

which is not zero unless $\rho_{\mathcal{E},\Gamma} = 0$, that is, unless $\mathcal{E}_i$ and Γ_i are mutually independent (so that the signal brings no information about the claim counts).

5.7.4   Multiple Signals

Assume that q signals, denoted as $S_{it}^{(j)}$, $j = 1, 2, \ldots, q$, are available in addition to the p features comprised in x_{it}. In case several signals are available, the insurer may either combine them into a single one and proceed as explained above. Another possibility is to extend the model from the previous section to the multivariate case by assuming a specific dynamics for each signal as explained next.

If the signals all consist in counts of different events then assumptions A1–A3 are supplemented with the following hypotheses:

A4   conditionally to $\mathcal{E}_i$, claim counts $N_{i1}, N_{i2}, \ldots$ are independent and independent of $\Gamma_i^{(j)}, S_{i1}^{(j)}, S_{i2}^{(j)}, \ldots$ for $j = 1, 2, \ldots, q$.

A5   conditionally to $\Gamma_i^{(j)}$, the signal counts $S_{i1}^{(j)}, S_{i2}^{(j)}, \ldots$ are independent and independent of $\mathcal{E}_i, N_{i1}, N_{i2}, \ldots$, and

$$S_{it}^{(j)} \sim \mathcal{P}oi\big(e_{it} \exp(\eta_{it}^{(j)} + \Gamma_i^{(j)})\big)$$

where

$\eta_{it}^{(j)}$ is the score for the jth signal based on a priori features x_{it} and

$\Gamma_i^{(j)}$ is Normally distributed with zero mean and represents the additional information contained in the jth signal about claim frequencies, corrected for the effect of the features x_{it}.

Also, the random vector $(\mathcal{E}_i, \Gamma_i^{(1)}, \Gamma_i^{(2)}, \ldots, \Gamma_i^{(q)})$ is multivariate Normally distributed with zero mean vector and variance-covariance matrix Σ.

A6 Given $(\mathcal{E}_i, \Gamma_i^{(1)}, \Gamma_i^{(2)}, \ldots, \Gamma_i^{(q)})$, all the observable random variables $N_{i1}, S_{i1}^{(1)}, S_{i1}^{(2)}, \ldots, N_{i2}, S_{i2}^{(1)}, S_{i2}^{(2)}, \ldots$ are independent.

5.7.5 Credibility Updating Formulas

In addition to accounting for overdispersion and serial correlation, the random effects

$$(\mathcal{E}_i, \Gamma_i^{(1)}, \Gamma_i^{(2)}, \ldots, \Gamma_i^{(q)})$$

allow for credibility updates. In the classical actuarial approach based on claim counts, only, past numbers of claims enter the credibility formulas in addition to observable features x_{i,T_i+1} to explain N_{i,T_i+1}. Formally, the experience used to update future premiums relates to past claims history

$$\mathcal{H}_{i,T_i}^{\text{claim}} = \{N_{it}, \quad t = 1, \ldots, T_i\}.$$

This information enters the predictive distribution, i.e. the conditional distribution of N_{i,T_i+1} given $\mathcal{H}_{i,T_i}^{\text{claim}}$. With experience rating, the a priori expectation

$$\mathrm{E}[N_{i,T_i+1}] = \lambda_{i,T_i+1}\mathrm{E}[\exp(\mathcal{E}_i)]$$

is replaced with the a posteriori expectation

$$\mathrm{E}[N_{i,T_i+1}|\mathcal{H}_{i,T_i}^{\text{claim}}] = \lambda_{i,T_i+1}\mathrm{E}[\exp(\mathcal{E}_i)|\mathcal{H}_{i,T_i}^{\text{claim}}]$$

as explained earlier in this chapter. The pricing structure is slow to adapt in personal lines because the a priori expected claim frequencies λ_{it} are generally small.

With telematics and IoT, the past claims history $\mathcal{H}_{i,T_i}^{\text{claim}}$ can be enriched with behavioral data. This allows the pricing structure to become much more reactive but requires the development of multivariate credibility models. In this case, the policy-specific history $\mathcal{H}_{i,T_i}$ gathers all the a posteriori information

$$\mathcal{H}_{i,T_i} = \mathcal{H}_{i,T_i}^{\text{claim}} \cup \mathcal{H}_{i,T_i}^{\text{signals}} = \{N_{it}, S_{it}^{(1)}, \ldots, S_{it}^{(q)}, \quad t = 1, \ldots, T_i\}.$$

The multivariate mixed/credibility model describes the joint dynamics of N_{it}, $S_{it}^{(1)}$, $\ldots$, $S_{it}^{(q)}$, given a priori features $\boldsymbol{x}_{it}$. The predictive distribution now corresponds to the conditional distribution of N_{i,T_i+1} given $\mathcal{H}_{i,T_i}$. The a priori expectation is replaced with an a posteriori one

$$E[N_{i,T_i+1}|\mathcal{H}_{i,T_i}] = \lambda_{i,T_i+1}E[\exp(\mathcal{E}_i)|\mathcal{H}_{i,T_i}].$$

The factor $E[\exp(\mathcal{E}_i)|\mathcal{H}_{i,T_i}]/E[\exp(\mathcal{E}_i)]$ is the credibility correction, i.e. the ratio between the a posteriori and the a priori expected numbers of claims.

5.7.6 Presentation of the Data Set

The case study in this section is based on real driving data recorded by GPS, collected by a Spanish insurance company within the framework of a new form of insurance cover. Under such policies, motor insurance premiums are determined by taking into account not only the traditional risk factors but also the number of kilometers driven over a period of time as well as information on the number of kilometers driven at night, the number of kilometers driven in an urban area, and the number of kilometers driven at excess speed. The information available is a panel that describes yearly records on the number of claims and the driving patterns for each driver measured thanks to telemetry.

Excess speed, night-time driving and urban driving are considered to be signals of the type of driving habits or skills. We treat these signals as entire numbers, by rounding excess speed, night-time driving and urban driving in natural units of 500 km. Specifically, the three signals at our disposal are as follows:

$S_{it}^{(1)}$ = distance travelled in the night (in multiples of 500 km)

$S_{it}^{(2)}$ = distance driven above the speed limit (in multiples of 500 km)

$S_{it}^{(3)}$ = distance travelled in urban zones (in multiples of 500 km).

The joint dynamics of the number of claims N_{it} reported by policyholder i during period t and the three signals $S_{it}^{(j)}$, $j = 1, 2, 3$, will be exploited to predict the future number of claims. The total distance driven per year (in kilometers) is considered as an exposure to risk and as such enters our models as an offset. To avoid large dispersion, distance driven is expressed in hundreds of kilometers.

Let us briefly comment on the choice of these three signals. Night-time driving is usually associated to more accidents than day-time, especially at young ages, and the first signal captures this effect. Vehicle speed is commonly considered as the major determinant of crash risk for young adults. It is generally considered that reducing the amount of time spent above the speed limit, holds the potential of dramatically reducing accidents. This is exactly the information captured by the second signal, time being here measured by the actual distance driven above the speed limits (integrating the total distance traveled by means of offset). Notice that the signal excess speed records the number of kilometers traveled at a speed in excess of the posted limit. However we do not possess information about the extent of excess, so we cannot distinguish between a driver who drives 10% faster or 20% than the posted limit. Finally, we note that urban areas are often congested and crash risk is higher there than in sub-urban or rural zones, because of heavy traffic. The third signal records the distance traveled in the accident-prone urban areas.

The sample is made up of $n = 2{,}494$ insured drivers followed over the three calendar years 2009–2011. All policyholders have been observed for three years (so that $T_i = 3$ for all i). The mean age of all drivers in the sample in 2009 is 25.17 years (standard deviation 2.44) because the policies that involve collecting telematics information were only offered to young drivers (the maximum age in the sample being 30 years). Our sample comprised 51.60% of male drivers and 48.40% of female drivers.

Our yearly responses are the number of claims together with the number of count units of excess speed, night-time driving and urban driving (rounded in 500s kilometers). Our measure of exposure-to-risk is the distance driven measured as a continuous variable in 100s km.

5.7.7 Fitted Models

Here, we assume that the joint dynamics of $(N_{it}, S_{it}^{(1)}, S_{it}^{(2)}, S_{it}^{(3)})$, $t = 1, 2, \ldots$, is described by the multivariate mixed Poisson model described above. The $\texttt{glmer}$ function included in the R package $\texttt{lme4}$ has been used to fit a GLMM which incorporates both fixed-effects parameters and random effects in a linear predictor, via maximum likelihood.

We know that the expression for the likelihood of a mixed-effects model involves an integral over all the random effects. In our case, the likelihood associated to the observations writes

$$
\mathcal{L} = \prod_{i=1}^{n} \int_{-\infty}^{\infty} \int_{-\infty}^{\infty} \int_{-\infty}^{\infty} \int_{-\infty}^{\infty} \prod_{t=1}^{3} \left(\exp\left(-e_{it} \exp(s_{it} + \epsilon)\right) \frac{\left(e_{it} \exp(s_{it} + \epsilon)\right)^{k_{it}}}{k_{it}!} \right.
$$

$$
\left. \prod_{j=1}^{3} \left(\exp\left(-e_{it} \exp(\eta_{it}^{(j)} + \gamma_j)\right) \frac{\left(e_{it} \exp(\eta_{it}^{(j)} + \gamma_j)\right)^{l_{it}^{(j)}}}{l_{it}^{(j)}} \right) \right)
$$

$$
f_{\Sigma}(\epsilon, \gamma_1, \gamma_2, \gamma_3) d\epsilon d\gamma_1 d\gamma_2 d\gamma_3
$$

where k_{it} denotes the observed number of claims for policyholder i in period t, where $l_{it}^{(1)}$, $l_{it}^{(2)}$, and $l_{it}^{(3)}$ denote the corresponding observed values for the three signals, and where f_{Σ} is the joint probability density function of the random vector $(\mathcal{E}_i, \Gamma_i^{(1)}, \Gamma_i^{(2)}, \Gamma_i^{(3)})$, corresponding to the assumed multivariate Normal distribution with zero mean vector and variance-covariance matrix Σ. Let us mention that to achieve convergence, some care is needed and appropriate control parameters must be selected in relation with the nonlinear optimizer. To ensure numerical stability of the optimization algorithms, policyholder's age has been rescaled (divided by 100). Gender is coded as 1 for male drivers and as 0 for female drivers. Also, different units have been tested for the three signals (in 100 and 1,000 km, without affecting the results).

The multivariate model considers claim counts and the three signals simultaneously. We fit the multivariate model at once following the approach proposed by Faraway (2016, Sect. 9.3). The idea is to define count and signal identifiers by means of a categorical feature signalName with 4 levels, N, S1, S2, and S3, say, treated as fixed effects and to introduce an interaction between the signals and the other fixed effects, as well as corresponding 4-dimensional policyholder-specific random effects. In order to get 4 correlated random effects between claim counts and the 3 signals, we need to specify the random effect structure as (-1+signalName|id) where id denotes the policy identifier (allowing the actuary to track the same contract over time) entering model formula.

To illustrate the relevance of the approach proposed in this section, we compare the multivariate credibility model described above, including past claims history as well as the three signals, with a classical credibility model, based on claim counts only. More precisely, we fit univariate mixed Poisson models for panel data, separately for each signal and the number of claims. The results for the univariate models can be considered as those obtained by replacing the covariance matrix $\boldsymbol{\Sigma}$ with a diagonal one, with marginal variances along the main diagonal. In the univariate modelling, the four responses N_{it}, $S_{it}^{(1)}$, $S_{it}^{(2)}$, and $S_{it}^{(3)}$ are thus considered to be mutually independent (but serial dependence for fixed i is taken into account in all four cases). In the univariate approach (i.e. considering claim counts, or each signal, in isolation), the random effects are included by means of the component (1|id) entering model formula. In this case, only past claim experience is used to update the expected number of claims in future years.

Table 5.5 presents the results of the univariate and the multivariate counts models (estimated with the three-year panel 2009–2011). The difference between the univariate approach and the multivariate approach is that the former only considers one of the signals at a time and it completely ignores the association between them. We see that age has an overall effect that is negative, meaning that the older the driver the less claims are expected. Here we included age as a continuous feature entering the score because the interval of ages is narrow for this sample of young drivers and there was no evidence supporting a non-linear effect of age on the score scale. Interactions between age and gender were tested but no significant cross-effects were found.

The joint dynamics of the number of claims N_{it} filed by policyholder i during period t and the three signals $S_{it}^{(1)}$, $S_{it}^{(2)}$ and $S_{it}^{(3)}$ is as follows. In the multivariate modeling, the correlation structure and the serial dependence are both taken into account for the four responses N_{it}, $S_{it}^{(1)}$, $S_{it}^{(2)}$, and $S_{it}^{(3)}$: precisely, given centered, multivariate Normally-distributed random effects $(\mathcal{E}_i, \Gamma_i^{(1)}, \Gamma_i^{(2)}, \Gamma_i^{(3)})$, the responses are Poisson distributed with respective conditional means

Table 5.5 Model results for panel data on claims and driving count signals, 2009-2011

	Multivariate Model	Univariate models			
		S1 (Night)	S2 (Speed)	S3 (Urban)	Claims
(Intercept)	−5.08***	−4.33***	−3.31***	−2.34***	−4.99***
	(0.29)	(0.14)	(0.15)	(0.09)	(0.31)
S1 (Night)	0.76*				
	(0.32)				
S2 (Speed)	1.78***				
	(0.32)				
S3 (Urban)	2.47***				
	(0.28)				
Age	−5.21***	−1.33*	−4.27***	−3.29***	−5.95***
	(1.13)	(0.55)	(0.61)	(0.36)	(1.22)
Gender	−0.11	0.38***	0.22***	0.03	−0.09
	(0.06)	(0.03)	(0.04)	(0.02)	(0.06)
S1 (Night):Age	3.84**				
	(1.25)				
S2 (Speed):Age	0.76				
	(1.29)				
S3 (Urban):Age	3.02**				
	(1.11)				
S1 (Night):Gender	0.49***				
	(0.07)				
S2 (Speed):Gender	0.34***				
	(0.07)				
S3 (Urban):Gender	0.14*				
	(0.06)				

*** $p < 0.001$, ** $p < 0.01$, * $p < 0.05$

$$E[N_{it}|\mathcal{E}_i] = d_{it} \exp\left(-5.08 - 5.21\text{age}_i - 0.11\text{I}[\text{gender}_i = \text{male}] + \mathcal{E}_i\right)$$

$$E[S_{it}^{(1)}|\Gamma_i^{(1)}] = d_{it} \exp\left((-5.08 + 0.76) + (-5.21 + 3.84)\text{age}_i\right.$$
$$\left. +(-0.11 + 0.49)\text{I}[\text{gender}_i = \text{male}] + \Gamma_i^{(1)}\right)$$

$$E[S_{it}^{(2)}|\Gamma_i^{(2)}] = d_{it} \exp\left((-5.08 + 1.78) + (-5.21 + 0.76)\text{age}_i\right.$$
$$\left. +(-0.11 + 0.34)\text{I}[\text{gender}_i = \text{male}] + \Gamma_i^{(2)}\right)$$

$$E[S_{it}^{(3)}|\Gamma_i^{(3)}] = d_{it} \exp\left((-5.08 + 2.47) + (-5.21 + 3.02)\text{age}_i\right.$$
$$\left. +(-0.11 + 0.14)\text{I}[\text{gender}_i = \text{male}] + \Gamma_i^{(3)}\right).$$

The estimated fixed effects are coherent between the multivariate and univariate models so that the estimated scores $\widehat{s}_{it}$ and $\widehat{\eta}_{it}^{(j)}$ are very similar in both cases. The main advantage of the multivariate model is to estimate the covariance matrix Σ of the random vector $(\mathcal{E}_i, \Gamma_i^{(1)}, \Gamma_i^{(2)}, \Gamma_i^{(3)})$ which connects claim counts N_{it} to corresponding signals $(S_{it}^{(1)}, S_{it}^{(2)}, S_{it}^{(3)})$.

The estimated covariance matrix $\widehat{\Sigma}$ is as follows. The marginal standard deviations are estimated to

$$\widehat{\sigma}_{\mathcal{E}} = 0.836$$
$$\widehat{\sigma}_{\Gamma,1} = 0.521$$
$$\widehat{\sigma}_{\Gamma,2} = 0.753$$
$$\widehat{\sigma}_{\Gamma,3} = 0.438.$$

The estimated correlation coefficients are given by

$$\widehat{\rho}_{\mathcal{E},\Gamma,1} = 0.019$$
$$\widehat{\rho}_{\mathcal{E},\Gamma,2} = -0.204$$
$$\widehat{\rho}_{\mathcal{E},\Gamma,3} = 0.602$$
$$\widehat{\rho}_{\Gamma,1,2} = 0.026$$
$$\widehat{\rho}_{\Gamma,1,3} = 0.058$$
$$\widehat{\rho}_{\Gamma,2,3} = -0.484.$$

We see that signal 1 (night-time driving) brings little information about claim counts in our data basis. The effects of signals 2 and 3 clearly dominate with respective correlations of about 20% and 60%, exhibiting opposite signs. Signal 3 (urban driving) appears to be the most informative, and negatively correlated to signal 2 (excess speed). This can be explained by traffic congestion, reducing speed in urban areas. On our data set, the estimated correlation between $\mathcal{E}_i$ and $\Gamma_i^{(2)}$ appears to be negative. This can be attributed to the way excess speeds have been recorded in the database, without distinctions between small and large violations of the posted speed limit.

This also suggests that those signals are not very informative and could have been replaced with appropriate geographic risk factors. This points out to a general issue with UB insurance: the proper definition of the signals to be included in the system is of utmost importance for the system to be successful.

5.7.8   A Posteriori Corrections

The multivariate model does not outperform the classical, univariate one on aggregate. This is easily seen from Table 5.5, by noticing that the estimated fixed effects are very similar for the claim count component of the multivariate model and the univariate model for claim counts, only. In fact, a simple Poisson GLM with an intercept would predict numbers of claims close to the observed ones, provided the portfolio experience is stationary. This at both the portfolio level and within sub-portfolios. This comes from the marginal totals constraints imposed by the Poisson likelihood equations under canonical, log-link. The added value of the multivariate model proposed in this section consists in refined individual premium corrections, as explained next.

The credibility approach consists in predicting the number of claims for next year using the conditional distribution of the response given the past experience. Here, past experience gathers the observed numbers of claims filed in the past for the univariate model. In the multivariate case, it also includes the history of the 3 signals.

The expected number of claims to be filed by policyholder i in year $T_i + 1$ given past numbers of claims $N_{it} = k_{it}, t = 1, 2, \ldots, T_i$, and past values of signals $S_{it}^{(j)} = l_{it}^{(j)}, t = 1, 2, \ldots, T_i, j = 1, 2, 3$, can be obtained as follows. As random effects are static, past experience is more conveniently summarized into the statistics

$$k_i = \sum_{t=1}^{T_i} k_{it} \text{ and } l_i^{(j)} = \sum_{t=1}^{T_i} l_{it}^{(j)}.$$

Also, we define

$$\lambda_i = \sum_{t=1}^{T_i} \lambda_{it} = \sum_{t=1}^{T_i} \exp(\ln e_{it} + s_{it}) \text{ and } \delta_i^{(j)} = \sum_{t=1}^{T_i} \exp(\ln e_{it} + \eta_{it}^{(j)}).$$

Then,

$$E\left[N_{i,T_i+1} \middle| N_{i1} = k_{it}, S_{it}^{(j)} = l_{it}^{(j)}, t = 1, 2, \ldots, T_i, j = 1, 2, 3\right]$$

$$= E\left[N_{i,T_i+1} \middle| N_{i1} + \ldots + N_{iT_i} = k_i, S_{i1}^{(j)} + \ldots + S_{iT_i}^{(j)} = l_i^{(j)}, j = 1, 2, 3\right]$$

$$= e_{i,T_i+1} \exp(s_{i,T_i+1}) E\left[\exp(\mathcal{E}_i) \middle| N_{i1} + \ldots + N_{iT_i} = k_i, S_{i1}^{(j)} + \ldots + S_{iT_i}^{(j)} = l_i^{(j)}, j = 1, 2, 3\right]$$

$$= e_{i,T_i+1} \exp(s_{i,T_i+1}) \frac{A}{B}$$

where

$$A = (k_i + 1) \int_{-\infty}^{\infty} \int_{-\infty}^{\infty} \int_{-\infty}^{\infty} \int_{-\infty}^{\infty} \exp\left(-\lambda_i \exp(\epsilon)\right) (\exp(\epsilon))^{k_i+1}$$
$$\prod_{j=1}^{3} \left(\exp\left(-\delta_i^{(j)} \exp(\gamma_j)\right) (\exp(\gamma_j))^{l_i^{(j)}}\right) f_{\Sigma}(\delta, \gamma_1, \gamma_2, \gamma_3) d\epsilon d\gamma_1 d\gamma_2 d\gamma_3$$

$$B = \int_{-\infty}^{\infty} \int_{-\infty}^{\infty} \int_{-\infty}^{\infty} \int_{-\infty}^{\infty} \exp\left(-\lambda_i \exp(\epsilon)\right) (\exp(\epsilon))^{k_i}$$
$$\prod_{j=1}^{3} \left(\exp\left(-\delta_i^{(j)} \exp(\gamma_j)\right) (\exp(\gamma_j))^{l_i^{(j)}}\right) f_{\Sigma}(\delta, \gamma_1, \gamma_2, \gamma_3) d\epsilon d\gamma_1 d\gamma_2 d\gamma_3.$$

The updating coefficient is thus given by A/B. The integrals involved in A and B can be computed numerically with quadrature formulas as implemented in the R package MultiGHQuad.

In the univariate case, we simply get the ratio of two Mellin transforms:

$$
\begin{aligned}
\mathrm{E}\Big[N_{i,T_i+1}\Big|N_{i1} = k_{it}, t = 1, 2, \ldots, T_i\Big] &= \mathrm{E}\Big[N_{i,T_i+1}\Big|N_{i1} + \ldots + N_{iT_i} = k_i\Big] \\
&= e_{i,T_i+1}\exp(s_{i,T_i+1})\mathrm{E}\Big[\exp(\mathcal{E}_i)\Big|N_{i1} + \ldots + N_{iT_i} = k_i\Big] \\
&= e_{i,T_i+1}\exp(s_{i,T_i+1})\frac{C}{D}
\end{aligned}
$$

where

$$
C = (k_i + 1)\int_{-\infty}^{\infty} \exp\big(-\lambda_i \exp(\epsilon)\big)\big(\exp(\epsilon)\big)^{k_i+1}\frac{1}{\sigma_{\mathcal{E}}\sqrt{2\pi}}\exp\left(-\frac{\epsilon^2}{2\sigma_{\mathcal{E}}^2}\right) d\epsilon
$$

$$
D = \int_{-\infty}^{\infty} \exp\big(-\lambda_i \exp(\epsilon)\big)\big(\exp(\epsilon)\big)^{k_i}\frac{1}{\sigma_{\mathcal{E}}\sqrt{2\pi}}\exp\left(-\frac{\epsilon^2}{2\sigma_{\mathcal{E}}^2}\right) d\epsilon.
$$

Let us now demonstrate the added value of the multivariate model by computing individual premium corrections. The boxplots of the values of $\mathrm{E}[\exp(\mathcal{E}_i)|\mathcal{H}_{i,3}]$ based on the multivariate model involving the three signals and of the values of $\mathrm{E}[\exp(\mathcal{E}_i)|\mathcal{H}_{i,3}^{\mathrm{claim}}]$ based on the univariate (i.e. the classical credibility construction) Poisson-LogNormal model for claim counts are displayed in Fig. 5.2. Apart from the common increasing trend according to the number of claims $N_{i1} + N_{i2} + N_{i3}$ filed during the observation period, we see that there is more dispersion in the $\mathrm{E}[\exp(\mathcal{E}_i)|\mathcal{H}_{i,3}]$ values compared to the $\mathrm{E}[\exp(\mathcal{E}_i)|\mathcal{H}_{i,3}^{\mathrm{claim}}]$ values, because of the variety in the signals.

Let us now compare the values of $\mathrm{E}[\exp(\mathcal{E}_i)|\mathcal{H}_{i,3}]$ based on the multivariate model involving the three signals to the values of $\mathrm{E}[\exp(\mathcal{E}_i)|\mathcal{H}_{i,3}^{\mathrm{claim}}]$ obtained from the univariate model for claim counts, only, according to the total number of claims $N_{i1} + N_{i2} + N_{i3}$ filed during the observation period. The numerical values are displayed in Figs. 5.3 and 5.4. For claim-free policyholders, we see that the univariate model always grants a discount whereas its multivariate counterpart may impose a penalty, depending on the experience with signals. When a single claim is reported, both univariate and multivariate models may still award a discount or induce a penalty. For the univariate model, it depends on the a priori features of the driver (a priori riskier drivers are less penalized when a claim is reported to the company). For the multivariate model, it depends on the a priori features as well as on the experience recorded on signals. When two claims are reported, the univariate model always imposes a penalty whereas its multivariate counterpart may still award a discount, based on a favorable experience related to signals. When three claims (or more) are reported, both the univariate and multivariate models impose a penalty, but its extent also depends on the signals in the multivariate case.

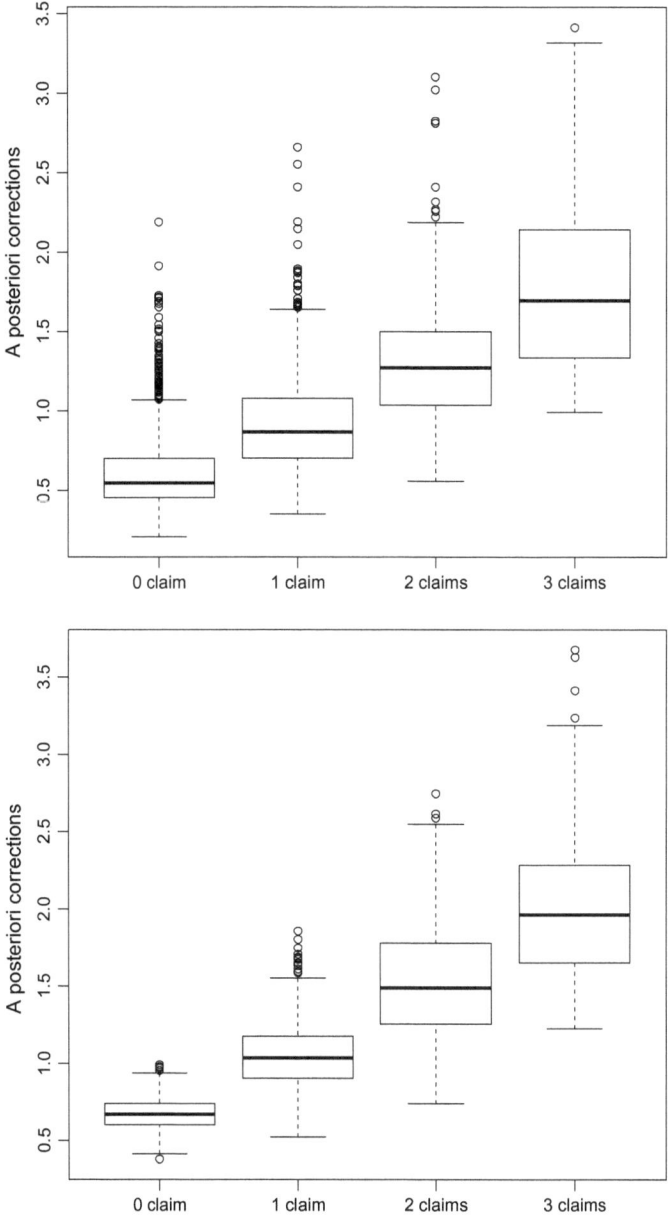

Fig. 5.2 Boxplots of the values of $E[\exp(\mathcal{E}_i)|\mathcal{H}_{i,3}]$ based on the multivariate model (upper panel) and of $E[\exp(\mathcal{E}_i)|\mathcal{H}_{i,3}^{\text{claim}}]$ (lower panel) obtained from the univariate model, according to the total number of claims $N_{i1} + N_{i2} + N_{i3}$ filed during the observation period

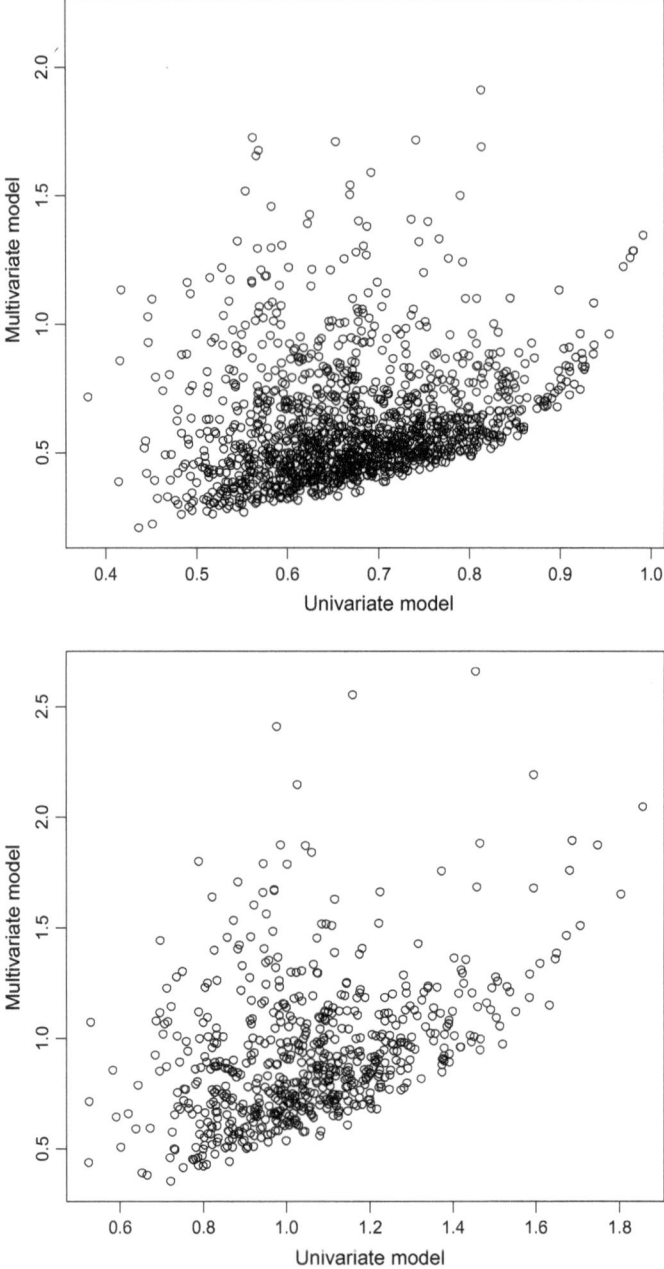

Fig. 5.3 Values of $E[\exp(\mathcal{E}_i)|\mathcal{H}_{i,3}]$ based on the multivariate model and of $E[\exp(\mathcal{E}_i)|\mathcal{H}_{i,3}^{\text{claim}}]$ obtained from the univariate model, according to the total number of claims filed during the observation period: $N_{i1} + N_{i2} + N_{i3} = 0$ (top panel) and $N_{i1} + N_{i2} + N_{i3} = 1$ (bottom panel)

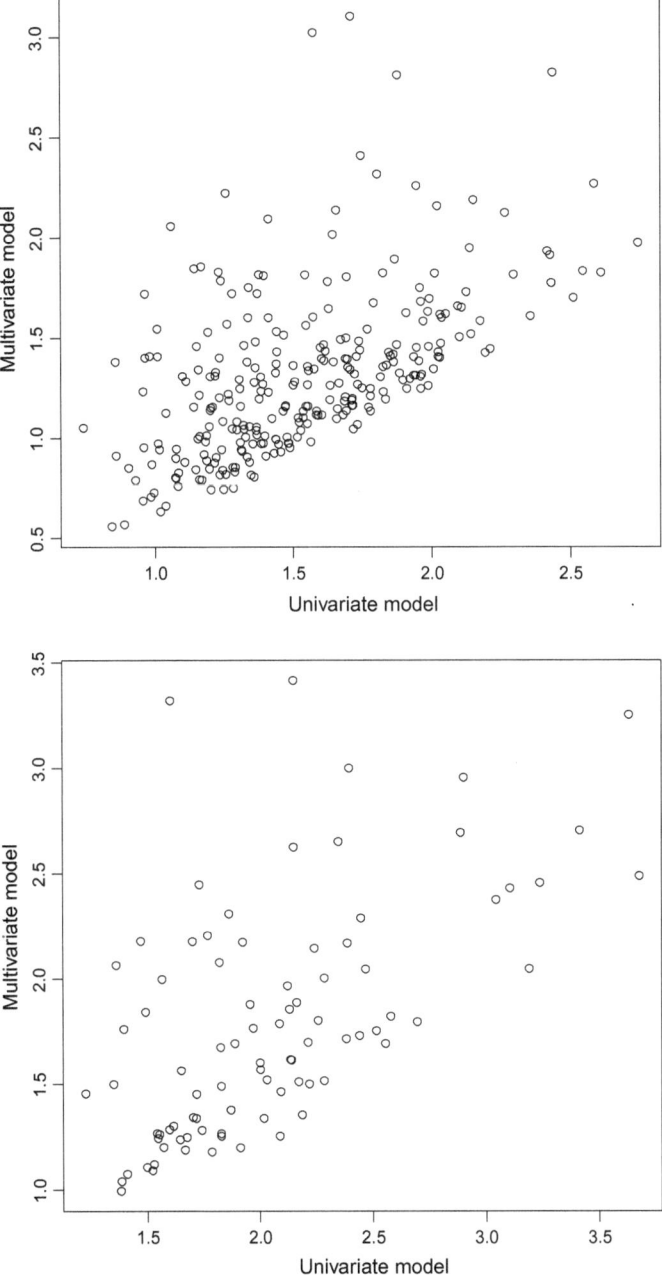

Fig. 5.4 Values of $E[\exp(\mathcal{E}_i)|\mathcal{H}_{i,3}]$ based on the multivariate model and of $E[\exp(\mathcal{E}_i)|\mathcal{H}_{i,3}^{claim}]$ obtained from the univariate model, according to the total number of claims filed during the observation period: $N_{i1} + N_{i2} + N_{i3} = 2$ (top panel) and $N_{i1} + N_{i2} + N_{i3} = 3$ (bottom panel)

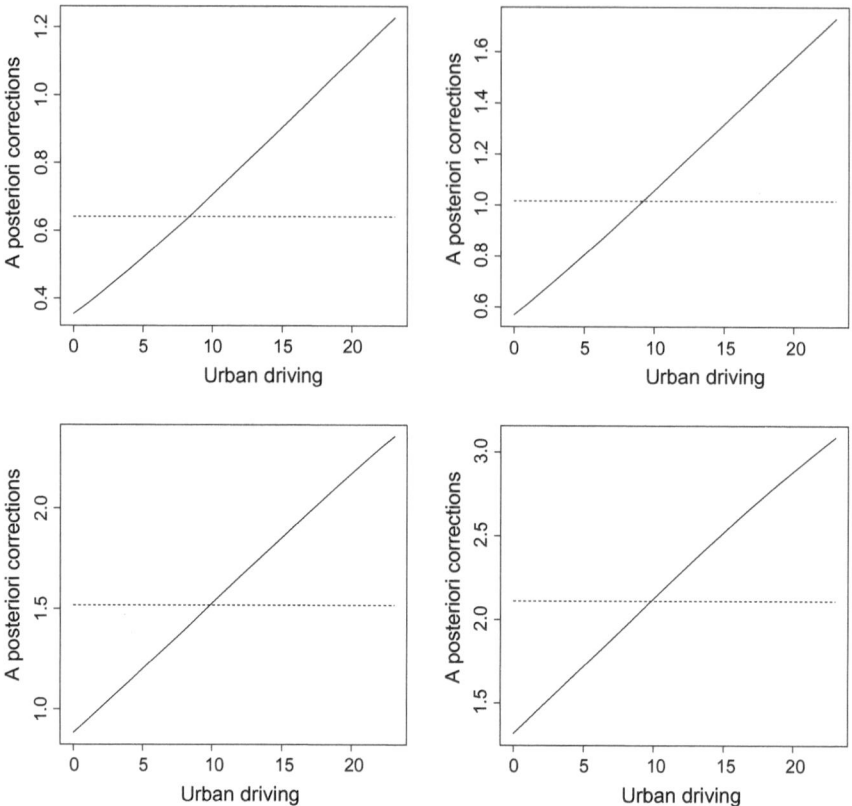

Fig. 5.5 Values of $E[\exp(\mathcal{E}_i)|\mathcal{H}_{i,3}]$ for an hypothetical male, mean-aged driver in function of the distance traveled in urban areas, based on the multivariate model (trending line) and of $E[\exp(\mathcal{E}_i)|\mathcal{H}_{i,3}^{claim}]$ obtained from the univariate model (horizontal line), according to the total number of claims filed during the observation period: $N_{i1} + N_{i2} + N_{i3} = 0$ (top left), $N_{i1} + N_{i2} + N_{i3} = 1$ (top right), $N_{i1} + N_{i2} + N_{i3} = 2$ (bottom left), and $N_{i1} + N_{i2} + N_{i3} = 3$ (bottom right)

Let us now consider a male policyholder with average age and driving the average annual distance. Also, we fix the signals 1 and 2 at their average value, but we let the third signal vary from 0 to its maximal value given the assumed total distance traveled (corresponding to the average number of kilometers driven in the database). Based on the number of claims reported during the three years, we compute the a posteriori corrections to assess the impact of the signal. The results are displayed in Fig. 5.5. For a policyholder without claim ($N_{i1} + N_{i2} + N_{i3} = 0$), we see that having a better driving style (small value of the signal) considerably increases the discount compared to the classical credibility correction based on past claims, only (represented by the horizontal line on the graph). However, a deteriorated signal may also reduce the discount, and even result in a penalty in the worst case. For a policyholder having reported a single claim ($N_{i1} + N_{i2} + N_{i3} = 1$), we see that depending on the value of the signal, the premium may increase or decrease (whereas it moderately increases using the classical credibility formula). Hence, the signal can compensate for the effect of a single claim. When two or three claims are reported, the policyholder suffers a penalty whatever the value of the signal, but the latter can attenuate the penalty compared to the classical credibility model based on past claim experience, only.

Remark 5.7.2 To end with, let us mention that the same approach can be used to model dependent counts, such as the number of claims with material damage, only, and the number of claims with bodily injuries reported by the same policyholder, for instance, or claim frequencies related to different insurance coverages.

5.8   Mixed Models, Penalized Log-Likelihood, and Regularization

Actuaries around the world are used to analyse large historical databases to identify the factors influencing the risk borne by the insurance provider, with the help of GLMs and related tools. Whereas the number of policyholders remained roughly stable over time, the amount of information available about each of them dramatically increased in the last decade. Indeed, web navigation and tools of the digital revolution such as smartphones and wearable devices provide actuaries with more data than ever before. Combining various sources of information, including banking or telematics, allows insurers to gather more and more data about their policyholders. Mining this increasing volume of data and developing innovative techniques to evaluate risks certainly offers great competitive advantages and the basic GLM tools have not been developed for that purpose.

GLMs are difficult to apply when the actuary deals with the kind of massive data sets that have now become routinely available in insurance studies. Precisely, they reveal unable to deal with the "large p" case: even if they can easily be applied in large portfolios (n running in millions is no more an issue), they have difficulties to fit many features (large p). Because of the "large p" inability, it is impossible to include all possible interactions in the GLM, letting it test for the relevant ones, because doing so dramatically increases the number of features (supplementing original ones with all possible products, say).

If n is much larger than p, estimates have low bias and low variance provided the GLM is correctly specified. Thus, the model performs well on test observations. However, if p becomes large with respect to n, there can be a lot of variability in the fit, resulting in overfitting and consequently poor predictions on observations not used in model training. Constraining, or shrinking the estimated regression coefficients often substantially reduces the variance at the cost of a negligible increase in bias. This leads to substantial improvements in prediction accuracy for observations not used in model training.

Model selection becomes problematic when many features are available, that is, when p is large. Indeed, it becomes rapidly difficult to consider all possible models: except for moderate values of p, testing all 2^p possible models becomes prohibitive. This is why penalties are more effective in a GLM context.

With penalized likelihood estimation, the objective function becomes

$$\text{log-likelihood} + \text{penalty}.$$

Such penalties are often used to impose smoothness in the estimates obtained with splines, as it will become clear from the next chapter. To ensure regularization, typical penalties are based on $\sum_{j=1}^{p} |\beta_j|^k$ multiplied with a tuning coefficient λ to be selected from the data. With $k = 0$ this essentially reduces to the number of parameters (as in the AIC), while $k = 1$ and $k = 2$ correspond to the lasso and ridge penalties, respectively.

Lasso is the acronym for "Least Absolute Selection and Shrinkage Operator". It consists in adding

$$\text{penalty} = \lambda \sum_{j=1}^{p} |\beta_j|$$

to the log-likelihood, where

- λ is a tuning parameter that can be determined by cross validation
- and the intercept is not subject to penalty.

Lasso has the effect of forcing some of the coefficient estimates to be exactly equal to 0 when the tuning parameter λ is sufficiently large. Precisely, lasso yields sparse models, i.e. models that involve only a subset of the available features:

- As λ increases, more coefficient estimates are set to zero (so that less covariates enter the model).
- With lasso, variable selection is built into the procedure.

In that respect, Lasso can be seen as a computationally feasible alternative to best subset selection, as well as to stepwise selection. It appears to be a powerful way to select the optimal set of predictors, by balancing goodness-of-fit and model complexity.

The lasso parameter λ can be selected by cross validation (that is, by training the model on a subset of the database, to predict the remaining data with the fitted model, and to evaluate the goodness of prediction on that basis). Basing the selection on out-of-sample performances, or predictive ability is the only valid approach because every features added to a model can only improve goodness-of-fit. However, adding unwarranted variables decreases the precision of parameter estimates. While a feature may be statistically significant, adding it to the model may not be worthwhile, because the improvement in fit may be outweighed by loss of estimation precision. This is appropriately measured by cross validation.

And here comes the link with the models studied in this chapter. Penalized likelihood is closely related to mixed models. Indeed, coming back to the general definition of GLMMs given in Remark 5.3.2, it becomes clear that the log-likelihood based on

$$f(y, b) = f(y|b) f(b)$$

is of the form

$$\ln f(y|b) + \ln f(b) = \text{ED log-likelihood} + \text{"Penalty"}.$$

Hence, the logarithm of the probability density function of the random effects plays the role of the penalty added to the log-likelihood. For instance, considering b as the outcome of a random vector B obeying the multivariate Normal distribution with independent $\mathcal{N}or(0, 1/\lambda)$ components, we recover the ridge penalty $\lambda \sum_{j=1}^{p} \beta_j^2$.

5.9 Bibliographic Notes and Further Reading

Cameron and Trivedi (1986) considered the NB p distributions having the mean λ_i but a variance of the form $\lambda_i + \alpha \lambda_i^p$. This kind of distribution can be generated with an heterogeneity factor following the Gamma distribution with unit mean and variance $\alpha \lambda_i^{p-2}$. Note that the variance of Θ_i now depends on the individual characteristics of the policyholders. When $p = 1$ and 2, we recover the NB1 and NB2 distributions, respectively. The NB1 model is interesting because the variance $\text{Var}[N_i] = \lambda_i + \alpha \lambda_i = \phi \lambda_i$ is the one corresponding to the Overdispersed Poisson GLM approach, or ODP. Winkelmann and Zimmermann (1991, 1995) proposed to treat p as an unknown parameter. This leads to a distribution called "Generalized Event Count".

We refer the reader to the two chapters by Katrien Antonio and Yanwei Zhang in Frees et al. (2014) for an overview of the estimation techniques for GLMMs, in the frequentist as well as Bayesian context, and to Fahrmeir et al. (2013) for an extensive treatment. Details about numerical methods, including PQL and quadratures as well as EM algorithm can be found there. The Bayesian approach to GLMMs and neural networks is presented in Chap. 3 of Hainaut et al. (2019).

Nelder and Verrall (1997) and Frees et al. (1999) demonstrate the connection of actuarial credibility models to statistical mixed models. This allows the actuary to resort to computational methods and softwares available for mixed models, to deal with the analysis of these actuarial models. See also Frees et al. (2001), Antonio and Beirlant (2007), and Ohlsson (2008). The cm procedure of the package actuR of the free software R allows to fit several standard actuarial credibility models.

GEEs have been proposed by Liang and Zeger (1986) as a computational tool for marginal models. This approach has been successfully applied in several insurance studies, including Brouhns et al. (2003).

The application to telematics data is taken from Denuit et al. (2019) where a modeling for continuous signals is also proposed. We also refer the interested reader to that paper for an extensive review of the literature.

Lasso and ridge penalties are also very useful in neural networks, as it will be seen in Hainaut et al. (2019). More details about the practical implementation of these techniques can be found there.

References

Antonio K, Beirlant J (2007) Actuarial statistics with generalized linear mixed models. Insur: Math Econ 40:58–76

Brouhns N, Guillen M, Denuit M, Pinquet J (2003) Bonus-malus scales in segmented tariffs with stochastic migration between segments. J Risk Insur 70:577–599

Cameron AC, Trivedi PK (1986) Econometric models based on count data: comparisons and applications of some estimators. J Appl Econ 46:347–364

Denuit M, Guillen M, Trufin J (2019) Multivariate credibility modeling for usage-based motor insurance pricing with behavioral data. Ann Actuarial Sci

Fahrmeir L, Kneib T, Lang S, Marx B (2013) Regression models, methods and applications. Springer

Faraway JJ (2016) Extending the linear model with R: generalized linear, mixed effects and non-parametric regression models, 2nd edn. CRC, Boca Raton, FL

Frees EW, Derrig RA, Meyers G (2014) Predictive modeling applications in actuarial science. Volume I: Predictive Modeling Techniques. International Series on Actuarial Science. Cambridge University Press

Frees E, Young V, Luo Y (1999) A longitudinal data analysis interpretation of credibility models. Insur: Math Econ 24:229–247

Frees E, Young V, Luo Y (2001) Case studies using panel data models. North Am Actuarial J 5:24–42

Hainaut D, Trufin J, Denuit M (2019) Effective statistical learning methods for actuaries—neural networks and unsupervised methods. Springer Actuarial Series

Liang K, Zeger S (1986) Longitudinal data analysis using generalized linear models. Biometrika 73:13–22

Nelder JA, Verrall RJ (1997) Credibility theory and generalized linear models. ASTIN Bull 27:71–82

Ohlsson E (2008) Combining generalized linear models and credibility models in practice. Scand Actuarial J 2008:301–314

Trufin J, Denuit M, Hainaut D (2019) Effective statistical learning methods for actuaries—tree-based methods. Springer Actuarial Series

Winkelmann R, Zimmermann KF (1991) A new approach for modeling economic count data. Econ Lett 37:139–143

Winkelmann R, Zimmermann KF (1995) Recent development in count data modelling: theory and application. J Econ Surv 9:1–24

Part III
Additive Models

Chapter 6
Generalized Additive Models (GAMs)

6.1 Introduction

We have seen in Chap. 4 that GLMs offer an effective solution to many problems in actuarial science. However, the linearity assumption about the score is sometimes questionable: nonlinear effects of continuous features such as policyholder's age may well need to be included on the score scale to reflect the actual experience. Continuous features can efficiently enter GLMs only if they are suitably transformed to reflect their true effect on the score scale. Unfortunately, it is not always clear how the variables should be transformed before inclusion in the score. It has been common practice in insurance companies to model possibly nonlinear effects with the help of polynomials. However, low-degree polynomials are often not flexible enough to capture the structure present in the data whereas increasing their degree produces unstable estimates, especially for extreme values of the features.

Another approach commonly used in practice consists in replacing the continuous feature with a categorical one obtained by dividing its domain into a moderate number of sub-intervals and to assume that the feature has a constant effect on the score within each interval. This means that the function giving the effect of the continuous feature on the score scale is assumed to be piecewise constant. Such a banding obviously results in a loss of information. Moreover, there is no general rule to determine the optimal choice of cut-offs so that banding may bias risk evaluation.

Generalized Additive Models (GAMs) have entered the actuary's toolkit to deal with continuous features in a flexible way. In this setting, the continuous features enter the model in a semi-parametric additive predictor. In Property and Casualty insurance, the effect of the policyholder's age, power of the vehicle or sum insured can be modeled by means of GAMs. GAMs also allow the actuary to analyze risk variations by geographic area, accounting for the possible interaction between continuous features (latitude and longitude giving the spatial location, in this case). Other interactions generally present in the data include age and power as well as gender and age in motor insurance. They can also be captured by GAMs.

© Springer Nature Switzerland AG 2019
M. Denuit et al., *Effective Statistical Learning Methods for Actuaries I*,
Springer Actuarial, https://doi.org/10.1007/978-3-030-25820-7_6

In life insurance, mortality statistics are often aggregated over groups of policyholders sharing the same characteristics (such as age, gender or type of product, for instance) in order to conduct a GLM analysis. Death counts originating from given exposures can then be studied using Poisson regression techniques. The key theoretical argument legitimating Poisson GLM is that under mild assumptions, the Poisson likelihood appears to be proportional to the true likelihood. Therefore, basing statistical inference on the Poisson likelihood is not restrictive. This is particularly interesting from a practical point of view since this allows actuaries to resort to the available statistical software performing Poisson regression to build life tables.

When continuous features are included in survival analysis, such as the sum insured for instance, the aggregation prior to GLM Poisson requires a preliminary, subjective banding. A clever use of GAMs avoids this problematic step and allows the actuary to analyze individual mortality data using Poisson regression. The necessary database contains one record per policy and per year (the same format as in motor insurance, for instance) with a binary variable indicating death and a reference number linking all the records related to the same policy. This sometimes requires to augment the data basis by splitting each individual observation into a set of independent realizations of Poisson counts sharing the same time-invariant explanatory variables, or letting these features evolve over time in case they are dynamic (such as attained age, for instance). This makes the GAM approach equally effective in life and nonlife insurance studies.

6.2 Structure of GAMs

6.2.1 Response Distribution

As for GLMs, we consider responses $Y_1, Y_2, \ldots, Y_n$ measured on n individuals. For each response, the actuary has a vector $\boldsymbol{x}_i = (1, x_{i1}, \ldots, x_{ip})^\top$ of dimension $p + 1$ containing the corresponding features (supplemented with a 1 in front when an intercept is included in the score). Features are now subdivided into p_{cat} categorical variables $x_{i1}, \ldots, x_{ip_{\text{cat}}}$ and $p_{\text{cont}} = p - p_{\text{cat}}$ continuous variables $x_{i,p_{\text{cat}}+1}, \ldots, x_{ip}$.

Given the information contained in $\boldsymbol{x}_i$, we assume again that the responses Y_i are independent with a conditional distribution in the ED family thoroughly studied in Chap. 2. Specifically, Y_i obeys the distribution (2.3) with specific canonical parameter θ_i and the same dispersion parameter ϕ. The way θ_i is linked to the information contained in $\boldsymbol{x}_i$ is explained next.

6.2.2 Additive Scores

In practice, many features are categorical and efficiently enter GLMs after coding by means of binary variables. However, some important rating variables are continuous: in motor ratemaking, age or power of the car are both continuous, for instance. In life and health insurance, age and sum insured are continuous features. In general, continuity refers here to a quantitative feature with many distinct, ordered values. For instance, age is generally recorded as age last birthday so that this feature assumes only integer values (18, 19, 20, ...in motor insurance, for instance). It is nevertheless considered as a continuous one as we expect a smooth progression of risk with age, without sudden jump.

Let us now explain the limitations of a GLM analysis if we fail to recognize that a continuous feature $x^\star$ may have a nonlinear effect on the score. Entering $x^\star$ directly in the GLM score boils down to assume a linear effect of $x^\star$ on the score scale: with the log-link function, this means that the mean response is constrained to vary exponentially with $x^\star$. However, such a monotonic exponential behavior may not be supported by the data as shown in the next example.

Example 6.2.1 Consider for example age in motor insurance ratemaking. Entering age directly in the GLM score implies a linear effect of age on the score scale. For instance, consider the number of claims Y_i reported by policyholder i and the Poisson GLM

$$Y_i \sim \mathcal{P}oi \left(e_i \exp \left(\beta_0 + \sum_{j=1}^{p} \beta_j x_{ij} + \beta_{\text{age}} \text{age}_i \right) \right),$$

where e_i is the exposure to risk and x_i summarizes all the information available about policyholder i, besides attained age. We see that the expected number of claims

$$\mu_i = e_i \exp \left(\beta_0 + \sum_{j=1}^{p} \beta_j x_{ij} + \beta_{\text{age}} \text{age}_i \right)$$

either increases (if $\beta_{\text{age}} > 0$) or decreases (if $\beta_{\text{age}} < 0$) with age in an exponential way. However, the impact of age on the expected number of claims in motor insurance has typically a U-shape (that may be distorted because of children borrowing their parents' car and causing accidents), conflicting with the exponential behavior implicitly assumed by GLMs.

Entering a continuous feature $x^\star$ in a linear way on the score scale may thus lead to a wrong risk evaluation, whereas treating it as a categorical feature with as many levels as distinct values of $x^\star$ may

- introduce a large number of additional regression parameters;
- fail to recognize the expected smooth variation of the mean response in $x^\star$ (as regression parameters corresponding to each level are no more related after coding by means of binary variables).

In general, this produces a lot of variability in the estimated regression coefficients associated to the different levels of the continuous feature $x^\star$ when it is treated as a categorical one.

Before GAMs entered the actuary's toolkit, the classical approach to deal with continuous features was purely parametric: every continuous feature $x^\star$ was transformed before being included in the score. In general, the function giving the impact of $x^\star$ on the score is linear, polynomial, or piecewise constant. So, the function is known up to a finite number of parameters to be estimated from the data and we are back to the GLM case. However, no matter what parametric form we specify, it will always exclude many plausible functions.

Actuaries often expect a smooth variation of the mean response with $x^\star$. For instance, the expected number of claims in motor insurance or the death rate in life insurance are expected to vary smoothly with age, without sudden jump from one age to the next. In this chapter, we explain how to fit a model with an additive score allowing for nonlinear effects of continuous features on the score scale that are learned from the available data. The following example in motor insurance explains the added value of GAMs in that respect.

Example 6.2.2 Let us continue with Example 6.2.1. Consider the data displayed in Fig. 6.1 The Poisson regression model has been fitted to the data (Y_i, e_i, x_i). We can see the risk exposures e_i (in policy-year) as well as the result obtained by treating age linearly on the score (straight line) or as a categorical features, with a specific regression coefficient for each age (circles). We see that none of these approaches is satisfying: the linear effect misses the U-shape age effect whereas the age-specific regression coefficients are not smooth enough.

This is why we switch to a score of the form

$$\beta_0 + \sum_{j=1}^{p} \beta_j x_{ij} + b(\text{age}_i)$$

for some unknown function $b(\cdot)$ of age. No assumption is made about $b(\cdot)$ except that it is a smooth function of age: it is not assumed to obey any rigid parametric form, like a polynomial for instance. In essence, $x \mapsto \widehat{b}(x)$ is a smoothed version of the regression coefficients $x \mapsto \widehat{\beta}_x$ obtained by treating age x as a categorical feature (so that age is coded by a set of binary features, one for each age observed in the data basis except the reference one). In our example, the $\widehat{\beta}_x$ are the circles visible in the upper right panel of Fig. 6.1. Specifically, we explain in this chapter how to fit the model

$$Y_i \sim \mathcal{P}oi\left(e_i \exp\left(\beta_0 + \sum_{j=1}^{p} \beta_j x_{ij} + b(\text{age}_i) \right) \right).$$

The result is displayed in the lower right panel of Fig. 6.1. We can see that $\widehat{b}(\cdot)$ nicely smooths the $\widehat{\beta}_x$ except at older ages where a lot of volatility is present in the data (because of small exposures, as it can be seen from the upper left panel in Fig. 6.1).

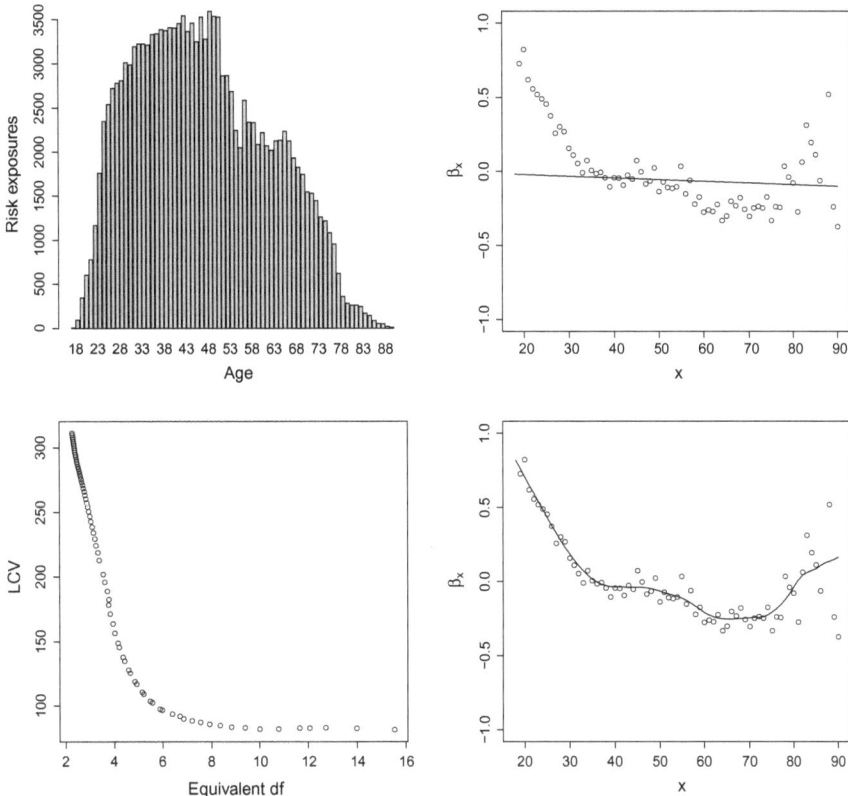

Fig. 6.1 Age effect on expected claim counts in motor insurance. Risk exposures (upper left panel), linear effect or categorical effects implicitly assumed by GLMs (upper right panel), Likelihood Cross-Validation, or LCV plot (bottom left panel) and corresponding GAM fit (lower right panel)

The criterion used to select the degree of smoothness from the data displayed in the lower left panel, called LCV for Likelihood Cross Validation, will be explained in the next section.

The same situation is encountered in life insurance. Considering Fig. 4.2, we see that death probabilities do not depend linearly on age over the whole age range. Death probabilities are rather high for newborns, then fall down rapidly to a lower mortality level for children and teenagers. For ages 20–25 the so-called accident hump is observed, which is particularly pronounced for males. From age 30 onwards, death probabilities increase almost linearly with age (on the score scale). At older ages, the one-year death probabilities remain increasing but with a different shape. The crude $\widehat{q}_x = D_x/l_x$ displayed in Fig. 4.2 can be seen as the output of a Binomial GLM where age x is treated as a categorical feature. The resulting estimated death probabilities are not connected so that erratic variations are clearly visible there.

Traditionally, age is included in the model by means of a polynomial. This has been done at older ages in Chap. 4 using the quadratic Perks formula. In order to represent the functional form of age for mortality rates usually a polynomial of order six or higher is required if the entire life span is considered. However, polynomials of higher order tend to show wild fluctuations in the tail regions (that is, for very young and very old ages) and individual observations in the tail regions might strongly influence the curve.

Another approach consists to estimate the functional form of age non-parametrically. In contrast to a parametric approach, the actuary does not specify a certain function in advance, but lets the data determine the optimal form. Methods used for non-parametric estimation include smoothing splines or local GLM methods, as explained in the next sections.

6.2.3 Link Function

We are now ready to state the definition of GAMs. As for the GLM, the response Y_i has a distribution in the ED family. The p_{cat} categorical features $x_{i1}, \ldots, x_{i,p_{cat}}$ are combined with the continuous features $x_{i,p_{cat}+1}, \ldots, x_{ip}$ to form the additive score

$$\text{score}_i = \beta_0 + \sum_{j=1}^{p_{cat}} \beta_j x_{ij} + \sum_{j=p_{cat}+1}^{p} b_j(x_{ij}).$$

Let g be the link function. The mean μ_i of Y_i is linked to the nonlinear score by the relationship

$$g(\mu_i) = \beta_0 + \sum_{j=1}^{p_{cat}} \beta_j x_{ij} + \sum_{j=p_{cat}+1}^{p} b_j(x_{ij}).$$

The smooth, unspecified functions b_j and the regression coefficients $\beta_0, \beta_1, \ldots, \beta_{p_{cat}}$ are estimated from the available data using the techniques presented in the next section (local GLMs or splines).

GLMs assume a specific shape for the functions b_j, such as linear or quadratic, that can be estimated parametrically. On the contrary, GAMs do not specify any rigid parametric form to b_j but let them unspecified. The only assumption made about b_j is smoothness. Translated into mathematical terms, this means that b_j is continuous with continuous first, second, third, ... derivatives. This restriction is far less limiting than the parametric approach. This assumption is often desirable in insurance studies where jumps or break rarely occurs. And this simple assumption appears to be enough to estimate b_j using semi-parametric approaches such as local GLMs or splines.

As little is required about the function b_j, we need to impose $E[b_j(X_{ij})] = 0$ to have an identifiable model. Otherwise, we could add and subtract constants into the one-dimensional functions b_j, leaving the score unchanged. This constraint is generally taken into account after the last step in estimation, by centering all the estimated function b_j (subtracting the average value of the transformed jth feature).

In addition to univariate functions, interaction terms of the form $b_{jk}(x_{ij}, x_{ik})$ or $b_{jkl}(x_{ij}, x_{ik}, x_{il})$ may also enter the score. For more than three features, it is generally impossible to fit higher-order interaction terms. Sometimes, it is useful to include constrained terms of the form $b_{jk}(x_{ij}x_{ik})$ with argument the product of the jth feature together with the kth one, especially when the sample size is moderate. The following examples illustrate the use of functions with two arguments in a GAM analysis.

Example 6.2.3 It has now become common practice to let the risk premium per unit exposure vary with geographic area when all other risk factors are held constant. In motor insurance for instance, most companies have adopted a risk classification according to the geographical zone where the policyholder lives (urban vs. non urban for instance). If the insurer wishes to implement a more accurate splitting of the country according to postcodes then the spatial variation related to geographic factors (i.e. any spatially structured effect, including socio-demographic factors) can be captured with the help of GAMs. The main assumption behind this approach is that claim characteristics tend to be similar in neighboring postcode areas (after other risk factors have been accounted for). GAMs exploit this spatial smoothness by allowing for information transfer to and from neighboring regions.

The geographic effect can be captured by the location of policyholder's residence on the map, represented by latitude and longitude. The score then includes a term of the form

$$\ldots + b_{\text{spat}}(\text{latitude}_i, \text{longitude}_i) + \ldots$$

where the function $b_{\text{spat}}(\cdot, \cdot)$ is assumed to be smooth and estimated from the data.

Example 6.2.4 An interaction commonly found in motor insurance involves age and power of the car. Often, powerful cars driven by young, unexperienced drivers appear to be particularly dangerous whereas the effect of power varies with age. This means that the score then includes a term of the form

$$\ldots + b(\text{power}_i, \text{age}_i) + \ldots$$

where the function $b(\cdot, \cdot)$ is assumed to be smooth and estimated from the data.

Interactions between categorical and continuous features can also be included in the score, as shown in the next example.

Example 6.2.5 To allow for a possible interaction between the policyholder's age and gender (gender$_i = 1$ if policyholder i is a male driver and gender$_i = 0$ otherwise), the predictor includes

$$b_1(\text{age}_i) + b_2(\text{age}_i)\text{gender}_i$$

In this model the function b_1 corresponds to the (nonlinear) effect of age for females, and b_2 to the deviation from this effect for the males. The effect for males is therefore given by the sum $b_1 + b_2$.

Compared to tree-based methods (Trufin et al. 2019) and neural networks (Hainaut et al. 2019), interactions must be structured before inclusion in GAMs. Thus, trees and neural networks are helpful to explore possible interaction effects present in the data to guide the actuary when specifying appropriate additive scores in GAMs.

To sum up, additive models are much more flexible than linear ones, but remain interpretable since the estimated function b_j can be plotted to visualize the marginal relationship between the continuous feature x_{ij} and the response. Compared to parametric transformations of the features, such as polynomials, additive models determine the best transformation b_j without any rigid parametric assumption beyond smoothness. GAMs thus offer a data-driven approach aiming at discovering the separate effect of each feature on the score, assumed to be additive.

6.3   Inference in GAMs

6.3.1   Local Polynomial GLMs

6.3.1.1   Intuitive Idea

To start with, let us assume that we have a single continuous feature x to model a response Y. Local regression is used to model the relation between the feature x and the response Y. Typically, x represents age in the application that we have in mind in this chapter, while Y may be related to mortality or morbidity, or be a claim count or a claim severity. As there is a unique feature, the link function only ensures that the estimated mean belongs to the set of admissible values and we can simply equate the canonical parameter to the unknown function we wish to estimate, i.e. work with the function $x \mapsto \theta(x)$. The central assumption is smoothness: the function $x \mapsto \theta(x)$ is assumed to be continuous, with continuous derivatives. This means that we expect a smooth progression of the mean response in the feature x. Once the function to be estimated has been assumed to be smooth, it can be expanded using Taylor formula to get the approximation

$$\theta(x) \approx \beta_0 + \beta_1 x + \ldots + \beta_k x^k.$$

However, such a polynomial approximation usually performs very poorly, especially in the tails. This is why alternatives have been developed.

With local polynomials, the polynomial approximation is not applied globally, but only in the neighborhood of the point of interest. This means that we recognize that the polynomial behavior suggested by Taylor expansion only holds locally, so that the coefficients of the approximating polynomial may vary according to the value of x. In its easiest form, the estimation method exploits the well-known result that any differentiable function can be approximated locally by its tangent line.

Having observed $(x_1, Y_1), (x_2, Y_2), \ldots, (x_n, Y_n)$, we assume that the response Y_i has a probability mass/density function (2.3) taken from the ED family with canonical parameter θ_i and unit weight, where $\theta_i = \theta(x_i)$ is a function of x_i (under canonical link function). In the GLM approach, it is assumed that $\theta(x)$ has a specific parametric form, for instance $\theta(x) = \beta_0 + \beta_1 x$ with the same intercept β_0 and slope β_1 whatever age x. With GAMs, no strong assumptions are made about the function $x \mapsto \theta(x)$, except that it is a smooth function that can be locally well approximated by simple parametric functions.

6.3.1.2   Smoothing Window

The local GLM approach no longer assumes that $\theta(x)$ has a rigid parametric form but fits a polynomial model locally within a smoothing window

$$\mathcal{V}(x) = (x - h(x), x + h(x)).$$

Recall from Taylor's theorem that any differentiable function can be approximated locally by a straight line, and any twice differentiable function can be approximated locally by a quadratic polynomial. Within the smoothing window $\mathcal{V}(x)$, the unknown function $\theta(\cdot)$ is then approximated by such a low-degree polynomial. Varying the window leads to another approximating polynomial. The coefficients of these local polynomials are then estimated by maximum likelihood in the corresponding GLM.

Only observations in the smoothing window $\mathcal{V}(x)$ play a role in the estimation of θ at some point x. A nearest neighbor bandwidth is generally used: the smoothing window is selected so that it contains a specified number of points. Precisely, the λ nearest neighbors of x are gathered in the set $\mathcal{V}(x)$, i.e.

$$\#\mathcal{V}(x) = \lambda.$$

The number λ of observations is often expressed as a proportion $\alpha = \frac{\lambda}{n}$ of the data set (α then represents the percentage of the observations comprised in every smoothing window). Then, $h(x)$ is the distance from x to the furthest of its λ nearest neighbors, i.e.

$$h(x) = \max_{i \in \mathcal{V}(x)} |x - x_i|,$$

The bandwidth $h(x)$ has a critical effect on the local regression. If $h(x)$ is too small, insufficient data fall within the smoothing window and a noisy fit results. On the other hand, if $h(x)$ is too large, the local polynomial may not fit the data well within the smoothing window, and important features of the mean function may be distorted or even lost.

6.3.1.3 Locally Weighted GLM Fit

The estimation of $\theta(x)$ is obtained from local GLM, on the basis of the observations in $\mathcal{V}(x)$. In order to estimate θ at some point x, the observations in $\mathcal{V}(x)$ are weighted in such a way that the largest weights are assigned to observations close to x and smaller weights to those that are further away. In many cases, the weight $v_i(x)$ assigned to (x_i, Y_i) to estimate $\theta(x)$ is obtained from some weight function $v(\cdot)$ chosen to be continuous, symmetric, peaked at 0 and defined on $[-1, 1]$. Precisely, the weights are obtained from the formula

$$v_i(x) = v\left(\frac{x_i - x}{h(x)}\right), \quad x_i \in \mathcal{V}(x). \tag{6.1}$$

Only observations in the smoothing window $\mathcal{V}(x)$ receive non-zero weights and play a role in the estimation of $\theta(\cdot)$ at x.

A common choice is the tricube weight function defined as

$$v(u) = \begin{cases} \left(1 - |u|^3\right)^3 & \text{for } -1 < u < 1, \\ 0 & \text{otherwise.} \end{cases} \tag{6.2}$$

Provided the sample size is large enough, the choice of the weight function is not too critical and the tricube weight function appears as a convenient choice. According to (6.1), weights are then given by

$$v_i(x) = \begin{cases} \left(1 - \left(\frac{|x-x_i|}{\max_{i \in \mathcal{V}(x)} |x-x_i|}\right)^3\right)^3 & \text{for } i \text{ such that } x_i \in \mathcal{V}(x), \\ 0 & \text{otherwise.} \end{cases}$$

Of course, several alternatives to the tricube weight function are available, including rectangular (or uniform), Gaussian or Epanechnikov, for instance. As a check of consistency, the actuary may vary the choice of the weight function to check that the resulting fit remains stable as it should.

A polynomial with a high degree can always provide a better approximation to the unknown function $\theta(\cdot)$ than a low-degree polynomial. But high-order polynomials have larger numbers of coefficients to estimate, and the result is an increased variability in the estimate. To some extent, the effects of the polynomial degree and bandwidth are confounded. It often suffices to chose a low-degree polynomial and to concentrate on selecting the bandwidth in order to obtain a satisfactory fit.

This is why within the smoothing window $\mathcal{V}(x)$, the unknown function $\theta(\cdot)$ is generally approximated by a low-degree polynomial:

$$\theta(x) \approx \begin{cases} \beta_0(x) \text{ (local constant regression)} \\[2mm] \beta_0(x) + \beta_1(x)x \text{ (local linear regression)} \\[2mm] \beta_0(x) + \beta_1(x)x + \beta_2(x)x^2 \text{ (local quadratic regression)} \end{cases}$$

with coefficients specific to x (as suggested by Taylor expansion). The local regression coefficients $\beta_j(x)$ are estimated by maximum likelihood in the corresponding GLM, with weights $v_i(x)$ specific to x. Notice that the local constant regression generalizes the KNN (or nearest neighbors) approach.

In its basic form, the local GLM assumes that $\theta_i = \theta(x_i)$ is linear in x_i, with intercept and slope specific to x_i. The likelihood is then maximized with respect to these coefficients. Precisely,

$$\theta(x_i) = \beta_0(x_i) + \beta_1(x_i)x_i$$

and the local log-likelihood at x is

$$L_x(\beta_0(x), \beta_1(x)) = \sum_{i=1}^{n} v_i(x) l(y_i, \beta_0(x) + \beta_1(x)x_i)$$

where l is the log of the ED density (2.3) given by

$$l(y_i, \beta_0(x) + \beta_1(x)x_i) = \frac{1}{\phi}\left(y_i(\beta_0(x) + \beta_1(x)x_i) - a(\beta_0(x) + \beta_1(x)x_i)\right) + \text{constant}.$$

These formulas are easily adapted to the local constant and local quadratic cases.

Example 6.3.1 Consider for instance the Poisson case with canonical, log link. Assume that the actuary specifies a local linear model. For a response Y with feature x and unit exposure, the unknown function $\theta(\cdot)$ is then given by

$$\theta(x) = \beta_0(x) + \beta_1(x)x.$$

The mean response is then

$$\exp\left(\theta(x)\right) = \exp\left(\beta_0(x) + \beta_1(x)x\right).$$

The estimated local intercept $\widehat{\beta}_0(x)$ and local slope $\widehat{\beta}_1(x)$ maximize the local linear Poisson log-likelihood at x given by

$$L_x(\beta_0(x), \beta_1(x)) = \sum_{i=1}^{n} v_i(x)\left(-e_i \exp\left(\beta_0(x) + \beta_1(x)x_i\right) + y_i\left(\ln e_i + \beta_0(x) + \beta_1(x)x_i\right)\right)$$

based on data (Y_i, x_i, e_i), $i = 1, 2, \ldots, n$, where e_i is the exposure to risk.

Here, $\widehat{\beta}_0(x)$ and $\widehat{\beta}_1(x)$ solve

$$\frac{\partial}{\partial \beta_0(x)} L_x = \sum_{i=1}^{n} v_i(x)\Big(y_i - e_i \exp\big(\beta_0(x) + \beta_1(x)x_i\big)\Big) = 0$$

$$\frac{\partial}{\partial \beta_1(x)} L_x = \sum_{i=1}^{n} v_i(x)x_i\Big(y_i - e_i \exp\big(\beta_0(x) + \beta_1(x)x_i\big)\Big) = 0.$$

Their numerical value can be obtained by the IRLS algorithm, exactly as with GLMs. Notice that with GAMs, because the weights assigned to the observations vary with the feature x, the IRLS algorithm has to be run repeatedly for every value of x under interest. Thus, it may be applied many times and not executed only once as in GLMs.

6.3.1.4 An Example with Mortality Graduation

Let us give an example in life insurance. Consider again the mortality statistics modeled using a Binomial GLM based on Perks formula (4.15). Instead of just one year, let us now consider three consecutive years and the whole life span, from birth to the oldest ages. The corresponding crude one-year death probabilities $\widehat{q}_x$ are displayed in Fig. 6.2, with three different colors for the three years under consideration.

Even if the general shape of the mortality curve is stable over the three consecutive calendar years under consideration, year-specific erratic variations remain. As long as these erratic variations do not reveal anything about the underlying mortality pattern, they should be removed before entering actuarial calculations. Graduation is used to remove the random departures from the underlying smooth curve $x \mapsto q_x$. As there is no simple parametric model able to capture the age-structure of mortality, we use the GAM regression model

$$D_x \sim \mathcal{B}in(L_x, q_x) \text{ with } \ln \frac{q_x}{1 - q_x} = \theta(x)$$

in order to smooth the crude estimates $\widehat{q}_x$ where the function $\theta(\cdot)$ is left unspecified, but assumed to be smooth.

The contribution of the observation at age x to the log-likelihood is

$$\begin{aligned} l(D_x, l_x) &= D_x \ln q_x + (l_x - D_x) \ln(1 - q_x) \\ &= D_x \ln \frac{q_x}{1 - q_x} + l_x \ln(1 - q_x). \end{aligned} \tag{6.3}$$

A local polynomial approximation for q_x is difficult since the inequalities $0 \leq q_x \leq 1$ must be fulfilled. Therefore, we prefer to work on the logit scale, defining the new parameter from the logit transformation

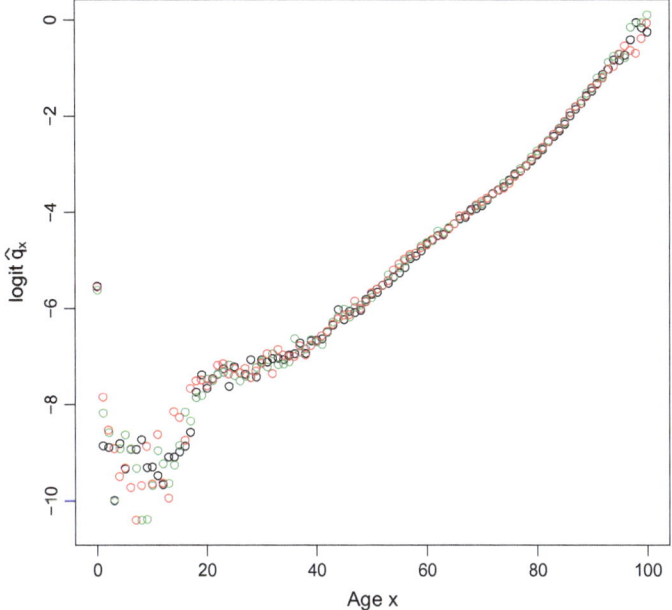

Fig. 6.2 Graph of the crude one-year death probabilities $\widehat{q}_x = D_x/l_x$ on the logit scale for three consecutive years. Data for the first year are printed in green, data for the central year are printed in black and data for the last year are printed in red

$$\theta(x) = \ln \frac{q_x}{1 - q_x}. \tag{6.4}$$

Now, $\theta(x)$ can assume any real value as q_x moves from 0 to 1. The local polynomial likelihood at x is then

$$\sum_{i=0}^{100} v_i(x)\Big(D_i\big(\beta_0(x) + \beta_1(x)i\big) - l_i \ln\big(1 + \exp(\beta_0(x) + \beta_1(x)i)\big)\Big). \tag{6.5}$$

The estimation of q_x is then obtained by inverting (6.4), that is,

$$\widehat{q}_x = \frac{\exp\big(\widehat{\beta}_0(x) + \widehat{\beta}_1(x)x\big)}{1 + \exp\big(\widehat{\beta}_0(x) + \widehat{\beta}_1(x)x\big)}. \tag{6.6}$$

Figure 6.3 illustrates this procedure graphically. First of all, a global GLM linear fit does not capture the age structure of the life table, as clearly illustrated in the upper left panel. The local GLM then proceeds as follows. To estimate q_5, let us first gather the observations close to age 5 (the red dot on Fig. 6.3), to be used in the local estimation of q_5. Here, we take $\lambda = 7$ so that

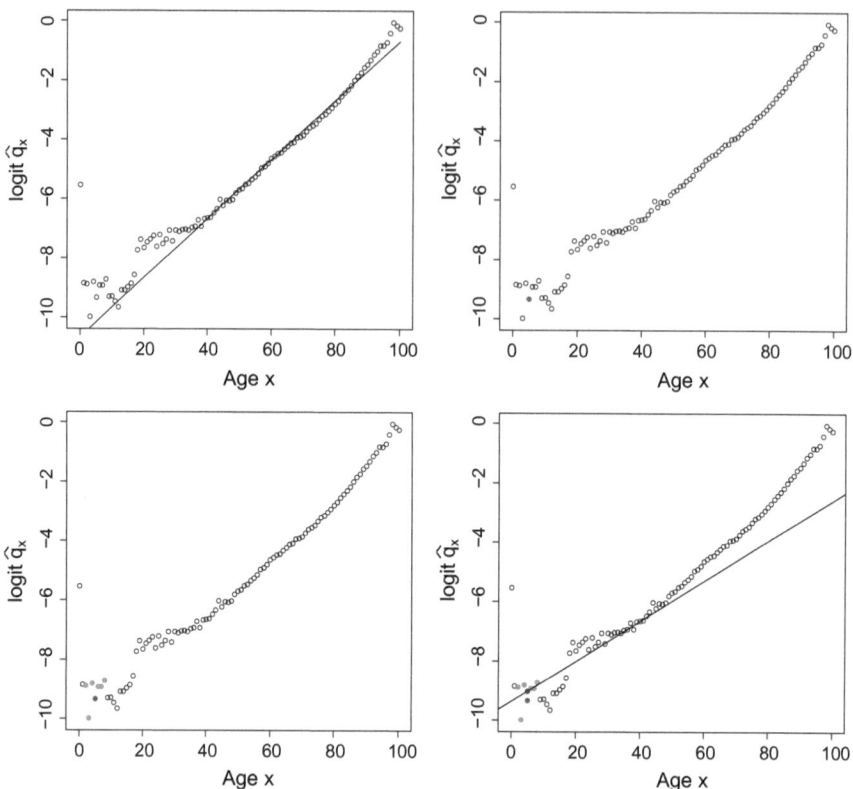

Fig. 6.3 Global linear fit (upper left panel) and local linear fit at age 5: the observation at age 5 (visible in the upper right panel) is supplemented with neighboring ones to form the set $\mathcal{V}(5)$ used to estimate q_5 (visible in the lower left panel) on which the GLM fit is performed, the resulting fit being linear (visible in the lower right panel)

$$\mathcal{V}(5) = \{2, 3, 4, 5, 6, 7, 8\}.$$

These are the green dots on Fig. 6.3. Then, we fit a Binomial GLM to the 7 observations in $\mathcal{V}(5)$, each receiving equal, unit weight. This results in the straight line that can be seen as the tangent line to the true life table at age 5. As the tangent line touches the curve at $x = 5$, we get the fitted $\widehat{q}_5$ represented as the blue dot on Fig. 6.3.

Repeating this procedure at different ages provides us with the results displayed in Fig. 6.4. The top panel repeats the procedure conducted for age 5 in Fig. 6.3 for a few other ages while the bottom panel shows the resulting fit (except for the three youngest ages and the three oldest ages because there are not enough observations to the left or to the right, respectively, to apply our basic strategy). We can see on the bottom panel of Fig. 6.4 that this very simple approach already produces rather reasonable results, except for the mortality at ages 0–15 where there is a lot of variability present in the observations.

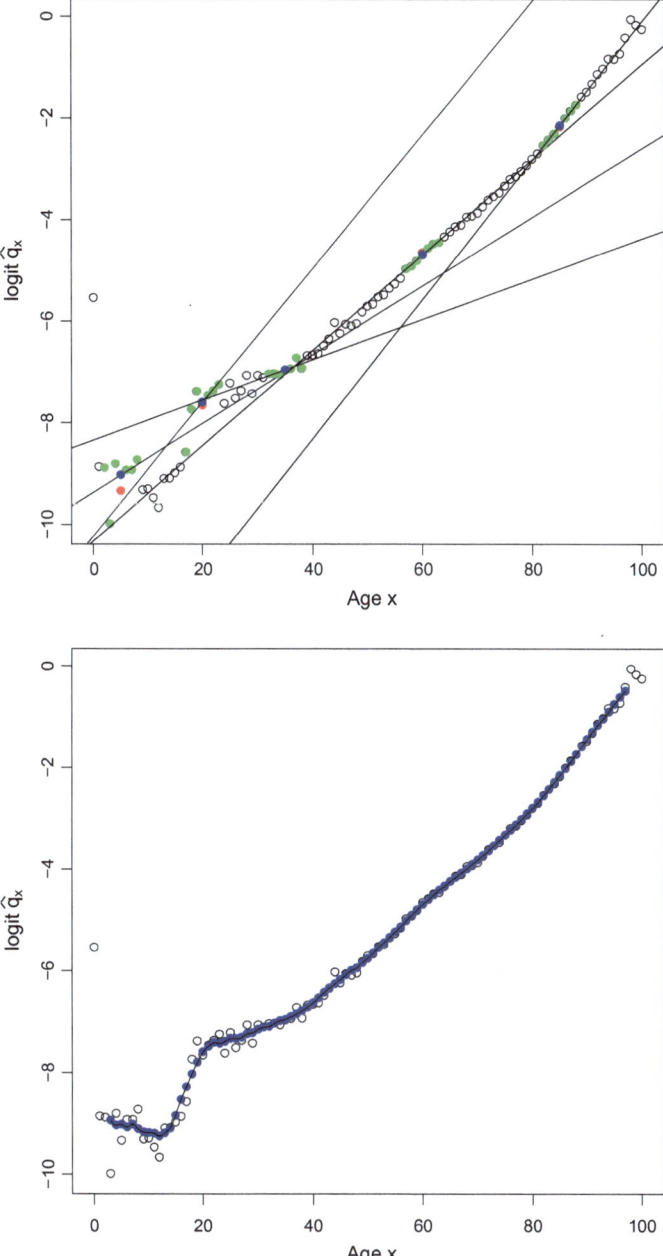

Fig. 6.4 Local linear fit. The procedure illustrated in Fig. 6.3 is repeated at each age (some examples are displayed in the upper panel), resulting in the fit appearing in the lower panel

6.3.1.5   Optimal Bandwidth

The goal in choosing the bandwidth is to produce an estimate that is as smooth as possible without distorting the underlying pattern of dependence of the response on the explanatory variables. In the application to mortality graduation for instance, this means that the actuary wants to remove all the erratic variations in the death probabilities, revealing the underlying life table. Of course, the bandwidth and the degree of the local polynomial interact. This is why we fix the degree of the local polynomial and focus on the choice of the bandwidth.

The number of equivalent degrees of freedom of a local fit provides a generalization of the number of parameters in a parametric approach. There are several possible definitions (see Sect. 4.5.5 for a possible approach to the calculation of the equivalent number of degrees of freedom), but they all purpose to measure model complexity. Intuitively speaking, they represent the number of parameters needed to obtain the same fit by means of a parametric GLM.

The optimal smoothing parameter λ (or α) is selected by cross-validation. In the Normal case, with responses

$$Y_i \sim \mathcal{N}or\left(b(x_i), \sigma^2\right), \quad i = 1, 2, \ldots, n,$$

for some unknown function $b(\cdot)$, the performance of the fit is assessed with the help of the mean squared error of prediction for an out-of-sample observations $(Y^\star, x^\star)$, that is, with the help of an empirical version of

$$\text{MSEP}(\widehat{b}) = \text{E}\left[\left(Y^\star - \widehat{b}(x^\star)\right)^2\right].$$

This indicator has already been considered in Sect. 4.5.5. It can be decomposed in three terms:

$$\text{MSEP}(\widehat{b}) = \text{Var}[\widehat{b}(x^\star)] + \left(b(x^\star) - \text{E}[\widehat{b}(x^\star)]\right)^2 + \text{Var}[Y^\star]$$

where

- the bias term
$$b(x^\star) - \text{E}[\widehat{b}(x^\star)]$$

measures the error introduced by approximating a complicated function b by a much simpler one. It is the difference between the truth and what can be discovered with the help of the model under consideration. The bias becomes high when we under-fit the data.

- the variance term
$$\text{Var}[\widehat{b}(x^\star)]$$

measures the change in the estimate if we use another training data set. If a model has high variance then small modifications in the training data can result in big changes in the resulting fit. A large variance suggests over-fitting the data.
- the last term $\text{Var}[Y^\star]$ often called process risk in insurance, is a constant not depending on the estimation procedure. It accounts for the variability inherent to the data under study.

The MSEP thus appears as a compromise between bias and variance, commonly referred to as bias-variance tradeoff.

6.3.1.6 Cross Validation

The cross-validation estimate of the MSEP is

$$\text{CV}(\widehat{b}) = \frac{1}{n} \sum_{i=1}^{n} \left(y_i - \widehat{b}^{(-i)}(x_i) \right)^2$$

where $\widehat{b}^{(-i)}(x_i)$ denotes the leave-x_i-out estimate of $b(x_i)$. The optimal smoothing parameter α minimizes CV.

In practice, each x_i is deleted from the dataset in turn, and the local regression estimate $\widehat{b}^{(-i)}$ is computed from the remaining $n-1$ data points:

$$\{(x_2, y_2), (x_3, y_3), \ldots, (x_n, y_n)\} \quad \rightarrow \quad \widehat{b}^{(-1)}$$
$$\rightarrow \quad \text{prediction } \widehat{y}_1 = \widehat{b}^{(-1)}(x_1)$$
$$\{(x_1, y_1), (x_3, y_3), \ldots, (x_n, y_n)\} \quad \rightarrow \quad \widehat{b}^{(-2)}$$
$$\rightarrow \quad \text{prediction } \widehat{y}_2 = \widehat{b}^{(-2)}(x_2)$$

$$\vdots \quad \vdots \quad \vdots$$

$$\{(x_1, y_1), (x_2, y_2), \ldots, (x_{n-1}, y_{n-1})\} \quad \rightarrow \quad \widehat{b}^{(-n)}$$
$$\rightarrow \quad \text{prediction } \widehat{y}_n = \widehat{b}^{(-n)}(x_n)$$

To ease the computations in large data sets, approximations to the CV criterion are available, avoiding to re-fit n times the model. The optimal smoothing parameter α is then obtained by a grid search.

For responses obeying other distributions in the ED family, we switch to likelihood (or deviance) cross-validation. The likelihood cross-validation is obtained by substituting the leave-x_i-out estimate $\widehat{\theta}^{(-i)}(x_i)$ in the log-likelihood, that is,

$$\text{LCV}(\widehat{\theta}) = -2 \sum_{i=1}^{n} l(Y_i, \widehat{\theta}^{(-i)}(x_i)),$$

where $\theta(\cdot)$ involves the unknown function $b(\cdot)$. Computing the n leave-x_i-out estimates $\widehat{\theta}^{(-i)}(x_i)$ can be time-consuming. Approximations for $\widehat{\theta}^{(-i)}(x_i)$ have been developed, based on an appropriate generalization of the influences of each observation.

In practice, the (L)CV plot provides graphical aid in choosing smoothing parameters. Specifically, (L)CV is computed for a range of candidate smoothing parameters. The (L)CV plot uses the number of equivalent degrees of freedom (edf) as the horizontal axis and the (L)CV statistic as the vertical axis. The number of edf of a local fit provides a generalization of the number of parameters involved in a parametric model. We refer the reader to Loader (1999, Sect. 4.4) for the precise definition of edf in local likelihood estimation. Letting the df appears along the horizontal axis helps interpretation (a too small fitted edf indicating a smooth model with very little flexibility, a too large fitted edf representing a noisy model) and comparability (as different polynomial degrees or smoothing methods can be added to the plot). The optimal smoothing parameter α corresponds to the minimum in the (L)CV plot.

It must be stressed that flat plots occur frequently: any model with a (L)CV score near the minimum is likely to have similar predictive power. In such a case, just minimizing (L)CV discards significant information provided by the whole (L)CV profile. If the (L)CV reaches a plateau after a steep descent then the smoothing parameter corresponding to the start of the plateau is a good candidate for being the optimal percentage α of data points comprised in every local neighborhood on which a separate GLM is estimated.

Example 6.3.2 (Poisson regression for counts) Let Y_i be the number of claims reported by policyholder i aged $x_i, i = 1, 2, \ldots, n$. Data consist in (y_i, x_i, e_i) where e_i is the corresponding exposure-to-risk. The model

$$Y_i \sim \mathcal{P}oi\big(e_i \exp(b(x_i))\big)$$

appears to be a good starting point for analyzing claim counts. The function b is assumed to be smooth but otherwise left unspecified. For t sufficiently close to x, b is approximated by a linear function, that is, $b(t) \approx \beta_0(x) + \beta_1(x)t$. Then, $\beta_0(x)$ and $\beta_1(x)$ are estimated by weighted (or local) maximum likelihood, assigning larger weights $v_i(x)$ to observations (y_i, x_i, e_i) corresponding to observations with ages x_i close to x. Notice that here, age x is not necessarily among those at which an observation is available. In fact, the same procedure applies, irrespective of the fact that x is comprised in the available database or not. In all cases, we first identify the observations close to age x of interest, assign them weights and fit the local GLM under consideration.

Cross-validation allows the actuary to select the optimal α, favoring the quality in predictions over goodness-of-fit. Let $\widehat{\beta}_0^{(-i)}(x_i)$ and $\widehat{\beta}_1^{(-i)}(x_i)$ be the estimated $\beta_0(x_i)$ and $\beta_1(x_i)$ based on the reduced data $\{(y_j, x_j, e_j), \ j \neq i\}$. The optimal bandwidth is selected on the basis of the likelihood cross-validation

$$LCV = -2 \sum_{i=1}^{n} \left(- e_i \exp \left(\widehat{\beta}_0^{(-i)}(x_i) + \widehat{\beta}_1^{(-i)}(x_i)x_i \right) \right.$$
$$\left. + y_i \left(\ln e_i + \widehat{\beta}_0^{(-i)}(x_i) + \widehat{\beta}_1^{(-i)}(x_i)x_i \right) \right).$$
$$+ \text{ constant.}$$

This is the procedure followed in Example 6.2.2. The bottom left panel in Fig. 6.1 displays the LCV plot. We select the optimal α from the beginning of the plateau visible in the right part of the LCV plot, corresponding to $\lambda = 19$ (which means that we take 9 ages to the left and 9 ages to the right of the age of interest, in the central part of the data).

6.3.1.7 Application to Mortality Graduation

Let us consider again the mortality data. Our naive approach produced the fit visible in the bottom panel of Fig. 6.4. Data pairs (y_i, l_i) where y_i is the observed number of deaths at age i, recorded among l_i individuals aged i, are modeled as follows:

$$Y_i \sim \mathcal{B}in\left(l_i, q_i\right) \text{ where } q_i = \frac{\exp\left(b(i)\right)}{1 + \exp\left(b(i)\right)}$$

for some smooth unspecified function b. In order to capture the high curvature in the observed data for ages 0–15, we use a local quadratic GLM. This means that for t sufficiently close to x, we approximate b by means of a quadratic polynomial:

$$b(t) \approx \beta_0(x) + \beta_1(x)t + \beta_2(x)t^2.$$

Then, regression coefficients $\beta_0(x)$, $\beta_1(x)$ and $\beta_2(x)$ are estimated by weighted (or local) maximum likelihood, assigning larger weights $w_i(x)$ to observations (y_i, l_i) corresponding to ages i close to x.

Let us use LCV to select the optimal α, In this approach, $\widehat{\beta}_0^{(-i)}(i)$, $\widehat{\beta}_1^{(-i)}(i)$ and $\widehat{\beta}_2^{(-i)}(i)$ are the estimated $\beta_0(i)$, $\beta_1(i)$ and $\beta_2(i)$ based on the reduced data $\{(y_j, l_j), \ j \neq i\}$. The optimal bandwidth is selected on the basis of the likelihood cross-validation

$$LCV = -2 \sum_{i=0}^{100} \left(y_i \left(\widehat{\beta}_0^{(-i)}(i) + \widehat{\beta}_1^{(-i)}(i)i + \widehat{\beta}_2^{(-i)}(i)i^2 \right) \right.$$
$$\left. - l_i \left(1 + \exp(\widehat{\beta}_0^{(-i)}(i) + \widehat{\beta}_1^{(-i)}(i)i + \widehat{\beta}_2^{(-i)}(i)i^2) \right) \right).$$
$$+ \text{ constant.}$$

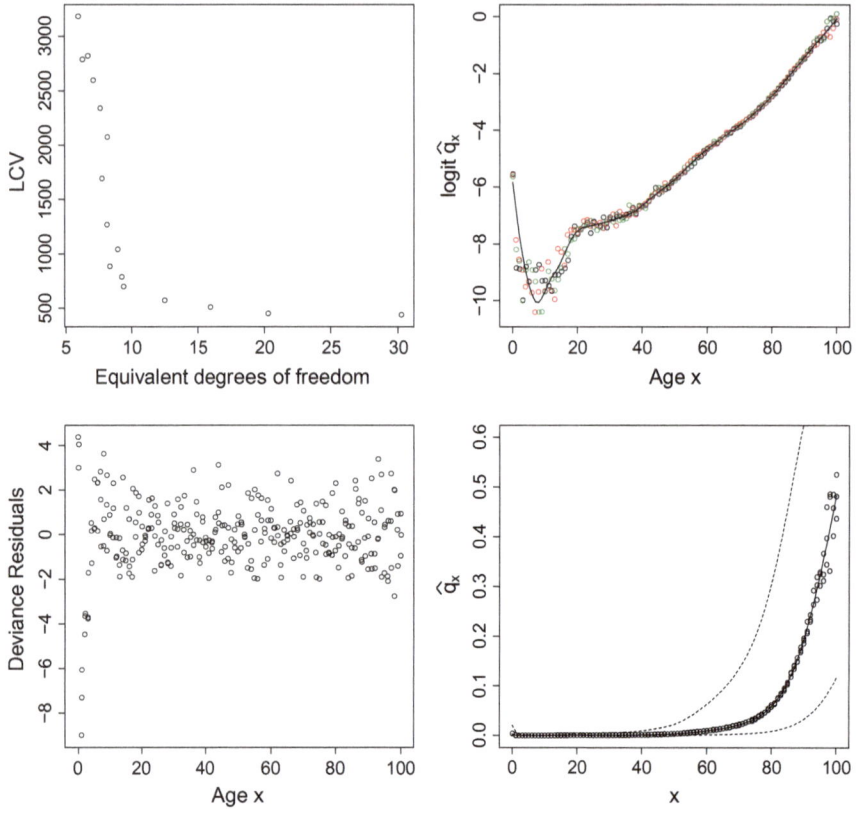

Fig. 6.5 LCV plot and corresponding graduated life table

Figure 6.5, top left panel, displays the LCV plot for different edf corresponding to values of α between 5 and 50%, by step of 3%. The optimal value for α is located at the beginning of the plateau. It is around 20%. The resulting fit is displayed in the top right panel. We can see there that the local quadratic Binomial GLM is able to capture the shape of mortality at all ages. We also display deviance residuals in the bottom left panel. We can see there that, except for very young ages, the model captures the data structure quite accurately. Finally, the resulting life table surrounded by 95% prediction intervals is displayed in the bottom right panel. We can see there that the error margin becomes quite substantial at older ages (at young ages, the death probabilities are so small that it is hard to see anything).

Remark 6.3.3 Adaptive smoothing could also have been used to get the right degree of smoothing at different ages (typically, the volatility is higher at young and old ages, requiring more smoothing compared to middle ages where mortality is much more stable). With adaptive smoothing, the smoothing window (or the percentage

α) is adapted to the local volatility present in the data instead of being kept constant at all ages.

6.3.1.8 Application to Geographic Ratemaking

Spatial models for insurance rating provide actuaries with a method for analyzing risk variation by geographic area. They are now common in insurance of domestic property lines, including motor and homeowner insurance. The spatial variation may be related to geographic factors (e.g. proximity to a fire brigade) or to socio-demographic factors (perhaps affecting theft rates in house insurance). Insurance rating for a potential policy then depends on its geographical location (contained in the postcode) in addition to standard factors.

Often, claim characteristics tend to be similar in neighboring postcode areas (after other factors have been accounted for). The idea of geographic rating models is to exploit this spatial smoothness and allow information transfer to and from neighboring regions. A convenient solution for estimating the geographic variations is to resort to local GLM regression models. This allows the actuary to estimate a geographic effect for each area, based on the insurer's experience.

As an example, consider the data displayed in Fig. 6.6. The upper left panel shows the average number of claims observed within each district, accounting for all other features available in the database. No clear pattern is visible there. Data points are mapped on the plane based on the respective latitude and longitude of the center of each district. Proximity is assessed by the Euclidean distance. The LCV plot is more difficult to interpret but suggests to use a rather small proportion α for the local polynomial fit. The resulting fit is displayed in the lower left panel of Fig. 6.6. It reveals the typical geographic variations in expected claim counts from motor third-party liability insurance. The smoothed spatial effects clearly put in evidence larger expected claim frequencies of urban areas: Brussels (in the middle), Antwerp (in the North) and Liège (in the West), three of the largest cities in Belgium. Deviance residuals are also mapped and do not reveal any remaining spatial pattern.

6.3.1.9 Application to the Mortality Surface

The dynamic analysis of mortality is often based on the modeling of the mortality surface depicted in Fig. 6.7, top panel. Mortality is studied there in an age-period framework. This means that two dimensions are used: age and calendar time. Both age and calendar time are discrete variables. In discrete terms, a person aged x, $x = 0, 1, 2, \ldots$, has an exact age comprised between x and $x + 1$. This concept is also known as "age last birthday" (that is, the age of an individual as a whole number of years, by rounding down to the age at the most recent birthday). Similarly, an event that occurs in calendar year t occurs during the time interval $[t, t + 1)$. Now, $q_x(t)$ is the probability that an x-aged individual in calendar year t dies before reaching age $x + 1$.

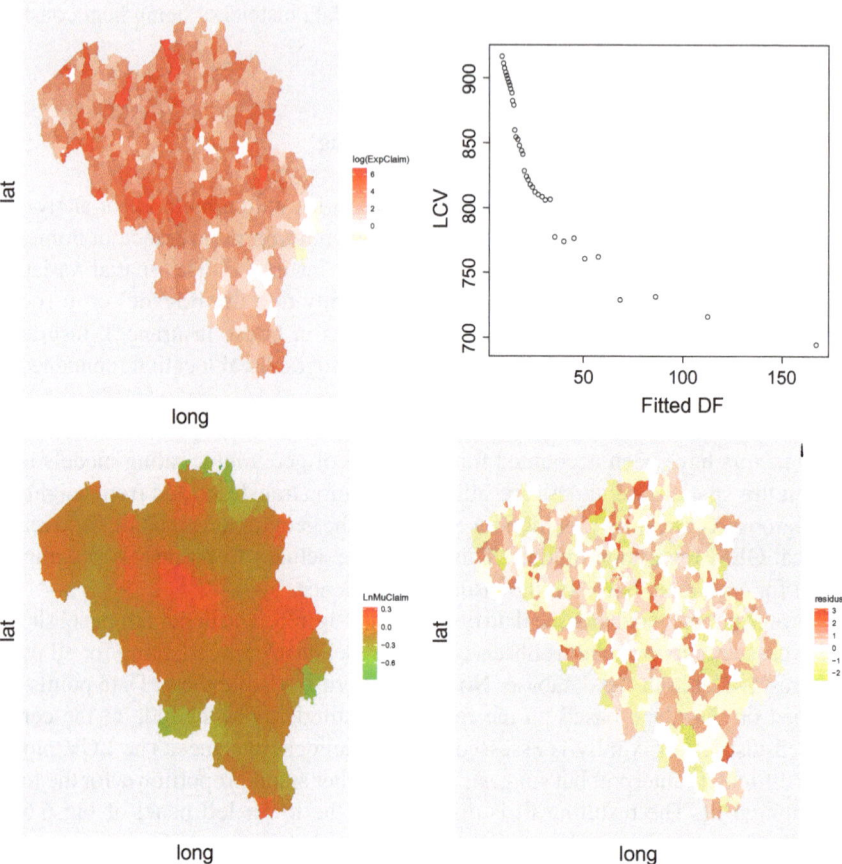

Fig. 6.6 Geographic variations in claim counts, LCV plot, corresponding smooth spatial effect, and deviance residuals

Let l_{xt} be the number of individuals aged x last birthday on January 1 of year t, and let Y_{xt} be the number of deaths recorded at age x last birthday during calendar year t, among these l_{xt} individuals. The one-year death probability $q_x(t)$ can then be estimated as Y_{xt}/l_{xt}.

The mortality surface consists of a 3-dimensional plot of the logit of $\widehat{q}_x(t)$ viewed as a function of both age x and time t. Fixing the value of t, we recognize the classical shape of a mortality curve visible on Fig. 6.2. Specifically, along cross sections of the mortality surface when t is fixed (or along diagonals when cohorts are followed), one observes relatively high mortality around birth, the well-known presence of a trough at about age 10, a ridge in the early 20's (which is less pronounced for females), and an increase at middle and older ages.

In order to remove erratic variations from the raw mortality surface, we perform a local Binomial GLM fit. The amount of smoothness is selected on the basis of the LCV plot which clearly shows a plateau. The corresponding fit displayed in the lower panel reveals the underlying, smooth mortality pattern.

6.3.2 Splines

Besides local GLMs, splines provide the second classical approach to incorporate non-linear effects of continuous risk factors on the score scale. Splines are piecewise polynomials specified by choosing a set of knots.

6.3.2.1 Piecewise Polynomials

A flexible approach for modeling nonlinear effects b_j comprised in GAM scores consists in using piecewise polynomials. Recall that banding consists in replacing the unknown function b_j with a piecewise constant one, to ease estimation. This is done by partitioning the domain of the jth feature in sub-intervals and by assuming that the function to be estimated is constant on each of them. In practice, this means that the actuary fits a 0-degree polynomial separately on each sub-interval. The partitioning is determined by the interval boundaries, called knots. Splines extend this basic approach by allowing for higher-degree polynomials which nicely connect to each other across sub-intervals (at the knots, thus) and by selecting appropriate knots partitioning the domain of the continuous features.

Precisely, the unknown functions b_j are approximated by polynomial splines, which may be regarded as piecewise polynomials with additional regularity conditions. With splines, the idea is to account for the remainder term in the Taylor expansion formula. Starting from the formula

$$b(x) = b(0) + b'(0)x + b''(0)\frac{x^2}{2} + b'''(0)\frac{x^3}{6} + \int_0^\infty b'''(t)\frac{(x-t)_+^3}{6}dt$$

it is natural to use the approximation

$$b(x) \approx \beta_0 + \beta_1 x + \beta_2\frac{x^2}{2} + \beta_3\frac{x^3}{6} + \sum_{j=1}^{p_\zeta}\beta_j\frac{(x-\zeta_j)_+^3}{6}. \tag{6.7}$$

The ζ_js entering the last formula are called knots. The approximation in the right-hand side of (6.7) appears to be a cubic polynomial between every pair of consecutive knots ζ_j and ζ_{j+1}. It is continuous and possesses continuous first and second derivatives at each knot ζ_j.

The idea is thus to replace the unknown contribution $b(\cdot)$ to the score with its approximation (6.7). Doing so, we recover a GLM score with the initial feature x and its transformations x^2, x^3, and $(x - \zeta_j)_+^3, j = 1, 2, \ldots, p_\zeta$. The IRLS algorithm can

Fig. 6.7 Mortality surfaces.
Raw data (top panel), LCV
plot (middle panel) and
resulting fit (bottom panel)

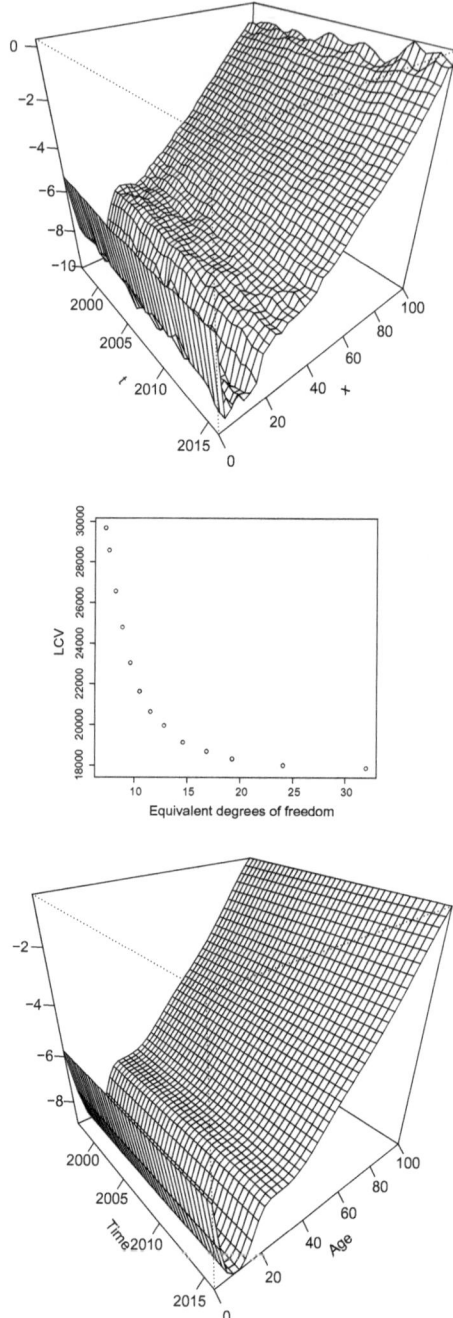

then be used to estimate the unknown parameters β_j entering (6.7) in the appropriate GLM.

The approximation (6.7) to b is a piecewise polynomial function: the domain of x is partitioned into sub-intervals by means of the knots ζ_j and the approximation consists in a different cubic polynomial on each interval. The resulting function, known as a cubic spline is continuous, and has continuous first and second derivatives at the knots. Notice that enforcing one more order of continuity would lead to a global cubic polynomial. Cubic splines are the lowest-order splines for which the knot-discontinuity is not visible to the human eye, so that there is seldom any good reason to go beyond cubic splines.

6.3.2.2 Cubic Splines

The limited Taylor expansion thus explains the use of splines: splines are piecewise polynomials joined together to make a single smooth curve. The core idea behind cubic splines is to replace the continuous feature x with new ones which are transformations of x, of the form x, x^2, x^3, and $(x - \zeta_j)^3_+$ based on Taylor expansion. The GLM machinery then applies to this new set of features.

When used to fit data, polynomials tend to be erratic near the boundaries and extrapolation may be meaningless. This unpleasant behavior is even exacerbated with splines. This is why natural cubic splines add further constraints, namely that the function is linear beyond the boundary knots. Consider the Normal regression model

$$Y_i \sim Nor\left(b(x_i), \sigma^2\right), \quad i = 1, 2, \ldots, n,$$

for some unknown function $b(\cdot)$. Natural cubic splines with knots at every observation x_i, that is, function (6.7) with $p_\zeta = n$ and $\zeta_j = x_j$, are solutions of the following optimization problem: among all functions b with continuous second derivatives, find the one which minimizes the penalized residual sum of squares

$$\sum_{i=1}^{n} \left(y_i - b(x_i)\right)^2 + \lambda \int \left(b''(x)\right)^2 dx$$

where $\lambda \geq 0$ is a smoothing parameter and the values of the continuous feature are assumed to have been ranked in ascending order, that is, $x_1 < x_2 < \ldots < x_n$. This penalized version of the least-squares objective function balances goodness-of-fit and smoothness of the estimated b: the first term imposes adherence to the data whereas the second one penalizes curvature (notice that $b'' = 0$ if b is linear). Here, λ is the smoothing parameter controlling the tradeoff between the two terms, goodness-of-fit and roughness penalty. The two extreme cases are as follows:

$\lambda = 0$ imposes no restriction and the resulting estimate interpolates the data (in fact, b is any function interpolating the data).

$\lambda \to \infty$ renders curvature impossible so that we are back to the linear regression model $b(x) = \beta_0 + \beta_1 x$ which fulfills $b'' = 0$.

Thus, a large λ corresponds to a smooth function b.

Remarkably, the minimizer is finite dimensional although the objective is a minimum over an infinite-dimensional space of functions. Precisely, the unique minimizer $\widehat{b}_\lambda$ is a natural cubic spline with knots located at the values x_i, that is, a function of the form (6.7) with $p_\zeta = n$ and $\zeta_j = x_j$. This means that $\widehat{b}_\lambda$ is a cubic polynomial on each interval $[x_i, x_{i+1})$ such that $\widehat{b}_\lambda$, $\widehat{b}'_\lambda$ and $\widehat{b}''_\lambda$ are continuous everywhere and $\widehat{b}''_\lambda(x_1) = \widehat{b}''_\lambda(x_n) = 0$. Knowing that the solution is a natural cubic spline, it can be estimated with the help of GLMs using the representation

$$\widehat{b}_\lambda(x) = \sum_{j=1}^{n} \beta_j B_j(x)$$

where the functions $B_j(\cdot)$ form a basis for such splines. The problem then reduces to fitting an ordinary linear regression model with transformed features $B_j(x)$. Denoting as $\boldsymbol{B}$ the design matrix with columns $(B_j(x_1), \ldots, B_j(x_n))$ and defining the matrix $\boldsymbol{\Omega}$ with element (j, k) given by

$$\Omega_{jk} = \int B''_j(z) B''_k(z) \mathrm{d}z,$$

the solution $\boldsymbol{\beta}$ minimizes $||\boldsymbol{Y} - \boldsymbol{B}\boldsymbol{\beta}||^2 + \lambda \boldsymbol{\beta}^\top \boldsymbol{\Omega} \boldsymbol{\beta}$ so that

$$\widehat{\boldsymbol{\beta}} = (\boldsymbol{B}^\top \boldsymbol{B} + \lambda \boldsymbol{\Omega})^{-1} \boldsymbol{B}^\top \boldsymbol{Y}.$$

Compared to formula (4.11), we see that $\lambda \boldsymbol{\Omega}$ is added to $\boldsymbol{B}^\top \boldsymbol{B}$ in the expression of $\widehat{\boldsymbol{\beta}}$. This is closely related to ridge regression (discussed in Chap. 5). The matrix $\lambda \boldsymbol{\Omega}$ serves as a ridge, or shrinkage matrix so that the estimates $\widehat{\boldsymbol{\beta}}$ are shrunken towards zero. This comes from the fact that for large λ, the expression $(\boldsymbol{B}^\top \boldsymbol{B} + \lambda \boldsymbol{\Omega})^{-1}$ becomes small. Also, the sum of squared residuals can be replaced with the log-likelihood of any ED distribution when non-Normal responses are considered.

As the fitted values $\widehat{y}_i = \widehat{b}_\lambda(x_i)$ can be obtained from

$$\widehat{\boldsymbol{Y}} = \boldsymbol{S}_\lambda \boldsymbol{Y} \text{ with } \boldsymbol{S}_\lambda = \boldsymbol{B}(\boldsymbol{B}^\top \boldsymbol{B} + \lambda \boldsymbol{\Omega})^{-1} \boldsymbol{B}^\top$$

the complexity of the estimated spline function can be deduced from the equivalent degrees of freedom $\mathrm{edf} = \mathrm{Tr}(\boldsymbol{S}_\lambda)$. Intuitively speaking, a smoothing spline with edf $= 5$ is as complex as a global polynomial of degree 4 (which has 5 parameters, including the intercept).

6.3.2.3 P-Splines, Smoothing Penalties

Once the knots ζ_j are fixed, we have a regression spline: the actuary has to select the number and location of these knots and then the GLM machinery applies. Regression splines differ from smoothing splines in that the number of knots is much smaller. The knots are generally located at quantiles of the feature x and not at every observed value x_i. In a simple regression spline approach the unknown regression coefficients β_j are estimated using standard algorithms for maximum likelihood in GLM, such as the IRLS algorithm.

The crucial point is the choice of the number of knots. We again face a bias-variance tradeoff here as using few knots results in a restrictive class of estimates (and thus high bias) whereas using many knots may overfit the data (and thus produces unstable estimates, i.e. an estimator with high variance). For a small number of knots, the resulting spline may be not flexible enough to capture the variability of the data. For a large number of knots, estimated curves tend to overfit the data and, as a result, too rough functions are obtained.

To overcome the difficulties of simple regression splines, it has become common practice to impose roughness penalties on the regression coefficients involved in the spline decomposition. This is the so-called P(enalized)-splines approach that has become one of the most effective method in applied statistics. The key idea behind P-splines is as follows. First, define a moderately large number of equally-spaced knots (usually between 20 and 40) to ensure enough flexibility of the resulting spline space. Then, define a roughness penalty based on the sum of squared first or second order differences of the coefficients β_j associated to adjacent $B_j(\cdot)$ to guarantee sufficient smoothness of the fitted curve. This leads to a penalized likelihood approach where the objective function is a compromise between goodness-of-fit and smoothness. Precisely, we combine the log-likelihood L corresponding to some ED distribution with roughness penalty to obtain

$$PL(\beta_0, \beta_1, \ldots, \beta_m) = L(\beta_0, \beta_1, \ldots, \beta_m) - \lambda \boldsymbol{\beta}^\top \boldsymbol{\Delta} \boldsymbol{\beta}$$

where λ is the smoothing parameter controlling the trade-off between adherence to data and smoothness, while $\boldsymbol{\beta}^\top \boldsymbol{\Delta} \boldsymbol{\beta}$ is the roughness penalty. Differences between adjacent β_j values are obtained from

$$\boldsymbol{P}\boldsymbol{\beta} = \begin{pmatrix} \beta_2 - \beta_1 \\ \beta_3 - \beta_2 \\ \vdots \end{pmatrix} \text{ with } \boldsymbol{P} = \begin{pmatrix} -1 & 1 & 0 & 0 & \cdots \\ 0 & -1 & 1 & 0 & \cdots \\ \vdots & \vdots & \vdots & \vdots & \ddots \end{pmatrix}.$$

Hence,

$$\sum_{j \geq 1} \left(\beta_{j+1} - \beta_j \right)^2 = \boldsymbol{\beta}^\top \boldsymbol{P}^\top \boldsymbol{P} \boldsymbol{\beta}$$

as

$$\boldsymbol{P}^\top \boldsymbol{P} = \begin{pmatrix} 1 & -1 & 0 & 0 & \cdots \\ 1 & 2 & -1 & 0 & \cdots \\ 0 & -1 & 2 & 1 & \cdots \\ \vdots & \vdots & \vdots & \vdots & \ddots \end{pmatrix}.$$

The IRLS algorithm can be adapted to obtain estimated regression coefficients $\beta_0, \beta_1, \ldots, \beta_m$ optimizing a penalized log-likelihood. Cross-validation can then be used to select the optimal λ. It is interesting to notice that the penalty $\lambda \boldsymbol{\beta}^\top \boldsymbol{\Delta} \boldsymbol{\beta}$ can be seen as the log-likelihood obtained by letting $\boldsymbol{\beta}$ be Multivariate Normal with zero mean and variance-covariance matrix $(\lambda \boldsymbol{\Delta})^{-1}$. This is in essence a mixed model as those considered in Chap. 5.

6.3.3 Extension to Multiple Continuous Features

6.3.3.1 Backfitting Algorithm

The regression parameters $\beta_0, \beta_1, \ldots, \beta_{p_{\text{cat}}}$ as well as the functions b_j have to be estimated from the available data. The fit of GLMs was performed as a sequence of fits in a Normal linear regression model by the IRLS algorithm. Similarly, the fit of GAMs is performed as a sequence of fits in a Normal additive regression model, based on working responses. At each step, the backfitting algorithm can be used to estimate the unknown functions b_j.

Backfitting proceeds as follows. All the function b_k, $k \neq j$, are put in the offset and the remaining one b_j is estimated by local likelihood or spline decomposition, as explained above. The iterative algorithm can be described as follows. We first set all the functions b_j equal to 0 and then repeat the following steps until convergence:

Step 1:   estimate $\beta_0, \beta_1, \ldots, \beta_{p_{\text{cat}}}$ in the GLM with mean response

$$g^{-1}\left(\beta_0 + \sum_{j=1}^{p_{\text{cat}}} x_{ij} \right),$$

that is, with the categorical features only, and offsets

$$\text{offset}_i = \sum_{j=p_{\text{cat}}+1}^{p} b_j(x_{ij}).$$

Step 2:   for $j = p_{\text{cat}} + 1, \ldots, p$, estimate b_j in the GAM with mean response

$$g^{-1}\big(b_j(x_{ij})\big),$$

that is, with the jth continuous feature only, and offset

$$\text{offset}_i = \beta_0 + \sum_{j=1}^{p_{\text{cat}}} x_{ij} + \sum_{k=p_{\text{cat}}+1, k \neq j}^{p} b_k(x_{ik}).$$

Sometimes, the function are taken to be linear in the initial step (so that the fit of the GAM starts from the corresponding GLM fit). Also, the linear effect of each feature may be included in the GLM part of the score and re-estimated at step 1. In such a case, the functions b_j estimate the departure from linearity for each continuous feature included in the score.

It is important to notice that we can use different smoothers for each continuous feature involved in the additive score, like local polynomials or splines, with differing amounts of smoothing.

6.3.3.2 Penalized Likelihood

In the penalized maximum likelihood approach, the log-likelihood is supplemented with a penalty for each smooth function b_j, penalizing its irregular behavior. To control the tradeoff between smoothness and adherence to data, each penalty is multiplied by an associated smoothing parameter, to be selected from the data.

Define a partition of the range of the jth feature into r_j non overlapping intervals. Then, b_j is written in terms of a linear combination of $m_j = r_j + l_j$ basis functions $B_{jk}^{l_j}$, i.e.

$$b_j(x_j) = \sum_{k=1}^{m_j} \beta_{jk} B_{jk}^{l_j}(x_j).$$

The vector of function evaluations

$$\boldsymbol{b}_j = (b_{j1}, \ldots, b_{jn})^\top$$

can be written as the matrix product of a design matrix $\boldsymbol{X}_j$ and the vector of regression coefficients $\boldsymbol{\beta}_j = (\beta_{j1}, \ldots, \beta_{jm_j})^\top$, i.e.

$$\boldsymbol{f}_j = \boldsymbol{X}_j \boldsymbol{\beta}_j.$$

The $n \times m_j$ design matrix $\boldsymbol{X}_j$ consists of the basis functions $B_{jk}^{l_j}$ evaluated at the observations. Precisely, the element in row i and column k of $\boldsymbol{X}_j$ is given by $B_{jk}^{l_j}(x_{ij})$.

Regression parameters are then estimated by maximizing a penalized log-likelihood of the form

$$PL(\boldsymbol{\beta}_1, \ldots, \boldsymbol{\beta}_p) = L(\boldsymbol{\beta}_1, \ldots, \boldsymbol{\beta}_p) - \lambda_1 \sum_{k=d+1}^{m_1} \left(\Delta^d \beta_{1k}\right)^2 - \cdots - \lambda_p \sum_{k=d+1}^{m_p} \left(\Delta^d \beta_{pk}\right)^2.$$

Here, PL is maximized with respect to $\boldsymbol{\beta}_1, \ldots, \boldsymbol{\beta}_p$. The index d indicates the order of differences. Usually $d = 1$ or $d = 2$ is used leading to first differences $\beta_{j,k} - \beta_{j,k-1}$ or second differences $\beta_{j,k} - 2\beta_{j,k-1} + \beta_{j,k-2}$, respectively. The trade off between fidelity to the data (governed by the likelihood term) and smoothness (governed by the p penalty terms) is controlled by the smoothing parameters λ_j. The larger the smoothing parameters the smoother the resulting fit. In the limit ($\lambda_j \to \infty$) we obtain a polynomial whose degree depends on the order of the difference penalty and the degree of the spline. For example for $d = 2$ and $l_j = 3$ (the most widely used combination) the limit is a linear fit. The choice of the smoothing parameters is crucial as we may obtain quite different fits by varying the smoothing parameters λ_j. Smoothing parameters can be selected with the help of cross-validation or with criteria such as generalized cross-validation (GCV), unbiased risk estimation (UBRE) or AIC that avoid re-fitting the model several times to predict out-of-sample data points.

6.4 Risk Classification in Motor Insurance

In this section, we work out a case study in motor insurance, to demonstrate the relevance of GAMs for risk evaluation in Property and Casualty insurance.

6.4.1 Description of the Data Set

The data set relates to a Belgian motor third-party liability insurance portfolio observed during the year 1997, comprising 162,468 insurance contracts.

6.4.1.1 Available Features and Responses

The variables contained in the database are as follows. First, we have information about the policyholder:

Gender   policyholder's gender, categorical feature with two levels Male and Female;

AgePh   policyholder's age last birthday, continuous feature measured in years, from 18 to 90;

District   postcode of the district in Belgium where the policyholder lives.

Next, we have information about the insured vehicle:

AgeCar   age of the vehicle (on January 1, 1997), rounded from below, in years, from 0 to 20;

Fuel   fuel oils of the vehicle, categorical feature with two levels Gasoline and Diesel;

PowerCat   power of the car, categorical feature with five levels labeled C1 to C5 of increasing power;

Use   use of the car, categorical feature with two levels Private and Professional.

Finally, we have information about the contract selected by the policyholder:

Cover   Extent of coverage selected by the policyholder, categorical feature with three levels,

– either third-party liability insurance only (level TP.Only),
– or limited material damage or theft in addition to the compulsory third-party liability insurance (level Limited.MD),
– or comprehensive coverage in addition to the compulsory third-party liability insurance (level Comprehensive);

Split   Number of premium payments per year, categorical feature with four levels,

– either premium paid once a year, labeled Yearly,
– or premium paid twice a year, labeled Half-Yearly,
– or premium paid four times a year, labeled Quarterly,
– or premium paid every month, labeled Monthly.

In addition to these features, the data basis also records

ExpoR   exposure to risk in years (based on the number of days the policy was in force during 1997);

Nclaim   the number of claims filed by each policyholder during 1997.

Cclaim   the resulting total claim amount. Costs of claims are expressed in Belgian francs as the Euro was introduced only later in Belgium: 1 Euro is approximately equal to 40 Belgian francs.

In this chapter, we consider Nclaim and postpone to Chap. 9 the analysis of Cclaim. This is because very expensive claims appear in the data basis, that cannot be properly described by ED distributions like the Gamma or Inverse-Gaussian ones, for instance. These large claims must first be isolated (and Extreme Value Theory offers a nice framework to this end) before the remaining ones can be modeled using GAMs.

6.4.1.2   Composition of the Portfolio with Respect to Features

The upper left panel in Fig. 6.8 displays the distribution of the exposure-to-risk in the portfolio. The majority (about 80%) of the policies have been observed during the whole year 1997. Considering the distribution of the exposure-to-risk, we see that policy issuances and lapses are randomly spread over the year. Remember that it is common practice to start a new record each time there is a change in one of the features (i.e. when policyholder moves or buys a new car, for instance) and this also results in curtailed exposures. The gender structure of the portfolio is also described in Fig. 6.8, as well as the composition of the data set with respect to vehicle age, power and fuel oils of the car as well as the extent of the coverage selected by the policyholders. In Fig. 6.9, we can see the composition of the portfolio with respect to premium payment, policyholder's age and use of the vehicle.

The total risk exposures per Belgian district (on the log-scale) are visible on Fig. 6.10. Precisely, we can see there the sum of the durations (in days) of all the policies whose holder lives in the district. Clearly, the company sells more policies in the center of the country. The exposure-to-risk is smaller in Southern districts. The sometimes large differences in exposure-to-risk between neighboring districts can be attributable to the way policies are sold in Belgium (mainly through independent brokers working with a few insurance companies, only).

6.4.2 Modeling Claim Counts

Separate analyses are conducted for claim frequencies and severities, including settlement expenses, to calculate a pure premium. This approach is particularly relevant in motor insurance, where the risk factors influencing the two components of the pure premium are usually different and where a separate model for claim numbers is required to build experience rating schemes (such as a bonus-malus scale, for instance).

Let us model the number of claims Y_i reported by policyholder i, $i = 1, \ldots, n$. Here, Y_i represents the number of accidents for which the company had to indemnify a third party during 1997 in relation with an accident for which policyholder i was recognized as liable. We have observations (y_i, e_i, x_i), $i = 1, 2, \ldots, n$, where y_i represents the observed claim count, e_i the exposure to risk and x_i the set of available features.

6.4.2.1 Marginal Impact of Features on Claim Counts

The top left panel in Fig. 6.11 displays the histogram of the observed claim numbers per policy. We can see there that the majority of the policies (142,902 out of the 162,468 contracts comprised in the portfolio, precisely) did not produce any claim. About 10% of the portfolio produced one claim (16,328 contracts). Then, 1,539 contracts produced 2 claims, 156 contracts produced 3 claims, 17 contracts 4 claims and there were 2 policies with 5 claims. This gives an average claim frequency equal to 13.93% of at the portfolio level (this is rather high compared to nowadays

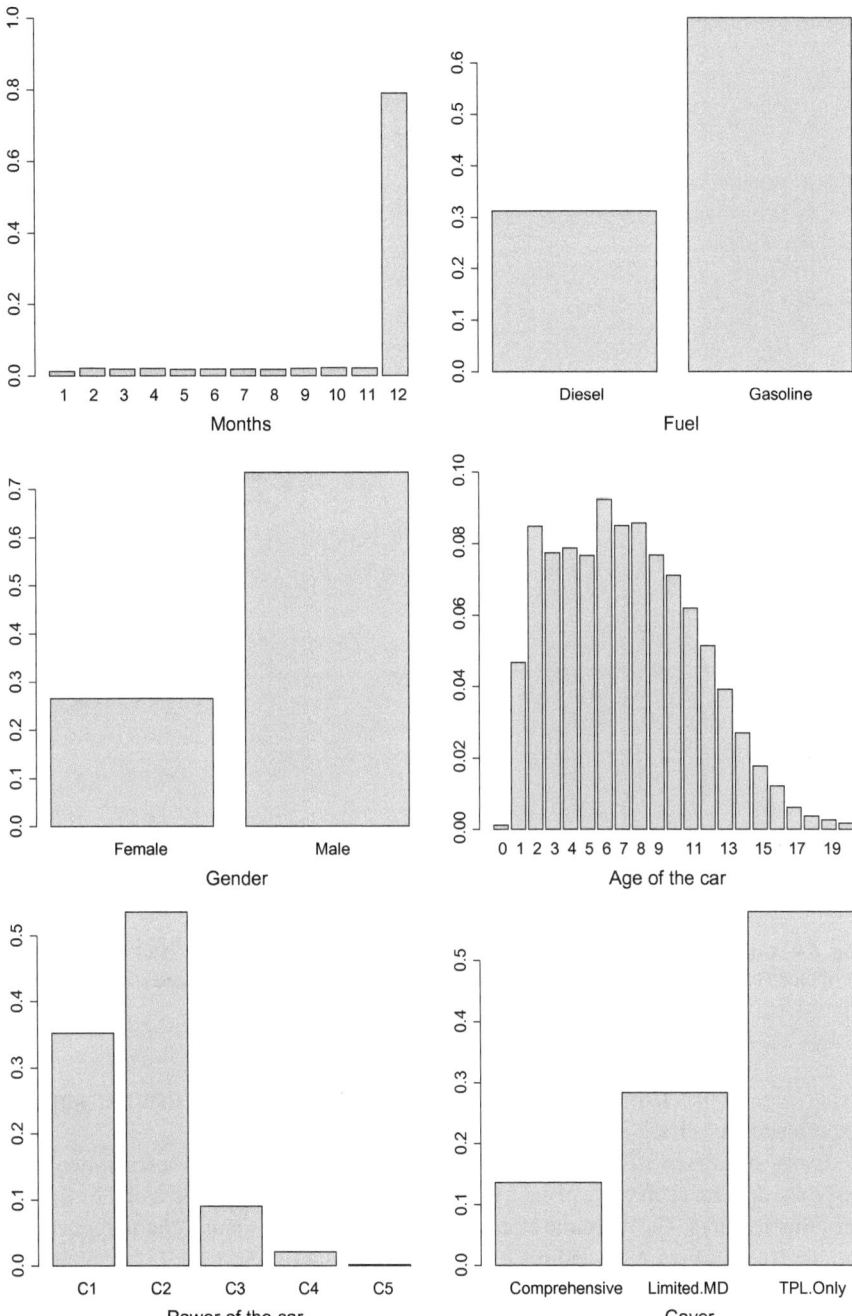

Fig. 6.8 Composition of the portfolio with respect to the available features. From upper left to lower right: exposures to risk (in months), fuel oils of the vehicle, policyholder's gender, vehicle age, power of the car, and cover extent

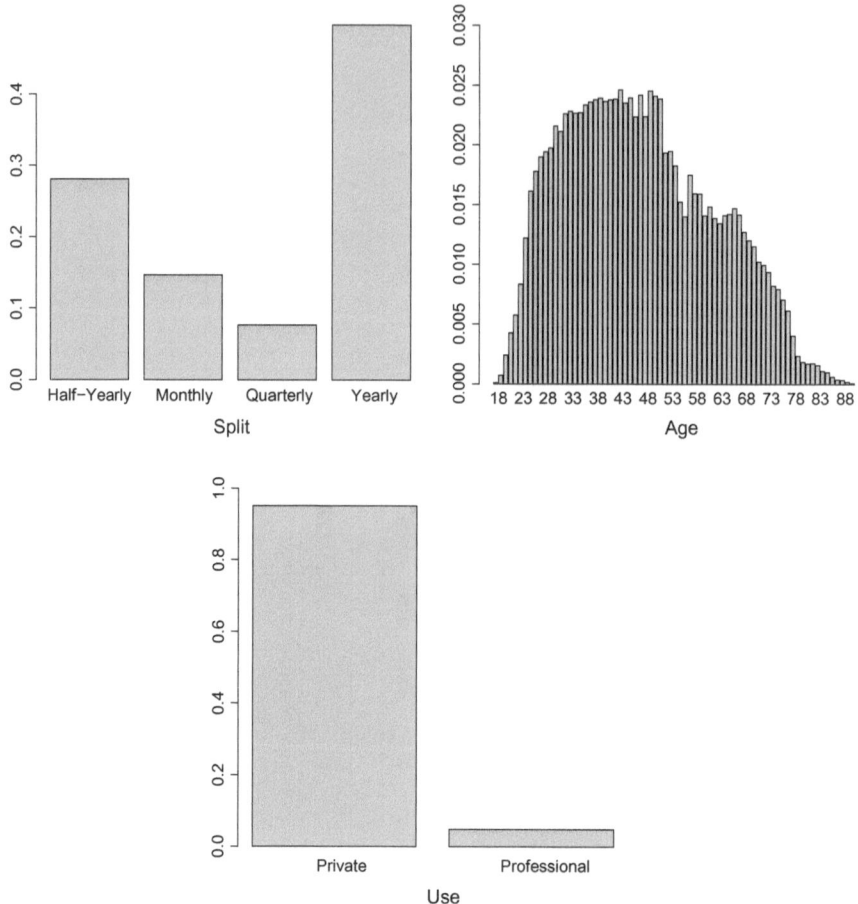

Fig. 6.9 Composition of the portfolio with respect to the available features (Ctd from Fig. 6.8). Periodicity of premium payment (upper left), policyholder's age (upper right), and use of the vehicle (lower)

experience, with claim frequencies about halved, but remember that the portfolio experience dates back to 1997).

Every insurance study starts with one-way, or marginal analyses summarizing response data for each value of each feature, but without taking account of the effect of the other features. The marginal effects of the features on the annual claim frequencies are depicted in Figs. 6.11 and 6.12, except polcyholder's age which is displayed in Fig. 6.13. We have supplemented point estimates of the mean claim frequency with 95% confidence intervals, to figure out whether the marginal difference is significant or not.

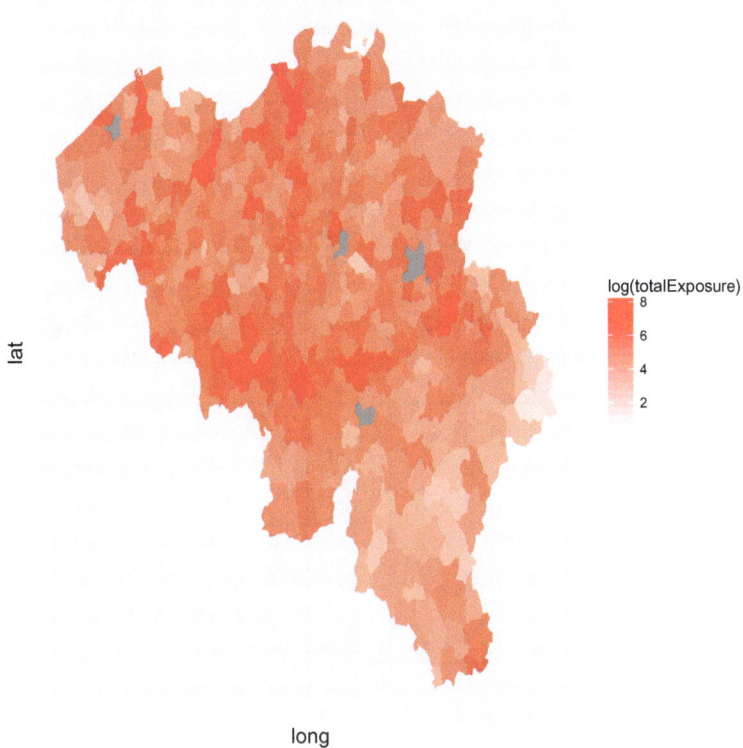

Fig. 6.10   Geographic variations of risk exposures (on the log scale)

We can see there that

– diesel vehicles tend to report more claims. This can be explained because such vehicles are more expensive but the fuel oils are cheaper compared to gasoline (at least this was the situation when the data have been recorded, drung the late 1990s in Belgium). Hence, policyholders opting for diesel vehicles typically drive higher annual mileage which generally increases the exposure to accidents.
– female drivers tend to report more claims compared to male ones (but this finding must be considered in relation to the interaction between policyholder's age and gender that is described in Fig. 6.13, which will refine this conclusion as explained below).
– new cars seem to be more risky, with a plateau for middle-aged cars, and a decrease in mean claim frequencies for older vehicles. This can be related to the compulsory annual mechanical check-up organized in Belgium, for all vehicles in use for more than 3 years. Policyholders driving longer distances typically cluster in the lower levels of this feature (as they tend to buy new cars, drive them intensively during three years before selling them as second-hand vehicles).

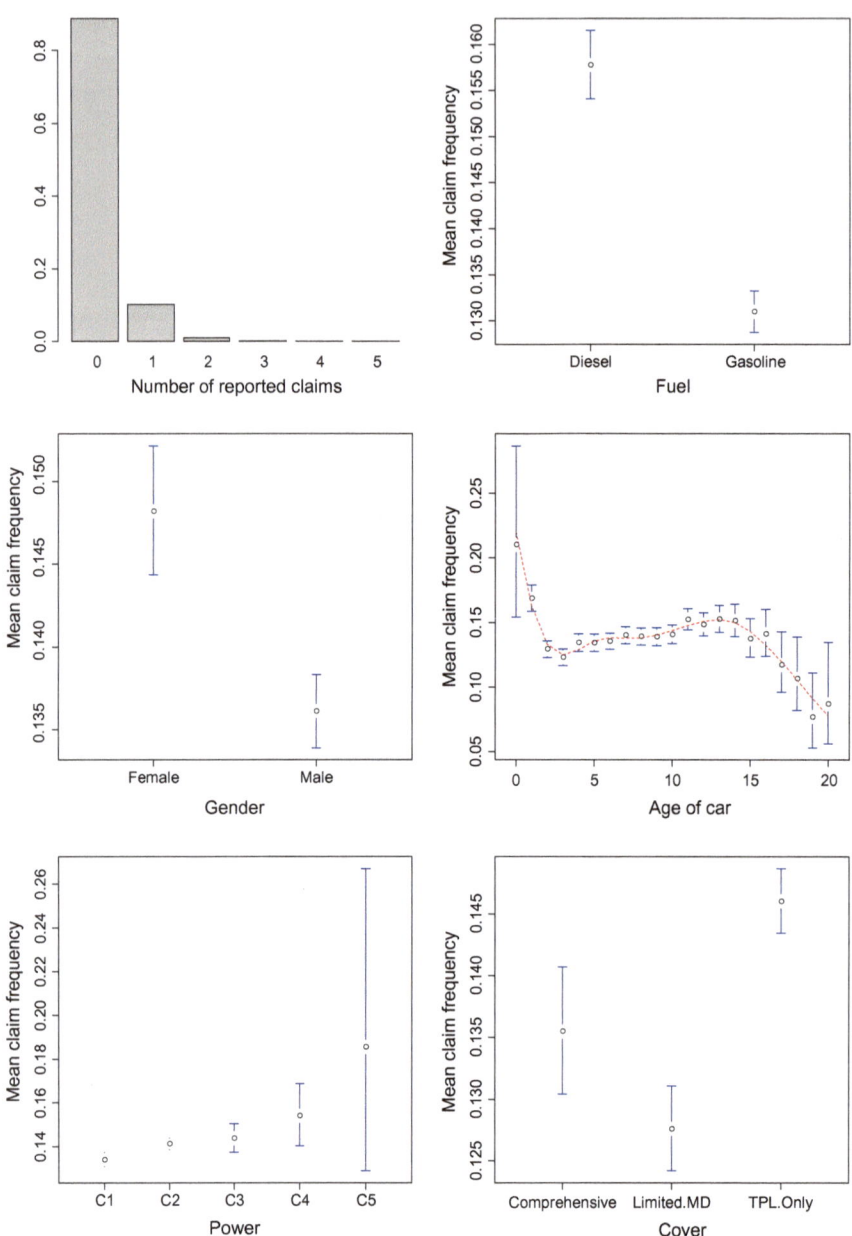

Fig. 6.11 Marginal impacts of the features on the number of claims. From upper left to lower right: distribution of the number of claims, fuel oils of the vehicle, policyholder's gender, vehicle age, power of the car, and cover extent

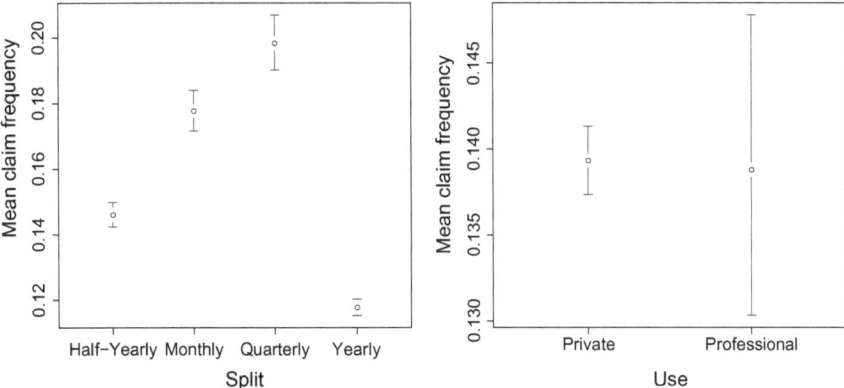

Fig. 6.12 Marginal impacts of the features on the number of claims (Ctd from Fig. 6.11). Periodicity of premium payment (left) and use of the vehicle (right)

- the mean claim frequency tends to increase with the power of the car. This effect is often visible in motor insurance, and generally explained by the fact that more powerful cars are often more difficult to drive. Notice that some refined analyses simultaneously resort to vehicle power and weight to account for the fact that heavier vehicles are generally equipped with more powerful engines.
- if optional guarantees are subscribed in addition to the compulsory third-party liability insurance then the mean claim frequencies are lower in the third-party liability product. This suggests that buying more guarantees reveals a lower risk profile (and rules out adverse selection).
- the mean claim frequencies are higher when premium payment is split across the year. This is usually attributed to a lower socio-economic profile when splitting premium payment results in an increase of the total amount of premium paid. Those policyholders paying the premium in several installments are thus penalized and this choice reveals strong budgetary constraints that may also delay some necessary repairs to the insured vehicle.
- there does not seem to be an effect of the use of the car on mean claim frequencies.

As policyholder's age generally interact with gender in motor insurance, in the sense that young male drivers are more dangerous compared to young female drivers, we display the observed claim frequencies separately for male and female divers in Fig. 6.13. In order to better visualize the possible interaction, we have estimated separately for male and female policyholders the average claim frequencies by age using a GAM with the single feature `AgePh` in interaction with `Gender`. This gives the last graph of Fig. 6.13. It is clear from these graphs that there is some interaction between these two features as the effect of gender clearly depends on age.

However, these graphs obtained from one-way analyses are at best difficult to interpret. Indeed, because of the limited exposure in some categories, we sometimes observe a rather jagged behavior. Moreover, these graphs depict the univariate effect

Fig. 6.13 Age-gender
interaction

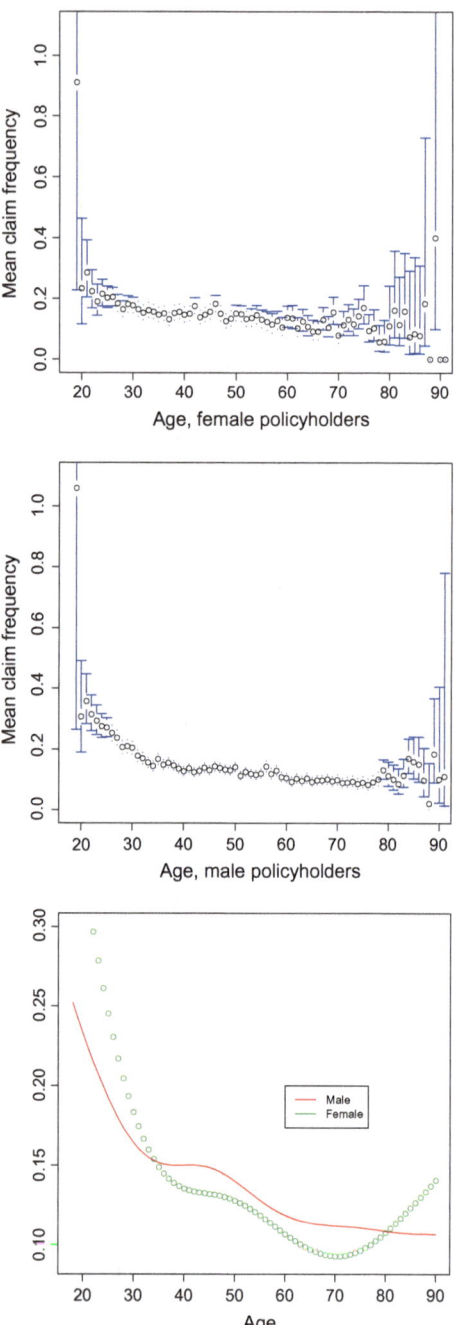

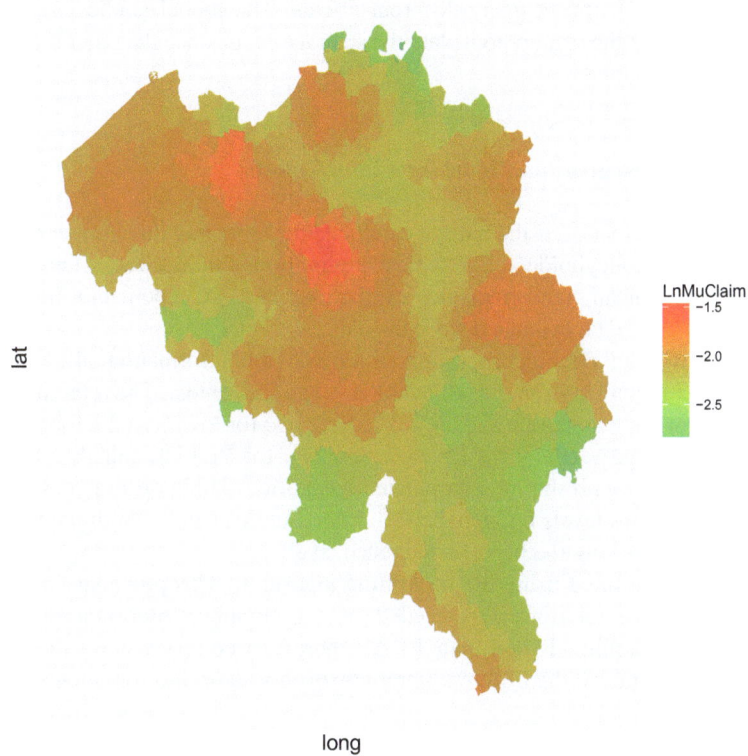

long

Fig. 6.14 Geographic variations of claim frequencies (on the log scale)

of the available features on the annual claim frequency. Because of possible correlations, seemingly high frequencies observed on Figs. 6.11 and 6.12 can be attributed
to confounding effects of other risk factors. For instance, higher frequencies for
new cars could be only apparent and result from the fact that new cars are driven
in majority by young, unexperienced policyholders. Therefore, we need to build a
multivariate model which will isolate the effect of each risk factor and smooth the
effects of continuous features over their respective ranges.

Figure 6.14 depicts the average claim frequency per Belgian district. A preliminary
smoothing has been performed because a map with crude claim frequencies is at best
difficult to interpret and can even be seriously misleading (due to the fact that the
crude frequencies tend to be far more extreme in regions with smaller risk exposures).
Hence regions with the least reliable data will typically draw the main visual attention.
This is one reason why it is difficult in practice to attempt any smoothing or risk
assessment "by eye".

No clear pattern emerges from the map, except that large cities are associated
with higher observed claim frequencies. Since the risk profile of policyholders living
in these districts may diverge to a large extent, no interpretation of such a map

is possible since it mixes many different effects. The model studied in the next subsection allows the actuary to isolate the spatial effects, once the other risk factors are taken into account.

6.4.2.2   GAM Regression Model for Claim Counts

The Poisson distribution is the natural candidate for modeling the number of claims reported by the policyholders. The typical assumption in these circumstances is that the conditional mean claim frequency can be written as an exponential function of an additive score to be estimated from the data.

We start with a model incorporating all the available information, allowing for a possible interaction between the policyholder's age and gender. The reference levels for the categorical features are as follows: Gasoline for Fuel, Male for Gender, C2 for Power, TPL.Only for Cover, Yearly for Split, and Private for Use. This in line with the most populated categories visible on Figs. 6.8 and 6.9. By default, R ranks levels by alphabetical order, but this can be modified using the command C specifying the base, or reference level.

The fit is performed using the backfitting algorithm. This procedure converged rapidly: only three iterations were needed, with as stopping criterion a relative difference in log-likelihoods less than 10^{-5}. Figure 6.15 compares the values of the parameters obtained at the different iterations of this algorithm for categorical fea-

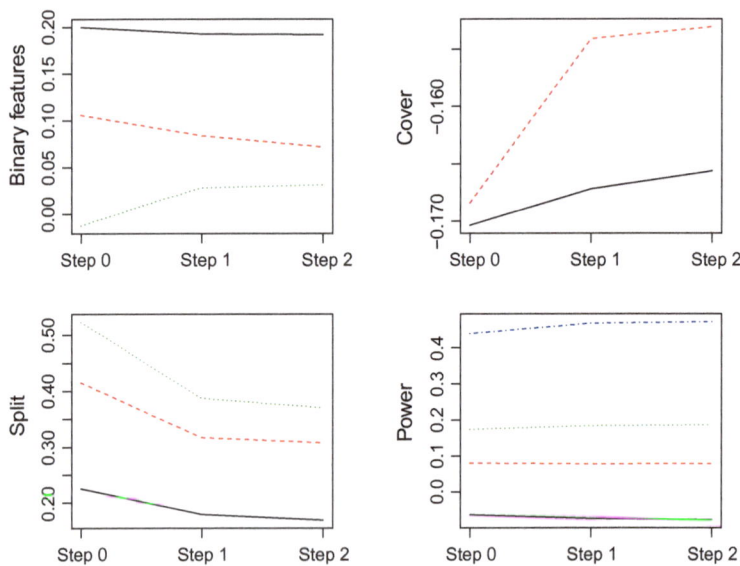

Fig. 6.15 Regression coefficients obtained at the different steps of the backfitting algorithm for categorical features

tures. Figure 6.16 displays the estimated effects for continuous features. We can see there that the initial estimates are already very close to the final ones. Figure 6.17 depicts the difference between the initial step and the final estimates per Belgian district. Again, we see that the variation between initial and final estimates are rather weak.

The small to moderate differences between the different steps of the backfitting algorithm legitimates the step-by-step procedure favored by many practitioners, who often just stop after the first cycle in the backfitting algorithm and merely include the effects in a progressive way, by freezing the previous estimations in the offset.

Let us briefly comment on the resulting fit. We begin with the effects of categorical features on the score. We see that subscribing more guarantees than just the compulsory TPL tends to reduce the expected number of claims (as it can be seen from the negative estimated regression coefficients). Also, driving a gasoline vehicle tends to be safer. Examining the nonlinear part of the score (see Fig. 6.16), we discover that the age-gender interaction is significant. Young males (below 35) tend to report more accidents than young females, as well as old males (above 80). In between, males appear less dangerous. This can be attributed to the fact that, because of the high premiums charged to young policyholders, it was common practice in Belgium at that time to ask older relatives (most often the mother) to purchase the policy. The peak of the age effect around 45 years is generally attributed to accidents caused by children behind the wheel. New cars appear more dangerous than older ones. The effect becomes almost constant after 4 years. Remember that in Belgium the car has not to undergo the annual mechanical check-up organized by the State during the first three years. So, drivers with high annual mileage often keep their car only for three years and then buy a new one. The geographic effect displayed in Fig. 6.18 is in line with our expectations, pointing out that large cities are more dangerous with respect to claim frequencies.

Let us now perform the analysis with the help of the gam function comprised in the R package mgcv. A penalized likelihood is implemented there and the optimal value of the smoothing parameters are selected by using the generalized cross-validation (GCV) criterion (given by $nD/(n - \text{edf})^2$ where D is the deviance and edf the equivalent number of degrees of freedom) or a re-scaled AIC.

The estimates of the linear part of the score, involving the categorical features, is displayed in Table 6.1. We can see there that the use of the vehicle does not significantly impact on the expected number of claims. This is why we exclude this feature from the model and re-fit the Poisson GAM. The resulting estimates of the linear part of the score are displayed in Table 6.2. We can see there that the omission of use leaves the other estimates almost unchanged.

Notice that some levels could be grouped together. For instance, the estimated regression coefficients for Limited.MD and Comprehensive are not significantly different given the associated standard errors, so that we could define a new feature Cover2 with only two levels, TPL.Only and TPL+, the latter combining the levels Limited.MD and Comprehensive. Similar groupings could be achieved for Split.

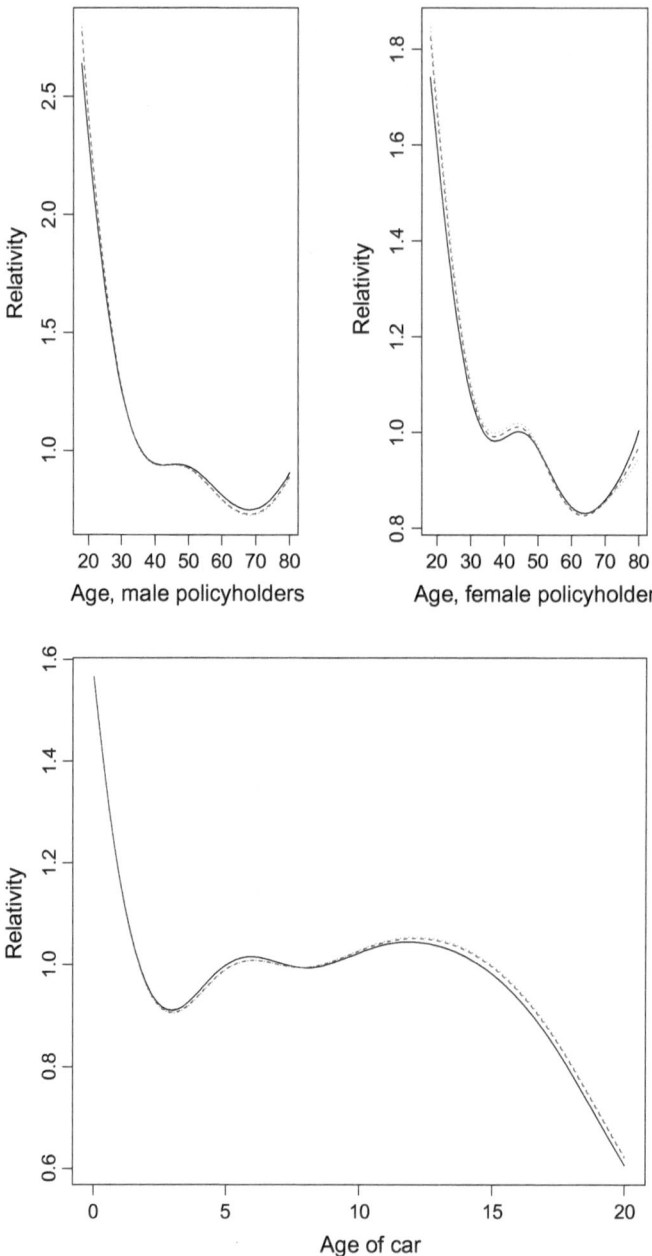

Fig. 6.16 Estimated nonlinear effects associated to age in interaction with gender and to car age obtained at the different steps of the backfitting algorithm for continuous features. Values at step 1 correspond to the broken line whereas final values (step 2) correspond to the continuous line

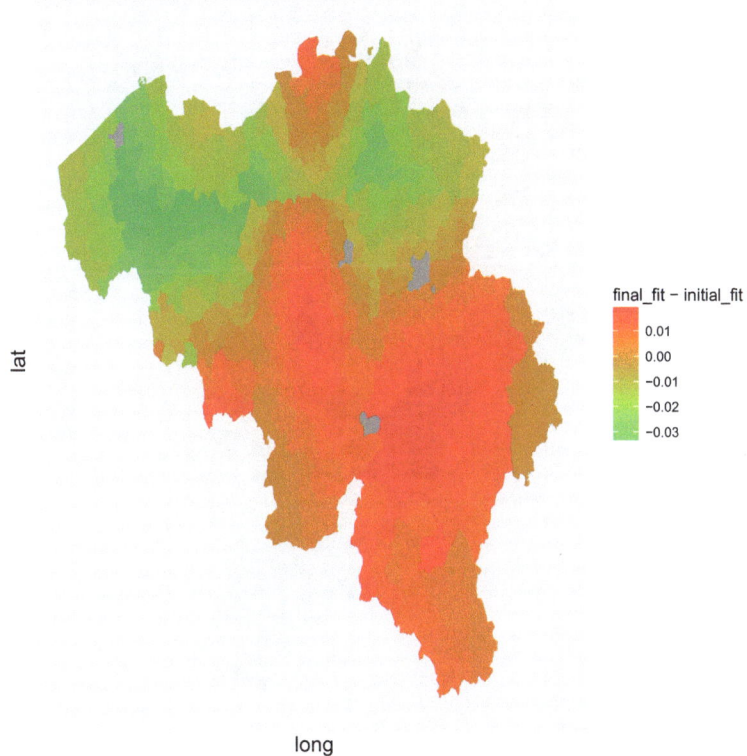

long

Fig. 6.17 Differences between the initial step and the final one of the backfitting algorithm per Belgian district

Let us now consider the fit of the non-linear part of the score. The effect of age for female drivers has an edf equal to 5.238 and is highly significant with a p-value less than 2×10^{-16}, whereas the edf associated to age for male drivers is equal to 6.132 with a p-value less than 2×10^{-16}. The other two continuous features are also highly significant, with an edf equal to 7.878 and a p-value of 1.14×10^{-13} for car age and and edf equal to 26.281 with a p-value less than 2×10^{-16} for the geographic effect. The estimated functions are displayed in Fig. 6.19. We can see that the obtained results are very similar to those in Fig. 6.16, except that the values are given on the log-scale here, and that the output for the geographic effect is not drawn on a map with gam.

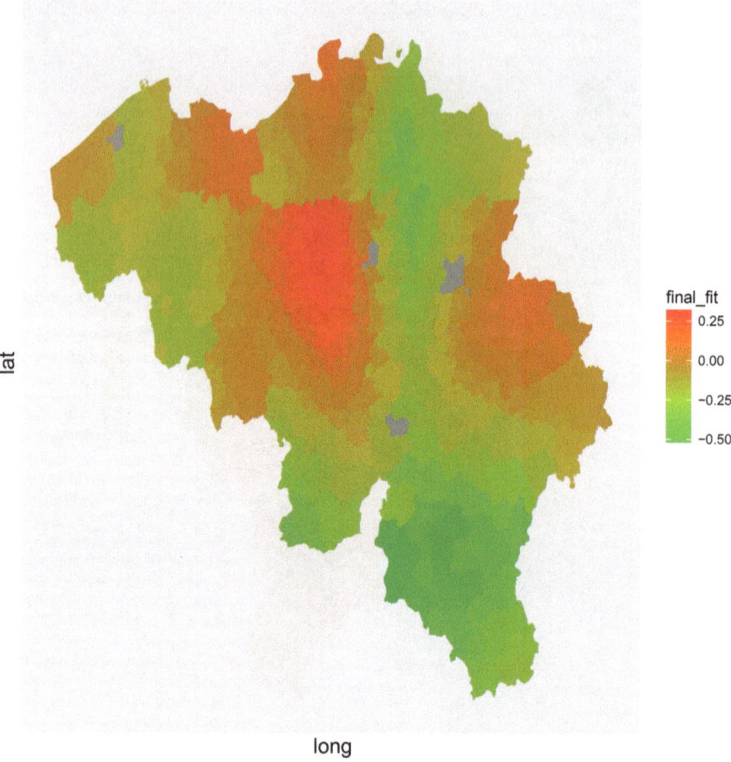

Fig. 6.18 Geographical effects for claim frequencies per Belgian district

Table 6.1 Output of the gam function of R package mgcv for the fit of th number of claims in motor third-party liability insurance

Coefficient	Estimate	Std. Error	z value	$\Pr(> |z|)$	
Intercept	−2.12986	0.01641	−129.812	$<2 \times 10^{-16}$	⋆ ⋆ ⋆
Fuel diesel	0.19219	0.01566	12.271	$<2 \times 10^{-16}$	⋆ ⋆ ⋆
Gender female	0.07499	0.01718	4.366	1.27×10^{-5}	⋆ ⋆ ⋆
Cover comprehensive	−0.16264	0.02540	−6.403	1.52×10^{-10}	⋆ ⋆ ⋆
Cover Limited.MD	−0.15170	0.01796	−8.444	$<2 \times 10^{-16}$	⋆ ⋆ ⋆
Split half-yearly	0.16614	0.01784	9.311	$<2 \times 10^{-16}$	⋆ ⋆ ⋆
Split monthly	0.30639	0.02212	13.854	$<2 \times 10^{-16}$	⋆ ⋆ ⋆
Split quarterly	0.36560	0.02545	14.365	$<2 \times 10^{-16}$	⋆ ⋆ ⋆
Use professional	0.03206	0.03349	0.957	0.338413	
PowerCat C1	−0.07913	0.01620	−4.885	1.04×10^{-6}	⋆ ⋆ ⋆
PowerCat C3	0.08000	0.02555	3.131	0.001741	⋆ ⋆
PowerCat C4	0.18747	0.04877	3.844	0.000121	⋆ ⋆ ⋆
PowerCat C5	0.47399	0.18631	2.544	0.010955	⋆

Table 6.2 Output of the `gam` function of R package `mgcv` for the fit of th number of claims in motor third-party liability insurance

Coefficient	Estimate	Std. Error	z value	Pr(> |z|)	
Intercept	−2.12876	0.01637	−130.075	$<2 \times 10^{-16}$	★★★
Fuel diesel	0.19320	0.01563	12.364	$<2 \times 10^{-16}$	★★★
Gender female	0.07484	0.01718	4.357	1.32×10^{-5}	★★★
Cover comprehensive	−0.16073	0.02532	−6.348	2.18×10^{-10}	★★★
Cover Limited.MD	−0.15142	0.01796	−8.430	$<2 \times 10^{-16}$	★★★
Split half-yearly	0.16565	0.01784	9.288	$<2 \times 10^{-16}$	★★★
Split monthly	0.30594	0.02211	13.837	$<2 \times 10^{-16}$	★★★
Split quarterly	0.36548	0.02545	14.360	$<2 \times 10^{-16}$	★★★
PowerCat C1	−0.07956	0.01619	−4.914	8.95×10^{-7}	★★★
PowerCat C3	0.08150	0.02550	3.196	0.00139	★★
PowerCat C4	0.19092	0.04864	3.926	8.65×10^{-5}	★★★
PowerCat C5	0.47702	0.18628	2.561	0.01045	★

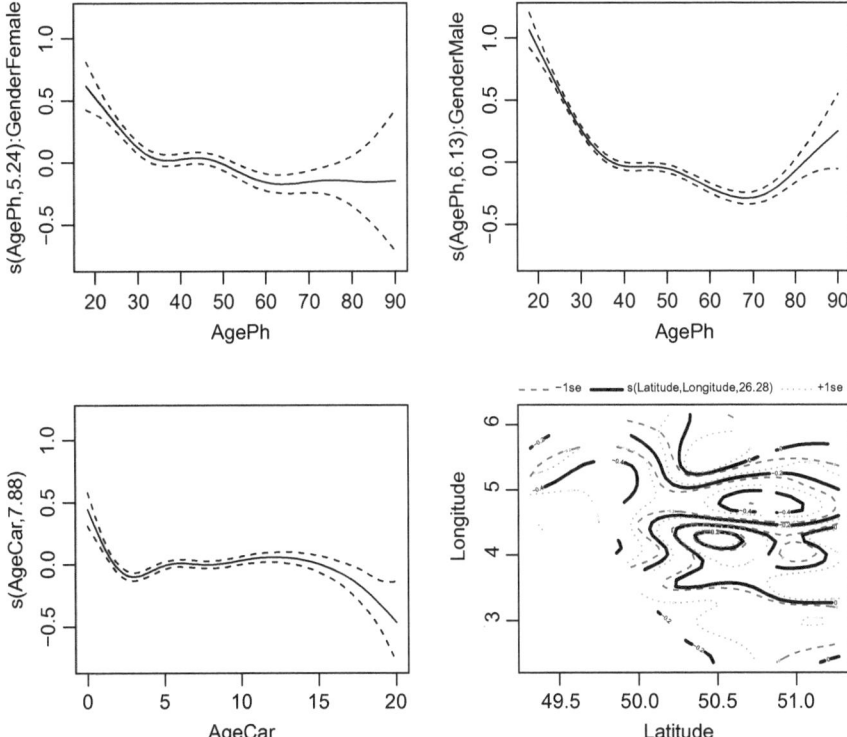

Fig. 6.19 Estimated nonlinear effects associated to age in interaction with gender, to car age and to the geographical location obtained with the help of the `gam` function included in R package `mgcv` for continuous features

6.5 Poisson Regression for Risk Classification in Life and Health Insurance

6.5.1 Life Table Modeling

Let $_\tau p_x$ be the τ-year survival probability for an individual aged x and let $_\tau q_x = 1 - {_\tau p_x}$ be the corresponding death probability. Henceforth, we assume that the function $\tau \mapsto {_\tau q_x}$ is differentiable almost everywhere and we denote the corresponding force of mortality at age $x + \tau$ as

$$\mu_{x+\tau} = \frac{1}{_\tau p_x} \frac{\partial}{\partial \tau} {_\tau q_x}.$$

Moreover, we assume throughout this section that the forces of mortality are piecewise constant on each unit interval, i.e.

$$\mu_{x+\xi} = \mu_x \text{ for } 0 \le \xi < 1 \text{ and integer } x. \tag{6.8}$$

This relationship is of course only approximately valid but is generally sufficiently accurate for practical purposes. Assumption (6.8) considerably simplifies the fitting of the statistical model. For instance, it can be shown that the identities

$$\mu_x = -\ln p_x$$

and

$$_\tau p_x = \exp(-\tau \mu_x) \ \ 0 \le \tau \le 1$$

both hold true when (6.8) is valid, for every integer age x.

6.5.2 Mortality Statistics

Assume that a life insurance portfolio has been observed during a given period of time (typically, 3–5 years). Policyholder i, $i = 1, 2, \ldots, n$, has been observed from age a_i to age b_i. Here, a_i may be the policyholder's age at the beginning of the observation period in case of an existing contract, or the age at entry in the portfolio for a new contract, issued during the observation period. The policyholder stopped being observed at age b_i, either because of death at that age, because the observation period terminated or because the policyholder left the portfolio due to policy cancellation or contract arriving at maturity. Notice that actuaries analyze mortality in function of attained age, so that age at death is the variable of interest.

    Let us first assume that we only use one year of observation and that we consider an homogeneous group of n individuals aged x. Hence, all the members of the group are subject to the same, unknown force of mortality μ_x to be estimated from their experience. The Binomial model does not apply anymore because we face an open group, with some arrivals in and departures from the portfolio. We will discuss the more general setting where the force of mortality may differ among individuals, because of observable features beyond gender and age, later in this section.

To each of these n individuals, we associate the indicator variable

$$d_i = \begin{cases} 1 \text{ if policyholder } i \text{ dies during the observation period,} \\ 0 \text{ otherwise,} \end{cases} \tag{6.9}$$

$i = 1, 2, \ldots, n$. The time spent in the portfolio by individual i during the observation year, or exposure to risk is

$$e_i = b_i - a_i \in (0, 1].$$

When working with mortality rates, the appropriate notion of risk exposure is the person-year of exposure, also called the (central) exposure-to-risk in the actuarial literature. Exposure-to-risk refers to the time, measured in years, a population was exposed to the risk of dying. In our case, the exposure-to-risk is

$$e_x = \sum_{i=1}^{n} e_i \in (0, n],$$

the total duration spent in portfolio by all policyholders aged x last birthday. By abuse of notation, we use the same letter "e" to denote the individual exposure to risk e_i, corresponding to individual i, and the aggregate exposure e_x, corresponding to the entire group aged x. The difference materializes in the subscript and it becomes clear from the context which the quantity is used.

6.5.3   Actual Likelihood

We assume that we have at our disposal independent and identically distributed observations (d_i, e_i) for each of the n individuals. The contribution of individual i to the likelihood is either

$$\exp(-e_i \mu_x)$$

in case of survival ($d_i = 0$) or

$$\exp(-e_i \mu_x)\mu_x$$

in case of death ($d_i = 1$). Therefore, the contribution of individual i to the likelihood can be written as

$$\exp(-e_i \mu_x)\left(\mu_x\right)^{d_i}.$$

Assuming mutual independence, the likelihood appears as the product over i of these individual contributions. Thus, the likelihood writes

$$\mathcal{L}(\mu_x) = \prod_{i=1}^{n} \exp(-e_i\mu_x)(\mu_x)^{d_i} = \exp\left(-e_x\mu_x\right)(\mu_x)^{d_x} \tag{6.10}$$

where

$$d_x = \sum_{i=1}^{n} d_i$$

is the number of deaths recorded among individuals aged x.

6.5.4   Poisson Likelihood

The key argument now consists in noticing that the actual likelihood $\mathcal{L}(\mu_x)$ is proportional to the Poisson likelihood

$$\mathcal{L}_{\mathcal{P}oi}(\mu_x) = \exp\left(-e_x\mu_x\right)\frac{\left(e_x\mu_x\right)^{d_x}}{d_x!} \tag{6.11}$$

which is the likelihood based on the distributional assumption $D_x \sim \mathcal{P}oi\left(e_x\mu_x\right)$.

Therefore, it is equivalent for conducting maximum-likelihood statistical inference to work on the basis of the "true" likelihood $\mathcal{L}(\mu_x)$ given in (6.10) or on the basis of the Poisson likelihood $\mathcal{L}_{\mathcal{P}oi}(\mu_x)$ given in (6.11). The great advantage of using the latter is that we then work in the GLM/GAM framework. Notice that we never formally assume that the number of deaths D_x obeys the Poisson distribution. In fact, the Poisson distribution is at best only a crude approximation to the exact distribution of D_x. For instance, the Poisson distribution puts a positive probability mass on every positive integer, so that under the Poisson assumption, D_x might well exceed n which is of course impossible in reality. Invoking the proportionality of $\mathcal{L}(\mu_x)$ and $\mathcal{L}_{\mathcal{P}oi}(\mu_x)$, we are allowed to do "as if" the number of death was Poisson distributed for the purposes of statistical inference. This is in contrast with the analysis of mortality at general population level where the number of deaths has been assumed to obey the Binomial distribution in Chap. 4 (general population being reasonably closed whereas insurance portfolios are in essence open groups). In both case, homogeneity is assumed (as the same force of mortality applies to individuals with the same gender and age).

6.5.5 Dual Modeling

Considering the expression (6.10) for the likelihood associated to the data, it can also be seen as proportional to a Gamma likelihood. Precisely, assume that given the observed number of deaths d_x, the total exposure to risk E_x obtained by summing the individual exposures E_i, $i = 1, 2, \ldots, n$, satisfies

$$E_x \sim \mathcal{G}am(y_x, \mu_x). \tag{6.12}$$

The Gamma likelihood associated to (6.12) writes

$$\mathcal{L}_{\mathcal{G}am}(\mu_x) = \exp(-e_x\mu_x)\frac{(e_x\mu_x)^{y_x}}{e_x\Gamma(y_x)} \tag{6.13}$$

which appears to be proportional to (6.10). The maximum likelihood estimates $\widehat{\mu}_x$ obtained from both likelihood functions are thus equal and statistical inference can be conducted on the basis of $\mathcal{L}_{\mathcal{G}am}$. Considering a Poisson likelihood $\mathcal{L}_{Poi}$, the parameter of interest is the force of mortality μ_x whereas in the dual Gamma likelihood $\mathcal{L}_{\mathcal{G}am}$, the parameter of interest becomes the force of vitality $1/\mu_x$.

Coming back to the individual modeling, the idea is to assume that the policyholder-specific exposure to risk E_i is the realization of a random variable obeying the distribution $\mathcal{G}am(d_x/n, \mu_x)$. By summing these individual exposures over the n individuals aged x, we get

$$E_x = \sum_{i=1}^{n} E_i = \sum_{j=1}^{d_x} G_j$$

where

$$G_j \sim \mathcal{G}am(1, \mu_x), \quad j = 1, \ldots, d_x,$$

are independent and identically distributed random variables. Again, we see that the Gamma distribution is not the exact distribution of the exposures to risk. The support of this distribution is the half positive real line whereas individual exposures belong to the unit interval and the aggregate one must be smaller than the number n of individuals under study. Again, we use the fact that the true likelihood is proportional to a Gamma one to take advantage of the GLM/GAM framework to conduct inference.

6.5.6 Poisson GLM with Aggregated Data

Until now, we have assumed that the group of individuals aged x was homogeneous, in that the same mortality rate μ_x applies to all of them. At this stage, the only continuous feature is age (the analysis is generally conducted separately for males and females so that gender also enters the analysis by subdividing the group into two sub-populations). As explained in the introduction to this chapter, there is often a considerable heterogeneity in the insured mortality. This heterogeneity can be accounted for by means of a regression model in the GLM framework as long as the actuary models the mortality of the population under consideration with respect to some reference life table (this is called a relational model in demography).

Specifically, let us now record the number of deaths Y_i from an exposure-to-risk e_i according to the value of a vector of $p + 1$ features $x_i = (1, x_{i1}, x_{i2}, \ldots, x_{ip})^\top$ including an intercept term and age, $i = 1, 2, \ldots, n$. Often, the analysis is performed with respect to some reference life table μ_x^{ref}. This can be a market life table, for instance. The logarithm of the reference death rate μ_x^{ref} is then included in the offset and the features x_{ij} position the mortality experience to the expected number of deaths according to the reference life table. The use of $\ln \mu_x^{\text{ref}}$ accounts for the effect of attained age and avoids the need for GAMs.

If a reference life table is used and all remaining features x_{ij} are categorical then we simply run a Poisson GLM to estimate the vector of unknown regression parameters $\boldsymbol{\beta} = (\beta_0, \beta_1, \ldots, \beta_p)^\top$ giving the effect of the features on mortality. Features are linked to the mortality rates with the help of a linear score and the logarithmic link function:

$$\ln \mu_i = x_i^\top \boldsymbol{\beta}.$$

The unknown parameters $\boldsymbol{\beta}$ can be estimated by Poisson maximum likelihood, i.e. under the working assumption $Y_i \sim \mathcal{P}oi(e_i \mu_i)$. The logarithm of the exposure-to-risk $\ln e_i$ is treated as an offset here. Once parameter estimates $\widehat{\boldsymbol{\beta}}$ for $\boldsymbol{\beta}$ are obtained, the fitted mortality rates for a given set of features can be calculated by $\exp(x_i^\top \widehat{\boldsymbol{\beta}})$.

Note that if the features are categorical and have been coded by means of binary variables x_{ij} then the likelihood equations have a clear interpretation. Equating to 0 the partial derivative of the log-likelihood with respect to β_j gives

$$\sum_{i|x_{ij}=1} y_i = \sum_{i|x_{ij}=1} e_i \mu_i.$$

This equation imposes that the model predicts exactly the total observed number of deaths for those cells with $x_{ij} = 1$. If an intercept is included in the linear score then also the total number of deaths is exactly fitted by the model. This can be related to the well-known method of marginal totals which predates Poisson regression in the non-life actuarial literature as explained in Chap. 4.

6.5.7 Poisson Regression with Individual Data

6.5.7.1 Score Decomposition

In addition to attained age, we have at our disposal a set of features denoted as x_i for policyholder i. The force of mortality is a function of attained age x and of x_i. It is denoted as $\mu(x|x_i)$ and is assumed to be piecewise constant with respect to age x, in accordance with (6.8). Thus, we assume that the force of mortality is constant over each year of age, but allowed to vary between ages, in line with actuarial practice. Let us nevertheless mention that we could use arbitrarily small age intervals (months, weeks or even days), subject to data availability.

The force of mortality is then typically decomposed in an additive way on the score scale, using a log link function, that is

$$\ln \mu(x|x_i) = \beta_0 + \sum_{j=1}^{p_{\text{cat}}} \beta_j x_{ij} + b_0(x) + \sum_{j=p_{\text{cat}}+1}^{p} b_j(x_{ij}) = \text{score}_i$$

for some smooth unspecified functions b_j. The regression parameters $\beta_0, \beta_1, \ldots, \beta_{p_{\text{cat}}}$ as well as the functions b_j are estimated from the available mortality data using the GAM machinery. The score includes nonlinear effects, estimated in a nonparametric way (by local polynomial models or spline decomposition, for instance). Interactions can also be included, for instance by using smooth functions with two arguments.

Example 6.5.1 Assume that we only have the sum insured z_i at our disposal ($p = 1$). Then, the following models could be considered. The general specification $\ln \mu(x|z) = b(x, z)$ allowing for all interactions between age x and sum insured z is often simplified into an additive decomposition of the form

$$\ln \mu(x|z) = b_0(x) + b_1(z) \tag{6.14}$$

where the functions b_0 and b_1 are left unspecified but assumed to be smooth. These functions can be estimated from the portfolio mortality experience, using local polynomial techniques or spline representations, for instance.

Often, mortality studies are conducted using a reference life table, corresponding to the market where the insurer operates. In such a case, the force of mortality is expressed as

$$\ln \mu(x|z) = b_0(\ln \mu_x^{\text{ref}}) + b_1(z) \tag{6.15}$$

in terms of a set of reference death rates μ_x^{ref} (treated as known constants satisfying (6.8)). Notice that this reference life table can be distorted by the function b_0 to better reflect portfolio experience. Even if there is, stricto sensu, no difference between the specifications (6.14)–(6.15), as μ_x^{ref} is itself a function of age, the second one performs generally much better in empirical illustrations as it suffices to distort the curve $x \mapsto \ln \mu_x^{\text{ref}}$ which looks similar to the portfolio experience life table.

Often, the estimated function b_0 in (6.15) appears to be approximately linear so that a linear hazard transform model can be used instead:

$$\ln \mu(x|z) = \beta_0 + \beta_1 \ln \mu_x^{\text{ref}} + b_1(z).$$

Calendar time t could be included in the features. However, portfolio analyses are generally conducted over relatively short periods of time (typically, 3–5 years). Therefore, time trends may be easier to detect using population or market data and the relational model (6.15) with reference mortality depending on calendar time might be a better alternative, as shown in the next example.

Example 6.5.2 Calendar time can also be included in the model when a projected life table is used as a reference. Precisely, the reference death rates are then made dynamic, with $\mu_{x,t}^{\text{ref}}$ representing the force of mortality at age x during calendar year t. Thus, the specification becomes

$$\ln \mu(x|t, z) = b_0(\ln \mu_{x,t}^{\text{ref}}) + b_1(z)$$

that can often be simplified into

$$\ln \mu(x|t, z) = \beta_0 + \beta_1 \ln \mu_{x,t}^{\text{ref}} + b_1(z).$$

6.5.7.2   Likelihood

Let d_i be the death indicator for policyholder i defined in (6.9). The contribution of policyholder i to the likelihood is then given by

$$\ell_i = \exp\left(-\int_{a_i}^{b_i} \mu(s|\boldsymbol{x}_i)ds\right)\left(\mu(b_i|\boldsymbol{x}_i)\right)^{d_i}. \tag{6.16}$$

Under assumption (6.8), the integrated force of mortality appearing in the exponential function can be further simplified into a sum over each year of age between a_i and b_i, as shown next.

Given a real number s, let $\lfloor s \rfloor$ denote s rounded from below and let $\lceil s \rceil$ denote s rounded from above. Precisely, $\lfloor s \rfloor$ is the largest integer that is smaller than, or equal to s and $\lceil s \rceil = \lfloor s \rfloor + 1$ is the smallest integer that is larger than, or equal to s. If $\lfloor a_i \rfloor < \lfloor b_i \rfloor$ then the integral appearing in ℓ_i can be splitted as follows:

$$\ell_i = \exp\left(-\int_{a_i}^{\lceil a_i \rceil} \mu(s|\boldsymbol{x}_i)ds\right)\exp\left(-\sum_{k=\lceil a_i \rceil}^{\lfloor b_i \rfloor - 1}\int_{k}^{k+1} \mu(s|\boldsymbol{x}_i)ds\right)$$
$$\exp\left(-\int_{\lfloor b_i \rfloor}^{b_i} \mu(s|\boldsymbol{x}_i)ds\right)\left(\mu(b_i|\boldsymbol{x}_i)\right)^{d_i}$$

with the convention that the sum over k is equal to 0 if $\lceil a_i \rceil = \lfloor b_i \rfloor$. Now, the force of mortality appearing in each integral is constant in accordance with our assumption (6.8), so that the contribution of policyholder i to the likelihood can be written as

$$\ell_i = \exp\left(-(\lceil a_i \rceil - a_i)\mu(\lfloor a_i \rfloor | \boldsymbol{x}_i)\right) \prod_{k=\lceil a_i \rceil}^{\lfloor b_i \rfloor - 1} \exp\left(-\mu(k | \boldsymbol{x}_i)\right)$$
$$\exp\left(-(b_i - \lfloor b_i \rfloor)\mu(\lfloor b_i \rfloor | \boldsymbol{x}_i)\right)\left(\mu(\lfloor b_i \rfloor | \boldsymbol{x}_i)\right)^{d_i},$$

with the convention that the product over k is equal to 1 if $\lceil a_i \rceil = \lfloor b_i \rfloor$. If $\lfloor a_i \rfloor = \lfloor b_i \rfloor$ then this contribution reduces to

$$\ell_i = \exp\left(-(\lceil a_i \rceil - a_i)\mu(\lfloor a_i \rfloor | \boldsymbol{x}_i)\right)\left(\mu(\lfloor a_i \rfloor | \boldsymbol{x}_i)\right)^{d_i}.$$

Assuming independent lifetimes, the likelihood is then obtained by multiplying the individual contributions ℓ_i over all policyholders $i = 1, \ldots, n$.

6.5.7.3 Independent Poisson Counts

Let us now relate the likelihood obtained in the preceding section to Poisson distributed random variables. To this end, define independent random variables D_{ik}, $k = \lfloor a_i \rfloor, \ldots, \lfloor b_i \rfloor$, that are assumed to be Poisson distributed with respective means

$$E[D_{ik}] = \begin{cases} (\lceil a_i \rceil - a_i)\mu(\lfloor a_i \rfloor | \boldsymbol{x}_i) \text{ for } k = \lfloor a_i \rfloor, \\ \mu(k | \boldsymbol{x}_i) \text{ for } k = \lceil a_i \rceil, \ldots, \lfloor b_i \rfloor - 1, \\ (b_i - \lfloor b_i \rfloor)\mu(\lfloor b_i \rfloor | \boldsymbol{x}_i) \text{ for } k = \lfloor b_i \rfloor. \end{cases}$$

Hence,

$$P[D_{ik} = 0] = \begin{cases} \exp\left(-(\lceil a_i \rceil - a_i)\mu(\lfloor a_i \rfloor | \boldsymbol{x}_i)\right) \text{ for } k = \lfloor a_i \rfloor, \\ \exp\left(-\mu(k | \boldsymbol{x}_i)\right) \text{ for } k = \lceil a_i \rceil, \ldots, \lfloor b_i \rfloor - 1, \end{cases}$$

and for $k = \lfloor b_i \rfloor$,

$$P[D_{i,\lfloor b_i \rfloor} = d_i] = \exp\left(-(b_i - \lfloor b_i \rfloor)\mu(\lfloor b_i \rfloor | \boldsymbol{x}_i)\right)\left((b_i - \lfloor b_i \rfloor)\mu(\lfloor b_i \rfloor | \boldsymbol{x}_i)\right)^{d_i}.$$

This shows that ℓ_i can be rewritten as

$$\ell_i = \left(\prod_{k=\lfloor a_i \rfloor}^{\lfloor b_i \rfloor - 1} P[D_{ik} = 0]\right) \frac{P[D_{i,\lfloor b_i \rfloor} = d_i]}{(b_i - \lfloor b_i \rfloor)^{d_i}}$$

with the convention that the product over k is equal to 1 if $\lfloor a_i \rfloor = \lfloor b_i \rfloor$. Hence, the contribution of each policyholder to the likelihood can be written as the product of Poisson probabilities, up to the factor $(b_i - \lfloor b_i \rfloor)^{-d_i}$. Therefore, we are allowed to perform inference using Poisson regression provided we convert the unique observation (a_i, b_i, d_i, x_i) related to policyholder i into a sequence of independent Poisson counts D_{ik}, $k = \lfloor a_i \rfloor, \ldots, \lfloor b_i \rfloor$, that are all equal to 0, except possibly the last one that equals $d_i \in \{0, 1\}$. Formulating the inference problem in terms of Poisson regression is important for practical purposes because tools performing GAM analyses are widely available and computationally efficient, not to mention that actuaries throughout the world are now used to conduct this kind of regression study.

In practice, the record (a_i, b_i, d_i, x_i) related to policyholder i in the available database is replaced with a block of $\lfloor b_i \rfloor - \lfloor a_i \rfloor + 1$ records (D_{ik}, d_i, x_i), $k = \lfloor a_i \rfloor, \ldots, \lfloor b_i \rfloor$. The Poisson regression analysis is then conducted on the responses D_{ik}, assuming their mutual independence. Typical actuarial mortality studies are performed on data gathered during 3 to 5 years so that the expanded database on which Poisson regression is conducted is three to five times bigger compared to the initial one. Even with large portfolios, this is not expected to be a problem as Poisson regression techniques can deal with very large data sets.

To be effective, life insurers should maintain databases similar to those encountered in Property and Casualty insurance. This means that there is one record per policy and per year (linked together with an identifiant like policy number), where in addition to the available features, we find the death indicator d_i.

6.6 Numerical Illustration

In order to illustrate the approach proposed in the previous section, let us consider the following hypothetical portfolio of life annuity contracts. It comprises 100,000 policies without duplicates. Annuitants are aged between 65 and 105. They are followed during three years. The available information consists in two continuous features (attained age and sum insured) together with a categorical feature with two levels (two possible sales channels for the contracts, denoted as SC1 and SC2, respectively). The sum insured is the yearly amount of annuity benefit.

Let us mention that this portfolio is realistic but not a real one (because of confidentiality issues). The mortality has been simulated according to the experience of the Belgian market. Specifically, reference death rates are taken from the Belgian MR regulatory life table (Makeham specification) whereas the effect of sum insured is inspired from Gschlossl et al. (2011).

Figure 6.20 describes the composition of the portfolio. The histograms displayed there show the age structure, the distribution of sum insured, and the number of policies according to sales channel.

The policy file comprises one record per contract, as displayed in the top panel of Table 6.3. Every individual record is then splitted into 1–3 records (depending on the remaining lifetime), as shown in the second panel of Table 6.3. The first annuitant

Fig. 6.20 Description of the
portfolio: Age structure
(top), distribution of the sum
insured (middle), and of the
sales channel (bottom, with 0
= SC1 and 1 = SC2)

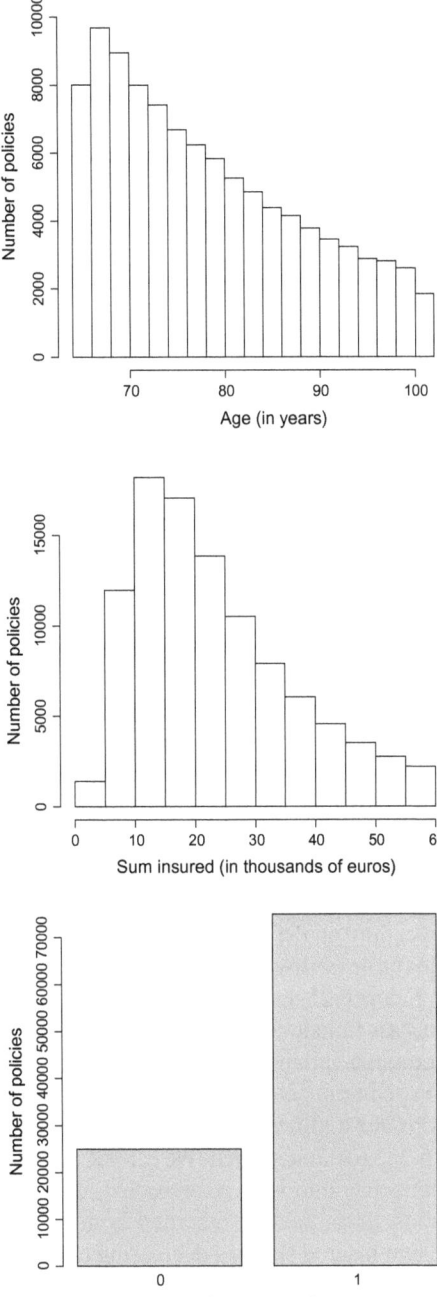

Table 6.3 Description of the initial portfolio and associated database to perform Poisson regression

Initial portfolio

Policy ID	Age (initial)	Sum insured	Sales channel	Remaining lifetime	
A	67	€14 000	SC1	≥ 3	
B	82	€36 000	SC2	2.0655	

Augmented database to perform Poisson regression

Policy ID	Age (attained)	Sum insured	Sales channel	Death indicator	Exposure-to-risk
A	67	€14 000	SC1	0	1
A	68	€14 000	SC1	0	1
A	69	€14 000	SC1	0	1
B	82	€36 000	SC2	0	1
B	83	€36 000	SC2	0	1
B	84	€36 000	SC2	1	0.0655

survives to the end of the observation period and is represented by means of three yearly records, each with a whole year of exposure and a zero death indicator. The second annuitant dies in the third year, being represented with three records. The last one accounts for death by means of the indicator, and the exposure is reduced to the survival time in the year of death. An annuitant dying in the second year would be represented similarly, with only two records. To end with, an individual dying in the first year would be represented by a single record in the augmented database. Notice that in Table 6.3, the sum insured (or any other feature) is allowed to vary in time, if appropriate. This may reflect bonuses awarded by the insurance company to the participating policies, or account for additional premiums paid by the policyholder, depending on the product under consideration. Moving to yearly records thus offers a lot of flexibility to the actuary analyzing the mortality experience.

Figure 6.21 displays the estimated effects of age and sum insured, obtained using the gam function of the mgcv package in the free statistical software R. Thin plate regression splines have been used, with equivalent numbers of degrees of freedom for each term selected from the data using generalized cross-validation. All model terms are significant, with equivalent degrees of freedom equal to 1.008 for age and to 5.383 for sum insured. We can see there point estimates surrounded with pointwise confidence intervals. As expected, the estimated age effect exhibits an increasing, almost linear pattern (inherited from the Makeham specification) whereas the effect of sum insured shows a decreasing trend, markedly non-linear.

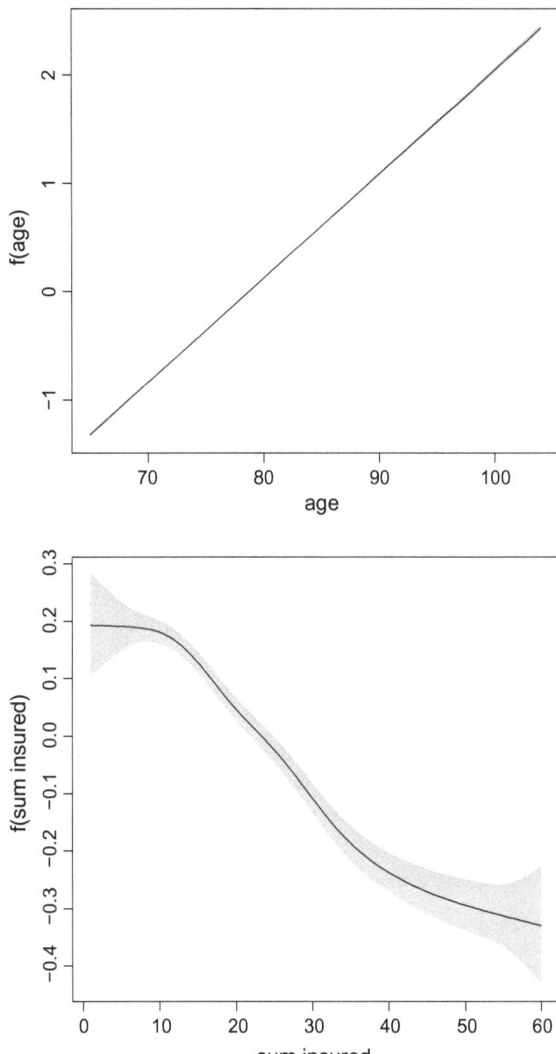

Fig. 6.21 Estimated effects for age (left) and sum insured (right)

6.7   Graduation of Morbidity Rates

6.7.1   Multi-state Modeling

Transition rates in multi-state models for health insurance can be analyzed in a way similar to mortality rates. The reason is that the multi-state model likelihood is proportional to a product of Poisson likelihoods when the transition rates are assumed to be piecewise constant, as shown next.

Specifically, assume that the states $s_0, s_1, \ldots,$ and s_m are needed to describe all benefits comprised in the insurance contract (related to sojourns in, or to transitions between states). For classical life insurance products, $m = 2$ and s_0 corresponds to "alive" whereas s_1 corresponds to "dead". In income protection insurance, we could use $m = 3$ states with $s_0 =$ "active", $s_1 =$ "disabled", and $s_2 =$ "dead", for instance, with transitions possible from s_0 to s_1 in both directions, and from these two states to s_2. In this case, premiums are paid by the policyholder as long as he or she occupies state s_0 while benefits are received in state s_1.

The stochastic process $X = \{X_t, \ t \geq 0\}$ with state space $\{s_0, s_1, \ldots, s_m\}$ describes the policyholder's trajectory across the m states $s_0, s_1, \ldots, s_m$, with X_t representing the state occupied by the policyholder at time t. By convention, time starts when the contract is issued. Time t thus measures the seniority of the policy (that is, the time elapsed since policy issue).

6.7.2 Transition Probabilities and Intensities

Assume for simplicity that the process X is Markovian. The calculations can then be performed using the transition probabilities

$$p_{ij}(s, t) = P[X_t = s_j | X_s = s_i] \text{ for } s < t \text{ and } i, j \in \{0, 1, \ldots, m\}.$$

In insurance applications, these transition probabilities can be assumed to be smooth in their two arguments, with a piecewise continuous first derivative. The transition intensities μ_{ij} for $s_i \neq s_j$ are then defined from

$$\mu_{ij}(t) = \lim_{\Delta t \searrow 0} \frac{P[X_{t+\Delta t} = s_j | X_t = s_i]}{\Delta t}.$$

In the 2-state "alive-dead" model, μ_{01} is the force of mortality used earlier in this chapter. In the 3-state model for income protection, μ_{01} is the disability rate, μ_{10} is the recovery rate, μ_{13} is the force of mortality for active individuals, and μ_{23} is the force of mortality for disabled individuals.

From the definition of $\mu_{ij}(t)$ as the limit of a transition probability divided by the length of the time interval where this transition must take place, we can write

$$P[X_{t+\Delta t} = s_j | X_t = s_i] \approx \mu_{ij}(t)\Delta t$$

where the approximation is accurate for sufficiently small Δt. The transition rates quantify the risk (or chance) for an individual in state s_i at time t to move to another state s_j at that time.

Exit intensities are defined as $\mu_{i\bullet} = \sum_{j=0|j \neq i}^{m} \mu_{ij}(t)$. Then, the sojourn probability is state s_i is simply given by

$$p_i^{(0)}(s, t) = P[X_{s+z} = s_i \text{ for all } 0 \le z \le t - s | X_s = s_i]$$

$$= \exp\left(\int_0^{t-s} \mu_{i\bullet}(s + z)\right) dz.$$

6.7.3 Likelihood Function

Let us now write the likelihood associated with observed individual trajectories across the sates $s_0, s_1, \ldots,$ and s_m. Consider an individual who has been observed from time t_0 to t_q. This individual is in state s_{i_0} at time t_0 and stays in state s_{i_j} from time t_j to time $t_{j+1}, j = 0, \ldots, q - 1$, where t_q is the end of the observation period. Each period of time spent in a given state s_{i_j} contributes a factor equal to the sojourn probability

$$p_{s_{i_j}}^{(0)}(t_j, t_{j+1}) = P[X_{t_j+z} = s_{i_j} \text{ for all } 0 \le z \le t_{j+1} - t_j | X_{t_j} = s_{i_j}]$$

$$= \exp\left(\int_0^{t_{j+1}-t_j} \mu_{i\bullet}(t_j + z)dz\right)$$

to the likelihood. Each transition from state s_{i_j} to state $s_{i_{j+1}}$ contributes a factor $\mu_{i_j i_{j+1}}$ to the likelihood. The likelihood associated to this individual trajectory then writes

$$\exp\left(-\int_{t_0}^{t_1} \mu_{i_0\bullet}(s)ds\right) \mu_{i_0 i_1}(t_1) \exp\left(-\int_{t_1}^{t_2} \mu_{i_1\bullet}(s)ds\right) \mu_{i_1 i_2}(t_2)$$

$$\ldots \mu_{i_{q-2} i_{q-1}}(t_{q-1}) \exp\left(-\int_{t_{q-1}}^{t_q} \mu_{i_{q-1}\bullet}(s)ds\right)$$

$$= \left(\prod_{j=1}^{q-1} \mu_{i_{j-1} i_j}(t_j)\right) \exp\left(-\sum_{j=1}^{q} \int_{t_{j-1}}^{t_j} \mu_{i_{j-1}\bullet}(s)ds\right).$$

If the process is Semi-Markovian then all transition intensities also depend on the time elapsed since the entry in the current state so that another continuous feature is included in the study.

We easily recognize a product of factors similar to ℓ_i defined in (6.16). This means that the analysis can be conducted separately for each pair of transitions, assuming that the response is Poisson distributed. The techniques proposed in the previous sections are thus also helpful to graduate transition rates in the presence of categorical (GLM) or continuous (GAM) features.

6.7.4 Application to Medical Expenses Cover

Such an insurance policy covers the hospitalization costs (in excess of the amount reimbursed by the Social Security regime). Let us consider a 2-state model with $s_0 = $ "active" and $s_1 = $ "at hospital". This means that we neglect mortality, which appears to be a safe strategy as long as the policy does not comprise benefits in case of death and reasonably accurate for young ages (at which forces of mortality are close to 0). The Markov assumption appears to be reasonable for short stays at hospital, excluding mental disorders or psychiatric cases, for instance.

For any integer age x and $0 \leq \xi < 1$, we assume that

$$\mu_{01}(x + \xi) = \mu_{01}(x) \text{ and } \mu_{10}(x + \xi) = \mu_{10}(x).$$

This assumption of piecewise constantness is in line with (6.8) stated for death rates.

Let us assume that we have observed for a group of policyholders aged x the number y_x^{01} of transitions from s_0 to s_1 for a total time spent in s_0 equal to $e_x^{(0)}$, and the number y_x^{10} of transitions from s_1 to s_0 for a total time spent in s_1 equal to $e_x^{(1)}$. The likelihood for x-aged people is then

$$\mathcal{L}(\mu_{01}(x), \mu_{10}(x)) = \exp\left(-\mu_{01}(x)e_x^{(0)}\right)\exp\left(-\mu_{10}(x)e_x^{(1)}\right)(\mu_{01}(x))^{y_x^{01}}(\mu_{10}(x))^{y_x^{10}}.$$

Interestingly, we can see that the likelihood factors into

$$\mathcal{L}(\mu_{01}(x), \mu_{10}(x)) = \mathcal{L}_{01}(\mu_{01}(x))\mathcal{L}_{10}(\mu_{10}(x))$$

where

$$\mathcal{L}_{01}(\mu_{01}(x)) = \exp\left(-\mu_{01}(x)e_x^{(0)}\right)(\mu_{01}(x))^{y_x^{01}}$$

and

$$\mathcal{L}_{10}(\mu_{10}(x)) = \exp\left(-\mu_{10}(x)e_x^{(1)}\right)(\mu_{10}(x))^{y_x^{10}}.$$

Each factor $\mathcal{L}_{01}$ and $\mathcal{L}_{10}$ entering the likelihood factorization only depends on a single transition rate. This allows the actuary to work separately on each set of transition rates.

Setting to zero the partial derivatives of the log-likelihood

$$\begin{aligned}
L(\mu_{01}(x), \mu_{10}(x)) &= \ln \mathcal{L}(\mu_{01}(x), \mu_{10}(x)) \\
&= \ln \mathcal{L}_{01}(\mu_{01}(x)) + \ln \mathcal{L}_{10}(\mu_{10}(x)) \\
&= -\mu_{01}(x)e_x^{(0)} + y_x^{01} \ln \mu_{01}(x) - \mu_{10}(x)e_x^{(1)} + y_x^{10} \ln \mu_{10}(x)
\end{aligned}$$

we get the likelihood equations

$$0 = \frac{\partial}{\partial \mu_{01}(x)} \ln \mathcal{L}_{01}\big(\mu_{01}(x)\big) = -e_x^{(0)} + \frac{y_x^{01}}{\mu_{01}(x)}$$

$$0 = \frac{\partial}{\partial \mu_{10}(x)} \ln \mathcal{L}_{10}\big(\mu_{10}(x)\big) = -e_x^{(1)} + \frac{y_x^{10}}{\mu_{10}(x)}.$$

Solving these equations produces the maximum-likelihood estimates of the piecewise constant transition intensities:

$$\widehat{\mu}_{01}(x) = \frac{y_x^{01}}{e_x^{(0)}} \quad \text{and} \quad \widehat{\mu}_{10}(x) = \frac{y_x^{10}}{e_x^{(1)}}.$$

Notice that the likelihood function $\mathcal{L}$ is proportional to that of two independent random variables Y_x^{01} and Y_x^{10} distributed as

$$Y_x^{01} \sim \mathcal{P}oi\big(e_x^{(0)}\mu_{01}(x)\big)$$
$$Y_x^{10} \sim \mathcal{P}oi\big(e_x^{(1)}\mu_{10}(x)\big).$$

This means that statistical inference can be conducted as if these assumptions were satisfied. The crude transition intensities $\widehat{\mu}_{01}(x)$ and $\widehat{\mu}_{10}(x)$ can therefore be graduated/smoothed using Poisson regression exactly as death rates.

Figure 6.22 displays the data available to graduate the transition rates $\widehat{\mu}_{01}(x)$. Figure 6.23 is the analog for $\widehat{\mu}_{10}(x)$. Observations displayed in Figs. 6.22 and 6.23 have been collected by a large insurance company operating in the European Union in the late 1990s. They relate to male policyholders. As the stays at hospital compensated by this insurance product are of rather short duration (policy conditions explicitly exclude hospitalizations related to mental disorders that may last for much longer), we see that $y_x^{01} \approx y_x^{10}$ for all ages x as sojourns at hospital starting during the calendar year have almost all ended during the same year. This explains why the middle panels of Figs. 6.22 and 6.23 look so similar.

Figures 6.24 and 6.25 displays the LCV plot, the corresponding local quadratic Poisson GLM fit as well as the deviance residuals. We recognize in the estimates $\widehat{\mu}_{01}$ the classical shape of a set of morbidity rates, with an age pattern similar to death rates but a higher magnitude. For female policyholders, a childbearing hump superposes to the accident hump at young adult ages. The graduated recovery rates $\widehat{\mu}_{10}$ globally decline with attained age, reflecting the increasing time spent at hospital when policyholders get older. The local minimum around age 20 corresponds to the accident hump in mortality/morbidity. There is a second local minimum around age 40 which could perhaps be attributed to midlife crisis but should certainly be submitted to medical doctors for validation.

Fig. 6.22 Exposures in s_0 (top), numbers of transitions (middle), and crude estimates of μ_{01} (bottom)

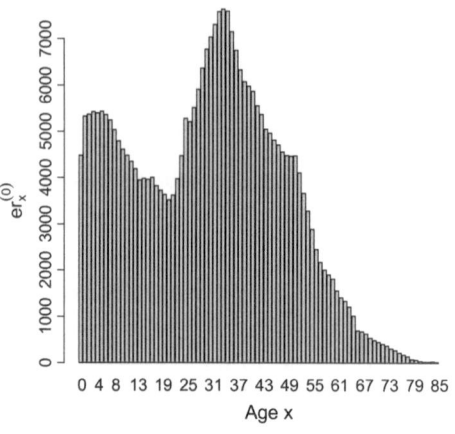

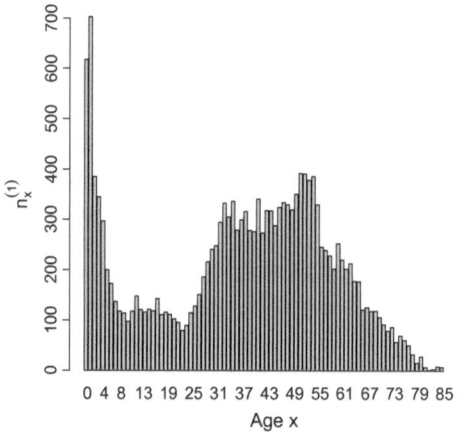

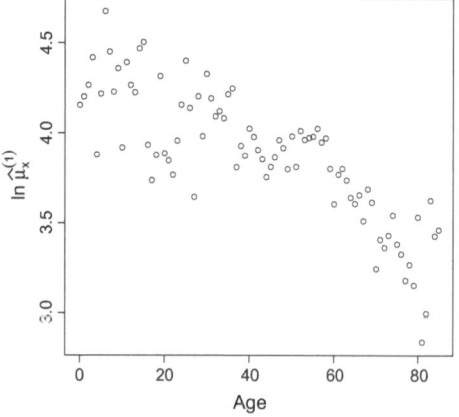

Fig. 6.23 Exposures in s_1
(top), numbers of transitions
(middle), and crude
estimates of μ_{10} (bottom)

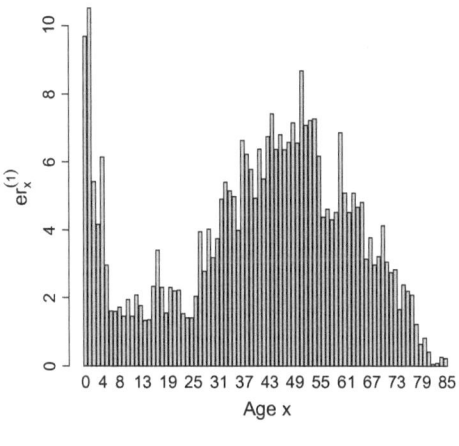

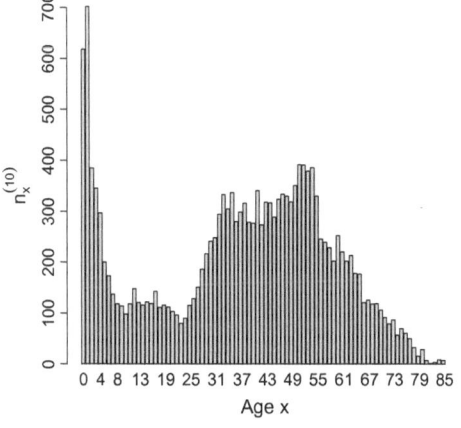

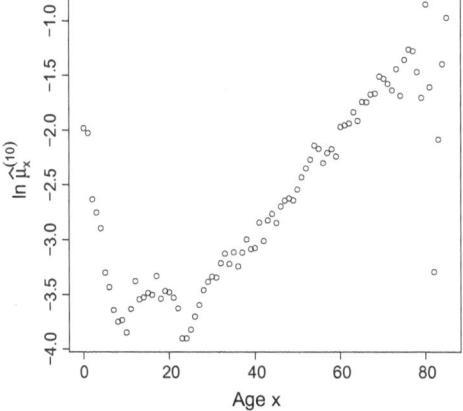

Fig. 6.24 LCV plot (top),
crude estimates $\widehat{\mu}_{01}(x)$ and
fitted values obtained with a
Poisson GAM (middle) and
deviance residuals (bottom)

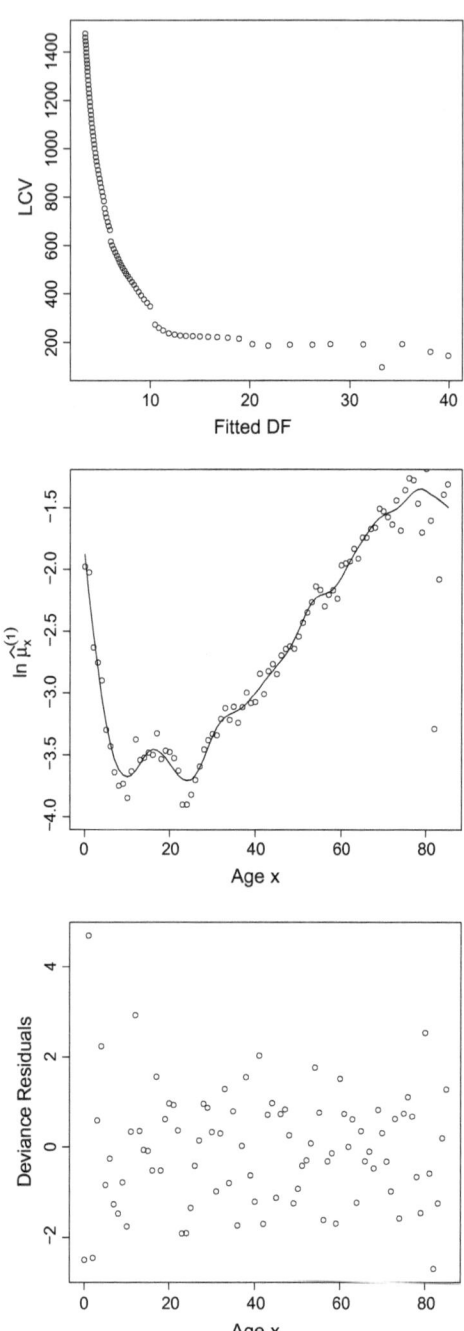

Fig. 6.25 LCV plot (top),
crude estimates $\widehat{\mu}_{10}(x)$ and
fitted values obtained with a
Poisson GAM (middle) and
deviance residuals (bottom)

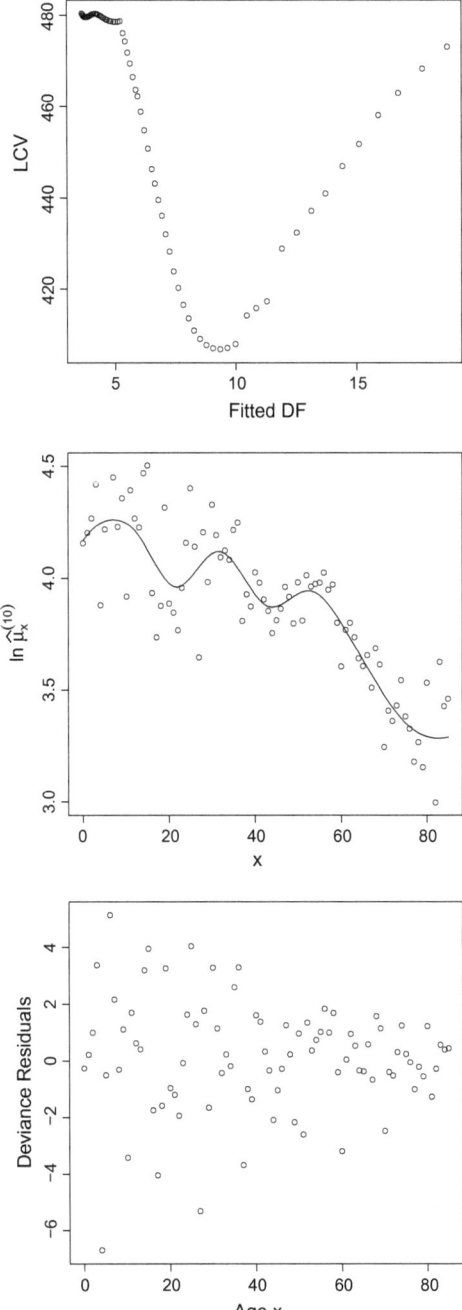

6.8   Boosting

This section offers a concise introduction to the concept of boosting. We refer the readers to Trufin et al. (2019) and Hainaut et al. (2019) for an extensive treatment, with applications to tree-based methods and neural networks, respectively.

6.8.1   General Principle

Boosting emerged from the field of machine learning. In this approach, the score is not estimated at once as in GLMs or GAMs but it is built sequentially, adding terms that weakly improve on the estimations obtained from the preceding step. Boosting algorithms allow the actuary to simultaneously select the relevant features and estimate their effect on the mean response. Instead of trying to obtain accurate first-order effects, as in GAMs, the basic idea is to boost the accuracy of weak learners by combining many of them to get a better estimation of the mean response.

The weight-judging competition studied in an article by Francis Galton published in 1907 can be seen as a precursor of boosting. The story is as follows. At a county fair, visitors were offered to estimate the weight of a fat ox with the promise to receive prizes in case of success. After the winners were announced by the organizers, Galton asked for a list of all estimates by the participants. He took this data set away and analyzed it. Arranging all the estimated ox weights from lowest to highest, he discovered that the correct weight of the ox was very close to the median, and even closer to the mean.

Participants to this contest were in majority lay people who knew very little about oxen. They can be seen as the prototype of weak learners involved in boosting algorithms. Even if each guess would be quite inaccurate, their combination reveals to be very close to the actual weight of the ox.[1]

AdaBoost is an early form of boosting for classification. With binary responses (considering the values 0 and 1 as two labels), a weak learner is defined to yield a correct classification rate just slightly better than random guessing (or flipping a coin to assign the observation to the 0 or to the 1 category). The machine learning community established that any such weak learner can be iteratively improved (or boosted) to lead to a powerful classification method. The central idea behind AdaBoost is to iteratively re-weight the observations so that observations that were misclassified in the preceding step get higher weights in the next step. This forces the algorithm to concentrate on the cases which are hard to classify, boosting its accuracy.

---

[1] It turns out that this phenomenon is not an isolated case and has led to a theory known as the Wisdom of Crowds.

6.8.2 Gradient Boosting

Later on, AdaBoost was interpreted as a functional gradient descent algorithm. This opened the door to more general responses obeying ED distributions and model fitting by a forward stage-wise procedure. When dealing with more general responses (not only binary ones), gradient boosting fits the base-learner not to re-weighted observations, as in AdaBoost, but to the negative gradient vector of the objective function evaluated at the previous iteration. With the squared error function (i.e. the Normal log-likelihood), these gradients reduce to the raw residuals. This is easily seen as follows. When $Y_i \sim \mathcal{N}or(b(\boldsymbol{x}_i), \sigma^2)$, maximizing the log-likelihood amounts to minimize the squared difference between the response Y_i and its mean $b(\boldsymbol{x}_i)$. The gradient of the objective (or loss) function to be minimized is then given by

$$\frac{\partial}{\partial b}(y - b)^2 = -2(y - b).$$

The (re-scaled) negative gradient $y - b$ are then taken as responses, starting from a constant score (including an intercept, only). In the Normal case, the boosting algorithm thus re-fits the residuals from the preceding step, adding the effect of the feature that best improves model performances.

In every boosting iteration, the base-learner is hence directly fitted to errors made in the previous iteration. Again, the performances of a simple base-learner are improved by iteratively shifting the focus towards problematic observations. With AdaBoost, this shift is done by up-weighting observations that were misclassified before. Gradient boosting with Normally distributed responses identifies problematic observations by large residuals computed in the previous iterations.

The base-learner used in boosting can be any regression technique. The boosting is for this reason called meta-algorithm or machine. Boosting is particularly effective when base-learners are trees of limited depth, as it will be seen in Trufin et al. (2019). But the base-learner can also be linear, or a penalized spline with a small number of degrees of freedom (to keep a weak learner) if the features might have a non-linear effect on the mean response. Also, only a small amount of the fit of the best-performing base-learner is added to the current additive score. This is achieved by multiplying the new effect entering the score with a shrinkage coefficient (a typical value is 0.1).

6.8.3 Likelihood-Based Boosting

Let us now stress the limitation of gradient boosting beyond the Bernoulli and Normal cases and advocate that likelihood-based boosting avoids this drawback. Instead of maximizing the log-likelihood, gradient boosting applies a least-squares principle on its gradients. Applied to Poisson responses, the actuary essentially uses a least-squares approach when resorting to gradient boosting. This is because when $Y_i \sim$

$\mathcal{P}oi(\exp(b(\boldsymbol{x}_i)))$, using minus the log-likelihood function as the objective function to be minimized means that the loss function is of the form

$$- \ln \mathcal{L}_{\mathcal{P}oi} = \sum_{i=1}^{n} \left(\exp \left(b(\boldsymbol{x}_i) \right) - y_i b(\boldsymbol{x}_i) \right) + \text{constant}.$$

Now,

$$\frac{\partial}{\partial b} \left(\exp(b) - yb \right) = \exp(b) - y.$$

Gradient boosting then consists in fitting the negative gradient $y - \exp(b)$ by least squares. However, we know that least-squares is outperformed by Poisson regression for low counts. This is why likelihood-based boosting described next is superior to gradient boosting in this case.

All models have been fitted so far by maximizing a likelihood. While such a likelihood can also be used to define a loss function in gradient boosting (by switching to minus the likelihood, to be minimized, or by using the deviance), a boosting approach could also be built on base-learners that directly maximize an overall likelihood in each boosting step. This is the underlying idea of likelihood-based boosting. Given a starting value or estimate from a previous boosting step, likelihood-based boosting approaches use base-learners for estimating a new effect correcting the previous estimation in a log-likelihood that contains the previous score as a fixed offset. For obtaining small improvements, similar to gradient boosting, a penalty term is attached to the log-likelihood. To speed the calculations, base learners are generally estimated using one step of the IRLS algorithm, using the current score as offset.

For Normally-distributed responses, including an offset is the same as fitting a model to the residuals from the previous boosting step, and likelihood maximization by a base-learner becomes standard least-squares estimation with respect to these residuals. In this special case, likelihood-based boosting thus coincides with gradient boosting for squared losses. For the kind of responses encountered in actuarial applications, such as claim frequencies and severities on individual policies, likelihood-based boosting improves on gradient boosting by recognizing the particular shape of the responses (whereas gradient boosting applies a least-squares principle, whatever the range of the gradients).

6.8.4 Feature Selection

In every boosting iteration, only the best-performing base-learner and hence the best-performing feature is included in the final model. This procedure effectively leads to data-driven variable selection during the model estimation. Boosting not only fits the model under consideration, but also automatically selects the relevant features in regression models. This allows the actuary to avoid problems with classical feature selection techniques.

To figure out the differences with the GAMs, recall that in the classical regression setting the actuary includes all the relevant features by monitoring the increase in log-likelihood, starting from the null model. Clearly, the objective function can only increase when more features are included. The question is then to detect when the improvement becomes small enough so that this feature does not warrant its inclusion in the model, the increase in model complexity being worse than the improvement in goodness of fit. The large-sample Chi-Square approximation to the deviance if generally used to this end.

With the more flexible boosting approach, the score is not optimized globally, but sequentially in a stage-wise fashion. This means that all the good properties of M-estimation listed in Sect. 4.11 are lost. In particular, we cannot use Chi-Square quantiles to monitor reductions in deviance. Instead, we evaluate the prediction capabilities of the model, no more the goodness of fit.

Often in practice, when regression models are used for prediction, the predictive accuracy is much worse compared to the goodness of fit in the training sample. With small samples (such as in the triangle-based loss reserving example or with life tables), cross validation can be achieved by omitting each observation in turn from the data set, fitting the model to the remaining $n - 1$ observations, predicting the value of the response for the omitted observation, and comparing the prediction with the actual observed value. This approach is known as the leave-one-out cross validation.

For larger data sets, this procedure rapidly becomes unfeasible. If the sample is large enough then it is possible to divide the sample randomly in several parts (or folds). Each part is then held out of the analysis and the model is fitted on the remaining data to predict the observed values of the response in the part set aside. This cross-validation of the model on the second independent sample gives a more realistic assessment of its predictive ability. Precisely, k-fold cross validation can be described as follows. With $k = 10$, the model is fit to 90% of the data and used to predict the remaining 10%. This is a convenient way to balance goodness of fit and model complexity: a model too close to the training set will often be worse for predictions, as it reproduces noise in the data (or over-fits the training data).

To prevent overfitting, cross-validation is used to stop the boosting algorithms when its prediction capabilities start to deteriorate. The optimal stopping iteration is hence the one which leads to the smallest average empirical loss on the out-of-sample test data. Hence, this process stops when predictive abilities start to deteriorate, as measured by cross-validation. Early stopping of statistical boosting algorithms plays a central role to ensure a sparse model with optimal performances on new data.

6.9 Projection Pursuit Regression

Projection pursuit regression (PPR) extends additive models in that the features are first projected in optimal directions before applying smoothing functions. Precisely, PPR creates optimal linear combinations of the features and include them in the mean modeling like in GAMs. The score in PPR is of the form

$$\beta_0 + \sum_{j=1}^{r} b_j (\boldsymbol{\beta}_j^\top \boldsymbol{x}_i)$$

where $\boldsymbol{\beta}_j$ are r vectors of length p which contain the unknown parameters, b_j are unknown smooth functions and r is a hyper-parameter. As r approaches infinity and with an appropriate set of functions b_j, the PPR model is a universal estimator able to approximate any continuous function with p arguments.

In general, additional components $(b_j, \boldsymbol{\beta}_j)$ are included in the score in a forward stage-wise fashion, like in boosting. Fitting PPR models is done as follows. Starting from an initial value for $\boldsymbol{\beta}_j$, the optimal b_j is determined as with GAMs, in a model including $\boldsymbol{\beta}_j^\top \boldsymbol{x}_i$ as unique feature. This can be done by local linear regression or spline decomposition, for instance. Then, once b_j has been estimated, it is kept fixed and the regression coefficients $\boldsymbol{\beta}_j$ are revised. This can be done by using a GLM with a user's defined link function. The updated $\boldsymbol{\beta}_j$ is then used to refit b_j, and so on. This alternating process is continued until $(b_j, \boldsymbol{\beta}_j)$ converges. The result is then included in the offset and the model proceeds with the next pair $(b_j, \boldsymbol{\beta}_j)$.

Beyond Normally-distributed responses Generalized PPR combines PPR with IRLS and a link function to estimate responses obeying ED distributions.

6.10 Regression Trees and Multivariate Adaptive Regression Splines

6.10.1 General Approach

In this section, regression trees and multivariate adaptive regression splines (MARS) are presented in a unified way. Such models complement GLMs and GAMs by identifying complex interaction effects that are often present in insurance data. This approach nicely connects with the industry practice, putting the output of GAM analysis in offset at the initializing step and correcting the GAM for the interaction effects it did not capture.

All insurance studies are performed under the assumption that the effects of the features on the mean response are smooth, in that they can be captured by continuous functions, possessing as many derivatives as needed. This somewhat conflicts with the approximation by means of step functions. Let us consider a function with a single argument to understand this point. Consider a real-valued function b with derivatives $b^{(k)}$, $k = 1, 2, \ldots$. If b is differentiable then the expansion

$$b(x) = b(0) - \int_{-\infty}^{0} \mathrm{I}[x \le t] b^{(1)}(t)\, \mathrm{d}t + \int_{0}^{+\infty} \mathrm{I}[x > t] b^{(1)}(t)\, \mathrm{d}t$$

shows that the function b can indeed be approximated by linear combinations of step functions

$$I[x \leq t] = \begin{cases} 1 \text{ if } x \leq t \\ 0 \text{ otherwise} \end{cases} \text{ and } I[x > t] = \begin{cases} 1 \text{ if } x > t \\ 0 \text{ otherwise.} \end{cases}$$

This is in essence the approach based on trees where the feature space in partitioned with the help of products of such indicator functions and the unknown mean function is approximated by a piecewise constant score.

Tree-based methods have been widely used in insurance studies, from single regression trees to random forests. Despite their attractiveness e.g. for identifying complex interaction effects, their output remains difficult to implement in insurance pricing because of the numerous irregularities in the resulting fit. This is the direct consequence of the additive decomposition of the score, as a sum of products of step functions.

If the function is more regular, then we can use higher-order Taylor's expansion of b to refine the approximation. Henceforth, denote as z_+ the positive part of z, i.e. $z_+ = \max\{z, 0\}$. We then get the following expansion formula

$$b(x) = \sum_{j=0}^{s-1} \frac{b^{(j)}(0)}{j!} x^j + (-1)^s \int_{-\infty}^{0} (t - x)_+^{s-1} \frac{b^{(s)}(t)}{(s-1)!} dt$$
$$+ \int_{0}^{+\infty} (x - t)_+^{s-1} \frac{b^{(s)}(t)}{(s-1)!} dt$$

for $s \geq 2$. For $s = 2$, we recover the MARS approach, where the unknown mean function is approximated by products of piecewise linear, or hockey stick functions.

The idea behind regression trees and MARS is to approximate the unknown mean function by linear combinations of products of reflected pairs $\left((t - x)_+^k, (x - t)_+^k\right)$ where x represents one of the continuous features available to explain the response, and t is a candidate for serving as knot (taken from the observed values of the feature). For $k = 0$, the reflected pairs reduce to $\left(I[x \leq t], I[x > t]\right)$ and we recover regression trees, while $k = 1$ corresponds to MARS. Even if larger values of k can be considered, this is generally not needed in practice so that we focus here on $k \in \{0, 1\}$. Notice that when t is the minimum value for the feature x, we essentially recover the monomial x^k.

6.10.2 Calibration

To fit regression trees ($k = 0$) and MARS ($k = 1$), an expansion in terms of piecewise polynomial functions $(t - x)_+^k$ and $(x - t)_+^k$ with knot t is performed. Having a response Y_i for individual i, we start from the null model

$$\mu_i = \overline{y} \text{ for all } i = 1, 2, \ldots, n.$$

The best reflected pair $\left((t - x)_+^k, (x - t)_+^k\right)$ is then selected and introduced into the score. The selection is made among all available features x, taking all their distinct values t as candidates for knot. The first step then gives a score of the form

$$\beta_0 + \beta_1(t_1 - x_{ij_1})_+^k + \beta_2(x_{ij_1} - t_1)_+^k \qquad (6.17)$$

where j_1 is the index of the feature producing the largest decrease in deviance for the knot t_1. Here, $\beta_0 = g^{-1}(\overline{y})$ and the pair (β_1, β_2) is estimated by IRLS. For $k = 1$, the score (6.17) is just a straight line with a different slope before and after t_1. For $k = 0$, it is a step function with different heights before and after t_1.

The model is now based on the two functions $(t_1 - x_{ij_1})_+^k$ and $(x_{ij_1} - t_1)_+^k$. The next step consists in multiplying one of them with a reflected pair $\left((t - x)_+^k, (x - t)_+^k\right)$, selected to maximize the decrease in the deviance. This means that we consider all candidate scores of the form

$$\beta_0 + \beta_{11}(t_1 - x_{ij_1})_+^k(t - x)_+^k + \beta_{12}(t_1 - x_{ij_1})_+^k(x - t)_+^k + \beta_2(x_{ij_1} - t_1)_+^k$$

or

$$\beta_0 + \beta_1(t_1 - x_{ij_1})_+^k + \beta_{21}(x_{ij_1} - t_1)_+^k(t - x)_+^k + \beta_{22}(x_{ij_1} - t_1)_+^k(x - t)_+^k$$

taken over all features x and knots t corresponding to their distinct values in the data basis. In the former case, $\beta_0 + \beta_2(x_{ij_1} - t_1)_+^k$ is taken as offset, whereas in the latter case, $\beta_0 + \beta_1(t_1 - x_{ij_1})_+^k$ is taken as offset.

Notice that the regression parameters involving the last reflected pairs included in the model are re-estimated at each step while the other ones are left unchanged. We then proceed in this way, repeating this step until some stopping criterion is met.

Most insurance studies also involve categorical features. The reflected pairs for a binary feature x_{ij}, with levels $x_{ij}^{[0]}$ and $x_{ij}^{[1]}$, say, is $\left(\mathrm{I}[x_{ij} = x_{ij}^{[0]}], \mathrm{I}[x_{ij} = x_{ij}^{[1]}]\right)$. For a categorical feature x_{ij} with k levels, we create all the binary cuts partitioning the k levels into two disjoint, non-empty subsets. This is effective when k is not too large. For larger values of k, some approximations can be used (such as considering the categorical feature as a continuous one equal to the average response value in each level). The reader is referred to Trufin et al. (2019) for extensive explanations.

Regression trees and MARS models typically overfit the data, so that a backward deletion procedure (called pruning) is applied. A boosting approach combining many regression trees or MARS with limited depth run on the responses themselves (not on gradients) offers a powerful alternative. This means that the actuary specifies a maximum interaction depth for every component of the score and each individual multivariate spline contribution is expanded up to this threshold.

The MARS approach imposes that every feature can appear at most once in a product. This prevents the formation of high powers of inputs which often increase or decrease too sharply near the boundary of the feature space. When combined with a GAM analysis for instance, taken as the initial step of the analysis, it appears reasonable to assume that the GAM captured all the marginal effects of the features, so that only possible interaction need to be identified. This means that each feature is allowed to enter only once in each individual multivariate spline and greatly simplifies the computations. At each step, the scores obtained from the preceding iterations are put in offset and the model estimates the multivariate spline best explaining the remaining effects. Classical stopping criteria are used to stop the fitting process, based on cross-validation.

6.11 Bibliographic Notes and Further Reading

As pointed out by Loader (1999), smoothing methods and local regression originated in actuarial science in the late 19th and early 20th centuries, in the problem of graduation. GAMs were developed in Hastie and Tibshirani (1990) to provide a convenient framework to overcome the linearity assumptions inherent to GLMs when smooth effects of continuous features need to be included in an additive predictor. A comprehensive reference about GAMs oriented towards applications is certainly Wood (2017) who contributed the R package mgcv that has been extensively used in this chapter. The key reference for local GLM is Loader (1999) who contributed the R package locfit.

In insurance, Denuit and Lang (2004) and Klein et al. (2014) demonstrated the usefulness of the GAM approach to deal with Property and Casualty insurance data. The Poisson modeling for death counts has been proposed by Brouhns et al. (2002) and is now widely adopted within the actuarial commmunity. The GLM analysis of mortality with aggregated data has been performed on German insurance statistics by Gschlossl et al. (2011). Denuit and Legrand (2018) extended these ideas to mortality and morbidity graduation with continuous features. The dual Gamma modeling for risk exposures has been proposed by Renshaw and Haberman (1996) and Renshaw et al. (1997). We refer the interested reader to Brouhns et al. (2002) for empirical evidence supporting the linearity of b_1 in (6.15).

The review papers by Mayr et al. (2014a, b) contain most of the material in Sect. 6.8. A detailed account of the analysis conducted by Francis Galton can be found in Wallis (2014). Likelihood-based boosting was introduced by Tutz and Binder (2006). These authors applied this principle to GAMs with B-spline base-learners. Likelihood-based boosting for GLMs was introduced by Tutz and Binder (2007). The best component for an update is then selected according to the deviance in each boosting step. An approach for mixed models was described by Groll and Tutz (2012).

Notice that gradient boosting extends beyond objective functions derived from maximum-likelihood estimation. We refer the reader to Ma and Huang (2005) for an application to the receiver operator characteristic (ROC) curve and the corresponding area under the ROC curve (AUC). An application to credit scoring can be found in Kraus and Kuchenhoff (2014). Wang (2011) introduced a gradient boosting approach to optimize the hinge-loss (HingeBoost). The hinge loss is the standard loss function for support vector machines (SVM) and its optimization is approximately equivalent to maximizing the AUC. Via HingeBoost it is also possible to incorporate unequal misclassification costs. This special feature could be of high relevance in insurance studies as covering a bad risk or declining to cover a good risk have very different consequences.

A popular traditional discriminatory measure for the evaluation of prediction models is the concordance index (C-index). We refer the reader to Denuit et al. (2019a, b) for goodness-of-fit measures based on concordance. See also Trufin et al. (2019). Chen et al. (2013) and Mayr and Schmid (2014) applied gradient boosting to optimize the C-index. Boosting has also been adapted to quantile (Fenske et al. 2011) and expectile (Sobotka and Kneib 2012) regression.

Boosting is particularly effective in insurance applications where weak learners are trees of limited depth. This is why we only provide a brief introduction here, to show how it connects to more traditional tools, deferring an extensive treatment to Trufin et al. (2019) where tree-based methods are comprehensively discussed.

PPR has been studied in Friedman and Stuetzle (1981). A methodological account of Generalized PPR can be found in Lingjarde and Liestol (1998). It is implemented in the gppr function of the R package gsg, for Bernoulli and Poisson responses. As it will be seen in Hainaut et al. (2019), neural networks models also project the features onto a one-dimensional hyperplane and then apply a nonlinear transformation of the input variables that are then added in a linear fashion. Thus, both PPR and neural networks follow basically the same steps. The main difference is that the functions b_j involved in PPR are estimated from the data whereas in neural networks these are all specified upfront.

MARS have been proposed by Friedman (1991). This extension of linear models automatically accounts for non-linearities and interactions between features. As explained in Friedman (1991), there is no reason to go beyond $k = 0$ or 1 in the applications. Because the term "MARS" has been trademarked and licensed to Salford Systems, many open source implementations of MARS are called "Earth" (including in R). Our presentation is inspired from Hastie et al. (2016, Sect. 9.4).

References

Brouhns N, Denuit M, Vermunt JK (2002) A Poisson log-bilinear approach to the construction of projected life tables. Insur Math Econ 31:373–393

Chen Y, Jia Z, Mercola D, Xie X (2013) A gradient boosting algorithm for survival analysis via direct optimization of concordance index. Comput Math Methods Med 13, Article ID 873595

Denuit M, Lang S (2004) Nonlife ratemaking with Bayesian GAMs. Insur Math Econ 35:627–647

Denuit M, Legrand C (2018) Risk classification in life and health insurance: extension to continuous covariates. Eur Actuar J 8:245–255

Denuit M, Mesfioui M, Trufin J (2019a) Bounds on concordance-based validation statistics in regression models for binary responses. Methodol Comput Appl Probab 21:491–509

Denuit M, Mesfioui M, Trufin J (2019b) Concordance-based predictive measures in regression models for discrete responses. Scand Actuar J

Fenske N, Kneib T, Hothorn T (2011) Identifying risk factors for severe childhood malnutrition by boosting additive quantile regression. J Am Stat Assoc 106:494–510

Friedman JH (1991) Multivariate adaptive regression splines. Ann Stat 19:1–67

Friedman JH, Stuetzle W (1981) Projection pursuit regression. J Am Stat Assoc 76:817–823

Groll A, Tutz G (2012) Regularization for generalized additive mixed models by likelihood-based boosting. Methods Inf Med 51:168–177

Gschlossl S, Schoenmaekers P, Denuit M (2011) Risk classification in life insurance: methodology and case study. Eur Actuar J 1:23–41

Hainaut D, Trufin J, Denuit M (2019) Effective statistical learning methods for actuaries–neural networks and unsupervised methods. Springer Actuarial Series

Hastie T, Tibshirani R (1990) Generalized additive models. Chapman and Hall, London

Hastie T, Tibshirani R, Friedman J (2016) The Elements of statistical learning: data mining, inference, and prediction. Springer Series in Statistics

Kraus A, Kuchenhoff H (2014) Credit scoring optimization using the area under the curve. J Risk Model Valid 8:31–67

Lingjarde OC, Liestol K (1998) Generalized projection pursuit regression. SIAM J Sci Comput 20:844–857

Loader C (1999) Local regression and likelihood. Springer, New York

Ma S, Huang J (2005) Regularized ROC method for disease classification and biomarker selection with microarray data. Bioinformatics 21:4356–4362

Mayr A, Binder H, Gefeller O, Schmid M (2014a) The evolution of boosting algorithms-from machine learning to statistical modelling. Methods Inf Med 53:419–427

Mayr A, Binder H, Gefeller O, Schmid M (2014b) Extending statistical boosting-an overview of recent methodological developments. Methods Inf Med 53:428–435

Mayr A, Schmid M (2014) Boosting the concordance index for survival data a unified framework to derive and evaluate biomarker combinations. PloS ONE 9:e84483

Renshaw AE, Haberman S (1996) Dual modelling and select mortality. Insur Math Econ 19:105–126

Renshaw AE, Haberman S, Hatzopoulos P (1997) On the duality of assumptions underpinning the construction of life tables. ASTIN Bull 27:5–22

Sobotka F, Kneib T (2012) Geoadditive expectile regression. Comput Stat Data Anal 56:755–767

Trufin J, Hainaut D, Denuit M (2019) Effective statistical learning methods for actuaries–tree-based methods. Springer Actuarial Series

Tutz G, Binder H (2006) Generalized additive modelling with implicit variable selection by likelihood based boosting. Biometrics 51:961–971

Tutz G, Binder H (2007) Boosting ridge regression. Comput Stat Data Anal 51:6044–6059

Wallis KF (2014) Revisiting Francis Galton's forecasting competition. Stat Sci 29:420–424

Wang Z (2011) HingeBoost: ROC-based boost for classification and variable selection. Int J Biostat 7:1–30

Wood SN (2017) Generalized additive models: an introduction with R, 2nd edn. CRC Press

Chapter 7
Beyond Mean Modeling: Double GLMs and GAMs for Location, Scale and Shape (GAMLSS)

7.1 Introduction

We know from Chap. 4 that GLMs impose a constant dispersion parameter ϕ. This means that the variance of the response is allowed to change with the risk factors only as a function of the mean. A constant ϕ appears to be restrictive with some distributions, such as those in the Tweedie subclass of the ED family. Moving to double GLMs, features can also enter the dispersion parameter which depends on a specific score. In addition to the mean modeling as in the GLM approach, there is thus also a dispersion modeling in double GLMs, with its own link function (generally, the log-link is used to ensure the positivity of the resulting dispersion parameter) and regression coefficients.

A Gamma GLM is used for dispersion modeling, run on the squared deviance or Pearson residuals. The double GLM fit is then obtained by iterating along a GLM for the mean followed by a Gamma GLM for the dispersion taking as response the contribution of each observation to the deviance. The connection between the two GLMs is as follows: the model for the mean produces residuals used as responses for the dispersion model, which in turn produces the dispersion for the mean model.

As ϕ enters the ED distribution through the ratio ϕ/ν_i where the weights ν_i are known constants, this means that the a priori weights ν_i cannot be modified by the model. If the dispersion parameter is allowed to vary with the features then the weights can be adapted to reflect the actuarial experience of each policy in the portfolio. A double GLM may thus improve the mean estimation in case certain classes of business appear to be more volatile compared to others. In this setting, the model is allowed to give less weight to the past observations of the volatile business and more weight to the stable business whose data is more informative. Hence, the model becomes capable of ignoring more noise when necessary and thus picking up more signal. The double GLM thus in a sense picks the linear score in the dispersion model to define optimal weights, maximizing goodness-of-fit. Sometimes, a feature that seemed to be irrelevant in the GLM setting becomes significant when the actuary moves to the double GLM framework.

© Springer Nature Switzerland AG 2019
M. Denuit et al., *Effective Statistical Learning Methods for Actuaries I*,
Springer Actuarial, https://doi.org/10.1007/978-3-030-25820-7_7

In the double GLM setting, features enter both ED parameters θ and ϕ, that is, features enter the mean and the dispersion parameters. The framework of generalized additive models for location, scale and shape (GAMLSS) extends this idea to more complex response distributions where not only the mean response but multiple parameters are related to additive scores with the help of suitable link functions. For instance, zero-inflated, skewed and zero-adjusted distributions can be embedded in this framework as special cases where all occurring parameters are related to specific scores.

In the GAMLSS approach, the ED distribution assumption for the response is relaxed so that the actuarial analysis is no more restricted to the distributions used in the classical GLM/GAM setting. The only restriction is that the individual contribution to the log-likelihood and its first two derivatives with respect to each of the parameters must be computable. This opens the door to numerous response distributions recognizing the specific nature of insurance data.

After a presentation of the methodological aspects of double GLM and GAMLSS, this chapter proposes an application to loss reserving approach, inspired from the collective model of risk theory. According to the collective paradigm, payments are not related to specific claims or policies but a frequency-severity setting is adopted, with a number of payments in every cell of the run-off triangle, together with the corresponding paid amounts. Compared to the Tweedie reserving model, that can be seen as a compound sum with Poisson-distributed number of terms and Gamma-distributed summands, more general severity distributions can be used, typically mixture models combining a light-tailed component with a heavier-tailed one, including inflation effects. The severity model is fitted to individual observations and not to aggregated data displayed into run-off triangles with a single value in every cell. In that respect, the modeling approach appears to be a powerful alternative to both the crude traditional aggregated approach based on triangles and the extremely detailed individual reserving approach developing each and every claim separately. A case study based on a motor third-party liability insurance portfolio illustrates the relevance of this approach.

7.2 Tweedie Compound Poisson Distributions

7.2.1 Compound Distributions

Sometimes, actuaries only have claim totals at their disposal. Such claim totals are zero with positive probability, but continuously distributed otherwise. A compound model is a good candidate to model such responses. Specifically, let N be a counting random variable representing the number of claims and let $C_1, C_2, \ldots, C_N$ be the respective amounts of these N claims. The total claim amount thus writes

$$Y = \sum_{k=1}^{N} C_k,$$

with the convention that the empty sum equals 0, i.e.

$$Y = \begin{cases} 0 \text{ if } N = 0, \\ C_1 + \cdots + C_N \text{ if } N \geq 1. \end{cases}$$

Therefore, $P[Y = 0] = P[N = 0] > 0$.

If the claim severities $C_1, C_2, \ldots$, are assumed to be independent and identically distributed, and independent of N, then, S is said to have a compound distribution. The mean and variance of Y are then respectively given by

$$E[Y] = E\Big[NE[C_1]\Big] = E[N]E[C_1]$$

and

$$\begin{aligned} \mathrm{Var}[Y] &= E\big[\mathrm{Var}[Y|N]\big] + \mathrm{Var}\big[E[Y|N]\big] \\ &= E\big[N\mathrm{Var}[C_1]\big] + \mathrm{Var}\big[NE[C_1]\big] \\ &= E[N]\mathrm{Var}[C_1] + \mathrm{Var}[N]\big(E[C_1]\big)^2. \end{aligned}$$

7.2.2 Compound Poisson Distributions

If $N \sim \mathcal{P}oi(\lambda)$ then Y obeys a compound Poisson distribution. In this case,

$$\begin{aligned} E[Y] &= \lambda E[C_1] \\ \mathrm{Var}[Y] &= \lambda \mathrm{Var}[C_1] + \lambda\big(E[C_1]\big)^2 \\ &= \lambda E[C_1^2]. \end{aligned}$$

Compound Poisson distributions play an important role in actuarial sciences as they provide a conservative approximation to the individual model in risk theory.

The moment generating function of the compound Poisson distribution is obtained from

$$\begin{aligned} m_Y(t) &= \sum_{k=0}^{\infty} P[N = k]E\left[\exp\left(t\sum_{j=1}^{k} C_j\right)\right] \\ &= \sum_{k=0}^{\infty} P[N = k]\big(m_C(t)\big)^k \text{ where } m_C(t) = E[\exp(tC_1)] \\ &= E\Big[\big(m_C(t)\big)^N\Big] = \varphi_N(m_C(t)) \text{ where } \varphi_N(t) = E[t^N]. \end{aligned}$$

Clearly, $\varphi_N(t) = E[t^N] = m_N(\ln t)$. Considering the Poisson moment generating function m_N obtained from Property 2.4.1, we finally get

$$m_Y(t) = \exp\Big(\lambda(m_C(t) - 1)\Big). \tag{7.1}$$

7.2.3 Tweedie and Compound Poisson with Gamma Summands

By cleverly choosing the claim frequency and severity distributions, we can see that the power variance function defining Tweedie distributions corresponds to a compound Poisson distribution belonging to the ED family provided $1 < \xi < 2$. Specifically, assume that

$$Y = \sum_{k=1}^{N} C_k \text{ with } N \sim \mathcal{P}oi(\lambda) \text{ and } C_k \sim \mathcal{G}am(\alpha, \tau),$$

all the random variables being independent. In order to connect this compound Poisson model to the ED family, let us identify the corresponding moment generating function to the canonical form derived in Property 2.4.1. To this end, we insert the Gamma moment generating function m_C into (7.1) to obtain

$$m_Y(t) = \exp\left(\lambda\left(\left(1 - \frac{t}{\tau}\right)^{-\alpha} - 1\right)\right).$$

Now, we write

$$\ln m_Y(t) = \lambda\left(\left(1 - \frac{t}{\tau}\right)^{-\alpha} - 1\right)$$

$$= \frac{1}{\phi}\Big(a(\theta + t\phi) - a(\theta)\Big)$$

where the second equality is valid for any distribution inside the ED family, according to Property 2.4.1. With the Tweedie power variance function $V(\mu) = \mu^\xi$, we know from Sect. 2.5 that

$$a(\theta) = \frac{\big((1 - \xi)\theta\big)^{\frac{2-\xi}{1-\xi}}}{2 - \xi}$$

so that

$$\ln m_Y(t) = \frac{1}{\phi} \frac{\left((1-\xi)(\theta+t\phi)\right)^{\frac{2-\xi}{1-\xi}} - \left((1-\xi)\theta\right)^{\frac{2-\xi}{1-\xi}}}{2-\xi}$$

$$= \frac{1}{\phi}\left((1-\xi)\theta\right)^{\frac{2-\xi}{1-\xi}} \frac{\left(1+\frac{(1-\xi)t\phi}{(1-\xi)\theta}\right)^{\frac{2-\xi}{1-\xi}} - 1}{2-\xi}.$$

The mean μ corresponding to the Tweedie subclass of ED distributions is

$$a'(\theta) = \left((1-\xi)\theta\right)^{\frac{2-\xi}{1-\xi}-1} = \left((1-\xi)\theta\right)^{\frac{1}{1-\xi}}$$

so that

$$\ln m_Y(t) = \frac{1}{\phi}\frac{\mu^{2-\xi}}{2-\xi}\left(\left(1+(1-\xi)t\phi\mu^{\xi-1}\right)^{\frac{2-\xi}{1-\xi}} - 1\right).$$

This finally leads to

$$\lambda = \frac{\mu^{2-\xi}}{\phi(2-\xi)}$$

$$\alpha = \frac{2-\xi}{\xi-1}$$

$$\frac{1}{\tau} = \phi(\xi-1)\mu^{\xi-1}.$$

To ensure $\alpha > 0$, the constraint $1 < \xi < 2$ must be fulfilled so that the exponent ξ is restricted to the interval $(1, 2)$.

Let us now identify the components of (2.3) for such a response Y. Clearly,

$$P[Y = 0] = P[N = 0] = \exp(-\lambda).$$

This is well of the form (2.3) with

$$\lambda = \frac{a(\theta)}{\phi}.$$

Now, given $N = k > 0$,

$$Y = C_1 + \cdots + C_k \sim \mathcal{G}am(\alpha k, \tau)$$

so that the probability density function of Y over $(0, \infty)$ is given by

$$f_Y(y) = \sum_{k=1}^{\infty} P[N = k] f_{C_1+\cdots+C_k}(y)$$

$$= \exp\left(-\tau y - \lambda\right) \sum_{k=1}^{\infty} \frac{\tau^{k\alpha}}{\Gamma(k\alpha)} y^{k\alpha-1} \frac{\lambda^k}{k!}, \qquad y > 0.$$

Since

$$\lambda \tau^\alpha = \frac{\mu^{2-\xi}}{\phi(2-\xi)} \left(\frac{1}{\phi(\xi-1)\mu^{\xi-1}} \right)^{\frac{2-\xi}{\xi-1}} = \frac{\left(\phi(\xi-1) \right)^{\frac{\xi-2}{\xi-1}}}{\phi(2-\xi)}$$

does not depend on μ, only on the parameter ϕ and the constant ξ, the series involved in the expression for f_Y depends on ϕ and y, but not on μ.

Define $c(y, \phi)$, $y > 0$, as the series appearing in f_Y, i.e.

$$c(y, \phi) = \sum_{k=1}^{\infty} \frac{\tau^{k\alpha}}{\Gamma(k\alpha)} y^{k\alpha-1} \frac{\lambda^k}{k!},$$

and $c(0, \phi) = 1$. Identifying

$$\exp\left(\frac{y\theta - a(\theta)}{\phi} \right) \text{ with } \exp\left(-\tau y - \lambda \right),$$

we get

$$-\tau = \frac{\theta}{\phi} \text{ and } \lambda = \frac{a(\theta)}{\phi}$$

which gives

$$\theta = -\tau\phi = \frac{1}{(1-\xi)\mu^{\xi-1}} \text{ and } a(\theta) = \lambda\phi = \frac{\mu^{2-\xi}}{2-\xi}.$$

This shows that the Tweedie distribution belongs to the ED family for fixed value of ξ. Also, we recover the power variance function $V(\mu) = \mu^\xi$ with $1 < \xi < 2$ as

$$\begin{aligned} \text{Var}[Y] &= E[N]E[C_1^2] \\ &= \lambda \left(\frac{\alpha}{\tau^2} + \left(\frac{\alpha}{\tau} \right)^2 \right) \\ &= \lambda \frac{\alpha(\alpha+1)}{\tau^2} \\ &= \phi\mu^\xi. \end{aligned}$$

7.2.4 Exponent Parameter

In actuarial applications, the exponent ξ appearing in the Tweedie variance function is generally confined to the interval $(1, 2)$ so that the corresponding Tweedie distributions correspond to compound Poisson sums with Gamma distributed severities. The Poisson ($\xi = 1$) and Gamma ($\xi = 2$) distributions can then be obtained as limits of this restricted Tweedie class.

In general, ξ is unknown and must be estimated from the data. By definition

$$\xi = \frac{\alpha + 2}{\alpha + 1}$$

is a function of the Gamma coefficient of variation. Values of ξ found in insurance modeling typically range between 1.5 and 1.8. When the Tweedie distribution is used in actuarial studies, the choice $\xi = 1.65$ appears to be convenient as a starting point.

7.2.5 Limitation of Compound Poisson Tweedie GLM

There is an implicit restriction when using the Compound Poisson Tweedie GLM, that is often disregarded by analysts. Recall from Sect. 2.5 that we have in this case

$$\mu = \lambda \frac{\alpha}{\tau}, \quad \xi = \frac{\alpha + 2}{\alpha + 1} \text{ and } \phi = \frac{\mu^{2-\xi}}{\lambda(2 - \xi)} = \frac{\lambda^{1-\xi} \left(\frac{\alpha}{\tau}\right)^{2-\xi}}{2 - \xi}.$$

With GLMs, the dispersion parameter ϕ is held constant so that

$$\phi = \text{constant} \Rightarrow \frac{\left(\frac{\alpha}{\tau}\right)^{2-\xi}}{\lambda^{\xi-1}} = \text{constant}. \tag{7.2}$$

Hence, any risk factor increasing the expected claim severity $\frac{\alpha}{\tau}$ must also increase the expected claim frequency λ to fulfill (7.2). Similarly, any risk factor decreasing the expected claim severity must also decrease the expected claim frequency for (7.2) to remain valid. In particular, every risk factor impacting on one of these two variables must also impact on the other one in the same direction.

However, this is often not the case in insurance applications. For instance in loss reserving, the expected number λ of payments typically decrease with the development lag whereas the corresponding average amounts increase. This phenomenon is clearly visible on the data displayed in Table 4.6. This means that the analysis of a single triangle with total amounts paid (obtained by multiplying both triangles cell by cell) using such a Tweedie model may distort the analysis.

In motor insurance, geographic factors typically have opposite effects on frequencies and severities, with higher frequencies but smaller severities in big cities and the opposite in rural areas. Here also, modeling the total losses using the Tweedie compound Poisson GLM does not recognize these two conflicting effects.

Using Tweedie model thus often requires to move to a double GLM setting where the dispersion parameter ϕ also depends on the information contained in x_i. In this case, we iterate between a GLM for the mean and a Gamma GLM for dispersion, run on deviance residuals (see the next Sect. 7.3 for more details).

7.3   Dispersion Modeling: Double GLMs

GLMs impose that the dispersion parameter ϕ must be the same for every individual. Hence, the variance $\text{Var}[Y_i] = \frac{\phi}{\nu_i} V(\mu_i)$ may change with the risk factors only as a function of the mean μ_i. Double GLMs avoids this restriction. Precisely, with double GLM, the dispersion parameter

$$\phi_i = \phi(\boldsymbol{x}_i)$$

also depends on the available features through a specific score, making it now specific to each observation. Thus, not only the mean response depends on available features, but also the dispersion coefficient. Precisely, besides the mean modeling $g(\mu_i) = \boldsymbol{x}_i^\top \boldsymbol{\beta}$ there is also a dispersion modeling

$$g_d(\phi_i) = \boldsymbol{x}_i^\top \boldsymbol{\gamma}$$

with its own link function g_d and regression coefficients $\boldsymbol{\gamma}$. Often, a log-link function is used for g_d, to ensure that $\widehat{\phi}_i$ remains positive. But other link functions may be useful in some applications (see for instance the application to mortality graduation in presence of duplicates discussed in the next section).

The regression coefficients $\boldsymbol{\gamma}$ are then estimated with the help of a Gamma GLM run on the squared deviance r_i^D or Pearson's r_i^P residuals defined in Sect. 4.9. Starting from $\phi_i = 1$ for all i at step 0, the regression coefficients $\boldsymbol{\beta}$ and $\boldsymbol{\gamma}$ are then estimated in turn, until convergence is achieved. Precisely, a double GLM fit is obtained by iterating along the following steps:

– fit a GLM for the mean response, with a constant ϕ for all observations, that is,

$$g(\text{E}[Y_i]) = \boldsymbol{x}_i^\top \boldsymbol{\beta}$$
$$\text{Var}[Y_i] = \frac{\phi}{\nu_i} V(\text{E}[Y_i]).$$

– calculate the contribution of each observation to the deviance and compute the squared Pearson or deviance residuals R_i^2.
– fit a GLM for dispersion, by taking as response the contribution R_i^2 of each observation to the deviance. The distribution is Gamma and there is no weight entering this step. The fitted values are the new dispersion parameter for each record. Precisely, the dispersion modeling uses a Gamma GLM with

$$g_d(\text{E}[R_i^2]) = \boldsymbol{x}_i^\top \boldsymbol{\gamma}$$
$$\text{Var}[R_i^2] = \tau (\text{E}[R_i^2])^2$$
$$\phi_i = g_d^{-1}(\boldsymbol{x}_i^\top \boldsymbol{\gamma}).$$

- fit the GLM for the mean, but this time using the specific dispersion parameter for each record (dividing the weight with the response-specific dispersion parameter obtained from the preceding step), that is,

$$g(\mathrm{E}[Y_i]) = x_i^\top \boldsymbol{\beta}$$

$$\mathrm{Var}[Y_i] = \frac{\phi_i}{v_i} V(\mathrm{E}[Y_i]).$$

- compute the squared Pearson or deviance residuals R_i^2 and repeat the preceding steps.

This iterative process continues until some stopping condition is fulfilled. The connection between the two GLMs is as follows: the model for the mean produces the response R_i^2 used for fitting the dispersion model, which in turn produces the dispersion for the mean model.

A double GLM may improve the mean estimation in case certain classes of business appear to be more volatile compared to others. In this setting, the model is allowed to give less weight to the past observations of the volatile business and more weight to the stable business whose data is more informative. Hence, the model becomes capable of ignoring more noise when necessary and thus picking up more signal. The double GLM thus in a sense determines the linear score to define optimal weights, maximizing goodness-of-fit. Sometimes, a feature that seems to be irrelevant in the GLM setting becomes significant when the actuary moves to the double GLM framework.

The double GLM is particularly relevant for Tweedie modeling, because of the implicit restriction inherent to these models discussed in Sect. 7.2.5. Consider again the application to loss reserving, with data displayed in triangular form indexed by accident year AY and development year DY. Assume that the amount appearing in the cell corresponding to AY$= j$ and DY$= k$ is Tweedie distributed with a mean expressed in function of factors related to accident year j and development k. Data displayed in run-off triangles generally exhibit the following two opposite effects:

- in general, most of the claims are reported early in development so that the frequency component has a decreasing trend throughout developments;
- the average cost per claim often increases through the development years so that the severity component exhibits an increasing trend.

Because frequencies and severities have opposite trends, models with constant dispersion are prone to errors. In the Tweedie case, a constant ϕ implies that the influence of j and k on both frequency and severity components must go in the same direction. This is why it is desirable to switch to a double GLM where the mean as well as the variance depend on the effects j and k.

7.4 Dispersion Modeling in Mortality Graduation with Duplicates

7.4.1 Duplicates

In practice, it is common for individuals to hold more than one policy and hence to appear more than once in the count of exposed to risk or deaths. In such a case, the portfolio is said to contain duplicates: it contains several policies concerning the same lives.

This phenomenon can easily be corrected at each company level but remains problematic when market data are collected and collated by some central agency. This is done by the Continuous Mortality Investigation (CMI) Bureau in the UK, or by the regulatory authorities (as the National Bank in Belgium). Even if the data are de-duplicated by each participant before submitting the mortality statistics to the central agency, that is all policies held by the same individual are consolidated into a single observation, this cannot be done across companies by the central agency in charge of data collection (because the data are anonymised by participating companies before being transmitted to the agency). Market data thus generally contain many duplicates so that deaths are confused with claims on policies: the death of a policyholder carrying m policies appears as m deaths in the data. In the absence of information about the distribution of policies per life insured, estimation of the death probabilities becomes more difficult.

When duplicates are present in the portfolio, the actuary knows the number of claims c_x, that is, the number of policies whose holder died at age x during the observation period, and not the actual number of deaths d_x recorded among the l_x individuals aged x. We know that when duplicates are present in mortality data the inequality $c_x \geq d_x$ holds valid. Also, the number n_x of policies whose holder is aged x is recorded in the data basis, with $n_x > l_x$, the lower bound corresponding to the absence of duplicates.

7.4.2 Numbers of Contracts, Lives, Deaths and Claims

Formally, let us denote as l_x the number of policyholders aged x and as D_x the number of deaths recorded among them. To ease the exposition, we assume that every policyholder has been covered for the whole year. The actuary thus deals with a closed group, assumed to be homogeneous. The number of deaths D_x is thus Binomially distributed with size l_x and probability q_x.

Further, define the random variables

$$N_{xi} = \text{number of contracts held by policyholder } i, \text{ aged } x$$

assumed to be independent and identically distributed

$$N_x = \text{number of contracts held by all policyholders aged } x$$

$$= \sum_{i=1}^{l_x} N_{xi}.$$

We denote as C_{xi} the number of claims corresponding to policyholder i, aged x. This random variable is zero if policyholder i survives and is equal to N_{xi} in case he or she dies.

7.4.3 Mean-Variance Relationship

We assume that the random variables C_{xi} are independent and identically distributed with

$$P[C_{xi} = 0] = p_x$$

and for $k \geq 1$,

$$P[C_{xi} = k] = q_x \psi_x(k)$$

where $\psi_x(k) = P[N_{xi} = k]$ denotes the probability that individual i aged x holds k policies, $k = 1, 2, \ldots$, with

$$1 = \sum_{k=1}^{\infty} \psi_x^{(k)}.$$

In words, $\psi_x(k)$ corresponds to the probability that a single death is recorded as k claims.

Define

$$\overline{\psi}_x^{[j]} = \sum_{k=1}^{\infty} k^j \psi_x(k) \text{ for } j \in \{1, 2\}.$$

Then,

$$E[C_{xi}] = q_x \sum_{k=1}^{\infty} k \psi_x(k) = q_x \overline{\psi}_x^{[1]}$$

and

$$\mathrm{Var}[C_{xi}] = E[C_{xi}^2] - \left(E[C_{xi}] \right)^2$$

$$= q_x \sum_{k=1}^{\infty} k^2 \psi_x(k) - \left(q_x \sum_{k=1}^{\infty} k \psi_x(k) \right)^2$$

$$= q_x \left(\overline{\psi}_x^{[2]} - q_x \left(\overline{\psi}_x^{[1]} \right)^2 \right).$$

Consider the total number of claims corresponding to policyholders aged x, given by

$$C_x = \sum_{i=1}^{l_x} C_{xi}.$$

We have

$$E[C_x] = l_x E[C_{x1}] = r_x q_x$$

with

$$r_x = l_x \overline{\psi}_x^{[1]}.$$

Here, r_x is the expected number of policies held by the l_x individuals aged x. Also,

$$\mathrm{Var}[C_x] = l_x \mathrm{Var}[C_{x1}] = \phi_x r_x q_x (1 - q_x)$$

with

$$\phi_x = \frac{1}{1 - q_x} \frac{\overline{\psi}_x^{[2]}}{\overline{\psi}_x^{[1]}} \left(1 - \frac{\left(\overline{\psi}_x^{[1]}\right)^2}{\overline{\psi}_x^{[2]}} q_x \right).$$

7.4.4 Overdispersed Binomial Distribution

When there are no duplicates in the data set, $\psi_x(1) = 1$ and $\psi_x(k) = 0$ for all $k \geq 2$. We then have

$$\phi_x = 1, \quad r_x = l_x \text{ and } C_x = D_x \sim \mathcal{B}in(l_x, q_x).$$

When there are duplicates, $\psi_x(k) > 0$ for at least some $k \geq 2$. The variance of the number of contracts per insured individual aged x is $\overline{\psi}_x^{[2]} - (\overline{\psi}_x^{[1]})^2$. The more variability in this number, the larger the variance and the smaller the ratio $(\overline{\psi}_x^{[1]})^2 / \overline{\psi}_x^{[2]}$. In the limiting case where each policyholder has only one contract, we get $\phi_x = 1$. This suggests to use the approximation

$$\phi_x \approx \frac{\overline{\psi}_x^{[2]}}{\overline{\psi}_x^{[1]}} > 1 \tag{7.3}$$

which appears to be accurate as long as q_x remains small. Since this is generally the case except at the oldest ages, the approximation (7.3) is effective for insurance studies. Under (7.3), C_x obeys an Overdispersed Binomial distribution.

7.4.5 Dispersion Modeling

Let $\widehat{\psi}_x(k)$ be the proportion of individuals aged x holding k policies. The variance ratios vr_x defined as

$$vr_x = \frac{\sum_{k \geq 1} k^2 \widehat{\psi}_x(k)}{\sum_{k \geq 1} k \widehat{\psi}_x(k)}$$

can be used as estimates for ϕ_x. Studies conducted on the UK market reveal variance ratios in the range $(1, 2)$.

In case no information is available about multi-detention, a possibility is to proceed as if vr_x was constant with respect to x (which is of course unrealistic) and use an Overdispersed Binomial model with constant dispersion parameter $\phi > 1$. A more effective approach is to supplement the Binomial GLM/GAM for death counts with a dispersion modeling where ϕ_x is learnt from the data.

When duplicates are present, mortality graduation proceeds in two stages, in the double GLM setting. In the first stage (mean modeling), the number of claims C_x is modeled according to the Overdispersed Binomial distribution with

$$E[C_x] = r_x q_x \text{ and } \text{Var}[C_x] = \phi_x r_x q_x (1 - q_x).$$

Alternatively, a Poisson approximation may be useful for open groups, replacing n_x with the corresponding exposure to risk based on policies and q_x with the force of mortality μ_x. In the second stage, the dispersion parameter ϕ_x is estimated in a Gamma regression model with the squared deviance residuals as responses. Because $\phi_x \in [1, 1 + \kappa)$ for some κ (with $\kappa = 1$ supported by the studies conducted on the UK market), the second-stage link function g_d should account for that particular range. For $\kappa = 1$, this is achieved with the translated complementary log-log link

$$\phi_x = 2 - \exp\left(-\exp(\text{score}_x)\right),$$

the translated logit link

$$\phi_x = \frac{1 + 2\exp(\text{score}_x)}{1 + \exp(\text{score}_x)},$$

or the translated probit link

$$\phi_x = 1 + \Phi(\text{score}_x)$$

where, as before, $\Phi(\cdot)$ denotes the distribution function of the $\mathcal{N}or(0, 1)$ distribution. The score involved in the dispersion modeling is generally a quadratic function of age x, that is,

$$\text{score}_x = \gamma_0 + \gamma_1 x + \gamma_2 x^2,$$

or a linear function of r_x, that is,

$$\text{score}_x = \gamma_0 + \gamma_1 r_x.$$

The great advantage of this approach is that the knowledge of the multi-detention probabilities $\psi_x(\cdot)$ are not necessarily needed to perform mortality graduation. The presence of duplicates is taken into account by the use of age-specific dispersion parameters ϕ_x replacing the empirical variance ratios vr_x, in a double GLM setting.

To end with, let us mention that the same approach can be used to graduate amounts at death, that is, considering the amount of benefits paid by insurance companies instead of the actual number of deaths.

7.5 Beyond Dispersion

7.5.1 The GAMLSS Framework

Generalized additive models for location, scale and shape (GAMLSS) extends GAMs to more complex response distributions where not only the expectation but multiple parameters are related to additive scores with the help of suitable link functions. This extends the double GLM approach where the dispersion parameter also depends on risk factors, not only the mean response. In particular, zero-inflated, skewed and zero-adjusted distributions can be embedded in the GAMLSS framework as special cases.

The main difference between GAMLSS and GAM is that GAMLSS do not only model the conditional mean of the response distribution but several of its parameters. For instance, in actuarial applications, the dispersion parameter and the no-claim probability may be modulated by a score, in addition to some location parameter. Several scores may then enter the pure premium, each one explaining a specific facet of the loss distribution.

GAMLSS still require a parametric distribution for the response but the ED assumption for the response Y is relaxed so that the actuarial analysis is no more restricted to the distributions used in the classical GLM setting. This opens the door to numerous response distributions in line with the characteristics of insurance data. The form of the distribution assumed for the response variable can be very general. The only restriction is that the individual contribution to the log-likelihood and its first two derivatives with respect to each of the parameters must be computable. Analytic expressions for these derivatives are preferable, but numerical derivatives can be used (resulting in reduced computational speed).

Known monotonic link functions relate the distribution parameters to explanatory variables. Specifically, distribution parameters $\vartheta_1, \vartheta_2, \ldots$ are given by $g_k(\vartheta_k) = \text{score}_k$ for some known link function g_k where each score is assumed to be additive with smooth, nonparametric effects of continuous features. Compared to the double GLM approach, scores may include nonlinear effects of the features in the GAMLSS setting. Also, considering the two-parameter ED distributions, both parameters are now functions of the available features, not only the mean response as in GLMs/GAMs.

7.5.2 GAMLSS for Count Data

7.5.2.1 Overview of Available Counting Distributions

Binomial-type response distributions fall in the GAMLSS framework, including the Binomial, the Beta-Binomial, and their inflated and zero-adjusted versions. Beta-Binomial models are mixed models where the random effect distribution is conjugate to the conditional distribution of the response. Hence, the posterior distribution of the response remains Beta-Binomial, with updated parameters.

Most of the other count data distributions are derived from the Poisson law. Many count data distributions have been proposed to remedy major problems often encountered when modeling count data using the Poisson distribution, including overdispersion and excess of zero values.

7.5.2.2 Mixed Poisson Distributions

Overdispersion has been extensively discussed in Chap. 5, in relation with extra variation in a count response which is not explained by the Poisson distribution alone. In addition to the type 1 and type 2 Negative Binomial distributions, the Poisson-Inverse Gaussian (PIG), the Sichel and the Delaporte mixed Poisson distributions are also interesting for insurance studies. The Sichel distribution has been found to provide a useful three-parameter model for over-dispersed Poisson count data exhibiting high positive skewness. This distribution is also known as the Generalized Inverse Gaussian Poisson (or GIGP) distribution. With the Sichel distribution, the actuary is able to model the mean, variance and skewness in terms of available features. The Delaporte distribution is a mixed Poisson distribution obtained with a shifted Gamma mixing distribution.

7.5.2.3 Zero-Inflated Distributions

The problem of excess of zero values occurs when the response variable has a higher probability of a zero value than a Poisson distribution. This is a phenomenon that occurs often in insurance practice. A solution to excess zero values in a particular discrete distribution is a zero-inflated discrete distribution. Zero-inflated distributions differ from zero-adjusted (or altered) ones offering a solution to excess and/or shortage of zero values in a particular discrete distribution.

In insurance studies, the number of observed zeroes is generally much larger than under the Poisson assumption. This motivates the use of a mixture of two distributions: a degenerated distribution for the zero case and a standard count distribution. Specifically, the probability mass function is given by

$$P[N = k] = \begin{cases} \rho + (1 - \rho)g(0) & \text{for } k = 0 \\ (1 - \rho)g(k) & \text{for } k = 1, 2, \dots \end{cases} \tag{7.4}$$

where $\rho \in (0, 1)$ and $g(\cdot)$ is the probability mass function corresponding to the standard distribution to be modified.

The number of claims N reported by a policyholder to the company can be represented as the product of an indicator variable J (equal to 1 if the policyholder reported at least 1 claim) and a non-negative counting variable K. Furthermore, J and K are assumed to be independent. Hence,

$$P[N = k] = P[JK = k] = \begin{cases} P[J = 0] + P[J = 1, K = 0] & \text{for } k = 0 \\ P[J = 1]P[K = k] & \text{for } k = 1, 2, \dots \end{cases} \tag{7.5}$$

which indeed corresponds to (7.4). Notice that here, K is allowed to be equal to 0 whereas $J \sim \mathcal{B}er(1 - \rho)$ so that $N = 0$ may be due either to $J = 0$ or to $K = 0$.

For instance, in the Zero-Inflated Poisson (ZIP) distribution, $K \sim \mathcal{P}oi(\lambda)$ so that g corresponds to the Poisson distribution with mean λ, that is, $g(k) = \exp(-\lambda)\frac{\lambda^k}{k!}$ for $k = 0, 1, \dots$. The two first moments of the ZIP distribution are

$$E[N] = E[J]E[K] = (1 - \rho)\lambda$$

and

$$\begin{aligned} \text{Var}[N] &= E[JK^2] - \left(E[J]E[K]\right)^2 \\ &= (1 - \rho)E[K^2] - (1 - \rho)^2\left(E[K]\right)^2 \\ &= E[N] + E[N]\left(E[K] - (1 - \rho)E[K]\right) \\ &= E[N] + E[N]\left(\lambda - E[N]\right). \end{aligned}$$

Since $\text{Var}[N] > E[N]$, ZIP models thus account for overdispersion. Notice that the ZIP model is a special case of a mixed Poisson distribution obtained with

$$\Lambda = \begin{cases} 0 \text{ with probability } \rho \\ \lambda \text{ with probability } 1 - \rho. \end{cases}$$

In some situations, even when the zero-count data are fitted adequately, overdispersion for non-zero count may be still present. Zero inflated Negative Binomial distributions can be used in such a case. Note that ZI models with overdispersion for the non-zero counts can be seen as particular cases of Poisson mixtures, obtained when Θ has a mixed distribution with a probability mass ψ at the origin.

7.5.2.4 Hurdle Models

Often, the vast majority of the insured drivers reports less than 2 claims per year. Consequently, a classification of the insured based on two processes turns out to be interesting. A dichotomic variable first differentiates insureds with and without claim. In the former case, another process then generates the number of reported claims. The most popular distribution implying the assumption that the data come from two separate processes is the hurdle count model. The simplest hurdle model is the one which sets the hurdle at zero, called the hurdle-at-zero model. This corresponds to a zero-adjusted (or zero-altered) distribution. Formally, given two probability mass functions g_1 and g_2, the hurdle-at-zero model has probability mass function

$$P[N = k] = \begin{cases} g_1(0) & \text{for } k = 0 \\ \frac{1-g_1(0)}{1-g_2(0)} g_2(k) = \rho g_2(k) & \text{for } k = 1, 2, \ldots \end{cases} \tag{7.6}$$

where

$$\rho = \frac{1 - g_1(0)}{1 - g_2(0)}$$

can be interpreted as the probability of crossing the hurdle (or more precisely in case of insurance, the probability to report at least one claim). Clearly, the model collapses to g if $g_1 = g_2 = g$.

The number of claims N can then be represented as the product of an indicator variable J (equal to 1 if the policyholder reported at least 1 claim, $J \sim \mathcal{B}er(1 - \rho)$) and a counting variable $K \geq 1$ (giving the number of claims reported to the company when at least 1 claim has been filed). Furthermore, J and K are assumed to be independent. Hence,

$$P[N = k] = P[JK = k] = \begin{cases} P[J = 0] & \text{for } k = 0 \\ P[J = 1]P[K = k] & \text{for } k = 1, 2, \ldots \end{cases} \tag{7.7}$$

which indeed corresponds to (7.6). The representation $N = JK$ is similar to the decomposition of the total claim amount in the individual model of risk theory. Compared to (7.5), we see that here $N = 0$ only if $J = 0$.

Let μ_2 be the expected value associated with the probability mass function g_2. The mean and variance corresponding to (7.6) are then given by

$$E[N] = \frac{1 - g_1(0)}{1 - g_2(0)} \sum_{k=0}^{\infty} k g_2(k)$$
$$= \rho \mu_2$$
$$\mathrm{Var}[N] = \rho \sum_{k=1}^{\infty} k^2 g_2(k) - \left(\rho \sum_{k=1}^{\infty} k^2 g_2(k) \right)^2$$
$$= P[N > 0]\mathrm{Var}[N|N > 0] + P[N = 0]E[N|N > 0].$$

Consequently, hurdle models can exhibit over or underdispersed, depending on the distributions g_1 and g_2. Many possibilities exist for g_1 and g_2. Nested models where g_1 and g_2 come from the same distribution, such as the Poisson or the Negative Binomial distributions, are often used.

The hurdle models are widely used in connection with health care demands where it is generally accepted that the demand for certain types of health care services depend on two processes: the decisions of the individual and the one of the health care provider. The hurdle model also possesses a natural interpretation for the number of reported claims. A reason for the good fitting of the zero-inflated models is certainly the reluctance of some insureds to report their accident (since they would then be penalized by some bonus-malus scheme implemented by the insurer). The behavior of the insureds may become different once a claim have reported in a year because of the bonus-malus scheme, with a worse claim experience once a claim is reported. Since the behavior of the insureds is likely to differ when they already have reported a claim, this suggests that two processes govern the total number of claims.

7.5.3   Finite Mixture Distributions

Finite mixture distributions consist in combinations of several distributions (say q, numbered 1 to q) with the help of a two-stage stochastic model. First a random number j is drawn from $\{1, \ldots, q\}$ according to some specific probabilities and then the jth component of the mixture distributions is considered. The components of the finite mixture may all be discrete or all be continuous, but they may also be of various types. When a continuous distribution is mixed with discrete distributions, this allows the actuary to create continuous distributions in which the support has been expanded to include some discrete values with non-zero probabilities. This is the case with the Beta distribution inflated at 0 and/or at 1, for instance.

Such a Beta distribution inflated at 0, at 1 or at 0 and 1, is appropriate to model material damage claims which are generally expressed as a percentage of the sum insured (such as the value of the insured vehicle, in motor insurance).

General finite mixture models can also be treated with the help of GAMLSS. We could for instance combine a Gamma distribution together with a Pareto one. The resulting mixture has both a light-tailed component and a heavy-tailed one. These finite mixtures may have no parameters in common or one or several parameters entering several scores in the model.

7.5.4 Zero-Augmented Distributions

Claim frequencies and claim severities are often analyzed separately. GAMLSS nevertheless allow the actuary to deal with total costs using the so-called zero-augmented models. Zero-augmented (ZA) distributions are useful for modeling a response valued in the interval $[0, \infty)$, such as a yearly claim amount. As the majority of policyholders do not report any claims, there is a high probability mass at zero but for policies with claims, the distribution of the yearly claim amount is defined on the positive real line $(0, \infty)$. Precisely, ZA models are finite mixtures with two components: a probability mass at zero and a continuous component which can be any parametric distribution as long as first and second derivatives of the log-likelihood can be computed.

ZA models are in line with the Tweedie distribution. Compound Poisson Tweedie distributions have a probability mass at zero and otherwise a continuous probability density function over $(0, \infty)$. When the actuary deals with claims totals, which are often zero and continuously distributed otherwise, this type of Tweedie distribution is therefore a natural candidate to model such responses. Notice that the probability mass at the origin is entirely determined by the Tweedie parameters whereas this key actuarial indicator is allowed to vary according to risk characteristics with ZA models.

ZA models represent the response Y as the product JZ where $J = I[Y > 0]$ accounts for zero responses and Z is distributed according to Y given $Y > 0$, these two random variables being mutually independent. The mean response is then

$$E[Y] = E[J]E[Z].$$

The expected outcome now appears to be a product of two quantities and not a single parameter in the analysis.

ZA models are in line with the claim amount representation in the individual model of risk theory. In this model, the total claim cost is decomposed into the product of an indicator J for the event "the policy produces at least one claim during the reference period" and a positive random variable Z representing the total claim amount produced by the policy when at least one claim has been filed. This exactly corresponds to the construction of the ZA models. Numerous powerful actuarial techniques have been developed for the individual model, which are thus directly applicable to the ZA modeling output and facilitates the actuarial analysis.

7.6   Beyond the Tweedie Loss Reserving Model

7.6.1   Motivation

In Property and Casualty insurance, we know that claims sometimes need several years to be settled. Meanwhile, insurers have to build reserves representing their estimate of outstanding liabilities for claims that occurred on or before the valuation date. As explained in Chap. 4, reserving calculation has traditionally been performed on the basis of aggregated data summarized in run-off triangles with rows corresponding to accident years and columns corresponding to development years. Such data exhibit three dimensions: for each accident (or occurrence, or underwriting) year AY and development period $DY = 1, 2, \ldots$, we read in cell (AY,DY) inside the triangle the total amount paid by the insurer in calendar year $CY = AY + DY - 1$ for claims originating in year AY. Techniques dealing with such aggregated triangular arrays of data with a single value in every observed cell go back to the pre-computer era, at a time where the available computing resources, data storage facilities and statistical methodologies were extremely limited. The Chain-Ladder approach is certainly the most popular technique falling in this category. Here, we go beyond this classical approach and deal with individual severities recorded in every cell of the run-off triangle.

7.6.2   Notation and Data

7.6.2.1   Accident and Development Indices

Henceforth, we denote as ω the time needed to settle all the claims occurred during a given accident year AY, i.e. these claims are closed in calendar year $AY + \omega - 1$ at the latest. For business lines with long developments, some claims for accident year 1 may still be open at the end of the observation period. If some claims of the first accident year are still open at the end of the observation period, the actuary must introduce a tail factor to account for the last developments before final settlement.

The data fill a triangle: accident year AY is followed from development 1 (corresponding to the accident year itself) to the last observed development $\omega - AY + 1$ (corresponding to the last calendar year for which observations are available, located along the last diagonal of the triangle).

7.6.2.2   The Data

The approach described in this chapter is applied on a data set extracted from the motor third-party liability insurance portfolio of an insurance company operating in the European Union. The observation period consists in calendar years 2004 till

2014. The available information concerns accident years 2004 to 2014 so that we have observed developments DY up to 11. Henceforth, we let AY belong to $\{1, \ldots, 11\}$ or range between 2004 and 2014, in order to make the numerical results easier to interpret.

There are 52, 155 claims in the data set. Among them, 4,023 claims are still open at the end of the observation period. Table 7.1 presents the information available for two claims of the database. Claim #16,384 corresponds to an accident occurred in 2009 that has been reported during the same calendar year. Payments have been made in years 2009–2013, but no payment has been recorded for 2014. We note that in our approach, all the payments related to the same claim are aggregated over the calendar year. At the end of the observation period, claim #16,384 is still open. Claim #20,784 corresponds to an accident occurred in 2010 that has been reported during the same calendar year. A payment has been made in 2010, there was no payment in 2011, and the claim has been closed in 2011, one year after its reporting to the insurer. Notice that in our data set, the declaration of a claim corresponds to the first time there is a payment or a positive case estimate for that claim. Hence, late reporting (i.e. at lags 3–4) is due here to the definition adopted for reporting as motor insurance contracts typically impose that policyholders rapidly file the claim against the company.

Table 7.2 displays descriptive statistics for payments per accident year and development period. It corresponds to Table 4.6 supplemented with some additional information. We can see there the number of payments, the proportion of claims with no payment, the average payments as well as the standard deviation and skewness per accident year and development period. Table 7.2 shows that the standard deviation is often about twice the mean while the large skewness values suggest highly asymmetric distributions. We also see there the typical increase of mean payments with development j, together with the corresponding decrease in the number of payments

Table 7.1 Information available for claims No 16, 384 and No 20, 784 in the data set. Claim No. 16, 384 is still open end of year 2014

Event	No	Year	Amount
Occurrence	16,384	2009	–
	20,784	2010	–
Declaration	16,384	2009	–
	20,784	2010	–
Payments	16,384	2009	5,022
	16,384	2010	67,363
	16,384	2011	903
	16,384	2012	6,295
	16,384	2013	13,850
	16,384	2014	0
	20,784	2010	1,605
	20,784	2011	0
Closure	16,384	Not settled	–
	20,784	2011	–

as described in the introduction. These opposite effects are known to invalidate the Tweedie specification with constant dispersion parameters (GLM setting).

Table 7.2 Descriptive statistics for payments per accident year 2004, ..., 2014 and development year 1, ..., 11, namely the number of payments (Num. pay.), the proportion of claims with no payment (% no pay.), the mean of the payments as well as the standard deviation and skewness

	1	2	3	4	5	6	7	8	9	10	11
2004											
Num. pay.	2,848	1,459	236	124	68	39	18	18	12	8	7
% no pay.	0.292	0.240	0.438	0.431	0.452	0.524	0.591	0.486	0.429	0.500	0.462
Mean	1,133	1,877	2,713	4,349	4,446	9,894	16,765	4,422	18,072	12,314	21,263
Std. dev.	2,378	5,317	4,861	9,405	7,918	26,576	27,037	9,768	24,203	14,436	50,490
Skewness	12.423	11.878	4.157	4.472	2.948	5.028	1.967	3.476	2.015	1.136	2.039
2005											
Num. pay	3,001	1,492	207	97	53	42	24	21	11	11	
% no pay.	0.284	0.229	0.423	0.484	0.579	0.475	0.529	0.382	0.500	0.214	
Mean	1,112	1,659	3,168	5,455	5,132	14,882	25,781	8,997	4,230	1,347	
Std. dev.	1,847	2,932	6,081	18,278	10,270	41,070	77,046	19,947	2,817	883	
Skewness	4.113	5.509	3.709	8.387	4.089	4.353	3.135	3.464	0.413	0.241	
2006											
Num. pay	3,007	1,659	268	117	61	41	21	10	10		
% no pay.	0.306	0.213	0.467	0.578	0.558	0.438	0.523	0.545	0.412		
Mean	1,164	1,624	5,799	4,494	7,287	6,055	6,141	4,688	12,205		
Std. dev.	2,972	2,932	49,737	7,632	22,190	12,682	9,173	4,594	26,907		
Skewness	23.651	6.020	15.840	3.104	4.485	3.248	1.624	0.506	2.440		
2007											
Num. pay	3,246	1,893	322	170	79	48	24	16			
% no pay.	0.316	0.257	0.542	0.377	0.423	0.385	0.400	0.385			
Mean	1,159	1,905	2,679	3,500	7,401	8,243	12,140	13,148			
Std. dev.	2,258	4,984	6,159	5,831	13,989	14,717	20,625	22,292			
Skewness	9.382	13.781	6.688	3.093	3.029	2.872	2.731	1.921			
2008											
Num. pay	3,574	1,816	304	125	71	37	22				
% no pay.	0.292	0.308	0.451	0.534	0.441	0.464	0.436				
Mean	1,104	1,720	2,189	4,203	4,611	7,775	6,310				
Std. dev.	1,837	3,644	3,524	8,791	9,908	12,249	7,275				
Skewness	5.521	7.533	3.851	4.451	4.774	2.352	0.983				
2009											
Num. pay	3,545	1,877	300	131	90	51					
% no pay.	0.314	0.318	0.543	0.518	0.379	0.311					
Mean	1,142	1,919	3,981	4,379	6,896	9,129					
Std. dev.	1,926	5,710	19,797	11,584	17,446	18,474					
Skewness	4.610	18.270	14.896	7.229	5.379	3.265					

(continued)

Table 7.2 (continued)

	1	2	3	4	5	6	7	8	9	10	11
2010											
Num. pay	2,874	2,072	338	161	75						
% no pay.	0.377	0.320	0.448	0.410	0.409						
Mean	1,663	1,984	3,637	5,147	14,935						
Std. dev.	4,012	5,832	11,419	13,420	60,912						
Skewness	18.229	16.910	11.563	5.037	5.849						
2011											
cre Num. pay	2,777	1,930	327	119							
% no pay.	0.368	0.311	0.443	0.526							
Mean	1,601	1,982	2,441	5,171							
Std. dev.	2,333	4,004	4,119	14,476							
Skewness	5.628	7.586	3.607	4.742							
2012											
Num. pay	2,860	1,749	282								
% no pay.	0.335	0.330	0.529								
Mean	1,716	2,328	4,390								
Std. dev.	4,587	10,085	31,803								
Skewness	36.917	31.363	16.083								
2013											
Num. pay	2,924	1,844									
% no pay.	0.358	0.357									
Mean	1,637	2,230									
Std. dev.	4,120	11,414									
Skewness	31.519	35.894									
2014											
Num. pay	2,723										
% no pay.	0.427										
Mean	1,662										
Std. dev.	2,360										
Skewness	7.018										

7.6.3   Collective Model for Losses

7.6.3.1   Compound Sum Decomposition

Assuming that the amount in cell (AY, DY) is Tweedie distributed with a mean expressed in function of factors related to accident year and development year is equivalent to assume that the number of payments in the cell is Poisson distributed, and independent of the amounts of each payment, assumed to be independent and Gamma distributed. This specification is thus intuitively appealing. Moreover, the Tweedie distribution belongs to the ED family, which facilitates the numerical treatment of the data with the help of GLM techniques. In particular, the loss prediction based on aggregate data in cell (AY, DY), i.e. the total payment corresponding to that cell, coincides with the result based on the knowledge of each and every individual payment (recall that in the GLM setting, all responses sharing the same features, AY and DY here, can be summed together without affecting the point estimates, provided the weights are adapted accordingly).

Actuaries are aware that

– in general, most of the claims are reported early in development so that the frequency component has a decreasing trend throughout developments;
– the average cost per claim often increases through the development years so that the severity component exhibits an increasing trend.

Because expected frequencies and severities have opposite trends, Tweedie GLMs with constant dispersion are prone to errors, making the double GLM appealing.

The equivalence between the aggregated approach and the detailed one, based on individual paid amounts, only holds in the Tweedie setting. Even if the Poisson assumption for the number of payments appears to be rather reasonable (at least as a starting point), the Gamma specification may lead to underestimation of the tail of the loss distribution. This is why we extend here the Tweedie approach to alternative distributions for the amounts paid by the insurer. Typically, we use a finite mixture model combining a light-tailed distribution with a heavier-tailed one. The parameters of this two-component mixture model are explained by means of AY, DY, and CY effects in a GAMLSS setting.

Henceforth, we have observations numbered as $i = 1, \ldots, n$ corresponding to accident year $AY(i)$ and development year $DY(i)$. Here, we decompose the total payment Y_i at development $DY(i)$ into the compound sum

$$Y_i = \sum_{k=1}^{N_i} C_{ik},$$

where N_i is the number of payments made at development $DY(i)$ for claims originating in accident year $AY(i)$, and the random variables C_{ik}s denote the corresponding amounts. All these random variables are assumed to be mutually independent. For each i, the random variables $C_{i1}, C_{i2}, \ldots$ are assumed to be identically distributed. Notice that here, payments related to individual policies are not tracked, only payments for the collective are modelled, and all payments related to the same claim are aggregated in C_{ik}.

The loss reserving model used in the present section is in line with the collective model of risk theory. Because we focus on the payments in cell (AY, DY), without reference to characteristics of the claims from which these payments originate, this means that these payments correspond to claims at various stages of development. The resulting heterogeneity can be taken into account by means of a mixture model combining light-tailed and heavier-tailed severity distributions. Notice that we model the number of payments and not the number of open claims (in order to avoid zero payments for some open claims at given developments, which complicates the mixed model for the severities).

Remark 7.6.1 Notice that the proposed approach does not explicitly include the number of reported, or open claims. Denoting as M_i the number of open claims at development $DY(i)$ originating in accident year $AY(i)$, and as Z_{il} the corresponding yearly payments, $l = 1, 2, \ldots, M_i$, we must account for the cases where no payments have been made for an open claim at that lag, i.e. Z_{il} must have a probability mass at 0. This is why we work here with the number of payments

$$N_i = \sum_{l=1}^{M_i} I[Z_{il} > 0].$$

Instead of modeling M_i and the probability mass of Z_{il} at zero, the approach proposed in this section directly targets N_i. As $E[N_i] = E[M_i]P[Z_{il} > 0]$ under suitable independence assumptions, the number of open claims implicitly appears in $E[N_i]$. The amounts C_{ik} then correspond to the kth aggregate yearly payment, among those open claim with a positive amount paid at development $DY(i)$.

7.6.3.2   Frequency and Severity Modeling

The observed counts N_i are displayed in Table 7.2. These data have been analyzed in Chap. 4 assuming that the random variables N_i are Poisson distributed. In line with the classical Chain-Ladder model, we use the multiplicative specification

$$E[N_{ij}] = \alpha_{AY(i)} \delta_{DY(i)}$$

subject to the usual identifiability constraint

$$\sum_{j=1}^{\omega} \delta_j = 1.$$

Notice that the number of reported claims for accident year 2004 is 4,196. Among them, 9 claims remain open at the end of the observation period, that is, at the end of 2014 so that $\omega > 11$. This is why the series of estimated δ_{DY} obtained by Poisson maximum-likelihood is extrapolated to $\omega = 13$. A simple linear extrapolation is supported by the decreasing trend exhibited by the estimated δ_{DY}. This gives $\widehat{\delta}_{12} = 0.00109$ and $\widehat{\delta}_{13} = 0.00055$.

Let us now turn to the modeling of severities. It is worth to mention that mixture distributions cannot be fitted to aggregated severity data displayed into a triangle, with a single value in every observed cell. Individual severities are needed for that purpose. Contrarily to individual loss reserving models which develop each and every claim over time, we work here with detailed severity data to fit a collective model inside each cell (AY,DY) without tracking individual claim development over time. Under the compound Poisson assumption, these cell-specific distributions can easily be aggregated by convolution to recover the outstanding claim distribution. This collective approach to loss development appears to outperform the Poisson-Gamma, or Tweedie model on the data used in our numerical illustrations (because of the mix between moderate and large severities). It offers enough flexibility to provide accurate answers to most problems encountered in practice as demonstrated in the numerical illustration.

In order to model the amounts of payments C_{ik}, $k = 1, \ldots, N_i$, in cell (AY(i), DY(i)) we need to accommodate for possibly large values. Therefore, we resort to a discrete mixture with two components:

- a lighter-tailed component with probability $1 - \rho_{DY(i)}$ such as Gamma or Inverse Gaussian distributions;
- a heavier-tailed component with probability $\rho_{DY(i)}$ with Pareto distribution.

The parameters (probabilities assigned to each component as well as distributional parameters for the light-tailed and the heavy-tailed components) are explained with the help of features related to AY, DY, and CY using appropriate regression models. Specifically, yearly payments are assumed to be mutually independent and explained by an inflation effect (in an hedonic approach) and a development effect. The first accident year is taken as the base year for inflation. If needed, the inflation effect may be structured (by specifying a constant inflation rate, for instance). Of course, other effects may also be included. Notice that working with single payments made by the insurer solves the severe identifiability issues faced in the aggregated triangle approach.

Precisely, we use a 2-component mixture with a Gamma and a Pareto distributions. The average payment is of the form

$$\delta_{1,DY(i)}(1 + g_1)^{CY(i)-1} \tag{7.8}$$

for the Gamma component and of the form

$$\delta_{2,DY(i)}(1 + g_2)^{CY(i)-1} \tag{7.9}$$

for the Pareto component. In these averages, g_1 (resp. g_2) can be interpreted as a constant inflation rate for the Gamma (resp. Pareto) component and $\delta_{1,DY}$ (resp. $\delta_{2,DY}$) models the development effect for the Gamma (resp. Pareto) component. The parameter δ_{DY} then represents the average amount paid at development DY, corrected for inflation.

Such a 2-component mixture of Gamma and Pareto distributions can be fitted with the help of the GAMLSS package of the statistical software R. The GAMLSS package supports many continuous, discrete and mixed distributions for modeling the response variable, including the 2-component Gamma-Pareto mixture used here.

Fig. 7.1 Estimated probabilities ρ_d by development $DY = d$ for the heavier-tailed Pareto component

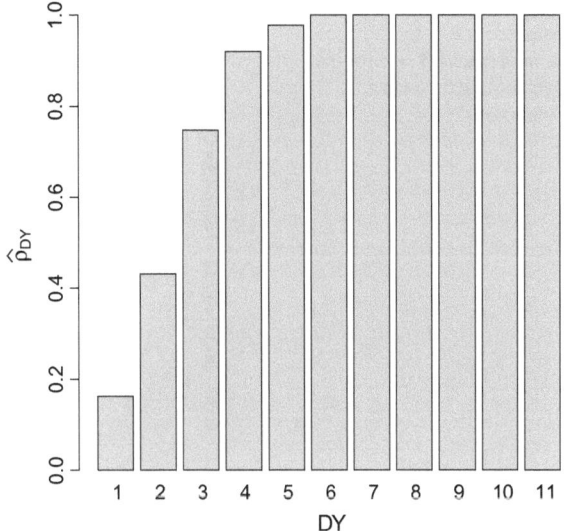

Development effects can also be smoothed using P-splines, cubic splines or loess[1] smoothing.

Considering the estimations displayed in Fig. 7.1, the weights for the Gamma components dominate the weights for the Pareto components at early developments, here $DY \in \{1, 2\}$. The weights assigned to the Gamma components rapidly decrease for $DY \in \{3, 4, 5\}$ to become negligible when $DY \geq 6$. Since $\omega = 13$ in our case, we still need to determine values for ρ_{12} and ρ_{13}. Obviously, we set the weights assigned to the Gamma component equal to 0 at developments 12 and 13, in line with the weights at previous lags displayed in Fig. 7.1.

The estimated inflation rates g_1 and g_2 are $\widehat{g}_1 = 1.91\%$ and $\widehat{g}_2 = 2.48\%$. The estimated $\delta_{1,DY}$ and $\delta_{2,DY}$ are obtained with the help of cubic splines. The Gamma component is present up to development 5 and the estimated mean parameters are

$$\widehat{\delta}_{1,1} = 678$$
$$\widehat{\delta}_{1,2} = 679$$
$$\widehat{\delta}_{1,3} = 214$$
$$\widehat{\delta}_{1,4} = 63$$
$$\widehat{\delta}_{1,5} = 17.$$

The estimated mean parameters for the Pareto component are displayed in Fig. 7.2.

The way these estimations vary with the development DY conforms with intuition. The mathematical expectations of the Gamma components decrease with DY whereas those of the Pareto component increase with DY, capturing the largest losses with

---

[1]Loess stands for Locally Weighted Scatterplot Smoother.

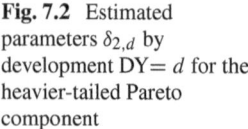

Fig. 7.2 Estimated parameters $\delta_{2,d}$ by development DY$= d$ for the heavier-tailed Pareto component

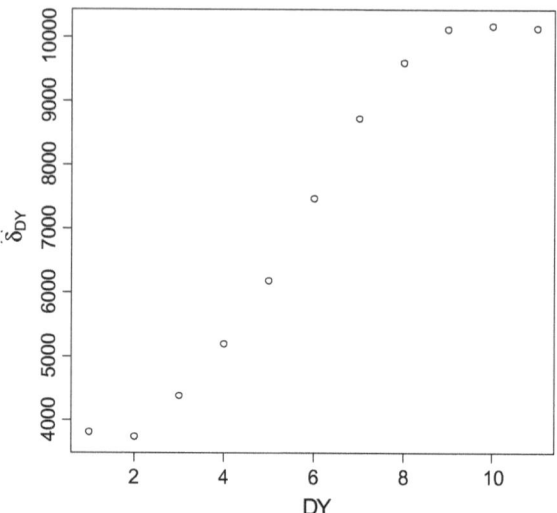

expensive payments and longer development to settlement. We also see that a plateau appears at the last developments. The tail factors $\delta_{2,12}$ and $\delta_{2,13}$ are set to 10, 150 because we see on Fig. 7.2 that the last estimations $\widehat{\delta}_{2,9}, \widehat{\delta}_{2,10}$ and $\widehat{\delta}_{2,11}$ tend to stabilize around that value.

The severity distribution combines a Gamma with a Pareto component, the weight attributed to the Pareto increasing with the development DY. This rules out the Gamma specification underlying the Tweedie reserving model (that can be seen as a compound Poisson sum with Gamma-distributed summands). Notice that the proposed model encompasses the Tweedie model which is thus rejected for the data set under consideration.

7.7 Bibliographic Notes and Further Reading

We refer the reader to Kaas et al. (2008) for an introduction to the individual and collective models of risk theory, that have been mentioned several times in this chapter. Also, arithmetic and geometric approaches to loss reserving are presented there.

Wuthrich (2003) proposed a reserving model assuming that the amount in each cell of the runoff triangle is Tweedie distributed with a mean expressed in function of factors related to accident year and development period. Boucher and Davidov (2011) proposed to switch to a double GLM setting. Precisely, these authors proposed a double GLM to let the mean as well as the variance depend on AY and DY in loss reserving. These authors justify this dispersion modeling by the existence of two opposite effects that invalidate the Tweedie specification. The collective loss reserving approach presented in this chapter is taken from Denuit and Trufin (2017).

Departing from aggregated run-off triangles, individual reserving methods allow for the development of individual claims in time. Individual reserving models describe how each claim evolves over time, from the occurrence of the accident until settlement of the claim. See for instance Pigeon et al. (2013, 2014). Denuit and Trufin (2018) adopted the new loss reserving approach proposed by Denuit and Trufin (2016) but instead of considering the whole set of claims as a collective, two types of claims are distinguished, those claims with relatively short development patterns and claims requiring longer developments. Only the latter are developed individually, whereas the former are modeled collectively. Thus, this hybrid approach bridges the gap between aggregated techniques, like Chain Ladder and fully individual reserving models.

Studies devoted to the consequences of duplicates on the estimation of death rates have been conducted early by UK actuaries (as it can be seen from the comments of the British CMI Committee as early as 1957). Renshaw (1992) proposed to apply a double GLM to deal with the presence of duplicate policies in mortality graduation.

Renshaw (1992) showed that it is possible to make proper adjustment for both duplicates when risk classification is performed by means of GLMs/GAMs. The question whether the existence of duplicates within a portfolio matters for risk management has been studied by several authors. As pointed out by Djeundje and Currie (2010), the answer might appear to be no, since the multiple deaths in the numerator are balanced by the additional exposed-to-risk in the denominator. In statistical terms, the estimate of mortality remains unbiased. However, the variance of such an estimate is too small, since it is based on more deaths than we have actually observed. There are other more subtle consequences. The results of simulating portfolio experience based on policies will be less volatile than they should be, since, for example, the lives with multiple policies should take all their sums assured with them when they die, and not just part of them. More formally, Denuit (2000) demonstrated how the presence of duplicates leads to a more dangerous portfolio in the sense of multivariate stochastic order relations for random vectors with different levels of multiplicity.

GAMLSS have been introduced by Rigby and Stasinopoulos (2005). See the website gamlss.org for exhaustive information. We refer the readers to the book by Stasinopoulos et al. (2017) for an extensive treatment of GAMLSS. Let us mention that Mayr et al. (2012) proposed a modified gradient boosting algorithm for GAMLSS, implemented in the R package gamboostLSS. For an application of GAMLSS to insurance ratemaking, we refer the reader to Klein et al. (2014). Many models considered in insurance studies can be analyzed with GAMLSS tools. Boucher et al. (2007a) compared different risk classification models for the annual number of claims reported to the insurer, including generalized heterogeneous, zero-inflated, hurdle and compound frequency models. Boucher et al. (2007b) extended hurdle models to panel count data with the help of independent random effects representing unexplained heterogeneity in both components (below and above the hurdle). Boucher et al. (2011) introduced correlated random effects in each process and used

Markov chain Monte Carlo simulations to compute posterior distributions of the random effects.

Another regression framework going beyond mean response modeling is the conditional transformation model approach by Hothorn et al. (2014), which can be seen as a semi-parametric competitor to GAMLSS. Let us also mention the quantile and expectile regression approaches. Quantile regression (see, for example Koenker (2005)) is based on the following representation for the quantile q_α: the quantile q_α of a random variable X can be defined as the minimizer of the piecewise linear loss function

$$\alpha E[(X - q)_+] + (1 - \alpha)E[(X - q)_-]$$

where $x_+ = \max\{x, 0\}$ and $x_- = \max\{-x, 0\}$. Newey and Powell (1987) introduced the expectiles as the minimizers of the piecewise quadratic loss function

$$\alpha E[(X - q)_+^2] + (1 - \alpha)E[(X - q)_-^2].$$

These quantities are closely related to risk measures, as pointed out by Bellini et al. (2014). Therefore, they reveal particularly interesting for insurance studies. Bellini et al. (2014) established for instance that expectiles are the only generalized quantiles that are coherent risk measures when $\alpha \geq 0.5$. While quantile regression can be seen as a generalization of median regression, expectiles are a generalized form of mean regression. An overview about methods focusing on estimation procedures regarding more features of the data than its centre (including semiparametric expectile and quantile regression) can be found in Kneib (2013).

References

Bellini F, Klar B, Muller A, Gianin ER (2014) Generalized quantiles as risk measures. Insur: Math Econ 54:41–48

Boucher JP, Davidov D (2011) On the importance of dispersion modeling for claims reserving: an application with the Tweedie distribution. Variance 5:158–172

Boucher J-P, Denuit M, Guillen M (2007a) Risk classification for claim counts: a comparative analysis of various zero-inflated mixed Poisson and hurdle models. North Am Actuarial J 11:110–131

Boucher J-P, Denuit M, Guillen M (2007b) Modelling of insurance claim count with hurdle distribution for panel data. In: Arnold BC, Balakrishnan N, Sarabia JM, Minguez R (eds) Advances in mathematical and statistical modeling. Statistics for Industry and Technology series, Birkhäuser, Boston, pp 45–59

Boucher J-P, Denuit M, Guillen M (2011) Correlated random effects for hurdle models applied to claim counts. Variance 5:68–79

Denuit M (2000) Stochastic analysis of duplicates in life insurance portfolios. Ger Actuarial Bull 24:507–514

Denuit M, Trufin J (2017) Beyond the Tweedie reserving model: the collective approach to loss development. North Am Actuarial J 21:611–619

Denuit M, Trufin J (2018) Collective loss reserving with two types of claims in motor third party liability insurance. J Comput Appl Math 335:168–184

Djeundje VAB, Currie ID (2010) Smoothing dispersed counts with applications to mortality data. Ann Actuarial Sci 5:33–52

Hothorn T, Kneib T, Buhlmann P (2014) Conditional transformation models. J Royal Stat Soc Ser B (Stat Methodol) 76:3–27

Kaas R, Goovaerts MJ, Dhaene J, Denuit M (2008) Modern actuarial risk theory using R. Springer

Klein N, Denuit M, Lang S, Kneib Th (2014) Nonlife ratemaking and risk management with Bayesian additive models for location, scale and shape. Insur: Math Econ 55:225–249

Kneib T (2013) Beyond mean regression (with discussion and rejoinder). Stat Modell 13:275–303

Koenker R (2005) Quantile regression. Cambridge University Press

Mayr A, Fenske N, Hofner B, Kneib T, Schmid M (2012) Generalized additive models for location, scale and shape for high-dimensional data - a flexible approach based on boosting. J Royal Stat Soc Ser C (Appl Stat) 61:403–427

Newey WK, Powell JL (1987) Asymmetric least squares estimation and testing. Econometrica 55:819–847

Pigeon M, Antonio K, Denuit M (2013) Individual loss reserving with the multivariate Skew Normal framework. ASTIN Bull 43:399–428

Pigeon M, Antonio K, Denuit M (2014) Individual loss reserving using paid–incurred data. Insur: Math Econ 58:121–131

Renshaw AE (1992) Joint modelling for actuarial graduation and duplicate policies. J Inst Actuaries 119:69–85

Rigby RA, Stasinopoulos DM (2005) Generalized additive models for location, scale and shape (with discussion). Appl Stat 54:507–554

Stasinopoulos MD, Rigby RA, Heller GZ, Voudouris V, De Bastiani F (2017) Flexible regression and smoothing: using GAMLSS in R. Chapman and Hall/CRC

Wuthrich MV (2003) Claims reserving using Tweedie's compound Poisson model. ASTIN Bull 33:331–346

Part IV
Special Topics

Chapter 8
Some Generalized Non-linear Models (GNMs)

8.1 Introduction

Generalized non-linear models (GNMs) are similar to GLMs except that the score also includes one or more non-linear terms. GNMs are still based on an ED distribution for the response, or equivalently a variance function, and on a link function mapping the mean response to the score scale. The score remains additive but it may now include some nonlinear effects of the unknown regression parameters. The Lee-Carter model for mortality projections can be included in this framework.

Regression models for count data displayed in contingency tables provide the typical example of GNMs. Assume that the response has been cross-classified according to two categorical features, a row one denoted as r and a column one denoted as c, say. The expected number of observations in cell (r, c) is denoted as μ_{rc}. The independence model assumes that μ_{rc} can be decomposed into a product of a row-specific effect and a column-specific effect (as in the Chain-Ladder model for loss reserving). On the log scale, this means that the following specification is adopted:

$$\ln \mu_{rc} = \beta_r + \gamma_c$$

Such a model thus falls into the GLM framework. If needed, row-column association can be included with the help of the GNM specification

$$\ln \mu_{rc} = \beta_r + \gamma_c + \delta_r \psi_c$$

that falls outside the scope of GLM because the score involves a nonlinear term $\delta_r \psi_c$, that is, the product of parameters.[1] Notice that if δ_r, or if ψ_c, is known, then we are back to the GLM setting and the model can easily be fitted. This property is very useful to fit such models by iteration with the help of simple GLM techniques.

To avoid tedious derivation of recursive algorithms in each and every particular case, the gnm package of R allows the actuary to fit a GNM to insurance data.

---

[1] Identifiability issues are not addressed here and are postponed to Sect. 8.2.3.

© Springer Nature Switzerland AG 2019
M. Denuit et al., *Effective Statistical Learning Methods for Actuaries I*,
Springer Actuarial, https://doi.org/10.1007/978-3-030-25820-7_8

It works very similarly to the glm function. In addition to the structured interaction terms discussed above, several other nonlinear effects are supported by gnm, such as $(\beta_0 + \beta_1 x)\gamma_{jk}$ where x denotes a continuous features and where j and k index levels of two categorical features. There is also the possibility to write a user-specific nonlinear function to be included in the score.

GNMs allow the actuary to deal with several problems in life and health insurance beyond contingency tables. In this chapter, models for mortality projection are extensively discussed. These models also apply to inflation modeling in health costs and for capturing joint trends in disability, recovery and mortality rates. Even if GNMs include many other possible specifications, we concentrate here on responses observed at different ages x during several years t with mean modeling involving scores of the bilinear form

$$\alpha_x + \beta_x \kappa_t. \tag{8.1}$$

The score (8.1) is often referred to as the Lee-Carter, or logbilinear decomposition in mortality studies. The specification (8.1) differs structurally from parametric models given that the dependence on age is nonparametric: age is treated as a factor in (8.1), with parameters α_x and β_x associated to each level x. The same approach is used for calendar time: time t is also treated as a factor in (8.1), with a parameter κ_t attached to each level t. This means that (8.1) does not assume any structured effect of age and time besides the factorization of the combined age and time effect into the product of an age effect β_x and a time effect κ_t. When the data fill a matrix with rows indexed by age and columns indexed by time, (8.1) is closely related to singular value decomposition (or SVD).

Regression models treating age and calendar time as factors are generally used to extract the α_x, β_x and κ_t from the available statistics. Different distributional assumptions have been proposed so far, including the Normal distribution for the logarithm of the estimated mortality or morbidity rates, or the Poisson, Binomial, and Negative Binomial distributions for the death counts, or more generally for numbers of transitions in a multi-state setting (counting the number of disabilities or recoveries, for instance). The maximum-likelihood estimates are easily found using iterative algorithms, that appear to converge very rapidly. The specifications listed above remain relevant for morbidity rates or claim frequencies in health insurance, as these quantities are still based on event counts. In general, they apply to numbers of transitions in a multistate setting (counting the number of disabilities or recoveries, for instance). The score (8.1) can also be useful to describe the dynamics of expected claim severities in health insurance, for instance. Turning to severities, discrete distributions are less appealing and the actuary could consider Normal, Gamma or Inverse-Gaussian specification, instead. Such a change can easily be handled in the GNM approach.

8.2 Mortality Projection

8.2.1 Observed Mortality Trends

8.2.1.1 Data and Assumptions

Here, mortality is studied in the age-period framework. This means that two dimensions enter the analysis: age and calendar time. Both age and calendar time can be either discrete or continuous variables. In discrete terms, a person aged x, $x = 0, 1, 2, \ldots$, has an exact age comprised between x and $x + 1$. This concept is also known as "age last birthday" (that is, the age of an individual as a whole number of years, by rounding down to the age at the most recent birthday). Similarly, an event that occurs in calendar year t occurs during the time interval $[t, t + 1)$.

Throughout this chapter, we assume that the age-specific forces of mortality evolve over time, remaining constant within bands of time, but allowed to vary from one band to the next. Specifically, given any integer age x and calendar year t, it is supposed that

$$\mu_{x+\tau}(t) = \mu_x(t) \text{ for } 0 \leq \tau < 1. \tag{8.2}$$

This is best illustrated with the aid of a coordinate system that has calendar time as horizontal axis and age as vertical axis. Such a representation is called a Lexis diagram after the German demographer who introduced it. Both time scales are divided into yearly bands, which partition the Lexis plane into rectangular segments. Model (8.2) assumes that the mortality rate is constant within each rectangle, but allows it to vary between rectangles.

In this section, two different sources are considered for mortality data: national statistical agency and international databases. We consider the case of Belgium but the approach described here equally applies to every industrialized country. Statistics Belgium is the official statistical agency for Belgium where a national population register serves as the centralizing database. This unique administrative record is used to provide official population figures for Belgium, including statistics on births and deaths. Besides official data, the often-used human mortality database (HMD) is also considered. HMD contains detailed mortality and population data to those interested in the history of human longevity. It has been put together by the Department of Demography at the University of California, Berkeley, USA, and the Max Planck Institute for Demographic Research in Rostock, Germany. It is freely available at http://www.mortality.org and provides a highly valuable source of mortality statistics. HMD contains original calculations of death rates and life tables for national populations, as well as the raw data used in constructing those tables. HMD includes life tables provided by single years of age up to 109, with an open age interval for 110+. These period life tables represent the mortality conditions at a specific moment in time.

In this chapter, we consider ages up to 99; older ages are considered in Chap. 9 where extreme survival is carefully studies on the basis of individual mortality data. Here, mortality statistics are available in aggregated format, grouping all individuals based on gender and attained age.

8.2.1.2 Death Rates

Figure 8.1 displays the crude death rates for Belgian males for four selected periods covering the last 150 years, according to HMD. For each period, death rates are relatively high in the first year after birth, decline rapidly to a low point around age 10, and thereafter rise, in a roughly exponential fashion (that is, linearly on the log-scale), before decelerating (or slowing their rate of increase) at the end of the life span. This is the typical shape of a set of death rates encountered in the preceding chapters.

From Fig. 8.1, it is obvious that dramatic changes in mortality have occurred over the 20th century. The striking features of the evolution of mortality are the downwards trends and the substantial variations in shape. We see that the greatest relative improvement in mortality during the 20th century occurred at the young ages, which has resulted largely from the control of infectious diseases. The hump in mortality around ages 18–25 has become increasingly important for young males. Accidents, injuries, and suicides account for the majority of the excess mortality of males under 45 (this is why this hump is referred to as the accident hump).

The dynamic analysis of mortality is often based on the modeling of the mortality surfaces that are depicted in Fig. 8.2. Such a surface consists of a 3-dimensional plot of the logarithm of the death rates viewed as a function of both age x and time t.

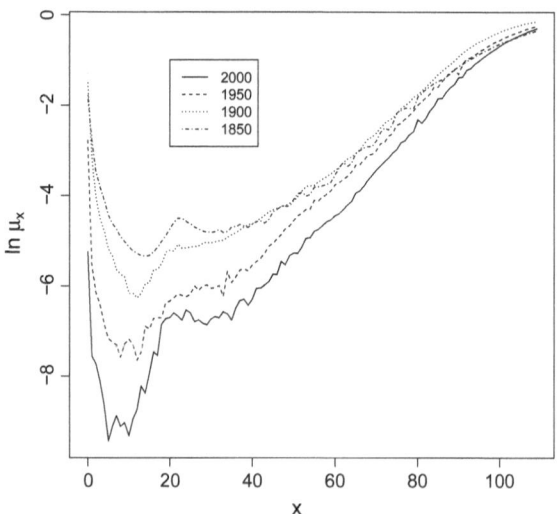

Fig. 8.1 Forces of mortality for some selected calendar years. Belgian males, HMD life tables

Fixing the value of t, we recognize the classical shape of a mortality curve visible on Fig. 8.1. Specifically, along cross sections when t is fixed (or along diagonals when cohorts are followed), one observes relatively high mortality rates around birth, the well-known presence of a trough at about age 10, a ridge in the early 20s, and an increase at middle and older ages.

Mortality does not vary uniformly over the age-year plane and the advantage of plots as in Fig. 8.2 is that they facilitate an examination of the way mortality changes over time and across cohorts, as well as with age. In addition to random deviation from the underlying smooth mortality surface, the surface is subject to period shocks corresponding to wars, epidemics, harvests, summer heat waves, etc. Roughness of the surface indicates volatility and ridges along cross sections at given years mark brief episodes of excess mortality. For instance, higher mortality rates are clearly visible for the years around World War II in the upper panel of Fig. 8.2 corresponding to HMD data going back to the 1850s. The lower panel of Fig. 8.2 corresponds to official mortality data published yearly by Statistics Belgium. It only starts later, in the 1950s. Comparing the two mortality surfaces, we can see marked differences at older ages. This is due to the data processing implemented in HMD at ages 85 and older, in order to ease the comparison of mortality across the industrialized countries covered by HMD. The mortality surface based on official Belgian data is a crude one, without treatment at older ages. The resulting higher volatility is clearly visible there. This is in contrast to HMD mortality data (visible in the upper panel of Fig. 8.2, where mortality statistics have been processed at older ages to ease international comparisons).

8.2.1.3 Life Expectancies

Life expectancy statistics are very useful as summary measures of mortality. Despite their obvious intuitive appeal, it is important to interpret their values correctly when their computation is based on period life tables. Period life expectancies are calculated using a set of age-specific mortality rates for a given period (either a single year, or a run of years), with no allowance for any future changes in mortality. Cohort life expectancies are calculated using a cohort life table, that is, using a set of age-specific mortality rates which allow for known or projected changes in mortality at later ages (in later years).

Period life expectancies are a useful measure of the mortality rates that have been actually experienced over a given period and, for past years, provide an objective means of comparison of the trends in mortality over time, between areas of a country and with other countries. Official life tables which relate to past years are generally period life tables for these reasons. Cohort life expectancies, even for past years, may require projected mortality rates for their calculation. As such, they are less objective because they are subject to substantial model risk and forecasting error.

Let $\overline{e}_x^{\uparrow}(t)$ be the period life expectancy at age x in calendar year t. Here, we have used a superscript "$\uparrow$" to recall that we work along a vertical band in the Lexis diagram, considering death rates associated with a given period of time. Specifically,

Fig. 8.2 Entire mortality surface according to HMD (top) and Statistics Belgium (bottom)

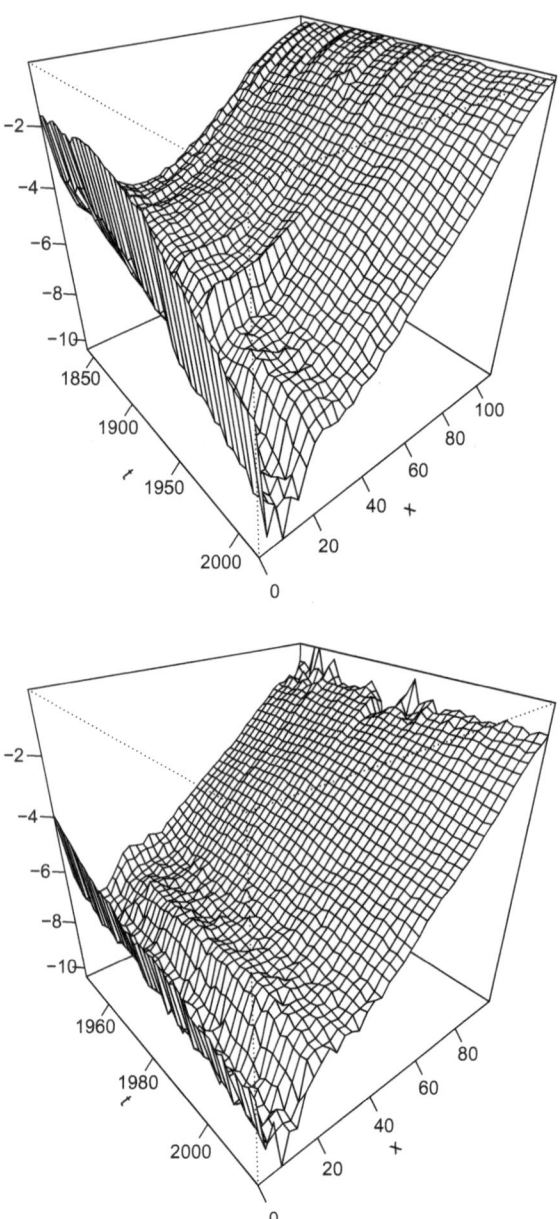

$\overline{e}_x^{\uparrow}(t)$ is computed from the period life table for year t, given by the set $\mu_{x+k}(t)$, $k = 0, 1, \ldots$. The formula giving $\overline{e}_x^{\uparrow}(t)$, under assumption (8.2), is

$$
\begin{aligned}
\overline{e}_x^{\uparrow}(t) &= \int_{\xi \geq 0} \exp\left(-\int_0^{\xi} \mu_{x+\eta}(t) d\eta\right) d\xi \\
&= \frac{1 - \exp\left(-\mu_x(t)\right)}{\mu_x(t)} \\
&\quad + \sum_{k \geq 1} \left(\prod_{j=0}^{k-1} \exp\left(-\mu_{x+j}(t)\right)\right) \frac{1 - \exp\left(-\mu_{x+k}(t)\right)}{\mu_{x+k}(t)}.
\end{aligned}
\tag{8.3}
$$

In this formula, the ratio $\frac{1-\exp(-\mu_{x+k}(t))}{\mu_{x+k}(t)}$ is the average fraction of the year lived by an individual alive at age $x + k$, and the product $\prod_{j=0}^{k-1} \exp(-\mu_{x+j}(t))$ is the probability $_k p_x^{\uparrow}(t)$ of reaching age $x + k$ computed from the period life table.

Figure 8.3 shows the trend in the period life expectancies at birth $\overline{e}_0^{\uparrow}(t)$ and at retirement age $\overline{e}_{65}^{\uparrow}(t)$. The period life expectancy at a particular age is based on the death rates for that and all higher ages that were experienced in that specific year. For life expectancies at birth, we observe a regular increase after 1950, with an effect due to World War II which is visible before that time. This is especially the case at the beginning and at the end of the conflict for $\overline{e}_0^{\uparrow}(t)$, and during the years preceding the conflict as well as during the war itself for $\overline{e}_{65}^{\uparrow}(t)$. Little increase was experienced from 1930 to 1945. It is interesting to note that period life expectancies are affected by sudden and temporary events, such as a war or an epidemic. The last panel of Fig. 8.3 shows that the life expectancy increased at all attained ages, not just at birth and age 65.

8.2.2 Log-Bilinear Decomposition

Lee and Carter (1992) proposed a simple model for describing the secular change in mortality as a function of a single time index. The method describes the logarithm of a time series of age-specific death rates as the sum of an age-specific component that is independent of time and another component that is the product of a time-varying parameter reflecting the general level of mortality, and an age-specific component that represents how rapidly or slowly mortality at each age varies when the general level of mortality changes. Precisely, it is assumed that the force of mortality $\mu_x(t)$ is expressed in terms of the bilinear score (8.1) using a log link, that is,

$$
\ln \mu_x(t) = \alpha_x + \beta_x \kappa_t.
\tag{8.4}
$$

Fig. 8.3 Evolution of period
life expectancy at birth (top),
at age 65 (middle) and
expected remaining lifetimes
in function of attained age
for some selected calendar
years (bottom) according to
HMD life tables

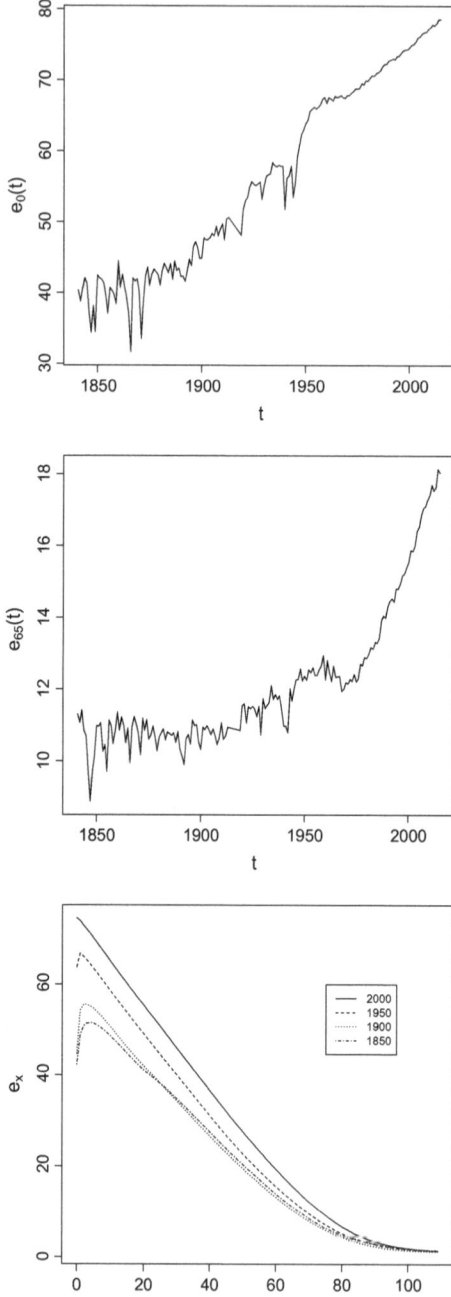

The age-specific parameters α_x and β_x are estimated together with the time-specific parameters κ_t in the regression model

$$\ln \widehat{\mu}_x(t) = \alpha_x + \beta_x \kappa_t + \mathcal{E}_x(t) \tag{8.5}$$

where $\widehat{\mu}_x(t)$ denotes the crude force of mortality at age x during year t and where the error terms $\mathcal{E}_x(t)$ are assumed to be independent and to obey the $\mathcal{N}or(0, \sigma_{\mathcal{E}}^2)$ distribution. It is interesting to notice that (8.5) can be equivalently stated as

$$\ln \widehat{\mu}_x(t) \sim \mathcal{N}or\left(\alpha_x + \beta_x \kappa_t, \sigma_{\mathcal{E}}^2\right).$$

Compared to models discussed previously in this book, we see that the responses are not direct observations here, but estimated parameters $\widehat{\mu}_x(t)$.

8.2.3 Identifiability Constraints

Let us assume that we have observed data for a set of calendar years $t = t_1, t_2, \ldots, t_n$ and for a set of ages $x = x_1, x_2, \ldots, x_m$. On the basis of these observations, we would like to estimate the corresponding parameters α_x, β_x and κ_t. However, this is not possible unless we impose additional constraints.

In (8.5), the parameters α_x can only be identified up to an additive constant, the parameters β_x can only be identified up to a multiplicative constant, and the parameters κ_t can only be identified up to a linear transformation. Precisely, if we replace β_x with $c\beta_x$ and κ_t with $\frac{\kappa_t}{c}$ for any $c \neq 0$ or if we replace α_x with $\alpha_x - c\beta_x$ and κ_t with $\kappa_t + c$ for any c, we obtain the same values for the death rates. This means that we cannot distinguish between the two parametrizations: different values of the parameters produce the same $\mu_x(t)$. In order to see that two constraints are needed to ensure identification, note that if (8.1) holds true, we also have

$$\ln \mu_x(t) = \widetilde{\alpha}_x + \widetilde{\beta}_x \widetilde{\kappa}_t$$

with

$$\widetilde{\alpha}_x = \alpha_x + c_1 \beta_x$$
$$\widetilde{\beta}_x = \beta_x / c_2$$
$$\widetilde{\kappa}_t = c_2 (\kappa_t - c_1).$$

Therefore, we need to impose two constraints on the parameters α_x, β_x, and κ_t in order to prevent the arbitrary selection of the parameters c_1 and c_2.

To some extent, the choice of these constraints is a subjective one, although some choices are more natural than others. In the literature, the parameters in (8.1) are usually subject to the constraints

$$\sum_{t=t_1}^{t_n} \kappa_t = 0 \text{ and } \sum_{x=x_1}^{x_m} \beta_x = 1 \tag{8.6}$$

ensuring model identification. Under this normalization, β_x is the proportion of change in the overall log mortality attributable to age x. We also note that other sets of constraints can be found in the literature, for instance, $\kappa_{t_n} = 0$ or $\sum_{x=x_1}^{x_m} \beta_x^2 = 1$.

Note that the lack of identifiability of the Lee-Carter model is not a real problem. It just means that the likelihood associated with the model has an infinite number of equivalent maxima, each one producing identical fits to the available data. Adopting the constraints (8.6) consists in picking one of these equivalent maxima. The important point is that the choice of constraints has no impact on the quality of the fit, or on forecasts of mortality.

8.2.4   Model Interpretation

Under (8.5), interpretation of the maximum-likelihood estimates of the parameters is quite simple, as shown next. In this case, the estimated parameters are obtained by minimizing the objective function

$$O(\boldsymbol{\alpha}, \boldsymbol{\beta}, \boldsymbol{\kappa}) = \sum_{x=x_1}^{x_m} \sum_{t=t_1}^{t_n} \left(\ln \widehat{\mu}_x(t) - \alpha_x - \beta_x \kappa_t \right)^2$$

where the sums are over all ages and calendar years included in the analysis. The differentiation of the objective function O with respect to α_x gives

$$\sum_{t=t_1}^{t_n} \ln \widehat{\mu}_x(t) = n\alpha_x + \beta_x \sum_{t=t_1}^{t_n} \kappa_t$$

where n denotes the number of calendar years. Assuming as in (8.6) that the parameters κ_t are centered, we see that the parameters α_x are estimated by

$$\widehat{\alpha}_x = \frac{1}{n} \sum_{t=t_1}^{t_n} \ln \widehat{\mu}_x(t).$$

The fitted values of α_x are thus equal to the average of $\ln \widehat{\mu}_x(t)$ over time t so that $\exp \alpha_x$ is the general shape of the mortality schedule. Define the matrix $\mathbf{Z}$ with element (x, t) equal to $z_{xt} = \ln \widehat{\mu}_x(t) - \widehat{\alpha}_x$. The estimated β_x and κ_t can then be obtained from a singular value decomposition of $\mathbf{Z}$ (picking the left and right eigenvectors associated to the largest eigenvalue to get the estimated β_x and κ_t).

Turning to the parameters β_x, their profile represents the age-specific patterns of mortality change. It indicates the sensitivity of the logarithm of the force of mortality at age x to variations in the time index κ_t. In principle, β_x could be negative at some ages x, indicating that mortality at those ages tends to rise when falling at other ages. In practice, this does not seem to happen over the long-run except sometimes at the oldest ages where volatility is very pronounced.

The time trend is represented by the parameters κ_t. The actual forces of mortality change according to an overall mortality index κ_t modulated by an age response β_x. The shape of the β_x profile tells which rates decline rapidly and which ones decline slowly over time in response of change in κ_t. When κ_t is linear in time, mortality at each age changes at its own constant exponential rate. Precisely, if $\kappa_t = c - \kappa_0 t$ then

$$\frac{d}{dt} \ln \mu_x(t) = \beta_x \frac{d}{dt} \kappa_t = -\kappa_0 \beta_x.$$

The error term $\mathcal{E}_x(t)$ involved in (8.5), with mean 0 and variance $\sigma_{\mathcal{E}}^2$ reflects particular age-specific historical influence not captured in the model.

Remark 8.2.1 If needed, the asymptotic level of mortality can also be controlled. Under the Lee-Carter setting, the death rates generally decline exponentially towards 0, i.e. $\widehat{\mu}_x(t) \to 0$ for all x as $t \to \infty$. For long-term predictions like those needed in actuarial practice, it appears reasonable to control the asymptotic level of mortality using limit life tables. Such life tables contain a set of μ_x^{limit} that can be considered as minimum values for death rates that could be attained once mortality improvements for all the causes of death would have been enforced. With such long-term constrains, the model is now

$$\ln \left(\widehat{\mu}_x(t) - \mu_x^{\text{limit}}\right) = \alpha_x + \beta_x \kappa_t + \mathcal{E}_x(t).$$

The interpretation of the parameters can be modified accordingly.

8.2.5 Calibration

8.2.6 Poisson Maximum Likelihood

The model described in Eq. (8.5) is not well suited to the situation of interest. This is because we are actually assuming that the errors are Normally distributed with constant variance across ages, which is quite unrealistic. The logarithm of the observed force of mortality is much more variable at older ages than at younger ages because of the much smaller absolute number of deaths at older ages. Allowing for dispersion modeling could be envisaged but switching to a more appropriate response than $\widehat{\mu}_x(t)$ and modifying the response distribution accordingly seems to be more promising, and easier to implement. As the crude death rates correspond to the ratio of the observed number of deaths D_{xt} to the corresponding exposure to risk e_{xt}, model (8.5) actually assumes that the death counts D_{xt} are LogNormally distributed, which is at best a very crude assumption.

Because the number of deaths is a counting random variable, the Poisson assumption appears to be plausible so that we now consider that

$$D_{xt} \sim \mathcal{P}oi\left(e_{xt}\mu_x(t)\right) \text{ with } \mu_x(t) = \exp\left(\alpha_x + \beta_x\kappa_t\right) \qquad (8.7)$$

where the parameters are still subjected to the constraints (8.6). The force of mortality is thus assumed to have the log-bilinear form $\ln \mu_x(t) = \alpha_x + \beta_x\kappa_t$ as assumed before in (8.4). The meaning of the parameters α_x, β_x, and κ_t is essentially the same as in the classical Lee-Carter model.

The Poisson assumption for the death counts is not restrictive. The reasoning is similar to the one developed in Chap. 6: proceeding along the same lines, it can easily be shown that the true likelihood is proportional to the likelihood based on the Poisson assumption, so that inference can be conducted on the Poisson likelihood. The true likelihood

$$\mathcal{L}(\alpha, \beta, \kappa) = \prod_{x,t} \exp\left(-e_{xt}\mu_x(t)\right)\left(\mu_x(t)\right)^{D_{xt}}$$

being proportional to the Poisson likelihood $\mathcal{L}_{\mathcal{P}oi}$ obtained under the assumption $D_{xt} \sim \mathcal{P}oi\left(e_{xt}\mu_x(t)\right)$, it is thus equivalent to work on the basis of the "true" likelihood or on the Poisson likelihood, once the assumption (8.2) has been made.

The parameters α_x, β_x and κ_t are we now estimated by maximum-likelihood in model (8.7). Formally, $\widehat{\alpha}$, $\widehat{\beta}$, and $\widehat{\kappa}$ maximize the log-likelihood

$$L_{\mathcal{P}oi}(\alpha, \beta, \kappa) = \sum_{x,t} \left(D_{xt}(\alpha_x + \beta_x\kappa_t) - e_{xt}\exp(\alpha_x + \beta_x\kappa_t)\right) + \text{constant}.$$

Because of the presence of the bilinear term $\beta_x\kappa_t$, it is not possible to estimate the proposed model with GLM machinery. The likelihood equations can be solved using an elementary Newton method. In iteration step $k + 1$, a single set of parameters is updated fixing the other parameters at their current estimates using the following updating scheme

$$\widehat{\theta}^{(k+1)} = \widehat{\theta}^{(k)} - \frac{\partial L^{(k)}/\partial\theta}{\partial^2 L^{(k)}/\partial\theta^2}$$

where $L^{(k)} = L(\widehat{\theta}^{(k)})$. Starting with $\widehat{\alpha}_x^{(0)} = 0$, $\widehat{\beta}_x^{(0)} = 1$, and $\widehat{\kappa}_t^{(0)} = 0$ (random values can also be used), the procedure is stopped if the gain in log-likelihood becomes smaller than a pre-determined threshold. This model specification is supported by the gnm package.

8.2.6.1 Binomial-Gumbel Bilinear Model

Let l_{xt} be the number of individuals aged x at the beginning of year t and denote as $q_x(t)$ the one-year death probability applying to these l_{xt} individuals. Provided the group can be considered as closed (because people leaving or entering the group have been properly taken into account in the determination of the deaths and exposures), the number of deaths D_{xt} at age x during year t obeys the Binomial distribution with parameters l_{xt} and $q_x(t)$. The log-bilinear specification adopted for $\mu_x(t)$ gives

$$
\begin{aligned}
q_x(t) &= 1 - p_x(t) \\
&= 1 - \exp\left(-\mu_x(t)\right) \text{ under (8.2)} \\
&= 1 - \exp\left(-\exp\left(\alpha_x + \beta_x \kappa_t\right)\right),
\end{aligned}
$$

for $x = x_1, ..., x_n$ and $t = t_1, ..., t_m$. It follows that

$$
D_{xt} \sim \mathcal{Bin}\left(l_{xt}, q_x(t)\right) \text{ with } q_x(t) = 1 - \exp\left(-\exp\left(\alpha_x + \beta_x \kappa_t\right)\right). \tag{8.8}
$$

To ensure identifiability, we adhere to the set of constraints (8.6) displayed above for the parameters α_x, β_x and κ_t. The log-likelihood is then given by

$$
\begin{aligned}
L(\boldsymbol{\alpha}, \boldsymbol{\beta}, \boldsymbol{\kappa}) &= \ln\left(\prod_t \prod_x \left(\binom{l_{xt}}{d_{xt}} \left(1 - \widehat{q}_{xt}\right)^{l_{xt} - d_{xt}} \widehat{q}_{xt}^{d_{xt}}\right)\right) \\
&= \sum_t \sum_x \left((l_{xt} - d_{xt}) \ln\left(1 - \widehat{q}_{xt}\right) + d_{xt} \ln \widehat{q}_{xt}\right) + \text{constant}.
\end{aligned}
$$

With general population data, a Binomial or a Poisson modeling for the number of deaths can often both be considered. Contrarily to the Binomial model, the Poisson model explicitly takes into account the exposure-to-risk, which may be considered as an advantage for the insurance applications. The Binomial model assumes that the group is closed.

8.2.6.2 Overdispersion and Negative Binomial Model

Model (8.7) induces equidispersion, that is,

$$
\mathrm{E}[D_{xt}] = \mathrm{Var}[D_{xt}] = \delta_{xt} \text{ where } \delta_{xt} = \mathrm{E}[D_{xt}] = e_{xt} \exp(\alpha_x + \beta_x \kappa_t).
$$

Here, δ_{xt} is the expected number of deaths. However, mortality is known to be influenced by income, wealth, marital status and educational attainment, for instance, so that we cannot assume that the same $\mu_x(t)$ applies to every individual aged x in year t. In general, these correlations work in the direction that individuals with higher socioeconomic status live longer than those in lower socioeconomic groups. Heterogeneity tends to increase the variance compared to the mean (a phenomenon termed as overdispersion and studied in Chap. 5). It rules out the Poisson specification (8.7) and favors a mixed Poisson model. This is why we now work out a Negative Binomial regression model for estimating the parameters α_x, β_x and κ_t.

Besides gender, age x and year t, there are many other factors affecting mortality. These factors are generally hidden when working with general population mortality data so that we take them into account by a random effect $\mathcal{E}_{xt}$ super-imposed to the predictor $\alpha_x + \beta_x \kappa_t$, exactly as in (8.5). More precisely, the Poisson model is replaced with a mixed Poisson model. Given $\mathcal{E}_{xt}$, the number of deaths D_{xt} is assumed to be Poisson distributed with mean

$$e_{xt} \exp(\alpha_x + \beta_x \kappa_t + \mathcal{E}_{xt}).$$

Unconditionally, D_{xt} obeys to a mixture of Poisson distributions. The random effects $\mathcal{E}_{xt}$ are assumed to be independent and identically distributed. This specification thus falls in the setting of mixed models studied in Chap. 5.

Taking D_{xt} mixed Poisson distributed leads to a variance of the form

$$\mathrm{Var}[D_{xt}] = \delta_{xt} + k\delta_{xt}^2 \geq \mathrm{E}[D_{xt}] = \delta_{xt}$$

where $k = \mathrm{Var}[\exp(\mathcal{E}_{xt})]$. Taking $\exp(\mathcal{E}_{xt})$ Gamma distributed, the death counts D_{xt} obey to the Negative Binomial law, that is,

$$P[D_{xt} = d_{xt}] = \frac{\Gamma\left(d_{xt} + \frac{1}{k}\right)}{\Gamma\left(\frac{1}{k}\right) d_{xt}!} \left(\frac{k\delta_{xt}}{1 + k\delta_{xt}}\right)^{d_{xt}} \left(\frac{1}{1 + k\delta_{xt}}\right)^{\frac{1}{k}}.$$

Having observed d_{xt} deaths from exposures e_{xt} for a set of ages x and calendar years t, the likelihood function is given by

$$\mathcal{L}(\boldsymbol{\alpha}, \boldsymbol{\beta}, \boldsymbol{\kappa}) = \prod_{x,t} \frac{\Gamma\left(d_{xt} + \frac{1}{k}\right)}{\Gamma\left(\frac{1}{k}\right) d_{xt}!} \left(\frac{k\delta_{xt}}{1 + k\delta_{xt}}\right)^{d_{xt}} \left(\frac{1}{1 + k\delta_{xt}}\right)^{\frac{1}{k}}.$$

The associated log-likelihood function is given by

$$L(\boldsymbol{\alpha}, \boldsymbol{\beta}, \boldsymbol{\kappa}, k) = \sum_{x,t} \left(\sum_{i=1}^{d_{xt}} \ln\left(\frac{1}{k} + d_{xt} - i\right) \right)$$
$$- \ln(d_{xt}!) - \left(d_{xt} + \frac{1}{k}\right) \ln(1 + k\delta_{xt}) + d_{xt} \ln(k\delta_{xt}). \quad (8.9)$$

Because of the presence of the bilinear term $\beta_x \kappa_t$, it is not possible to estimate the proposed model with statistical packages that implement Negative Binomial regression (like glm.nb). The algorithm implemented to solve the likelihood equations is the uni-dimensional Newton method. It consists in updating each parameter at a time, fixing the other parameters at their current estimates.

Mortality data from the life insurance market often exhibit overdispersion because of the presence of duplicates. It is common for individuals to hold more than one life insurance policy and hence to appear more than once in the count of exposed to risk or deaths. In such a case, the portfolio is said to contain duplicates, that is, the portfolio contains several policies concerning the same lives. A double GLM, supplementing the mean modeling with a dispersion modeling, has been proposed in Chap. 7 to deal with the presence of duplicate policies in graduation. We know that the variance gets inflated in the presence of duplicates. Consequently, even if the portfolio (or one of its risk class) is homogeneous, the presence of duplicates increases the variance and causes overdispersion. The Negative Binomial model for estimating the Lee-Carter parameters is thus appealing for actuarial applications.

8.2.6.3 GLM Fit to Bilinear Models

Before the Lee-Carter model becomes the standard approach to mortality forecasting, GLMs were also used for mortality projection. The effect of age and time was then modeled with the help of polynomials. Legendre polynomials are defined by

$$(n + 1)L_{n+1}(x) = (2n + 1)x L_n(x) - n L_{n-1}(x) \text{ for } n = 1, 2, \ldots,$$

starting from $L_0(x) = 1$ and $L_1(x) = x$. The force of mortality at age x at time t was then expressed as

$$\mu_x(t) = \exp\left(\beta_0 + \sum_{j=1}^{s} \beta_j L_j(x') + \sum_{j=1}^{r} \beta_{s+j}(t')^j + \sum_{j=1}^{r}\sum_{k=1}^{s} \gamma_{jk} L_j(x')(t')^k\right)$$

where x', t' denote the age and time variables which have been transformed linearly and mapped on to the range $[-1, 1]$. Such a model is linear in the features $L_j(x')$, $j = 1, \ldots, s$, $(t')^j$, $j = 1, \ldots, r$, and $L_j(x')(t')^k$, $j = 1, \ldots, r$, $k = 1, \ldots, s$. It can thus be estimated with the help of standard GLM machinery, based on IRLS algorithm.

It is interesting to stress that the Lee-Carter model can also be fitted with the help of the IRLS algorithm for the Poisson or (Negative) Binomial distributions. This is because the Newton scheme used to fit bilinear models fixes some parameters at their current values to revise the estimates of the other parameters. This renders the bilinear model linear at each step of the algorithm. This is because the Lee-Carter model is in the GLM framework if either β_x or κ_t is known.

Precisely, the procedure can be described as follows. First, select starting values for the parameters β_x (take for instance $\beta_x = 1$ for all x). Then, enter the updating cycle:

Step 1   given β_x, estimate the parameters α_x and κ_t in the appropriate GLM (Poisson, Binomial, or Negative Binomial);

Step 2   given κ_t, estimate the parameters α_x and β_x in the appropriate GLM;

Step 3   compute the deviance.

Repeat the updating cycle until the deviance stabilize. Revert back to the Lee-Carter constraints $\sum_x \beta_x = 1$ and $\sum_t \kappa_t = 0$ in (8.6) once convergence is attained. It has been observed empirically that this simple approach produces almost the same estimates for α_x, β_x and κ_t than the more formal procedure based on Newton-Raphson algorithm described before.

8.2.6.4   Application to Belgian Mortality Statistics

Let us now fit the log-bilinear model (8.4) by the least-squares method, Poisson and Binomial maximum likelihood. We consider here Belgian males, calendar years 1950–2015 and ages 0–99. Data come either from HMD or from Statistics Belgium (remember that the mortality data at high ages have been processed in the HMD, so that we expect to see some differences at these ages).

Figure 8.4 plots the estimated parameters α_x, β_x and κ_t. The estimated α_x appear in the top panels. They exhibit the typical shape of a set of death rates on the log-scale, with relatively high values around birth, a decrease at infant ages, the accident hump at late teenager and young adult ages, and finally the increase at adult ages with an ultimately concave behavior. We can see there some differences between the estimations obtained from HMD (on the left) and Statistics Belgium (on the right) data. We can see at the oldest ages an effect of the closure mechanism applied to HMD life tables: the estimates based on Statistics Belgium crude data exhibit some curvature at the very end of the age range, with some volatility. We see on Fig. 8.4 that the estimated α_x obtained with the three error structures closely agree.

The estimated parameters β_x appear in the middle panels. They decrease with age, suggesting that most of the mortality decreases are concentrated on the younger ages. Again, a difference between the estimates obtained from HMD (left panel) and Statistics Belgium (right panel) is visible at older ages. Notice that the last few $\widehat{\beta_x}$ become negative. This is unfortunate because the parameters β_x represent the sensitivity of each age to a change in the level of general mortality. Therefore, they should all have the same sign (positive, without loss of generality). As we are going to structure the age and time effects, we do not address this issue now but come back to it later on.

The estimated parameters κ_t are displayed in the bottom panels. They exhibit a gradually decreasing underlying trend. There is a clear break visible in the estimated time index κ_t, occurring in the early 1970s. The presence of such a break is well documented, for all industrialized countries.

Considering the estimated β_x and κ_t displayed in Fig. 8.4, the Binomial and Poisson models deliver almost the same values, but the Normal approach produces somewhat different results for young and old ages. Because the Normal error structure does

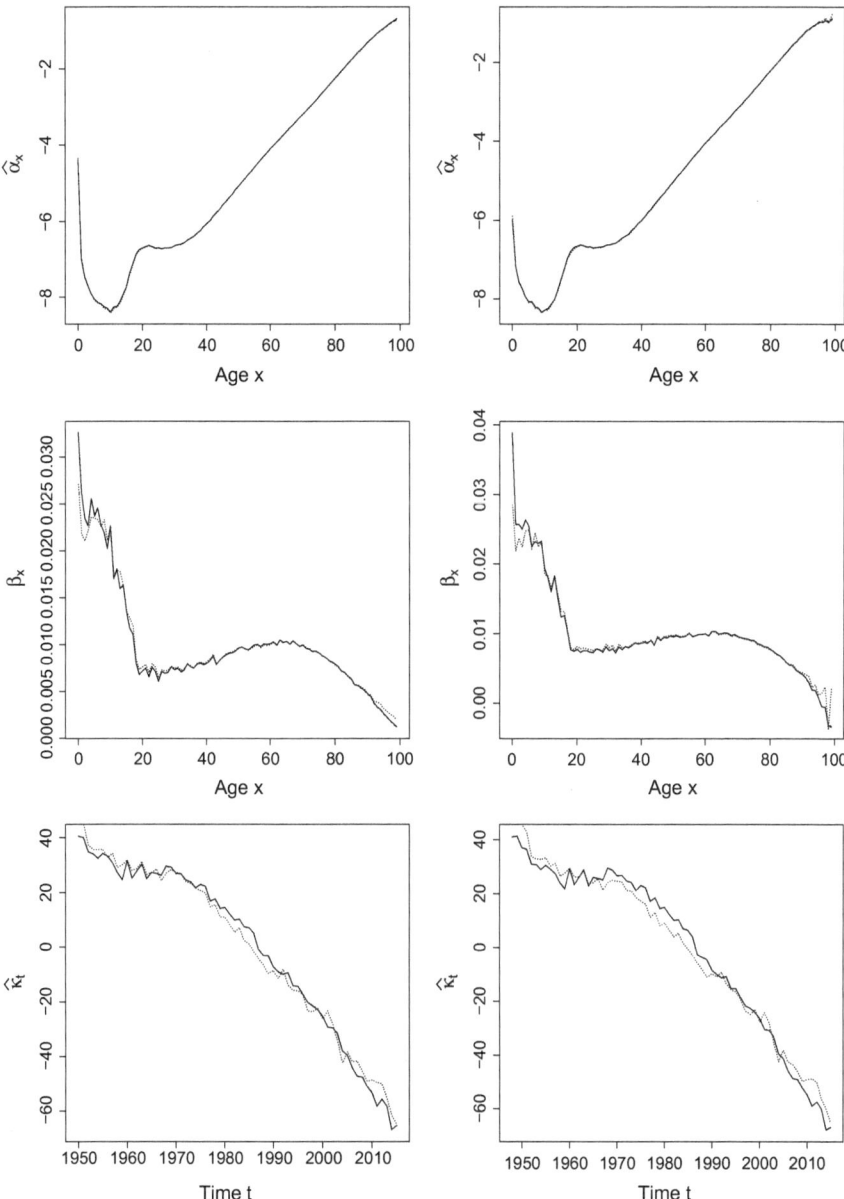

Fig. 8.4 Estimated α_x, β_x and κ_t (from top to bottom), $x = 0, 1, \ldots, 99$, $t = 1948$, 1949, ..., 2015, obtained in the Binomial, Poisson and Normal models with HMD data (left) and Statistics Belgium data (right)

not recognize the integer nature of the response, we favor the Binomial and Poisson models because they both target the observed numbers of deaths D_{xt} whereas the Normal error structure applies to the estimated parameter $\mu_x(t)$. Also, because of the processing of the data at older ages, we continue our analysis with death counts and corresponding exposures published by Statistics Belgium. Let us now compare the performances of the Binomial and Poisson regression models. The AIC for the Binomial model is 75,436.5 and the AIC for the Poisson model is 75,762.31. This shows that the Binomial error structure outperforms the Poisson one on this data set.

We know from Fig. 8.4 that the estimated κ_t are markedly linear over the last part of the observation periods, starting in the 1970s. Let us now concentrate on the period for which the estimated mortality index parameter κ_t is linear: here we restrict the observation period to 1975–2015 and replace the unstructured time index κ_t with a linear function of t. We are then back to the GLM setting using a linear effect for the continuous feature calendar time with an interaction with the categorical feature age, together with an age-specific term. Even if this approach produces almost the same projections as when the time index κ_t is modeled by a time series (because the random walk with drift model adopted for the time index in the vast majority of empirical studies possesses a linear mean trajectory), there is a difference when measuring the uncertainty in mortality forecasts. The resulting confidence intervals are quite narrow when the linear trend in the parameters κ_t has been imposed.

Now, we assume that D_{xt} is Binomially distributed, with

$$q_x(t) = 1 - \exp\left(-\exp\left(\alpha_x + \beta_x(\bar{t} - t)\right)\right)$$

where $\bar{t}$ is the average of the calendar years 1975–2015 included in the analysis. Here, age is treated as a factor and the α_x correspond to regression coefficients associated to each age 0–99. The effect of the linear feature $\bar{t} - t$ on the score is then nested in age, so that an age-specific slope is estimated. To compare with the values obtained from the Lee-Carter model, the GLM slopes are normalized so that they sum to unity and the feature $\bar{t} - t$ is multiplied by the sum of the estimated slopes.

Figure 8.5 plots the estimated parameters α_x and β_x. The parameters κ_t estimated without constraints are also compared with their linear version. The estimated α_x appearing in the top panel still exhibit the typical shape of a set of death rates on the log-scale. We see on Fig. 8.5 that the estimated α_x obtained with and without imposing linearity to κ_t closely agree. The estimated parameters β_x appear in the middle panel. They keep the same shape as those obtained before. The linear version of the estimated parameters κ_t is displayed in the bottom panel. We can see there that imposing linear κ_t is clearly supported by the data once the observation period is restricted to calendar years posterior to 1975.

The estimated parameters β_x displayed in Fig. 8.5 exhibit an irregular pattern. This is due to the fact that age is treated as a factor in the model, so that the parameters β_x are not linked together. Working with unstable $\widehat{\beta}_x$ may be regarded as undesirable from an actuarial point of view, since the resulting projected life tables will also show some erratic variations across ages. Irregularities in the life tables then propagate to

Fig. 8.5 Estimated α_x, β_x and κ_t (from top to bottom), $x = 0, 1, \ldots, 99$, $t = 1975, 1976, \ldots, 2015$, obtained in the Binomial model with Statistics Belgium data

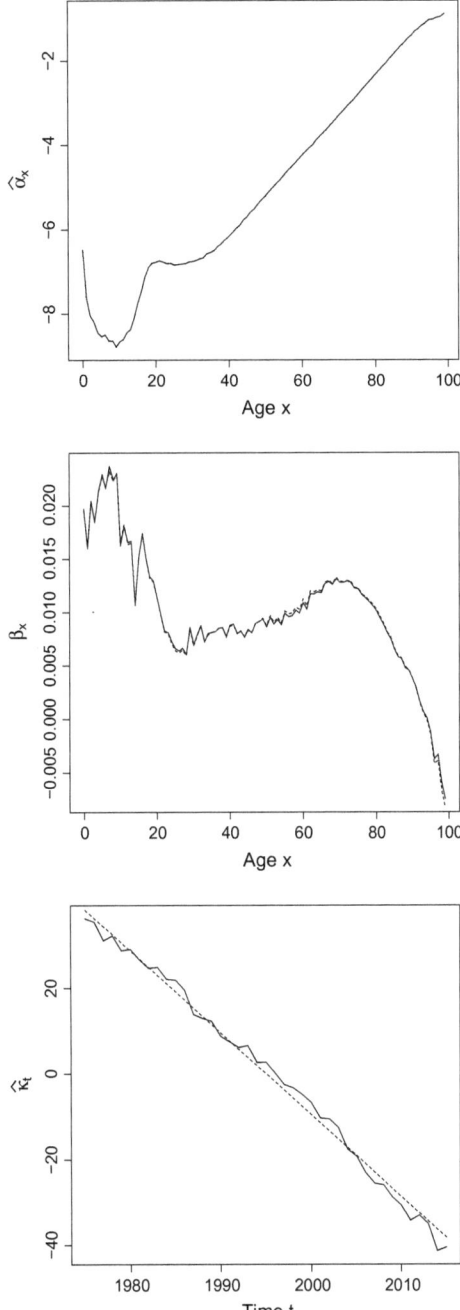

the price list and to balance sheets, which is not desirable. Therefore, as long as these irregularities do not reveal particular features of the risk covered by the insurer, but are likely to be caused by sampling errors, actuaries prefer to resort to statistical techniques producing smooth life tables.

Often in practice, mortality surfaces are first smoothed as explained in Chap. 6. The output of this procedure is then used to produce a smoothed version of the data set on which the Lee-Carter model is fitted. This approach may however produce undesirable results because the volatility of death counts is artificially reduced and local correlation is induced by the preliminary smoothing.

An alternative solution consists in smoothing the estimated parameters β_x, without modifying the estimated α_x. This may however produce large discrepancies between observations and model predictions. To avoid this problem, the smoothing of the parameters β_x is implemented by running the following GAM: we assume here that D_{xt} is Binomially distributed, with

$$q_x(t) = 1 - \exp\left(-\exp\left(\alpha_x + b(x)(\bar{t} - t)\right)\right)$$

where the smooth function $b(\cdot)$ replacing the set of parameters β_x is assumed to be smooth. Notice that the estimated parameters α_x are usually very smooth, since they represent an average effect of mortality at age x over the observation period. The estimated κ_t have been constrained to be linear here (in the Lee-Carter model, they are often rather irregular, but only the projected κ_t enter the actuarial calculations, the latter being obtained from some time series model, they are smooth). Hence, we only need to smooth the parameters β_x in order to get projected life tables with mortality varying smoothly across ages.

Such a decomposition into the product of a smooth age effect and the continuous feature corresponding to minus the centered calendar year can be implemented in the gam function of the R package mgcv using the by statement. The result is displayed in Fig. 8.6. Compared to the estimated β_x displayed in Fig. 8.5, we see that their smoothed version removes erratic variations.

Since we work in a regression framework, it is essential to inspect the residuals. Model performance is assessed in terms of the randomness of the residuals. A lack of randomness would indicate the presence of systematic variations, such as age-time interactions. If the residuals r_{xt} exhibit some regular pattern, this means that the model is not able to describe all of the phenomena appropriately. In practice, looking at $(x, t) \mapsto r_{xt}$, and discovering no structure in those graphs ensures that the time trends have been correctly captured by the model.

When the parameters are estimated by least-squares, Pearson residuals have to be inspected. With a Poisson, Binomial or Negative Binomial random component, it is more appropriate to consider the deviance residuals in order to monitor the quality of the fit. These residuals are defined as the signed square root of the contribution of each observation to the deviance statistics. These residuals should also be displayed as a function of time at different ages, or as a function of both age and calendar year.

Fig. 8.6 Estimated function $b(\cdot)$ replacing the parameters $\beta_x, x = 0, 1, \ldots, 99$, $t = 1975, 1976, \ldots, 2015$, obtained in the Binomial model with Statistics Belgium data

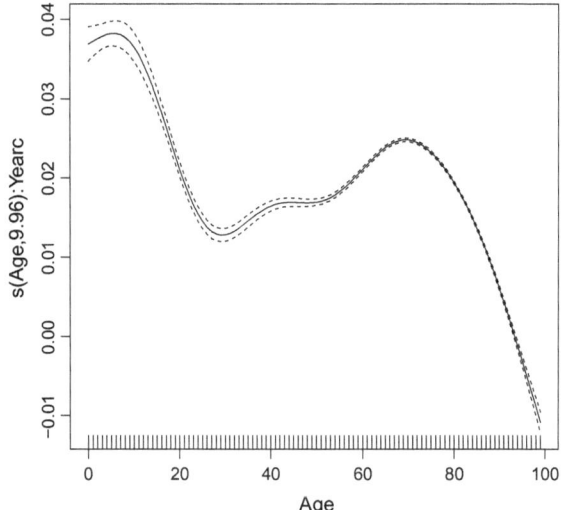

Residuals can be displayed as a function of both age and calendar time, and inspected with the help of a surface or a map as displayed in Fig. 8.7. The residuals appear to be unstructured, except for some moderate cohort effects.

8.2.7 Cairns-Blake-Dowd Model

8.2.7.1 Specification

In this section, we use an alternative mortality projection model, known as the Cairns-Blake-Dowd one (named after its inventors). It can be seen as a dynamic version of the Perks formula used in Chap. 4. Recall that the empirical analysis conducted in Chap. 4 suggests that $\ln q_x(t)/p_x(t)$ is reasonably linear in x beyond age 60, with a small degree of curvature in the plot of x versus $\ln q_x(t)/p_x(t)$. Precisely, formula (4.15) is made dynamic and the one-year death probability $q_x(t)$ at age x in calendar year t is modeled as

$$\ln \frac{q_x(t)}{1 - q_x(t)} = \beta_0^{(t)} + \beta_1^{(t)}(x - \overline{x}) + \beta_2^{(t)}(x - \overline{x})^2$$

$$\Leftrightarrow q_x(t) = \frac{\exp\left(\beta_0^{(t)} + \beta_1^{(t)}(x - \overline{x}) + \beta_2^{(t)}(x - \overline{x})^2\right)}{1 + \exp\left(\beta_0^{(t)} + \beta_1^{(t)}(x - \overline{x}) + \beta_2^{(t)}(x - \overline{x})^2\right)} \qquad (8.10)$$

Fig. 8.7 Residuals obtained
in the Binomial model with
Statistics Belgium data,
$x = 0, 1, \ldots, 99,$
$t = 1975, 1976, \ldots, 2015$

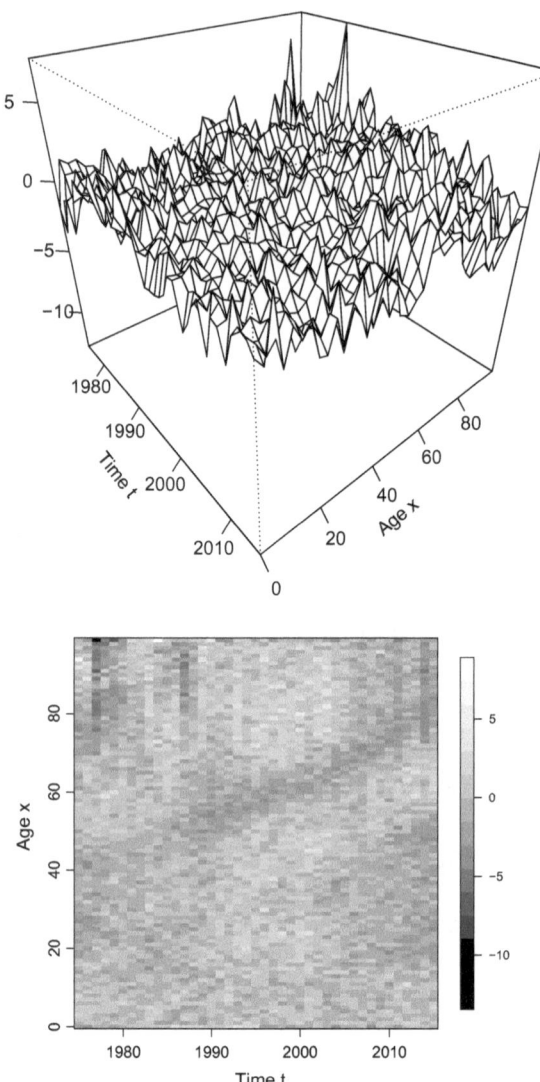

where $\bar{x}$ denotes the mean age and where the intercept $\beta_0^{(t)}$, slope $\beta_1^{(t)}$ and curvature $\beta_2^{(t)}$ now depend on calendar time, hence the superscript "(t)". Thus, the
Cairns-Blake-Dowd model takes advantage of the approximate quadratic behavior
in age (on the logit scale) at higher ages to provide a parsimonious representation of
one-year death probabilities. Notice that this specification does not suffer from any
identifiability problems so that no constraints need to be specified.

The Cairns-Blake-Dowd model includes three time factors (whereas only one time
factor drives the future death rates in the Lee-Carter case) with a smoothing of age

effects using a logit transformation of one-year death probabilities. Specifically, the logit of the one-year death probabilities is modeled as a quadratic function of age, with intercept, slope and curvature parameters following some stochastic process. The Cairns-Blake-Dowd model falls into the GLM setting in that age enters the Cairns-Blake-Dowd model as a continuous feature whereas the effect of calendar time is captured by three factors (time-varying intercept, slope and curvature parameters).

Once the Cairns-Blake-Dowd model has been fitted to historical mortality data, the resulting estimates for the time-varying parameters are then projected using a multivariate time series model. From this forecast of the future intercept, slope and curvature parameters, the future one-year death probabilities are computed in combination with the linear and quadratic age effects. Here, we rather follow the regression approach and we structure the time-varying parameters so that they can easily be extrapolated to the future.

8.2.7.2   Calibration

We assume that we have observed data for a set of calendar years $t = t_1, t_2, \ldots, t_n$ and for a set of ages $x = x_1, x_2, \ldots, x_m$. On the basis of these observations, we would like to estimate the time-varying intercept, slope and curvature parameters. If no exposure is available, only the set of crude one-year death probabilities $\widehat{q}_x(t)$, this can be done by least-squares in the regression model

$$\ln \frac{\widehat{q}_x(t)}{\widehat{p}_x(t)} = \beta_0^{(t)} + \beta_1^{(t)}(x - \overline{x}) + \beta_2^{(t)}(x - \overline{x})^2 + \mathcal{E}_x(t) \tag{8.11}$$

is fitted to the observations of calendar year t, where the error terms $\mathcal{E}_x(t)$ are independent and Normally distributed, with mean 0 and constant variance $\sigma_{\mathcal{E}}^2$. Notice that, in contrast to the Lee-Carter case, where the estimated time index κ_t depends on the observation period, the time-varying parameters $\beta_0^{(t)}$, $\beta_1^{(t)}$ and $\beta_2^{(t)}$ are estimated separately for each calendar year t in the Cairns-Blake-Dowd model. Hence, reducing the observation period to its last part does not modify their fitted values.

The Cairns-Blake-Dowd model can also be calibrated in a number of alternative ways, as it was the case for the Lee-Carter model. For instance, a Poisson regression model can be specified by assuming that the observed death counts are independent and Poisson distributed, with a mean equal to the product of the exposure-to-risk e_{xt} times the population force of mortality of the form

$$\mu_x(t) = -\ln(1 - q_x(t)) = \ln\left(1 + \exp\left(\beta_0^{(t)} + \beta_1^{(t)}(x - \overline{x}) + \beta_2^{(t)}(x - \overline{x})^2\right)\right) \tag{8.12}$$

inherited from (8.10). Estimation based on a Binomial or Negative Binomial error structure can also be envisaged.

8.2.7.3 Application to Belgian Mortality Statistics

As for the implementation of the Lee-Carter approach, we fit the Cairns-Blake-Dowd model to the mortality data published by Statistics Belgium, considering Belgian males from the general population. We fit the Cairns-Blake-Dowd model to these mortality data with the help of Binomial GLMs estimated separately for each calendar year t, assuming that the number of deaths D_{xt} at age x during year t are obey the Binomial distribution with parameters l_{xt} and $q_x(t)$ of the form (8.10).

Remember that the Cairns-Blake-Dowd model is based on the Perks formula and was thus never designed to cover all ages, certainly not down to age 0. This is why we consider the restricted age range $60, 61, \ldots, 99$. As for the Lee-Carter, we use calendar years 1950–2015. The results are displayed in Fig. 8.8.

We see there that the intercept period term $\beta_0^{(t)}$ is decreasing over time, expressing the fact that mortality rates have been decreasing over time at all ages. The estimated slope $\beta_1^{(t)}$ and curvature $\beta_2^{(t)}$ parameters share the same U-shape, and appear to be ultimately increasing over time. Despite having the same overall shape, notice that their magnitude is quite different, the estimated $\beta_2^{(t)}$ parameters being smaller. This expresses the fact that the upward-sloping plot of the logit of death probabilities against age is shifting downwards over time. If during the fitting period, the mortality improvements have been greater at lower ages than at higher ages, the slope period term $\beta_1^{(t)}$ would be increasing over time. In such a case, the plot of the logit of death probabilities against age would be becoming more steep as it shifts downwards over time. Starting from 1980, the estimated $\beta_1^{(t)}$ and $\beta_2^{(t)}$ tend to increase over time, indicating that mortality improvements have been comparatively greater at younger ages over this period.

The trend over the last years seems to be markedly linear for all the three time-varying parameters and this can be exploited to produce mortality forecasts.

8.3 Projection Models for Health Expenses

8.3.1 Context

Morbidity rates and mortality rates often share a very similar age pattern, with higher values around birth and at young adult ages (near the so-called accident hump), and then monotonically increasing at older ages, first exponentially before switching to an increasing concave behavior. The same structure is often found for the expected number of claims in sickness or medical insurance, with some peculiarities (such as the hump induced by childbearing for young women). Corresponding yearly insurance claim costs, being influenced by their frequency component, also exhibit a similar age shape. This suggests that models developed to describe the age structure of mortality can be useful for these related quantities, too.

Fig. 8.8 Estimated intercept $\beta_0^{(t)}$, slope $\beta_1^{(t)}$ and curvature $\beta_2^{(t)}$ (from top to bottom) for ages $x = 60, 61, \ldots, 99$ and calendar years 1950–2015

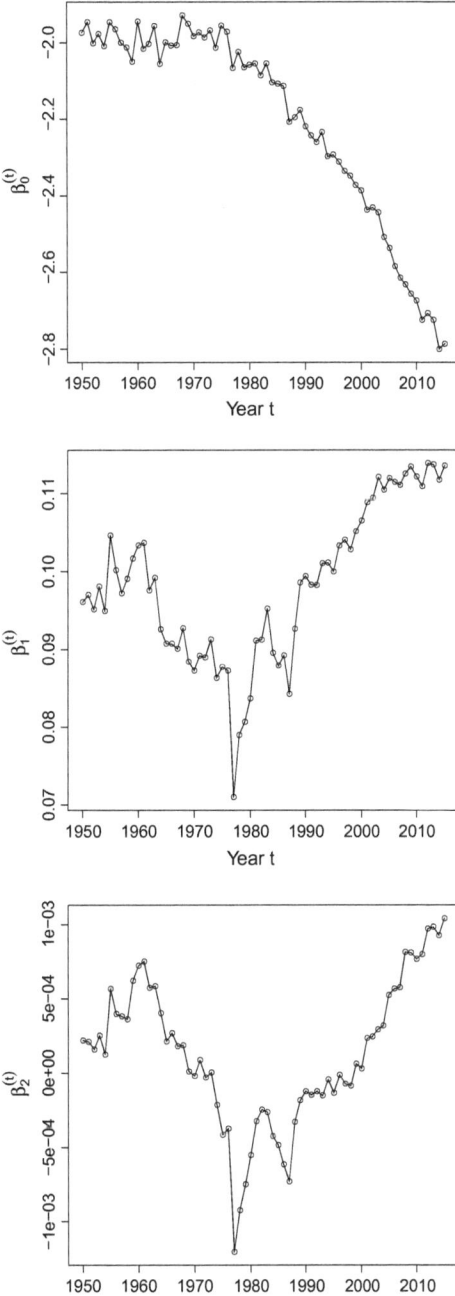

In this section, the response $Y_x(t)$ represents the average yearly medical inpatient cost for a set of calendar years t at ages x. Thus, $Y_x(t)$ is indexed by attained age x and calendar time t. Let us first briefly explain how the expected values $E[Y_x(t)]$ are modeled in practice. The average yearly claim costs are often decomposed into

$$E[Y_x(t)] = \gamma_x(t)E[Y_{x_0}(t)] \Leftrightarrow \gamma_x(t) = \frac{E[Y_x(t)]}{E[Y_{x_0}(t)]}$$

for a fixed reference age x_0, say $x_0 = 40$. The expected amount at age x and time t is split into a product of an expected amount at some reference age x_0 modulated by the age pattern $\gamma_x(t)$.

German private health insurance companies assume that $\gamma_x(t)$ is approximately constant for at least a short period of time, i.e. $\gamma_x(t) = \gamma_x(t_0)$ for all t in a small interval around t_0. By defining $\beta_x = \gamma_x(t_0)$ and $\kappa_t = E[Y_{x_0}(t)]$, we obtain the model

$$E[Y_x(t)] = \beta_x \kappa_t. \tag{8.13}$$

The future expected annual medical costs κ_t at reference age x_0 are estimated by linear regression, based on the assumption that

$$\kappa_t = \kappa_{t_0} + \theta(t - t_0) \tag{8.14}$$

for some real number θ. This is referred to as Rusam method on the German market.

8.3.2 Data Description

The data used in this section are those published by the German Federal Financial Supervisory Authority (BaFin). The data set covers the period 1995–2011. Henceforth, the response $Y_x(t)$ is indexed by attained age x and calendar time t. It describes the average yearly medical inpatient cost for year $t = 1995, \ldots, 2011$ at age $x = 20, 21, \ldots, 80$. Here, we only consider data for German males.

The procedure adopted by BaFin for creating the data tables can be summarized as follows. BaFin requires data from all German insurance companies, each year, according to different tariffs and types of benefits. The collected data are crude. For each specific table (e.g. inpatient, double room), BaFin smooths the collected crude data with Whittaker-Henderson (with possibly different smoothing parameters for different age groups). There is no further adjustment of the data beyond smoothing.

The observed $y_x(t)$ are displayed in Fig. 8.9. The shape of the data is similar to a mortality surface, with the increase at young adult ages paralleling the accident hump followed by a linear increase after age 40 similar to the Gompertz part of the mortality schedule, before an ultimate concave behavior at oldest ages. Surprisingly, the accident hump temporarily vanishes in the calendar years 1999–2001. The effect of

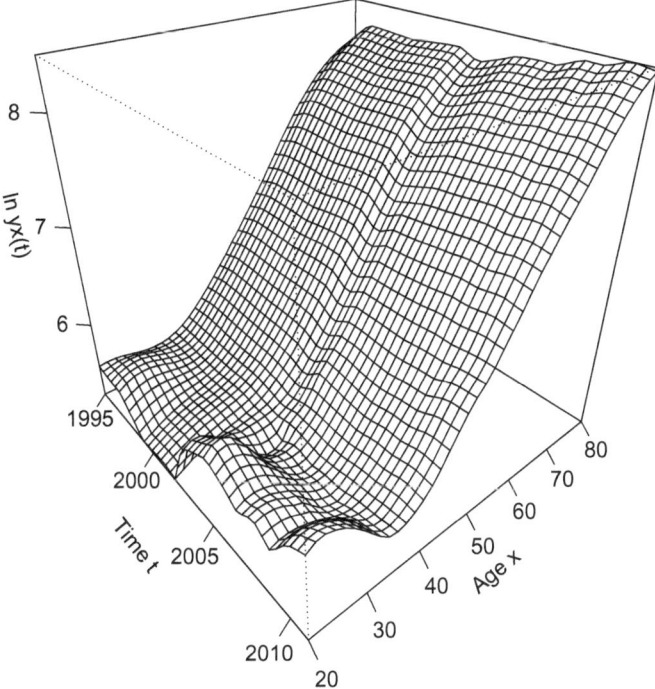

Fig. 8.9 Observed inpatient costs $y_x(t)$ for German males, $t = 1995, \ldots, 2011$, $x = 20, 21$, $\ldots, 80$, log scale

medical inflation is also clearly visible, causing an increase in yearly health expenses as time passes.

Figure 8.10 displays the number of policies according to policyholder's age for each calendar year 2007–2011. This information was missing for years up to 2006. Apart from a moderate aging effect, the available age-specific volumes appear to be relatively stable over time. Therefore, unavailable volumes for specific years were inferred from observed volumes. Specifically, the age structure is henceforth supposed to be constant over 1995–2006, as suggested by available data.

8.3.3 Modeling Approach

To relax the linear trend assumption (8.14) underlying the Rusam approach, we replace (8.13) with the specifications

$$E[Y_x(t)] = \alpha_x + \beta_x \kappa_t \tag{8.15}$$

Fig. 8.10 Number of
policies according to
policyholder's age for each
calendar year 2007–2011

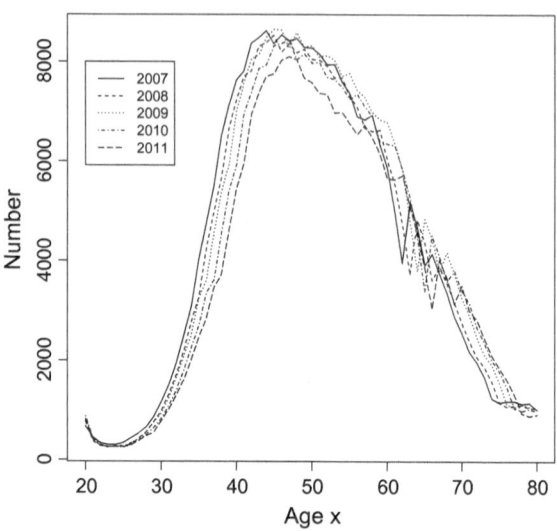

and

$$E[Y_x(t)] = \exp(\alpha_x + \beta_x \kappa_t) \tag{8.16}$$

involving the bilinear score (8.1) commonly used in mortality studies with an identity
or a log link function, respectively. Similar to mortality projection, the decomposition
into a age-specific pattern α_x and a time pattern κ_t (trend) modulated by an age
response β_x is easy to interpret.

We estimate age profiles α_x and β_x in combination with a time profile κ_t based
on all past data. The parameters κ_t are next projected with an appropriate time series
model. In contrast to the deterministic trend model (8.14), the time series approach
for κ_t better captures empirically observed patterns and moreover allows to quantify
the forecasting uncertainty.

In our case study we consider four different models. All models share similarity
with the log-bilinear structure (8.1). However, we assume continuous distributions
for the response variable because we do not have to deal with event counts but average
claim severities instead. The models are from the class of GNMs in the sense that:

- The response is generated from a distribution of the ED family.
- We specify a link function (logarithm or identity function).
- The score is not a linear function of the unknown parameters.

8.3.4 Model Specifications

Henceforth, we consider the following specifications to estimate the parameters α_x,
β_x and κ_t appearing in (8.15)–(8.16):

Model M0:

Setting α_x equal to zero, the first model is in line with the Rusam method (8.13) with $Y_x(t)$ Normally distributed with mean

$$E[Y_x(t)] = \beta_x \kappa_t$$

and constant variance

$$\text{Var}[Y_x(t)] = \sigma^2,$$

that is,

$$Y_x(t) \sim \textit{Nor}\left(\beta_x \kappa_t, \sigma^2\right).$$

For this model, we only normalize the parameters β_x by imposing the second constraint of (8.6). Different from the Rusam method, here β_x and κ_t do not refer to a specific year or age. Moreover, for κ_t we allow departures from the linear trend (8.14).

Model M1:

This model is based on (8.15) with $Y_x(t)$ Normally distributed, i.e.

$$Y_x(t) \sim \textit{Nor}\left(\alpha_x + \beta_x \kappa_t, \sigma^2\right).$$

This model is close to model M0 except that the set of α_x captures an average level over the observation period.

Model M2:

This model is based on (8.16) with $Y_x(t)$ Gamma distributed with mean

$$E[Y_x(t)] = \exp\left(\alpha_x + \beta_x \kappa_t\right),$$

that is,

$$Y_x(t) \sim \textit{Gam}\left(\exp\left(\alpha_x + \beta_x \kappa_t\right), \tau\right).$$

This model differs from M0–M1 by the distribution assumption (Gamma vs. Normal) as well as by the link function (log-link vs. identity link).

Model M3:

The last model is based on (8.16) with $Y_x(t)$ Inverse Gaussian, i.e.

$$Y_x(t) \sim \textit{IGau}\left(\exp\left(\alpha_x + \beta_x \kappa_t\right), \tau\right).$$

This model thus differs from M2 by a different distributional assumption for the response variable (Inverse Gaussian vs. Gamma) but not by the link function (log-link in both cases).

8.3.5 Model Selection

Because of the temporary disappearance of the accident hump, we restrict our study
to calendar years 2002–2011. Figure 8.11 displays the estimated effects for each
model M0–M3. For all models, the overall shape of $\widehat{\kappa}_t$ looks very similar. Combined
with positive estimated parameters β_x, this shows that average costs were indeed

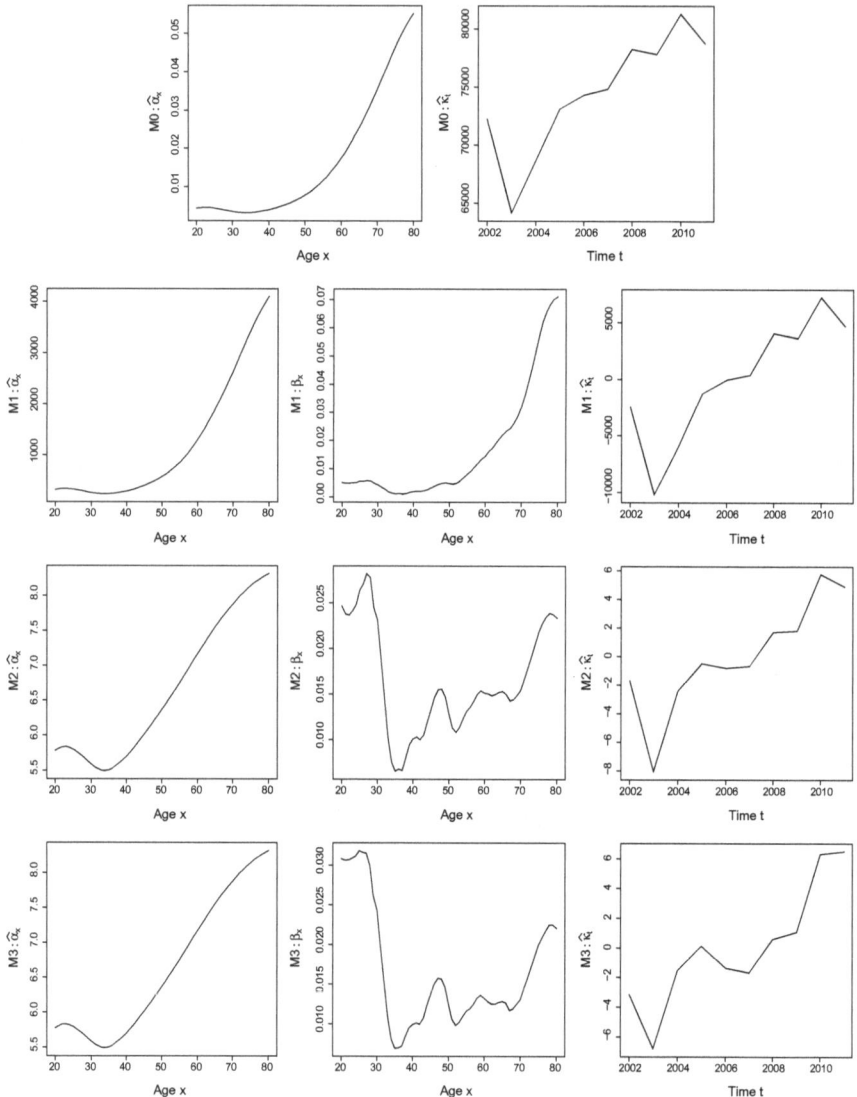

Fig. 8.11 Estimated effects for models M0–M3 (from top to bottom)

increasing at all ages because of medical inflation over the observation period. Even if the estimated parameters α_x look similar for models M1–M3, the estimated β_x profile for models M2–M3 look surprising. Models M0–M1 exhibit similar estimated β_x profile.

Since we work in a regression framework, it is essential to inspect the residuals. Model performance is assessed in terms of the randomness of the residuals. Here, we use Pearson residuals for all models M0–M3, computed from $r_{xt} = y_{xt} - \widehat{y}_{xt}$ suitably standardized. If the residuals r_{xt} exhibit some regular pattern, this means that the model is not able to describe all of the phenomena appropriately. In practice, looking at $(x, t) \mapsto r_{xt}$, and discovering no structure in those graphs ensures that the time trends have been correctly captured by the model.

Figures 8.12 and 8.13 display the residuals obtained for models M0–M3 as a function of both age and calendar time. The structure in the residuals can be attributed to the preliminary smoothing procedure implemented by BaFin. This induces similar

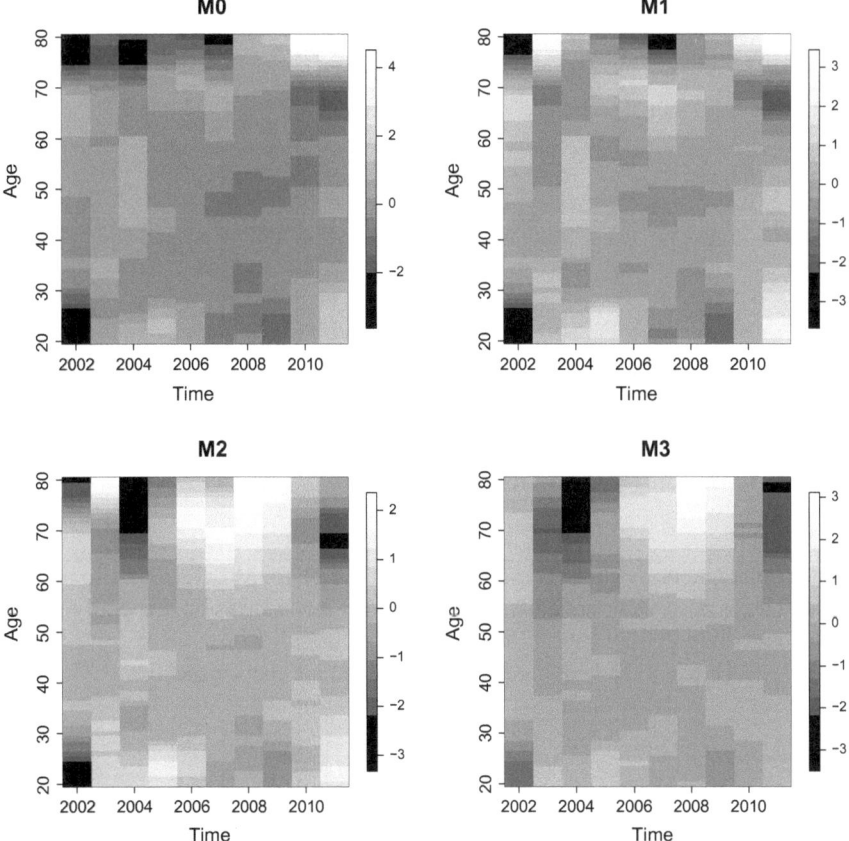

Fig. 8.12 Image plots for standardized residuals r_{xt} corresponding to models M0–M3

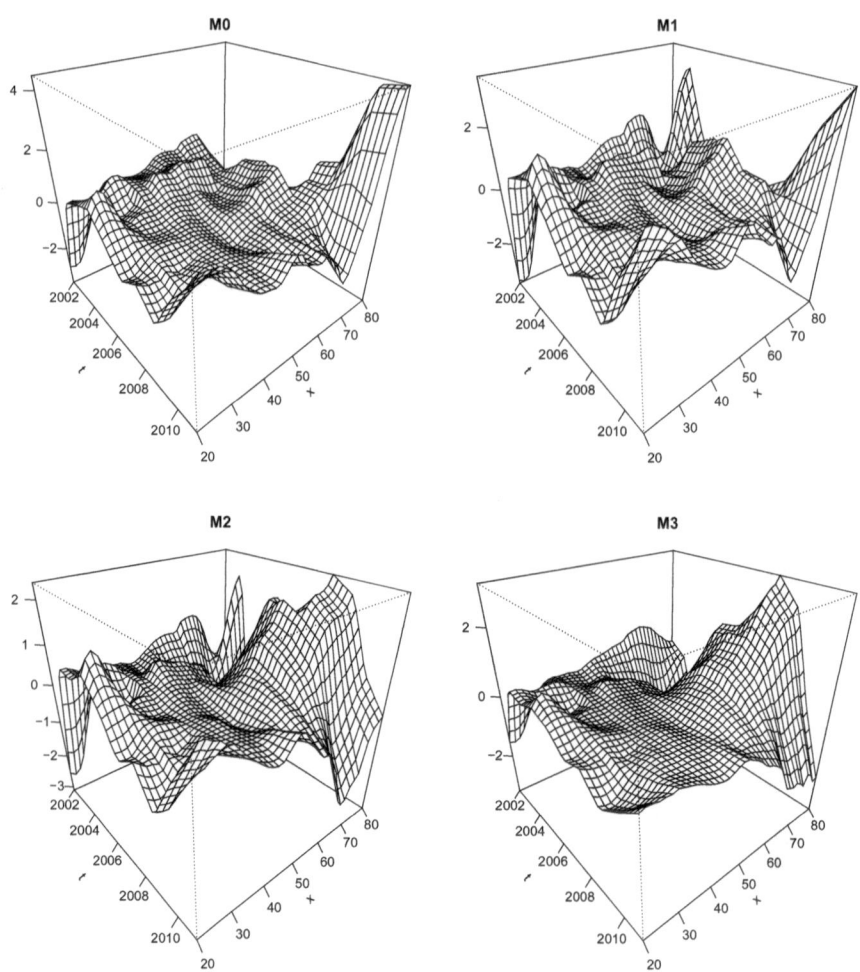

Fig. 8.13 Standardized residuals r_{xt} for models M0–M3

values for neighboring cells. This is especially the case for M0. Residuals sometimes assume large positive or negative values near data boundaries but generally remain in the interval $[-2, 2]$.

In order to select the best model, we now compare their respective AIC values. This gives

$$6,204 \text{ for } M0$$
$$5,964 \text{ for } M1$$
$$22,664,312 \text{ for } M2$$
$$24,101,837 \text{ for } M3.$$

Based on this performance measure, model M1 appears to be the best one. The linear trend in the time index κ_t can be exploited to produce forecasts of the average costs in the next few years.

8.4 Bibliographic Notes and Further Reading

The statistical analysis of log-bilinear models is originally traced back to Goodman (1979), in relation to contingency tables. The log-bilinear decomposition (8.1) plays a central role in dynamic mortality modeling, after the pioneering work by Lee and Carter (1992). See also Lee (2000) as well as Pitacco et al. (2009) for a textbook treatment of the topic. Different distributional assumptions have been proposed so far. Lee and Carter (1992) initially suggested to use least-squares. Brouhns et al. (2002a, b) and Renshaw and Haberman (2003) embedded (8.1) in a Poisson regression model, which is perfectly suited for age-gender-specific mortality rates. The Binomial (Cossette et al. 2007) and Negative Binomial (Delwarde et al. 2007a) distributions have been proposed afterwards. The maximum-likelihood estimates are easily found using iterative algorithms. In practice, they appear to converge very rapidly. The gnm package of the statistical software R (Turner and Firth 2007) can be used to fit GNMs with a score of the form (8.1), avoiding to develop such algorithms case by case (we refer the interested reader to the aforementioned papers for the derivation of these algorithms in the Normal, Poisson, Binomial, and Negative Binomial cases). The numerical illustrations proposed in this chapter have been performed with gnm.

In empirical studies, the estimated parameters β_x often exhibit an irregular pattern. This issue has been solved by adopting the GAM setting in this chapter, after having linearized the time index κ_t. A more general way to impose some smoothness on the estimated parameters β_x has been proposed by Delwarde et al. (2007b). It consists in introducing a roughness penalty with the help of penalized least-squares or penalized maximum-likelihood using the counting distributions listed above. The estimated parameters can be obtained with the help of the Newton-Raphson method and the optimal smoothing parameter can be selected by cross-validation. As explained earlier in this chapter, the estimated parameters α_x are usually very smooth, since they represent an average effect of mortality at age x over the observation period. The estimated parameters κ_t are often rather irregular, but the projected values of κ_t, obtained from some time series model (like ARIMA), will be smooth. Hence, we only need to smooth the parameters β_x in order to get projected life tables with mortality varying smoothly across ages. Notice that smoothness can also be incorporated by an appropriate choice of a priori distributions in the Bayesian version of the Poisson log-bilinear model for mortality projection proposed by Czado et al. (2005).

The presence of a break in the trend of the Lee-Carter index of mortality is well documented for post-1950 mortality data in industrialized countries. Structural changes in the rate of decline in overall mortality have been found by Coelho and Nunes (2011) for almost every country, especially in male populations. See also

Van Berkum et al. (2016). Borger and Schupp (2018) proposed an alternative trend process where trend changes occur randomly over time, and also the trend change magnitude is stochastic.

Hunt and Blake (2014) presented the variety of age-period-cohort mortality projection models developed after the Lee-Carter one in a unified way. Many projection models decompose the death rates (on the log scale) or the one-year death probabilities (on the logit scale) into a linear combination of a limited number of time factors. Precisely,

$$\ln \mu_x(t) \text{ or } \text{logit} q_x(t) = \ln \frac{q_x(t)}{p_x(t)}$$

is decomposed into

$$= \alpha_x \text{ static reference life table}$$

$$+ \sum_{j=1}^{p} \beta_x^{(j)} \kappa_t^{(j)} \text{ time trends interacting with age}$$

$$+ \gamma_x \lambda_{t-x} \text{ cohort effect}$$

with suitable identifiability constrains. Here are some examples:

Clayton-Schifflers model $\ln \mu_x(t) = \alpha_x + \kappa_t + \lambda_{t-x}$

Lee-Carter model $\ln \mu_x(t) = \alpha_x + \beta_x \kappa_t$

Renshaw-Haberman model $\ln \mu_x(t) = \alpha_x + \beta_x \kappa_t + \gamma_x \lambda_{t-x}$

Debonneuil model $\ln \mu_x(t) = \alpha_x + \beta_x \kappa_t + \lambda_{t-x}$

Cairns-Blake-Dowd model $\text{logit}(q_x(t)) = \kappa_t^{(1)} + \kappa_t^{(2)}(x - \overline{x})$
                                $(+\text{quadratic effects of age})$

Plat model $\ln \mu_x(t) = \alpha_x + \kappa_t^{(1)} + \kappa_t^{(2)}(x - \overline{x})$
                                $+ \kappa_t^{(3)}(\overline{x} - x)_+ + \lambda_{t-x}$

Mortality projections are generally obtained as follows. In a first step, regression techniques are used to extract the time factors from the available mortality data. In a second step, the time factors are intrinsically viewed as forming a time series to be projected to the future. The actual age-specific death rates are then derived from this forecast using the estimated age effects. This in turn yields projected life expectancies, for instance.

In the first step of the two-step model calibration procedure, the random nature of the unobservable time factor is disregarded, and this may bias the analysis. As possible incoherence may arise from this two-step procedure, Czado et al. (2005)

integrated both steps into a Bayesian version of the Lee-Carter model in order to avoid this deficiency. After Czado et al. (2005), Pedroza (2006) formulated the Lee-Carter method as a state-space model, using Gaussian error terms and a random walk with drift for the mortality index. See also Girosi and Ging (2008), Kogure et al. (2009, 2010) and Li (2014) for related works. However, the practical implementation of Bayesian methods often requires computer-intensive Markov Chain Monte Carlo (MCMC) simulations. This is why Schinzinger et al. (2016) proposed a simple credibility model ensuring robustness over time while keeping the computational issues relatively easy and allowing for the flexibility of time series modeling. Treating the time factor as such, and not as a parameter to be estimated from past mortality statistics using regression techniques before entering time series models, recognizes its hidden nature and intrinsic randomness.

Several authors suggested to target improvement rates to forecast future mortality, instead of the death rates. While the time dependence structure of death rate models are dominated by the continuing downward trend, the improvement rates are already trend adjusted. See, e.g. Mitchell et al. (2013) and Denuit and Trufin (2016). Precisely, the mortality improvement rates $\mu_x(t)/\mu_x(t-1)$ can be specified as

$$\ln \mu_x(t) - \ln \mu_x(t-1) = \alpha_x + \sum_{j=1}^{p} \beta_x^{(j)} \kappa_t^{(j)} + \gamma_x \lambda_{t-x}.$$

Both approaches are related as under the Lee-Carter model for instance

$$\ln \mu_x(t) - \ln \mu_x(t-1) = \alpha_x + \beta_x \kappa_t - \left(\alpha_x + \beta_x \kappa_{t-1}\right)$$

$$= \beta_x(\kappa_t - \kappa_{t-1})$$

so that trends are removed. Schinzinger et al. (2016) studied the dynamics of these mortality improvement rates and proposed an evolutionary credibility model to describes their joint dynamics through time in several populations. Dynamic random effects ensure the necessary smoothing across time, as well as the learning effect. Last but not least, they also serve to stabilize successive mortality projection outputs, avoiding dramatic changes from one year to the next.

As pointed out by Booth et al. (2002), the original approach by Lee and Carter (1992) makes use of only the first term of the singular value decomposition of the matrix of centered log death rates. In principle, the second and higher order terms could be incorporated in the model. The full expanded model is

$$\ln \widehat{m}_x(t) = \alpha_x + \sum_{j=1}^{r} \beta_x^{[j]} \kappa_t^{[j]}, \tag{8.17}$$

where r is the rank of the $\ln m_x(t) - \alpha_x$ matrix. In this case, $\beta_x^{[j]} \kappa_t^{[j]}$ is referred to as the jth order term of the approximation. Any systematic variation in the residuals from fitting only the first term would be captured by the second and higher terms. In their empirical illustration, Booth et al. (2002) find a diagonal pattern in the residuals that was interpreted as a cohort-period effect. Brouhns et al. (2002a) have tested whether the inclusion of a second log-bilinear term significantly improves the quality of the fit, and this was not the case in their empirical illustrations.

Hyndman and Ullah (2007) extend the principal components approach by adopting a functional data paradigm combined with nonparametric smoothing (penalized regression splines) and robust statistics. Univariate time series are then fitted to each component coefficient (or level parameter). The Lee-Carter method then appears to be a particular case of this general approach.

An important aspect of the Lee-Carter and the Cairns-Blake-Dowd models, as well as of the other models discussed in this chapter, is that the time factors are intrinsically viewed as time series. Their forecasts in turn yield projected age-specific mortality rates, life expectancies and single premiums for life annuities. We do not enter the details here and refer to Pitacco et al. (2009) for more details. In this chapter, we favor a parametric approach, replacing the time index with linear functions of calendar time. See e.g. Denuit and Goderniaux (2005) for a comparison of both approaches in the Lee-Carter setting.

The Lee-Carter model has also been successfully applied to project disability rates, in addition to mortality ones, in Christiansen et al. (2012) as well as in Levantesi and Menzietti (2012). For another approach based on parametric models, see Renshaw and Haberman (2000). It appears to provide a effective starting point for the analysis of morbidity rates.

Section 8.3 is based on Christiansen et al. (2018). Piontkowski (2019) further exploited the similarities between mortality and health claim costs and adopted the functional data paradigm proposed by Hyndman and Ullah (2007) for mortality modeling.

Since the empirically observed time trend is quite volatile, confidence interval estimates rather than point estimates should be used for predictions. Different bootstrap techniques were applied for creating prediction intervals. See e.g. Brouhns et al. (2005) and Koissi et al. (2006). The longevity fan charts proposed by Dowd et al. (2010) offer convenient representations of the uncertainty surrounding mortality projections.

References

Booth H, Maindonald J, Smith L (2002) Applying Lee-Carter under conditions of variable mortality decline. Popul Stud 56:325–336

Borger M, Schupp J (2018) Modeling trend processes in parametric mortality models. Insur Math Econ 78:369–380

Brouhns N, Denuit M, Van Keilegom I (2005) Bootstrapping the Poisson log-bilinear model for mortality projection. Scand Actuar J 2005:212–224

Brouhns N, Denuit M, Vermunt JK (2002a) A Poisson log-bilinear approach to the construction of projected life tables. Insur Math Econ 31:373–393

Brouhns N, Denuit M, Vermunt JK (2002b) Measuring the longevity risk in mortality projections. Bull Swiss Assoc Actuar 2002:105–130

Christiansen MC, Denuit MM, Lazar D (2012) The Solvency II square-root formula for systematic biometric risk. Insur Math Econ 50:257–265

Christiansen M, Denuit M, Lucas N, Schmidt J-Ph (2018) Projection models for health expenses. Ann Actuar Sci 12:185–203

Coelho E, Nunes LC (2011) Forecasting mortality in the event of a structural change. J R Stat Soc Ser A (Stat Soc) 174:713–736

Cossette H, Delwarde A, Denuit M, Guillot F, Marceau E (2007) Pension plan valuation and dynamic mortality tables. North Am Actuar J 11:1–34

Czado C, Delwarde A, Denuit M (2005) Bayesian Poisson log-bilinear mortality projections. Insur Math Econ 36:260–284

Delwarde A, Denuit M, Eilers P (2007a) Smoothing the Lee-Carter and Poisson log-bilinear models for mortality forecasting: a penalized log-likelihood approach. Stat Modell 7:29–48

Delwarde A, Denuit M, Partrat Ch (2007b) Negative Binomial version of the Lee-Carter model for mortality forecasting. Appl Stoch Models Bus Ind 23:385–401

Denuit M, Goderniaux A-C (2005) Closing and projecting life tables using log-linear models. Bull Swiss Assoc Actuar 2005:29–49

Denuit M, Trufin J (2016) From regulatory life tables to stochastic mortality projections: the exponential decline model. Insur Math Econ 71:295–303

Dowd K, Blake D, Cairns AJG (2010) Facing up to uncertain life expectancy: the longevity fan charts. Demography 47:67–78

Girosi F, King G (2008) Demographic forecasting. Princeton University Press

Goodman LA (1979) Simple models for the analysis of association in cross-classifications having ordered categories. J Am Stat Assoc 74:537–552

Hunt A, Blake D (2014) A general procedure for constructing mortality models. North Am Actuar J 18:116–138

Hyndman RJ, Ullah MS (2007) Robust forecasting of mortality and fertility rates: a functional data approach. Comput Stat Data Anal 51:4942–4956

Kogure A, Kitsukawa K, Kurachi Y (2009) A Bayesian comparison of models for changing mortalities toward evaluating longevity risk in Japan. Asia Pac J Risk Insur 3:1–22

Kogure A, Kurachi Y (2010) A Bayesian approach to pricing longevity risk based on risk-neutral predictive distributions. Insur Math Econ 46:162–172

Koissi MC, Shapiro AF, Hognas G (2006) Evaluating and extending the Lee-Carter model for mortality forecasting: bootstrap confidence intervals. Insur Math Econ 38:1–20

Lee RD (2000) The Lee-Carter method of forecasting mortality, with various extensions and applications. North Am Actuar J 4:80–93

Lee RD, Carter L (1992) modeling and forecasting the time series of US mortality. J Am Stat Assoc 87:659–671

Levantesi S, Menzietti M (2012) Managing longevity and disability risks in life annuities with long term care. Insur Math Econ 50:391–401

Li J (2014) An application of MCMC simulation in mortality projection for populations with limited data. Demogr Res 30:1–48

Mitchell D, Brockett P, Mendoza-Arriaga R, Muthuraman K (2013) Modeling and forecasting mortality rates. Insur Math Econ 52:275–285

Pedroza C (2006) A Bayesian forecasting model: predicting US male mortality. Biostatistics 7:530–550

Piontkowski J (2019) Forecasting health expenses using a functional data model. Ann Actuar Sci (in press)

Pitacco E, Denuit M, Haberman S, Olivieri A (2009) Modelling longevity dynamics for pensions and annuity business. Oxford University Press

Renshaw AE, Haberman S (2000) Modeling the recent time trends in UK permanent health insurance recovery, mortality and claim inception transition intensities. Insur Math Econ 27:365–396

Renshaw A, Haberman S (2003) Lee-Carter mortality forecasting: a parallel generalized linear modelling approach for England and Wales mortality projections. Appl Stat 52:119–137

Schinzinger E, Denuit M, Christiansen MC (2016) A multivariate evolutionary credibility model for mortality improvement rates. Insur Math Econ 69:70–81

Turner H, Firth D (2007) gnm: a package for generalized nonlinear models. R News 7:8–12

Van Berkum F, Antonio K, Vellekoop M (2016) The impact of multiple structural changes on mortality predictions. Scand Actuar J 2016:581–603

Chapter 9
Extreme Value Models

9.1 Introduction

As in other disciplines sensitive to extreme observations, like for instance hydrology and climatology, standard statistical techniques fail to analyze large claims located far in the tail of the severity distribution. This is because the actuary wants to make inference about the extremal behavior of claim distributions, that is, in an area of the sample where there are very few data points, if any.

Extreme Value Theory (EVT, in short), and its "excesses over threshold" model together with the Generalized Pareto distribution, offers an unified approach to the modeling of the tail of a claim distribution. This method is not solely based on the data at hand but includes a probabilistic argument concerning the behavior of the extreme sample values, allowing for extrapolation beyond the range of the data, that is, in areas where there are no observations at all.

Let us briefly discuss typical situations where actuaries are concerned with extreme values. In third-party liability insurance, actuaries often face difficulties when they analyze claim severities because

- of the limited number of observations, as only claiming policies provide information about claim costs.
- a long time is sometimes needed to settle the claim so that the exact claim cost is unknown to the company during that period.
- the amounts paid by the company depend to a large extent on the third party characteristics, which are unknown when the premium is computed.

Moreover, a few large claims hitting a portfolio often represent a significant part of the insurance benefits paid by the company. These extreme events are therefore of prime interest for actuaries. They also form the statistical material

- for the pricing of reinsurance agreements such as excess-of-loss reinsurance treaties (under which the reinsurer has to pay for the excess of a claim above a fixed threshold named the deductible).
- for the estimation of high quantiles (also called Value-at-Risk in financial risk management).
- for the derivation of the probable maximal loss (PML, in short), to be used as a working upper bound for the claim size in the calculation of risk quantities.

© Springer Nature Switzerland AG 2019
M. Denuit et al., *Effective Statistical Learning Methods for Actuaries I*,
Springer Actuarial, https://doi.org/10.1007/978-3-030-25820-7_9

When the main interest is in the tail of loss severity distributions, it is essential to have a good model for the largest claims. Distributions providing a good overall fit, such as ED ones, can be particularly bad at fitting the tails. EVT precisely focuses on the tails, being supported by strong theoretical arguments. The idea behind EVT is to conduct the analysis based on that part of the sample which carries the information about the extremal behavior, i.e. only the largest sample values. This chapter recalls basics of EVT, useful for actuarial applications.

Throughout this chapter, two applications are used to illustrate the interest of EVT in insurance studies: claim severities in motor third-party liability insurance and mortality at the oldest ages. In the former case, identifying the large claims which require a separate, Pareto-type modeling is of prime importance in risk management. In the latter case, the close examination of the extreme lifetimes provides the actuaries with a meaningful closure procedure for the life table.

9.2   Basics of EVT

9.2.1   Sample Averages and Maxima

Considering a sequence of independent and identically distributed random variables (claim severities or remaining lifetimes, say) $Y_1, Y_2, Y_3, \ldots$, most classical results from probability and statistics that are relevant for insurance are based on sums $S_n = \sum_{i=1}^{n} Y_i$. Let us mention the law of large numbers and the central limit theorem, for instance.

The central limit theorem identifies real sequences a_n and b_n such that the normalized partial sums $\frac{S_n - a_n}{b_n}$ converge to the standard Normal distribution as n grows. Specifically, denoting as $\mu = E[Y_1]$ and $\sigma^2 = Var[Y_1] < \infty$, we know that the identity

$$\lim_{n \to \infty} P\left[\frac{\sum_{i=1}^{n} Y_i - a_n}{b_n} \le x \right] = \Phi(x)$$

holds true with $a_n = n\mu$ and $b_n = \sqrt{n}\sigma$.

Another interesting yet less standard statistics for the actuary is

$$M_n = \max\{Y_1, \ldots, Y_n\}$$

the maximum of the n random variables $Y_1, \ldots, Y_n$. EVT addresses the following question: how does M_n behave in large samples (i.e. when n tends to infinity)? Of course, without further restriction, M_n obviously tends to ω, where ω is the right-endpoint of the support of Y_1, precisely defined as

$$\omega = \sup\left\{ y \in (-\infty, \infty) \,\middle|\, F(y) < 1 \right\}, \text{ possibly infinite,}$$

where F denotes the common distribution function of the random variables Y_1, Y_2, Y_3, ... under consideration. We know that

$$P[M_n \leq y] = P[Y_i \leq y \text{ for all } i = 1, \ldots, n] = \left(F(y)\right)^n.$$

But in practice, the exact distribution of M_n is of little interest as for all $y < \omega$,

$$\lim_{n \to \infty} P[M_n \leq y] = \lim_{n \to \infty} \left(F(y)\right)^n = 0.$$

If $\omega < \infty$ then for all $y \geq \omega$

$$\lim_{n \to \infty} P[M_n \leq y] = \lim_{n \to \infty} \left(F(y)\right)^n = 1.$$

Moreover, a small estimation error on F can have dramatic consequences on the estimation of high quantiles of M_n.

9.2.2 Extreme Value Distribution

Once M_n is appropriately centered and normalized, however, it may converge to some specific limit distribution (of three different types, according to the thickness of the tails associated to F). Together with the normalizing constants, this limit distribution determines the asymptotic behavior of the sample maxima M_n. If there exist sequences of real numbers $c_n > 0$ and $d_n \in (-\infty, \infty)$ such that the normalized sequence $(M_n - d_n)/c_n$ converges in distribution to H, i.e.

$$\lim_{n \to \infty} P\left[\frac{M_n - d_n}{c_n} \leq y\right] = \lim_{n \to \infty} \left(F(c_n y + d_n)\right)^n = H(y) \qquad (9.1)$$

for all points of continuity of the limit H, then H is a Generalized Extreme Value Distribution. Notice that (9.1) stands in contrast to the central-limit theorem which is about sums S_n.

The Generalized Extreme Value (GEV) distribution function with tail index ξ appearing as the limit in (9.1) is of the form

$$H_\xi(y) = \begin{cases} \exp\left(-(1 + \xi y)_+^{-1/\xi}\right) & \text{if } \xi \neq 0, \\ \exp\left(-\exp(-y)\right) & \text{if } \xi = 0, \end{cases}$$

where $z_+ = \max\{z, 0\}$ is the positive part of z. The support of H_ξ is

$$(-1/\xi, \infty) \text{ if } \xi > 0$$
$$(-\infty, -1/\xi) \text{ if } \xi < 0$$
$$(-\infty, \infty) \text{ if } \xi = 0.$$

Here, the parameter ξ controlling the right tail is called the tail index or the extreme value index. The three classical extreme value distributions are special cases of the GEV family:

if $\xi > 0$ then we have the Frechet distribution,

if $\xi < 0$ then we have the Weibull distribution, and

if $\xi = 0$ then we have the Gumbel distribution.

The GEV distribution appears to be the only non-degenerate limit distribution for appropriately normalized sample maxima, as formally stated in the next section.

9.2.3   Fisher-Tippett Theorem

If the appropriately normalized sample maxima M_n converge in distribution to a non-degenerate limit H in (9.1), the distribution function F is said to be in the domain of attraction of the GEV distribution H. The class of distributions for which such an asymptotic behavior holds is large: all commonly encountered continuous distributions fulfill (9.1).

The results broadly described so far can be summarized in the following theorem.

Theorem 9.2.1 (Fisher-Tippett Theorem) *If there exist sequences of real constants c_n and d_n such that (9.1) holds for some non-degenerate (i.e. not concentrated on a single point) distribution function H then H must be a GEV distribution, i.e.*

$$H = H_\xi \text{ for some } \xi.$$

The parameter ξ is known as the extreme value index (or Pareto index, or tail index).

The Pareto distribution is the typical example of element belonging to the Frechet class, as shown in the next example.

Example 9.2.2 $(\mathcal{P}ar(\alpha, \tau) \Rightarrow \xi > 0)$ Assume that $Y_1, Y_2, Y_3, \ldots$ are independent and all obey the $\mathcal{P}ar(\alpha, \tau)$ distribution, with distribution function

$$F(y) = 1 - \left(\frac{\tau}{\tau + y}\right)^\alpha, \quad \alpha, \tau > 0, \quad y \geq 0.$$

Considering the normalizing constants

$$c_n = \frac{\tau n^{1/\alpha}}{\alpha} \text{ and } d_n = \tau n^{1/\alpha} - \tau$$

we get for

$$c_n y + d_n \geq 0 \Leftrightarrow 1 + \frac{y}{\alpha} \geq n^{-1/\alpha}$$

that

$$\left(F(c_n y + d_n)\right)^n = \left(1 - \left(\frac{\tau}{\tau + y\frac{\tau n^{1/\alpha}}{\alpha} + \tau n^{1/\alpha} - \theta}\right)^\alpha\right)^n$$

$$= \left(1 - \frac{1}{n}\left(1 + \frac{y}{\alpha}\right)^{-\alpha}\right)^n$$

$$\to \exp\left(-\left(1 + \frac{y}{\alpha}\right)^{-\alpha}\right) = H_{1/\alpha}(y) \text{ as } n \to \infty.$$

Thus, the extreme value index of the $Par(\alpha, \tau)$ distribution is $\xi = 1/\alpha$. Notice that the constraint $1 + \frac{y}{\alpha} \geq n^{-1/\alpha}$ becomes $1 + \frac{y}{\alpha} > 0$ as $n \to \infty$ so that the domain is well of the form $(-1/\xi, \infty)$, as announced. This shows that the $Par(\alpha, \tau)$ distribution belongs to the domain of attraction of the Frechet distribution as the associated extreme value index $\xi = 1/\alpha$ is positive.

Of course, the Frechet class ($\xi > 0$) does not reduce to the Pareto distribution. It can be described in broad terms as follows. The Fisher-Tipett theorem holds with H_ξ, $\xi > 0$, if, and only if, the representation

$$1 - F(y) = y^{-1/\xi}\ell(y) \tag{9.2}$$

holds for some slowly varying function $\ell(\cdot)$, i.e. for a function $\ell(\cdot)$ such that

$$\lim_{y \to \infty} \frac{\ell(ty)}{\ell(y)} = 1 \text{ for all } t > 0.$$

In words, this essentially means that if $1 - F$ decays like a power function then the distribution is in the domain of attraction of the Frechet distribution. The formula (9.2) is often taken as a definition of heavy-tailed distributions. In addition to the Pareto distribution whose tails are polynomially decreasing, examples of distributions belonging to the Frechet class include Burr, Log-Gamma, Cauchy and Student t distributions. Not all moments are finite.

The appropriateness of heavy-tailed distributions in the Frechet class to model catastrophe risks manifests in the asymptotic relation

$$\lim_{t \to +\infty} \frac{P[M_n > t]}{P[S_n > t]} = 1.$$

This relation describes a situation where the sum S_n of n claims gets large if, and only if, its maximum M_n gets large. Distributions satisfying this condition are usually referred to as sub-exponential in the actuarial literature.

Heavy-tailed distributions are most often encountered in nonlife insurance applications. Distributions with lighter tails (typically with exponential decrease in the tails) belong to the Gumbel class, as shown in the next example.

Example 9.2.3 ($\mathcal{E}xp(\tau) \Rightarrow \xi = 0$) Assume that $Y_1, Y_2, Y_3, \ldots$ are independent and all obey the $\mathcal{E}xp(\tau)$ distribution, with distribution function

$$F(y) = 1 - \exp(-\tau y), \quad \tau > 0, \quad y \geq 0.$$

With $c_n = \frac{1}{\tau}$ and $d_n = \frac{\ln n}{\tau}$, we obtain for

$$c_n y + d_n \geq 0 \Leftrightarrow y \geq -\ln n$$

that

$$\begin{aligned}
\left(F(c_n y + d_n)\right)^n &= \left(1 - \exp\left(-\tau\left(\frac{1}{\tau}y + \frac{\ln n}{\tau}\right)\right)\right)^n \\
&= \left(1 - \frac{1}{n}\exp(-y)\right)^n \\
&\to \exp(-\exp(-y)) = H_0(y), \quad y \in (-\infty, \infty) \text{ as } n \to \infty.
\end{aligned}$$

Notice that the constraint $y \geq -\ln n$ is not binding anymore at the limit so that the domain extends to the whole real line $(-\infty, \infty)$. Thus, the $\mathcal{E}xp(\tau)$ distribution belongs to the domain of attraction of the Gumbel distribution with zero associated extreme value index.

The characterization of the Gumbel class $\xi = 0$ is more complicated. Roughly speaking, it contains distributions whose tails decay exponentially towards 0 (light-tailed distributions). All moments are finite. Examples include the Normal, LogNormal, and Gamma (in particular, the Negative Exponential) distributions.

The Weibull class ($\xi < 0$) appears to be particularly useful in life insurance, as its elements have a bounded support. This behavior will be studied in the next sections devoted to the modeling of remaining lifetimes at older ages.

9.2.4   Extreme Lifetimes

Let us now re-phrase the central result of EVT in the life insurance setting, for mortality modeling at the oldest ages. Responses Y typically represent lifetimes in a life insurance context and are henceforth denoted as T. Consider a sequence of independent individual lifetimes $T_1, T_2, T_3, \ldots$ with common distribution function

$$F(x) = {}_xq_0 = P[T_i \leq x], \quad x \geq 0,$$

for $i = 1, \ldots, n$, satisfying ${}_0q_0 = 0$. Here, T_i represents the total lifetime, from birth to death. In many insurance applications, it may also represent the remaining lifetime after a given initial age α, with common distribution function $F_\alpha(x) = {}_xq_\alpha$.

In words, M_n represents in this setting the oldest age at death observed in an homogeneous group of n individuals subject to the same life table $x \mapsto {}_xq_0$. EVT studies the asymptotic behavior of M_n when n tends to infinity, provided some mild technical conditions on $x \mapsto {}_xq_0$ are fulfilled. Of course, without further restriction, M_n obviously approaches the upper limit of the support

$$\omega = \sup\{x \geq 0 | {}_xq_0 < 1\}, \text{ possibly infinite.}$$

In life insurance, ω is referred to as the ultimate age of the life table. As before, this is easily seen from

$$P[M_n \leq x] = \left({}_xq_0\right)^n \rightarrow \begin{cases} 0 \text{ if } x < \omega \\ 1 \text{ if } x \geq \omega \end{cases} \quad \text{as } n \rightarrow \infty.$$

Once M_n is appropriately centered and normalized, however, it may converge to some specific limit distribution, as shown in (9.1). When $\xi > 0$, we face lifetimes with heavy tails which contradicts empirical evidence available for human lifetimes (in this case, forces of mortality decrease with attained age). Thus, the cases $\xi = 0$ and $\xi < 0$ are of interest for life insurance applications. Notice that if (9.1) holds with $\xi < 0$ then $\omega < \infty$, so that a negative value of ξ supports the existence of a finite ultimate age ω.

Recall that

$$\mu_x = \lim_{\Delta x \searrow 0} \frac{P[x < T \leq x + \Delta x | T > x]}{\Delta x} = \frac{\frac{d}{dx}{}_xq_0}{{}_xp_0}$$

is the force of mortality at age x. It quantifies the risk to die instantaneously for an individual alive at age x. A sufficient condition for (9.1) to hold is

$$\lim_{x \rightarrow \omega} \frac{d}{dx} \left(\frac{1}{\mu_x}\right) = \xi. \tag{9.3}$$

Intuitively speaking, $\frac{1}{\mu_x}$ can be considered as the force of resistance to mortality, or force of vitality at age x. The resistance to mortality must thus stabilize when $\xi = 0$ or becomes ultimately linear. A negative ξ indicates that the resistance ultimately decreases at advanced ages. For $\xi < 0$ we have $\omega < \infty$ and condition (9.3) implies

$$\lim_{x \rightarrow \omega} \left((\omega - x)\mu_x\right) = -\frac{1}{\xi}.$$

9.3 Excess Over Threshold and Mean Excess Function

9.3.1 Excess Distribution

Given a sequence of independent and identically distributed random variables $Y_1, Y_2, \ldots$ and a threshold u, supposed to be large, actuaries are often interested in the excesses $Y_i - u$ over u given that $Y_i > u$. Notice that here, we restrict ourselves to the elements Y_i such $Y_i > u$ is fulfilled, i.e. Y_i reaches the threshold u under consideration.

Let F_u stand for the distribution function of the excess of the random variable Y over the threshold u, given that the threshold u is reached, that is,

$$F_u(y) = P[Y - u \le y | Y > u]$$

$$= \frac{F(y + u) - F(u)}{1 - F(u)}$$

$$= 1 - \frac{\overline{F}(u + y)}{\overline{F}(u)}, \quad y \ge 0,$$

where $\overline{F} = 1 - F$ is the excess (or de-cumulative, or survival) function associated to the (cumulative) distribution function F.

9.3.2 Mean Excess Function

Provided $E[Y] < \infty$, the mean excess function $e(\cdot)$ associated to Y is defined as

$$e(u) = E[Y - u | Y > u] = \int_0^\infty \overline{F}_u(t) dt$$

where $\overline{F}_u = 1 - F_u$. Thus, $e(u)$ represents the expected excess over the threshold u given that the threshold u is reached. It is easy to prove that if Y is exponentially distributed, its mean excess function is constant, that is

$$Y \sim \mathcal{E}xp(\tau) \Leftrightarrow e(u) = E[Y] = \frac{1}{\tau}.$$

This is a direct consequence of the memoryless property characterizing the Negative Exponential distribution. Consequently, the plot of $e(u)$ against the threshold will be an horizontal line. Short-tailed distributions will show a downward trend. On the contrary, an upward trend will be an indication of heavy-tailed behavior. A downward trend is apparent for distributions with finite supports.

Usually the mean excess function $e(\cdot)$ is not known but it can be easily estimated from a random sample and this empirical estimator $\widehat{e}_n$ can be plotted. Precisely, the mean excess function can be estimated from a random sample $\{y_1, y_2, \ldots, y_n\}$ by

$$\widehat{e}_n(u) = \frac{\sum_{i=1}^n y_i I[y_i > u]}{\#\{y_i | y_i > u\}} - u = \frac{\sum_{i=1}^n (y_i - u)I[y_i > u]}{\#\{y_i | y_i > u\}}$$

where $\#\{y_i | y_i > u\}$ denotes the number of observations exceeding u and $I[\cdot]$ is the indicator function. In words, $e(u)$ is estimated by the sum of all the excesses over the threshold u divided by the number of data points exceeding the threshold u.

Usually, the mean excess function is evaluated in the observations of the sample. Precisely, denote the sample observations arranged in ascending order (or order statistics) as

$$y_{(1)} \leq y_{(2)} \leq \cdots \leq y_{(n)}.$$

We then have

$$\widehat{e}_n(y_{(k)}) = \frac{1}{n-k} \sum_{j=1}^{n-k} \left(y_{(k+j)} - y_{(k)}\right).$$

9.3.3 Illustration in Nonlife Insurance

Let us consider the claim severities observed in the Belgian motor third-party liability insurance portfolio considered in Chap. 6. The information about claim amounts in the database has been recorded 6 months after the end of the observation period. Hence, most of the "small" claims are settled and their final cost is known. However, for the large claims, we work here with incurred losses (payments made plus reserve). All costs are expressed in Belgian francs, the currency in force in the country in the late 1990s (one euro is about forty Belgian francs).

Descriptive statistics for claim costs are displayed in Table 9.1. We have at our disposal 18,042 observed individual claim costs, ranging from 1 to over 80,000,000 Belgian francs, with a mean of 72,330.7 Belgian francs. We see in Table 9.1 that 25% of the recorded claim costs are below 5,850 Belgian francs, that half of them are smaller than 23,242 Belgian francs, and that 90% of them are less than 121,902 Belgian francs. The remaining 10% of observed claim severities thus extend to over 80 millions, which suggests that the data set contains some very large amounts far in the tail of the distribution.

As the data are considerably skewed to the right, we opt for the log-scale to represent the histogram of claim severities. The histogram displayed in Fig. 9.1 shows that the right part of the distribution reveals the presence of very large losses even on the log-scale. The tail behavior of the data is then further examined with the help of the graph of $\widehat{e}_n$. The empirical mean excess function displayed in Fig. 9.2 demonstrates the heavy-tailed character of the claim severities that materializes in

Table 9.1 Descriptive statistics of the claim costs (only strictly positive values) for a Belgian motor third-party liability insurance portfolio observed in the late 1990s, all cost values expressed in Belgian francs

Number of observations	18,042
Minimum	1
Maximum	80,258,970
Mean	72,331
Standard deviation	710,990.5
25th percentile	5,850
Median	23,242
75th percentile	58,465
90th percentile	121,902
95th percentile	170,530.9
99th percentile	794,900.9

Fig. 9.1 Histogram of the logarithm of claim severities in motor insurance for a Belgian motor third-party liability insurance portfolio observed in the late 1990s

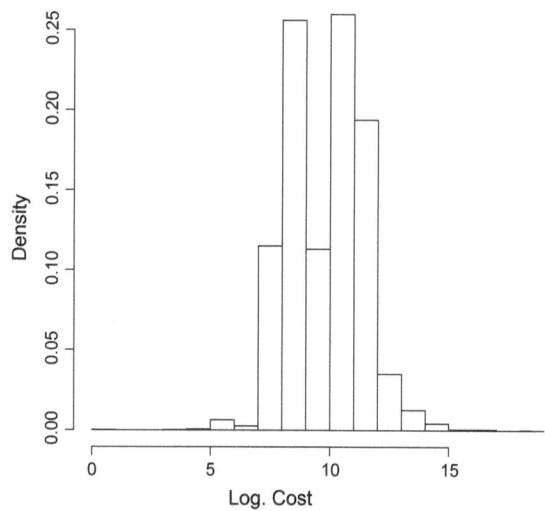

the visible upward trend. This graph has been obtained with the `mrlplot` function of the R package POT.

We know that the mean excess function is constant if the claim severities obey the Negative Exponential distribution. Short-tailed distributions show a downward trend whereas an upward trend indicates a heavy-tailed behavior. The increasing trend of $\widehat{e}_n$ that is clearly visible on Fig. 9.2 supports a heavy-tailed behavior. Therefore, a positive value of tail index ξ is expected.

The Negative Exponential distribution also serves as benchmark in the Exponential QQ-plot. The interpretation of such a plot is easy. If the data comply with the Negative Exponential distribution, the points should lie approximately along a straight line. A convex departure from the linear shape indicates a heavier tailed distribution in the sense that empirical quantiles grow faster than the theoretical ones.

Fig. 9.2 Empirical mean
excess function plot of claim
severities for a Belgian motor
third-party liability insurance
portfolio observed in the late
1990s, all cost values
expressed in Belgian francs

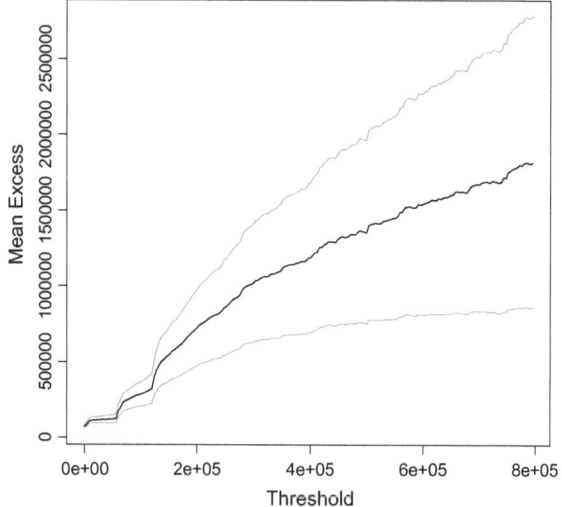

Fig. 9.3 Exponential
QQ-plot of claim severities
for a Belgian motor
third-party liability insurance
portfolio observed in the late
1990s, all cost values
expressed in Belgian francs

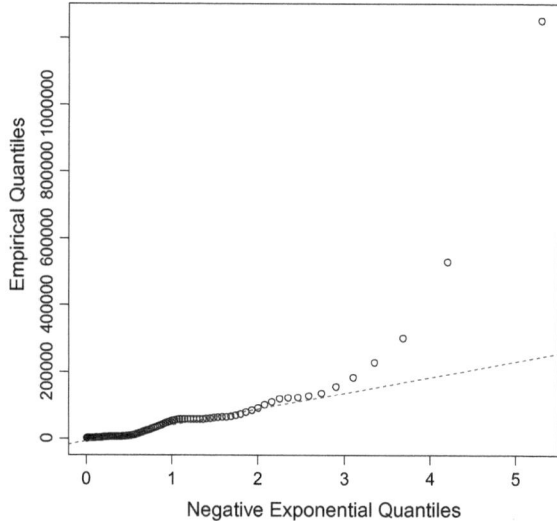

On the contrary, concavity indicates a shorter tailed distribution. The convex pattern
of the Exponential QQ-plot displayed in Fig. 9.3 confirms the heavy-tailed behavior
of the claim severity distribution.

9.3.4   Illustration in Life Insurance

EVT techniques are also useful to study the very end of life tables. Insurers need reliable estimate of mortality at oldest ages to price life annuities and reverse mortgages, for instance. Parametric models have often been used by actuaries to fit mortality statistics, including

– Gompertz model:
$$\mu_x = a \exp(bx) \text{ with } a, b > 0;$$

– Makeham model:
$$\mu_x = c + a \exp(bx) \text{ with } a, b, c > 0.$$

The exponential increase in the force of mortality assumed under the Gompertz and Makeham models is generally appropriate for adult and early old ages but the mortality increases tend to slow down at older ages, typically around 85, so that these specifications usually fail to describe the end of the life table (overestimating mortality at the oldest ages, which appears to be problematic for life annuities, for instance). Alternative models have been proposed to account for the mortality deceleration observed at older ages in industrialized countries. The plateau that appears at these advanced ages is often attributed to the earlier selection of the more robust individuals in heterogeneous cohorts. This empirical finding supports the Logistic model obtained by including a Gamma distributed frailty coefficient accounting for the heterogeneity in the Gompertz model, as explained next.

A frailty is defined as a non-negative random variable whose level expresses the unobservable risk factors affecting individual mortality. The underlying idea is that those people with a higher frailty die on average earlier than others. Let Z denote the frailty corresponding to the lifetime T. Given $Z = z$, the force of mortality in the multiplicative frailty model is then defined as

$$\mu_x(z) = z\mu_x$$

where μ_x represents the force of mortality for an individual with $Z = 1$; μ_x is considered as the standard force of mortality. If $z < 1$, then $\mu_x(z) < \mu_x$, which suggests that the person is in good conditions; vice versa if $z > 1$. Now, assume that $Z \sim \mathcal{G}am(\delta, \theta)$. Referring to adult ages, we can assume the Gompertz model for describing the standard force of mortality so that

$$\mu_x(z) = za \exp(bx).$$

The population force of mortality is then given by

$$\frac{a\delta \exp(bx)}{\left(\theta - \frac{a}{b}\right) + \frac{a}{b} \exp(bx)} = \frac{1}{\theta - \frac{a}{b}} \frac{a\delta \exp(bx)}{1 + \frac{a}{b\theta - a} \exp(bx)}.$$

Table 9.2 Descriptive statistics for the observed ages at death above 95, cohorts born in Belgium from 1886 to 1904

	Males	Females
Number of observations	10,050	36,616
Mean	97.35	97.75
Standard deviation	2.05	2.37
25th percentile	95.77	95.91
Median	96.79	97.12
75th percentile	98.36	98.99
Maximum	111.47	112.58

Defining $\frac{a\delta}{\theta - \frac{a}{b}} = a'$ and $\frac{a}{b\theta - a} = \delta'$, we get the logistic model

$$\frac{a' \exp(bx)}{1 + \delta' \exp(bx)}$$

for the population force of mortality, which produces the mortality plateau mentioned previously.

When the main interest is in the mortality at the oldest ages, inference is conducted in an area of the sample where there is a very small amount of data. Moreover, extrapolation beyond the range of the data is often desirable, a procedure known as the closure of the life table. In such a case, it is essential to have a good model for the longest lifetimes: EVT thus appears as the natural candidate for analyzing mortality at the oldest ages. Also, EVT is closely linked to limiting residual life distributions and offers a unified approach to the modeling of the right tail of a lifetime distribution. For instance, the mean excess function corresponds to the expected remaining lifetime viewed as a function of attained age.

Considering the data analyzed in this section, we have at our disposal the ages at death for every individual who died at age 95 or older in Belgium, plus a list of survivors aged 95 and older at the end of the observation period. There are 46,666 observations relating to individuals born in Belgium who died after age 95 since 1981. The data basis comprises 22% males and 78% females. For each individual, we know the exact birth and death dates.

Basic descriptive statistics of the data sets are given in Table 9.2. The average age at death for women is greater than the one for men, as expected. However, the differences are rather modest, compared to the differences observed for the total lifetime from birth, except for the much higher number of females reaching age 95 compared to males.

Cohort-specific data are displayed in Table 9.3, for each gender separately. We can read there the initial number of individuals included in the analysis, for each extinct cohort born between 1886 and 1904 (that is, the number L_{95} of individuals born in calender year $c \in \{1886, \ldots, 1904\}$ still alive at age 95) as well as the age at death $M_{L_{95}}$ of the last survivor for each cohort. We can see there that L_{95} increases as time passes, as expected from the combined effect of the increasing population size and decreasing mortality at younger ages over time. Also, the number of individuals

Table 9.3 Initial size for each cohort L_{95}, observed highest age at death $M_{L_{95}}$ and mean age at death $95 + \widehat{e}(95)$ for each cohort born in calendar year c from 1886 to 1904

Cohort	Males			Females		
c	L_{95}	$M_{L_{95}}$	$95 + \widehat{e}(95)$	L_{95}	$M_{L_{95}}$	$95 + \widehat{e}(95)$
1886	359	108.17	97.35	1045	107.78	97.70
1887	418	105.13	97.41	1130	110.45	97.64
1888	412	106.33	97.20	1242	110.32	97.68
1889	462	105.58	97.53	1208	110.16	97.69
1890	425	107.70	97.50	1311	112.58	97.60
1891	428	105.81	97.35	1368	109.72	97.70
1892	444	105.44	97.29	1510	110.89	97.72
1893	492	110.29	97.46	1544	107.75	97.63
1894	498	106.19	97.22	1760	107.41	97.71
1895	526	106.62	97.24	1852	109.38	97.86
1896	545	106.27	97.43	1894	109.79	97.85
1897	568	106.43	97.33	2009	109.85	97.72
1898	572	105.74	97.26	2149	110.89	97.71
1899	630	106.88	97.35	2301	111.60	97.72
1900	576	111.47	97.27	2460	111.70	97.78
1901	635	103.77	97.23	2787	110.36	97.89
1902	632	106.79	97.42	2829	112.36	97.82
1903	669	104.71	97.46	2980	109.96	97.81
1904	759	106.15	97.32	3237	110.18	97.74

entering the database is larger for females compared to males because of the lower mortality levels for women compared to men at all ages. The maxima $M_{L_{95}}$ are relatively stable over cohorts, and are higher for females compared to males. Notice that for each cohort, the last survivor was a female. The highest ages at death observed in Belgium for these cohorts are 112.58 for females and 111.47 for males. Recall[1] that Jeanne Calment died at the age of 122 years in 1997. After this record, only a single person, Sarah Knauss has lived for 119 years, and died in 1993. Since then, three women lived more than 117 years and three others are still alive at 116 years old. Given the limited size of the Belgian population, the maxima contained in the database are in line with these world records.

We also see from Table 9.3 that there is no visible trend in the average age at death, or shifted life expectancy $95 + \widehat{e}(95)$ for individuals alive at age 95. It even remains stable around 97 for all male and female cohorts. The observation that $\widehat{e}(95)$ did not improve across the 19 birth cohorts is quite remarkable. This seems to suggest no

---

[1] Even if some controversy arises about her age at death, we keep here Jeanne Calment as the world record of longevity.

Fig. 9.4 Histograms of the observed ages at death above 98, cohorts born in Belgium from 1886 to 1904, for males (top panel) and females (bottom panel)

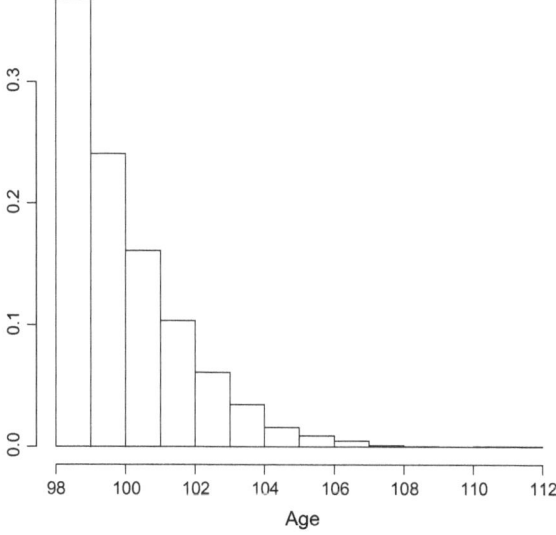

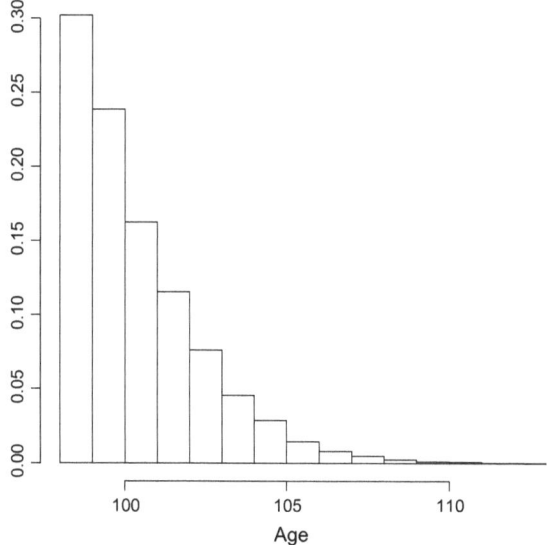

potential gain of extreme longevity through generations at these advanced ages. This can be confirmed using a formal test (as performed in Gbari et al. 2017a).

Let us now restrict the data set to ages 98 and over. The reason is as follows: market mortality statistics are available from the Belgian National Bank, acting as the insurance regulator, up to age 98 (the last category is open, gathering ages 99 and over). Because the aim here is to provide actuaries with an extrapolation of the market life tables to older ages, it is therefore natural to concentrate on ages

Fig. 9.5 Empirical expected
remaining lifetimes
$x \mapsto \widehat{e}(x)$ for males (upper
panel) and females (lower
panel)

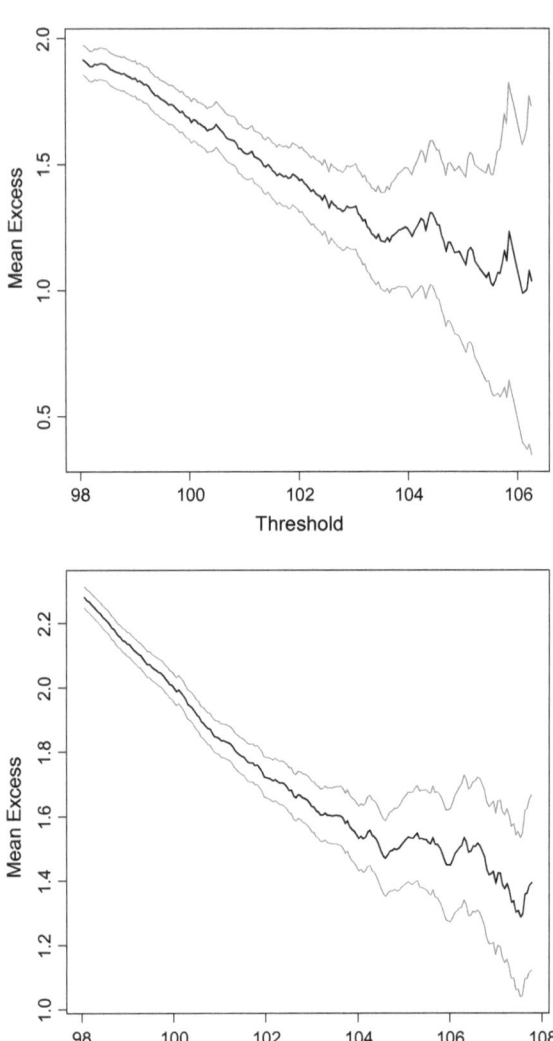

not covered by the market life tables. Histograms are displayed in Fig. 9.4. These
graphs have been obtained from the observed ages at death $\{t_1, t_2, \ldots, t_n\}$ with $t_i \geq$
98 comprised in the database. We can see there that these histograms suggest that
remaining lifetimes beyond age 98 possess a decreasing probability density function,
with some observations located far in the tail of the distributions.

The empirical version of the remaining life expectancy $\widehat{e}(x)$ viewed as a function
of attained age x is displayed in Fig. 9.5. We know that if the remaining lifetimes
obey the Negative Exponential distribution, then the mean excess function is constant.
Consequently, the plot of $e(x)$ versus age x will be an horizontal line. Short-tailed

Fig. 9.6 Negative
Exponential QQ-plot for
lifetimes for males (upper
panel) and females (lower
panel)

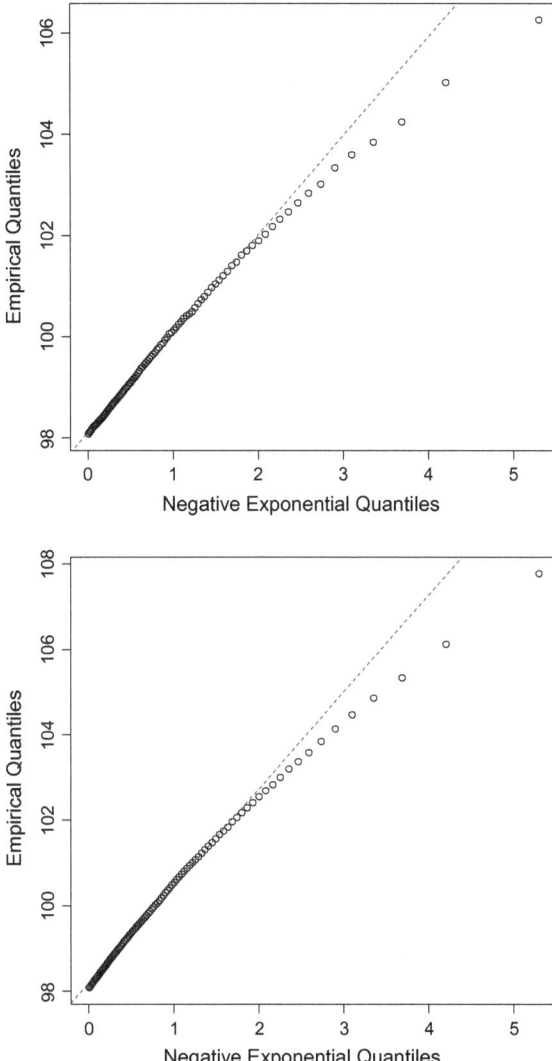

distributions will show a downward trend. On the contrary, an upward trend will be an indication of heavy-tailed behavior. The decreasing shape of $\widehat{e}(x)$ visible in Fig. 9.5 contradicts the Negative Exponential behavior of the remaining lifetime. A downward trend is clearly visible on Fig. 9.5, supporting short-tailed behavior. Therefore, a negative value of tail index ξ is expected, resulting in a finite ultimate age ω.

The comparison to Negative Exponentially distributed lifetimes can be performed with the help of an Exponential QQ-plot, as explained before with motor insurance claim severities. If the data are an independent and identically distributed sample from a Negative Exponential distribution, a linear trend should be visible. A convex departure from this shape indicates a heavier tailed distribution in the sense that empirical quantiles grow faster than the theoretical ones. On the contrary, concavity indicates a shorter tailed distribution. The concave pattern of the Exponential QQ-plot visible in Fig. 9.6 confirms the short-tailed behavior of the lifetime distribution for both genders.

9.4 Generalized Pareto Distribution

9.4.1 Definition

Generalized Pareto distributions play an important role in EVT, in relation with the distribution of excesses over high thresholds. This family of continuous probability distributions involves two parameters: a scale parameter τ and a shape parameter ξ. Sometimes it also allows for a location parameter. Precisely, the distribution function $G_{\xi;\tau}$ of the Generalized Pareto distribution $\mathcal{GPar}(\xi, \tau)$ is given by

$$
G_{\xi;\tau}(y) = \begin{cases} 1 - \left(1 + \xi\frac{y}{\tau}\right)_+^{-1/\xi} & \text{if } \xi \neq 0 \\ 1 - \exp\left(-\frac{y}{\tau}\right) & \text{if } \xi = 0 \end{cases}
$$

where $\tau > 0$ and

$$
y \in (0, \infty) \text{ when } \xi \geq 0
$$

$$
y \in \left(0, -\frac{\tau}{\xi}\right) \text{ when } \xi < 0.
$$

As particular cases of the Generalized Pareto distributions, we find the Pareto distribution when $\xi > 0$, the type II Pareto distribution when $\xi < 0$ and the Negative Exponential distribution when $\xi = 0$. The corresponding probability density function is

$$
g_{\xi;\tau}(y) = \begin{cases} \frac{1}{\tau}\left(1 + \xi\frac{y}{\tau}\right)_+^{-1/\xi-1} & \text{if } \xi \neq 0 \\ \frac{1}{\tau}\exp\left(-\frac{y}{\tau}\right) & \text{if } \xi = 0. \end{cases}
$$

9.4.2 Properties

9.4.2.1 Mixture of Negative Exponential Distributions

When $\xi > 0$, the Generalized Pareto distribution can be obtained as a mixture of Negative Exponential distributions, with a Gamma mixing. Precisely, assume that given $\Lambda = \lambda$, the random variable Y obeys the Negative Exponential distribution $\mathcal{E}xp(\lambda)$. Further, assume that the random variable Λ is Gamma distributed $\mathcal{G}am(\alpha, \beta)$. Then, unconditionally, Y obeys the Generalized Pareto distribution with parameters $\xi = 1/\alpha$ and $\tau = \beta/\alpha$. Notice however, that since the parameter α for the Gamma distribution must be positive, this construction only applies to the case $\xi > 0$. This extends the corresponding property of the Pareto distribution to the Generalized Pareto distribution with positive shape parameter ξ.

9.4.2.2 Moments

Provided $\xi < 1$, the mean value for $Y \sim \mathcal{G}Par(\xi, \tau)$ is

$$E[Y] = \frac{\tau}{1 - \xi}.$$

If $Y \sim \mathcal{G}Par(\xi, \tau)$ and $\xi > 0$ then

$$E[Y^k] = \infty \text{ when } k \geq \frac{1}{\xi}.$$

Whatever $\xi > 0$, some moments of a response obeying the Generalized Pareto distribution become infinite when the power becomes sufficiently large.

9.4.2.3 Threshold Stability

If $Y \sim \mathcal{G}Par(\xi, \tau)$ then the excess function over the threshold u corresponding to $F = G_{\xi;\tau}$ is given by

$$\overline{F}_u(y) = \frac{\left(1 + \xi \frac{y+u}{\tau}\right)^{-1/\xi}}{\left(1 + \xi \frac{y}{\tau}\right)^{-1/\xi}}$$
$$= \left(\frac{\tau + \xi(y + u)}{\tau + \xi u}\right)^{-1/\xi}$$
$$= \left(1 + \xi \frac{y}{\tau + \xi u}\right)^{-1/\xi}$$

so that

$$F_u(y) = G_{\xi, \tau(u)}(y) \text{ with } \tau(u) = \tau + \xi u$$

where

$$y \geq 0 \text{ when } \xi \geq 0 \text{ and } 0 \leq y \leq -\frac{\tau}{\xi} - u \text{ when } \xi < 0.$$

In words, if the excesses of Y over u obey the $\mathcal{GP}ar(\xi, \tau)$ distribution then the excesses of Y over a larger threshold $v > u$ still conform to the Generalized Pareto distribution with the same shape parameter ξ but an updated scale parameter $\tau + \xi(v - u)$.

In particular, if $Y \sim \mathcal{E}xp(\tau) = \mathcal{GP}ar(0, \tau)$, that is, $F(y) = 1 - \exp(-y/\tau)$, then we find the identity

$$F_u(x) = F(x) \text{ for all } u > 0.$$

We thus recover the memoryless property which characterizes the Negative Exponential distribution.

9.4.2.4   Mean-Excess Function

As the excesses over a threshold remain Generalized Pareto distributed, the mean excess function can easily be obtained from the formula for the expected value associated to the Generalized Pareto distribution. Precisely, the $\mathcal{GP}ar(\xi, \tau)$ distribution with $\xi < 1$ has a mean excess function of the form

$$e(u) = \frac{\tau(u)}{1 - \xi} = \frac{\tau}{1 - \xi} + \frac{\xi}{1 - \xi} u$$

where $u \geq 0$ if $0 \leq \xi < 1$ and $0 \leq u \leq -\frac{\tau}{\xi}$ if $\xi < 0$. Therefore, the linearity of the mean excess function characterizes the Generalized Pareto distribution.

Based on the linear behavior of the Generalized Pareto mean excess function, a natural idea to detect such a behavior is to determine on the basis of the graph of the empirical mean excess function $\widehat{e}_n$ a region (u, ∞) where $\widehat{e}_n(t)$ becomes approximately linear for $t \geq u$. This indicates that the excesses obey the $\mathcal{GP}ar(\xi, \tau)$ distribution in the tail area above u. This also suggests graphical estimates of $\xi < 1$ and τ from the slope and the intercept of the ultimate linear trend of $\widehat{e}_n$.

9.5   Peak Over Threshold Approach

9.5.1 *Principle*

The traditional approach to EVT is based on extreme value limit distributions. Here, a model for extreme losses is based on the possible parametric form of the limit distribution of maxima. A more flexible model is known as the "Peak Over

Threshold" (POT in short) method. This approach appears as an alternative to maxima analysis for studying extreme behaviors. Essentially, POT analyzes the series of excesses over a high threshold u.

It turns out that the Generalized Pareto distribution provides the actuary with an approximation to the excess distribution F_u over sufficiently large thresholds.

Theorem 9.5.1 (Pickands-Balkema-de Haan Theorem) *The Fisher-Tippett theorem holds with H_ξ if, and only if, we can find a positive function $\tau(u)$ such that*

$$\lim_{u \to \omega} \sup_{0 \leq y \leq \omega - u} \left| F_u(y) - G_{\xi, \tau(u)}(y) \right| = 0.$$

Theorem 9.5.1 indicates that the $\mathcal{GP}ar(\xi, \tau)$ distribution provides a good approximation to the distribution of the excesses over sufficiently large thresholds u. In practice, for some function $\tau(u)$ and some Pareto index ξ depending on F, we can use the approximation

$$\overline{F}_u(y) \approx 1 - G_{\xi; \tau(u)}(y), \quad y \geq 0,$$

provided that u is sufficiently large. We then obtain the useful approximation

$$\overline{F}(u + z) \approx \overline{F}(u) \left(1 - G_{\hat{\xi}; \hat{\tau}}(z) \right)$$

that appears to be accurate for a sufficiently large threshold u and any $z \geq 0$. Tail formulas based on the Generalized Pareto distributions such as those derived in the next section thus appear to be useful when the actuary deals with extreme values of the response.

9.5.2 Tail Formulas

If the tail distribution of Y is Generalized Pareto then the actuary is able to derive a host of useful identities, as shown next. If $F_u(y) = G_{\xi, \tau}(y)$ for $0 \leq y < \omega - u$ as suggested by the Pickands-Balkema-de Haan theorem then for $y \geq u$,

$$
\begin{aligned}
P[Y > y] &= P[Y > u]P[Y > y | Y > u] \text{ for } x \geq u \\
&= \overline{F}(u)P[Y - u > y - u | Y > u] \\
&= \overline{F}(u)\overline{F}_u(y - u) \\
&= \overline{F}(u) \left(1 + \xi \frac{y - u}{\tau} \right)^{-1/\xi}.
\end{aligned}
$$

This formulas is useful for the calculation of high quantiles of the response. The quantile function of the $\mathcal{GP}ar(\xi, \tau)$ distribution is given by

$$G_{\xi;\tau}^{-1}(p) = \frac{\tau}{\xi}\left((1-p)^{-\xi} - 1\right), \quad 0 < p < 1.$$

Now, for $p \geq F(u)$, the quantile of Y at probability level p (often called the Value-at-Risk in risk management applications) is obtained as the solution z to the equation

$$1 - p = \overline{F}(u)\left(1 + \xi\frac{z-u}{\tau}\right)^{-1/\xi}.$$

This gives

$$F_Y^{-1}(p) = u + \frac{\tau}{\xi}\left(\left(\frac{1-p}{\overline{F}(u)}\right)^{-\xi} - 1\right).$$

The mean response when the Value-at-Risk has been exceeded is also a very useful indicator in risk management, to quantify the losses in the adverse scenarios. This indicator is known as the conditional tail expectation. If $\xi < 1$ then for $p \geq F(u)$, the conditional tail expectation is given by

$$
\begin{aligned}
E\left[Y \mid Y > F_Y^{-1}(p)\right] &= F_Y^{-1}(p) + E\left[Y - F_Y^{-1}(p) \mid Y > F_Y^{-1}(p)\right] \\
&= F_Y^{-1}(p) + E\left[Y - u - \left(F_Y^{-1}(p) - u\right) \mid Y - u > F_Y^{-1}(p) - u\right] \\
&= F_Y^{-1}(p) + \frac{\tau + \xi\left(F_Y^{-1}(p) - u\right)}{1 - \xi} \\
&= \frac{F_Y^{-1}(p)}{1 - \xi} + \frac{\tau - \xi u}{1 - \xi}
\end{aligned}
$$

where $F_Y^{-1}(p)$ has been derived in the previous formula.

9.5.3   Tail Estimators

Pickands-Balkema-de Haan theorem shows that (provided u is sufficiently large) a potential estimator for the excess-of-loss distribution $F_u(x)$ is $G_{\hat{\xi};\hat{\tau}}(x)$. The selection of the appropriate threshold u is discussed in the next sections. So, $G_{\hat{\xi};\hat{\tau}}(x)$ approximates the conditional distribution of the losses, given that they exceed the threshold u. Quantile estimators derived from this curve are conditional quantile estimators which indicate the scale of losses which could be experienced if the threshold u were to be exceeded. When estimates of the unconditional quantiles are of interest, it is necessary to relate the unconditional cumulative distribution function F to $G_{\hat{\xi};\hat{\tau}}$ through F_u.

Denote the number of claims above the threshold u as

$$N_u = \sum_{i=1}^{n} \mathrm{I}[Y_i > u] \sim \mathcal{B}in\left(n, \overline{F}(u)\right).$$

Provided we have a large enough sample, we can accurately estimate $\overline{F}(u)$ by its empirical counterpart N_u/n. For $x > u$, $\overline{F}(x) = \overline{F}(u)\overline{F}_u(x - u)$ so that we can estimate $\overline{F}(x)$ with the help of

$$\widehat{\overline{F}}(x) = \frac{N_u}{n}\left(1 + \widehat{\xi}\frac{x - u}{\widehat{\tau}}\right)^{-1/\widehat{\xi}}.$$

High quantiles contain useful information for insurers about the distribution of the claim amounts. Usually quantiles can be estimated by their empirical counterparts but when we are interested in the very high quantiles, this approach is not longer valid since estimation based on a low number of large observations would be strongly imprecise. The Pickands-Balkema-de Haan theorem suggests the following estimates. For $p \geq F(u)$, the quantile at probability level p can be estimated from

$$\widehat{F}^{-1}(p) = u + \frac{\widehat{\tau}}{\widehat{\xi}}\left(\left(\frac{n(1 - p)}{N_u}\right)^{-\widehat{\xi}} - 1\right).$$

If $\xi < 1$ then the conditional tail expectation can be estimated from

$$\widehat{\mathrm{E}}[Y|Y > F^{-1}(p)] = \frac{\widehat{F}^{-1}(p)}{1 - \widehat{\xi}} + \frac{\widehat{\tau} - \widehat{\xi}u}{1 - \widehat{\xi}}$$

where we insert the aforementioned expression for $\widehat{F}^{-1}(p)$.

9.5.4 Applications to Remaining Lifetimes

The analysis of residual lifetimes at high ages is in line with the POT method where the threshold u corresponds to some advanced age x. Transposed to the life insurance setting, the remaining lifetime $T - x$ at age x, given $T > x$, is distributed according to

$$s \mapsto {}_s q_x = \mathrm{P}[T - x \leq s | T > x].$$

It may happen that for large attained ages x, this conditional probability distribution stabilizes after a normalization, that is, there exists a positive function $\tau(\cdot)$ such that

$$\lim_{x \to \omega} \mathrm{P}\left[\frac{T - x}{\tau(x)} > s \,\middle|\, T > x\right] = 1 - G(s), \quad s > 0, \tag{9.4}$$

where G is a non-degenerate distribution function. Only a limited class of distribution functions are eligible in (9.4), namely the Generalized Pareto $G_{\xi,\tau}$ ones. When $\xi = 0$, the remaining lifetimes at high ages become ultimately Negative Exponentially distributed so that the forces of mortality stabilize.

For some appropriate function $\tau(\cdot)$, the approximation

$$_sq_x \approx G_{\xi;\tau(x)}(s) \text{ for } s \geq 0 \qquad (9.5)$$

holds for x large enough. The approximation (9.5) is justified by the Pickands-Balkema-de Haan theorem. In view of (9.5) the remaining lifetimes at age x can be treated as a random sample from the Generalized Pareto distribution provided x is large enough.

If $\xi < 0$, so that $\omega < \infty$, then a suitable transformation of the extreme value index ξ possesses an intuitive interpretation. Recall that the mean-excess function

$$e(x) = \mathrm{E}\left[T - x | T > x\right]$$

coincides with the remaining life expectancy at age x, denoted as e_x. It can be shown that for $\xi < 0$, (9.1) is equivalent to

$$\lim_{x \to \omega} \mathrm{E}\left[\frac{T - x}{\omega - x} \middle| T > x\right] = \lim_{x \to \omega} \frac{e(x)}{\omega - x} = -\frac{\xi}{1 - \xi} = \alpha$$

The parameter $\alpha = \alpha(\xi)$ is referred to as the perseverance parameter. The intuitive interpretation of the perseverance parameter is as follows. Let us consider an individual who is still alive at some advanced age x. The ratio $\frac{T-x}{\omega-x}$ represents the percentage of the actual remaining lifetime $T - x$ to the maximum remaining lifetime $\omega - x$. This percentage stabilizes, on average when $x \to \omega$ and converges to α, which thus appears as the expected percentage of the maximum possible remaining lifetime effectively used by the individual.

Provided the selected threshold age $x^\star$ is sufficiently large, a potential estimator for the remaining lifetime distribution $_sq_{x^\star}$ is $G_{\widehat{\xi};\widehat{\tau}}(s)$. Quantile estimators derived from this curve are conditional quantile estimators which indicate the potential survival beyond the threshold age $x^\star$ when it is attained. If $\widehat{\xi} < 0$ then individuals possess bounded remaining lifetimes with estimated ultimate age

$$\widehat{\omega} = x^\star - \frac{\widehat{\tau}}{\widehat{\xi}}.$$

When estimates of the unconditional quantiles are of interest, we can relate the unconditional distribution function $_xq_0$ to $G_{\widehat{\xi};\widehat{\tau}}$ for

$$x^\star < x \leq \widehat{\omega} = x^\star - \frac{\widehat{\tau}}{\widehat{\xi}},$$

through

$$x q_0 = 1 - {}_x p_0$$
$$= 1 - {}_{x^\star} p_0 \times {}_{x-x^\star} p_{x^\star}$$
$$\approx 1 - {}_{x^\star} p_0 \left(1 - G_{\widehat{\xi};\widehat{\tau}}(x - x^\star)\right).$$

High-level quantiles then correspond to solutions z to the equations

$$z q_0 = 1 - \epsilon$$

where the approximation just derived is useful for small probability levels ϵ.

Provided the sample size is large enough, we can estimate ${}_{x^\star} p_0$ by its empirical counterpart. In the example considered in this chapter, we start with individuals aged 95. The survival probability to age x such that $x^\star - \frac{\widehat{\tau}}{\widehat{\xi}} \geq x > x^\star$ can then be estimated by

$$_x \widehat{p}_{95} = \frac{L_{x^\star}}{L_{95}} \left(1 - G_{\widehat{\xi};\widehat{\tau}}(x - x^\star)\right)$$

where $L_{x^\star}$ and L_{95} are the number of survivors at the threshold age $x^\star$ and at age 95 (i.e., the total number of individuals under study for a given cohort), respectively. The high quantiles are then estimated as

$$\widehat{F}^{-1}(\epsilon) = x^\star + \frac{\widehat{\tau}}{\widehat{\xi}} \left(\left(\frac{L_{95}}{L_{x^\star}}(1 - \epsilon)\right)^{-\widehat{\xi}} - 1\right). \tag{9.6}$$

Furthermore, for ages x such that $x^\star - \frac{\widehat{\tau}}{\widehat{\xi}} \geq x > x^\star$, the force of mortality can be obtained from the Generalized Pareto formula

$$\widehat{\mu}_x = \frac{1}{\widehat{\tau} + \widehat{\xi}(x - x^\star)}.$$

Notice that $\widehat{\mu}_x$ diverges to ∞ when age x approaches the finite endpoint $\widehat{\omega}$ because no individual is allowed to survive beyond age ω.

9.5.5 Selection of the Generalized Pareto Threshold

9.5.5.1 Principle

Selecting the appropriate threshold u beyond which the Generalized Pareto approximation applies is certainly a very difficult task. In this section, we provide the reader with some general principles in that respect, referring to the literature for particular approaches applying in specific situations. Often, the techniques for threshold selection only provide a range of reasonable values, so that simultaneous application of them is highly recommended in order to get more reliable results.

Two factors have to be taken into account in the choice of an optimal threshold u beyond which the Generalized Pareto approximation holds for the distribution of the excesses:

– A too large value for u yields few excesses and consequently volatile upper quantile estimates. The actuary also loses the possibility to estimate smaller quantiles.
– A too small value for u implies that the Generalized Pareto character does not hold for the moderate observations and it yields biased quantile estimates. This bias can be important as moderate observations usually constitute the largest proportion of the sample.

Thus, our aim is to determine the minimum value of the threshold beyond which the Generalized Pareto distribution becomes a reasonable approximation to the tail of the distribution under consideration.

We know that the Generalized Pareto distribution enjoys the convenient threshold stability property which ensures that

$$Y \sim \mathcal{GP}ar(\xi, \tau) \Rightarrow \mathrm{P}[Y - u > t | Y > u] = 1 - G_{\xi, \tau + \xi u}(t) \qquad (9.7)$$

with the same index parameter ξ, for any $u > 0$. This basically says that provided Y obeys the Generalized Pareto distribution, the excesses of Y over any threshold remain distributed according to the Generalized Pareto. This property is exploited by the majority of the procedures to select the optimal threshold u beyond which the Generalized Pareto approximation applies.

The stability property (9.7) of the Generalized Pareto distribution ensures that the plot of the estimators $\widehat{\xi}$ computed with increasing thresholds reveals estimations that stabilize when the smallest threshold for which the Generalized Pareto behavior holds is reached. This can be checked graphically and allows the actuary to determine the smallest threshold u above which Generalized Pareto distribution provides a good approximation of the tail.

9.5.5.2   Application to Claim Severities

In line with market practice, we concentrate on severities exceeding 1,000,000 Belgian francs (approximately 25,000€), which corresponds to 13.83 times the observed mean claim severity. The 97.5% quantile of the Gamma distribution with mean and variance matching their empirical counterparts is 360,481.7 Belgian francs, whereas the 99% and 99.5% quantiles are respectively equal to 1,937,007 and to 4,004,094 Belgian francs. The corresponding values for the Inverse Gaussian distribution are 409,671.3 Belgian francs, 1,403,344 Belgian francs, and 2,952,146 Belgian francs. Thus, we see that ED distributions used to model moderate losses extend well beyond

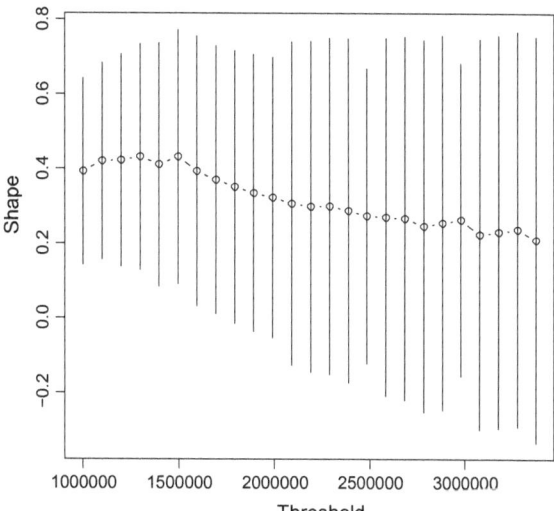

Fig. 9.7 Pareto index plot for the claim severities from a Belgian motor third-party liability insurance portfolio observed in the late 1990s, all cost values expressed in Belgian francs: the estimated shape parameter $\widehat{\xi}$ is plotted against the threshold used to isolate the observations entering the estimation, starting from 1,000,000 Belgian francs

the smallest threshold considered if the POT analysis starts at 1,000,000 Belgian francs.

The Pareto index plot for claim severities in motor insurance is displayed in Fig. 9.7. It has been produced with the help of the `tcplot` function of the R package `POT`. We can see there that estimated indices appear to be relatively stable as they should according to the Pickands-Balkema-de Haan theorem. The confidence intervals are however quite wide. Overall, this graph does not provide the actuary with a lot of guidance for selecting the threshold.

The Gertensgarbe plot is another graphical tool that can be used to estimate the threshold defining the large losses. It is based on the assumption that the optimal threshold can be found as a change point in the series of spacings between ordered claim costs. The key idea is that it may be reasonably expected that the behavior of the differences corresponding to the extreme observations will be different from the one corresponding to the non-extreme observations. So there should be a change point if the POT analysis applies, and this change point can be identified with the help of a sequential version of the Mann-Kendall test as the intersection point between a normalized progressive and retrograde rank statistics. It is implemented in `ggplot` function of the R package `tea`. Precisely, given the series of differences $\Delta_i = y_{(i)} - y_{(i-1)}$, the starting point of the extreme region will be detected as a change point of the series $\{\Delta_i, \ i = 2, 3, \ldots, n\}$. In this test, two normalized series U_p and U_r are determined, first based on the series $\Delta_1, \Delta_2, \ldots$ and then on the series of the differences from the end to the start, $\Delta_n, \Delta_{n-1}, \ldots$, instead of from the start to the

Fig. 9.8 Gertensgarbe plot
for claim severities from a
Belgian motor third-party
liability insurance portfolio
observed in the late 1990s.
Normalized progressive U_p
and retrograde U_r rank
statistics and associated
intersection point

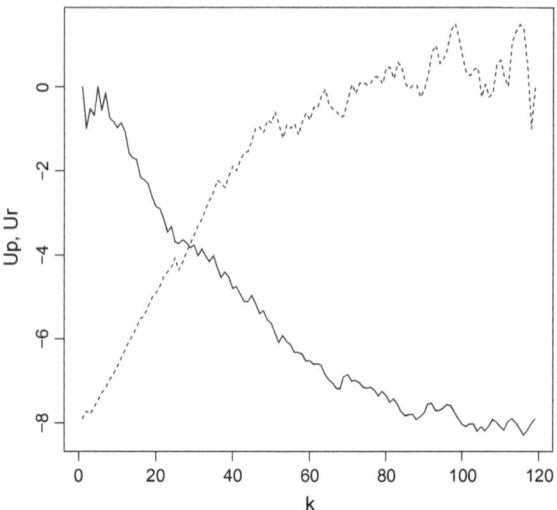

end. The intersection point between these two series determines a candidate change
point that will be significant if it exceeds a high Normal percentile.

The Gertensgarbe plot is displayed in Fig. 9.8. It gives a threshold of 2,471,312
Belgian francs (so that 29 claims qualify as being large losses). The p-value of the
Mann-Kendall test is equal to 1.299018×10^{-4}, so that the result is significant. This
is in line with the conclusion drawn from the Pareto index plot displayed in Fig. 9.7 so
that we select a threshold $u = 2,471,312$ for the claim severities in motor insurance.

Now that the threshold value has been selected, there are several methods to estimate the corresponding Generalized Pareto parameters. Let us mention maximum-
likelihood (with or without penalty), (un)biased probability weighted moments,
moments, Pickands', minimum density power divergence, medians, and maximum
goodness-of-fit estimators, all available from the `fitgpd` function of the R package `POT`. Here, we adopt the maximum-likelihood approach. To fit the Generalized
Pareto model to the excesses over the threshold u, we thus maximize the likelihood
function

$$\mathcal{L}(\xi, \tau) = \prod_{i|y_i > u} \left(\frac{1}{\tau} \left(1 + \frac{\xi}{\tau}(y_i - u) \right)^{-\frac{1}{\xi} - 1} \right)$$

or the corresponding log-likelihood function

$$L(\xi, \tau) = \ln \mathcal{L}(\xi, \tau) = -N_u \ln \tau - \left(1 + \frac{1}{\xi} \right) \sum_{i|y_i > u} \ln \left(1 + \frac{\xi}{\tau}(y_i - u) \right)$$

where $N_u = \#\{y_i|y_i > u\}$ is the number of large claims observed in the portfolio as introduced before.

This optimization problem requires numerical algorithms and appropriate starting values for the parameters ξ and τ. The mean and the variance of the Generalized Pareto distribution are respectively given by $\frac{\tau}{1-\xi}$ provided $\xi < 1$, and by $\frac{\tau^2}{(1-\xi)^2(1-2\xi)}$ provided $\xi < 1/2$. Initial values by the method of moments are

$$\widehat{\xi}_0 = \frac{1}{2}\left(1 - \frac{\bar{y}^2}{s^2}\right) \text{ and } \widehat{\tau}_0 = \frac{1}{2}\bar{y}\left(\frac{\bar{y}^2}{s^2} + 1\right),$$

where $\bar{y}$ and s^2 are the sample mean and variance. The values of τ and ξ coming from the linear fit to the right part of the graph of the empirical mean excess function could also be used.

With the threshold selected at 2,471,312 Belgian francs, we get $\widehat{\xi} = 0.2718819$ so that we face a heavy-tailed distribution. The corresponding $\widehat{\tau} = 7,655,438$. Notice that $\widehat{\xi}$ indeed corresponds to the plateau visible on the Pareto index plot displayed in Fig. 9.7.

9.5.5.3 Application to Life Insurance

Let us now consider the modeling of extreme lifetimes. The Pareto index plots for lifetimes are displayed in Fig. 9.9, separately for males and females. The graphical tools are more difficult to interpret compared to claim severities. We refer the reader to Gbari et al. (2017a) for the determination of the threshold age $x^\star$ such that the approximation (9.5) is sufficiently accurate for $x \geq x^\star$, by means of several automated procedures. According to their conclusions, we can set $x^\star$ to 98.89 for males and to 100.89 for females. These values appear to be reasonable considering the Pareto index plots displayed in Fig. 9.9. Maximum-likelihood estimates of the Generalized Pareto parameters can be found in Table 9.4 together with standard errors.

Table 9.4 Maximum likelihood estimates for the Generalized Pareto parameters τ and ξ, together with standard errors

Parameters	Male	Female
$x^\star$	98.89	100.89
$L_{x^\star}$	1,940	4,104
$\widehat{\xi}$	−0.132	−0.092
s.e $(\widehat{\xi})$	0.015	0.014
$\widehat{\tau}$	2.098	2.019
s.e $(\widehat{\tau})$	0.057	0.042

Fig. 9.9 Pareto index plot:
the estimated shape
parameter $\widehat{\xi}$ is plotted against
the threshold used to isolate
the observations entering the
estimation. Males appear in
the upper panel, females in
the lower panel

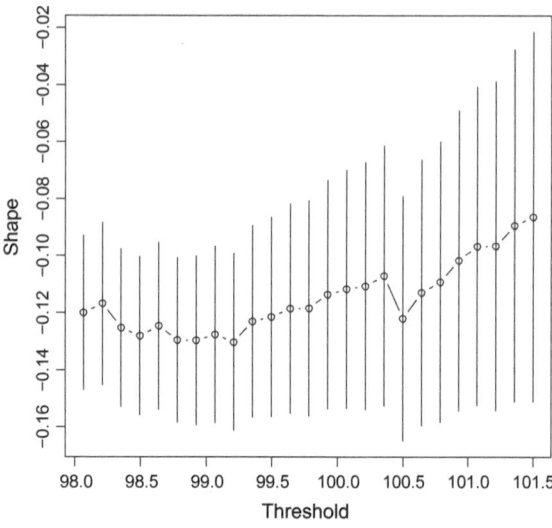

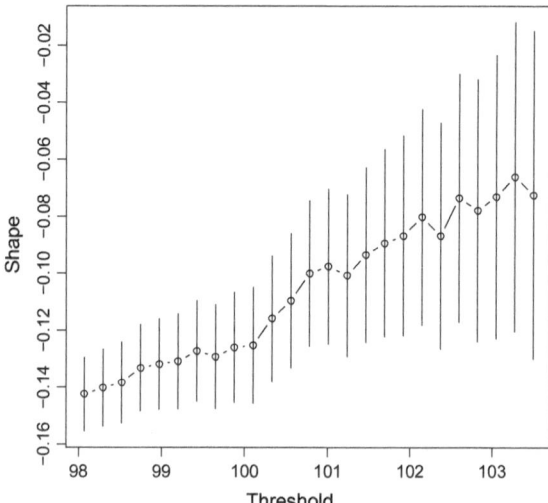

The estimated ultimate age ω is given by

$$\widehat{\omega} = x^{\star} - \frac{\widehat{\tau}}{\widehat{\xi}} = \begin{cases} 114.82 \text{ for males,} \\ \\ 122.73 \text{ for females.} \end{cases} \tag{9.8}$$

The female population has the highest estimated ultimate age. Estimations are in line
with the highest ages at death 112.58 recorded for females and 111.47 recorded for
males for the considered cohorts born in Belgium. These results are consistent with

the worldwide maximum observed ages at death. Interestingly, the obtained ultimate age for females is close to Jeanne Calment record of 122.42 (122 years and 164 days, being born in Arles, France, on February 21, 1875, and died at the same place on August 4, 1997).

Every goodness-of-fit measure can be used in order to check whether the data comply with the Generalized Pareto distribution beyond the selected threshold. The Generalized Pareto QQ-plot is often used to that end. In this plot the empirical quantiles are represented against the estimated Generalized Pareto quantiles. If the Generalized Pareto distribution effectively fits the data under consideration then the plotted pairs must be located close to the 45-degree line. The qqgpd function of the R package tea can be used to plot the empirical observations above a given threshold against the theoretical quantiles of a Generalized Pareto distribution. The Generalized Pareto QQ-plots for the extreme lifetimes are displayed in Fig. 9.10. The marked linear pattern in the QQ-plot confirms that the Generalized Pareto model adequately describes the remaining lifetime distribution above x^*.

To conclude this application to life insurance, let us discuss the difference with the analyses conducted on abridged mortality data. In demography, mortality levels are usually assessed based on statistical data aggregated by attained age, and not individual ages at death. This is especially the case with general population data. The actuary only knows the observed numbers L_x of individuals reaching age x, the corresponding death counts $D_x = L_x - L_{x+1}$ and the corresponding exposure E_x at this age. The available data are thus now those displayed in Table 9.5.

The Generalized Pareto model can be estimated on aggregated data (L_x, D_x), $x \geq \lceil x^* \rceil$, where $\lceil x^* \rceil$ is the lowest integer greater or equal to x^*. Considering ages $x \geq x^*$, the one-year survival probabilities are obtained from the approximation (9.5), which gives

$$p_x = p_x(\xi, \tau) = \left(1 + \frac{\xi}{\tau + \xi(x - x^*)}\right)^{-1/\xi}.$$

The Generalized Pareto parameters can be estimated in the conditional model

$$D_x \sim \mathcal{B}in(L_x, q_x) \text{ where } q_x = q_x(\xi, \tau) = 1 - p_x(\xi, \tau).$$

Precisely, we maximize the Binomial log-likelihood function

$$\mathcal{L}(\xi, \tau) = \sum_{x \geq x^*} \left(L_x \ln p_x(\xi, \tau) + D_x \ln q_x(\xi, \tau)\right).$$

This gives

$$\widehat{\xi} = \begin{cases} -0.131 \text{ with standard error } 0.016 \text{ for males,} \\ \\ -0.096 \text{ with standard error } 0.015 \text{ for females} \end{cases}$$

and

Fig. 9.10 Generalized
QQ-plots for the remaining
lifetimes at age $x^\star = 98.89$
for males (upper panel) and
$x^\star = 100.89$ for females
(lower panel)

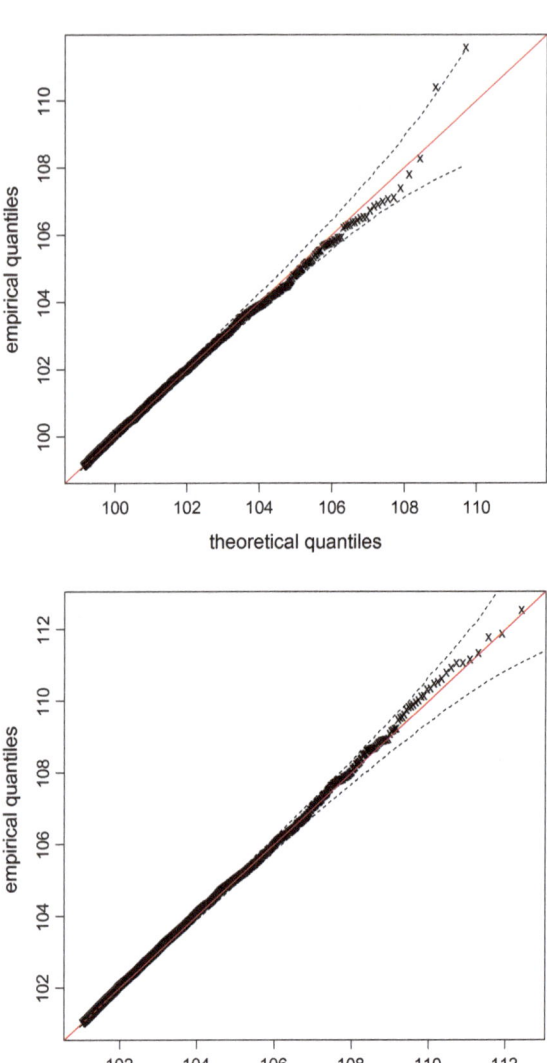

$$\widehat{\tau} = \begin{cases} 2.100 \text{ with standard error } 0.061 \text{ for males,} \\ \\ 2.034 \text{ with standard error } 0.046 \text{ for females.} \end{cases}$$

These values are close to those reported in Table 9.4. The estimated ultimate age ω
is now

Table 9.5 Aggregated mortality data

Age	Male			Female		
x	L_x	E_x	D_x	L_x	E_x	D_x
99	1,836	1,468.636	691	–	–	–
100	1,145	886.959	465	–	–	–
101	680	517.208	307	3,883	3,064.211	1,508
102	373	275.967	184	2,375	1,827.112	1,002
103	189	138.175	102	1,373	1,045.293	592
104	87	59.808	45	781	581.049	373
105	42	29.027	23	408	299.518	187
106	19	10.819	14	221	160.216	101
107	5	3.707	2	120	87.512	59
108	3	2.17	1	61	45.129	31
109	2	2	0	30	22.838	15
110	2	1.288	1	15	10.605	10
111	–	–	–	5	3.46	3

$$\widehat{\omega} = \begin{cases} 114.90 \text{ for males,} \\ \\ 122.13 \text{ for females,} \end{cases} \tag{9.9}$$

which is in line with the values in (9.8) obtained with individual ages at death. The main difficulty when working with aggregated data is to select the appropriate threshold age beyond which the Generalized Pareto behavior becomes apparent.

9.6 Poisson-Generalized Pareto Model for Excesses Over a High Threshold

9.6.1 Distribution of the Maximum of a Generalized Pareto Sample of Poisson Size

The number N_u of excesses over the threshold u obeys the $Bin\left(n, \overline{F}(u)\right)$ distribution. Provided u and n are large enough, this Binomial distribution can be approximated by the Poisson one. Hence, it is interesting to consider a sample with random, Poisson-distributed size. This is precisely the topic investigated in this section.

Consider a random variable N distributed according the Poisson law with mean λ, and let $Y_1, \ldots, Y_N$ be a sequence of N independent and identically distributed random variables with common cumulative distribution function $G_{\xi;\tau}$. Define

$$M_N = \max\{Y_1, Y_2, \ldots, Y_N\}.$$

Also, define the location-scale family $H_{\xi;\mu,\psi}$ corresponding to the GEV distribution function H_ξ by

$$H_{\xi;\mu,\psi}(y) = H_\xi\left(\frac{y-\mu}{\psi}\right), \quad \mu \in (-\infty, \infty), \quad \psi > 0.$$

The next result gives the distribution of M_N in the Poisson-Generalized Pareto setting.

Property 9.6.1 *Consider a sequence* $Y_1, Y_2, \ldots$ *of independent random variables with common* $\mathcal{G}Par(\xi, \tau)$ *distribution, independent of* $N \sim \mathcal{P}oi(\lambda)$. *We then have*

$$P[M_N \le y] = H_{\xi;\mu,\psi}(y), \tag{9.10}$$

with

$$\mu = \tau\xi^{-1}(\lambda^\xi - 1) \text{ and } \psi = \tau\lambda^\xi.$$

Proof This result can be established as follows. For $\xi \ne 0$,

$$\begin{aligned}
P[M_N \le y] &= P[\max\{Y_1, \ldots, Y_N\} \le y] \\
&= \sum_{n=0}^{\infty} P[\max\{Y_1, \ldots, Y_N\} \le y]P[N = n] \\
&= \sum_{n=0}^{\infty} \left(P[Y_1 \le y]\right)^n P[N = n] \\
&= \sum_{n=0}^{\infty} \frac{\left(\lambda G_{\xi,\tau}(y)\right)^n}{n!} \exp(-\lambda) \\
&= \exp\left(-\lambda\left(1 - G_{\xi,\tau}(y)\right)\right) \\
&= \exp\left(-\lambda\left(1 + \xi\frac{y}{\tau}\right)_+^{-1/\xi}\right) \\
&= \exp\left(-\left(1 + \xi\frac{y - \xi^{-1}\tau(\lambda^\xi - 1)}{\tau\lambda^\xi}\right)_+^{-1/\xi}\right) \\
&= H_{\xi;\tau\xi^{-1}(\lambda^\xi-1),\tau\lambda^\xi}(y),
\end{aligned}$$

which proves the announced result. For $\xi = 0$, it suffices to notice that

$$P[M_N \leq y] = \sum_{n=0}^{\infty} \frac{(\lambda(1 - \exp(-y/\tau))^n}{n!} \exp(-\lambda)$$

$$= \exp\left(-\lambda \exp(-y/\tau)\right)$$

$$= \exp\left(-\exp\left(-\frac{1}{\tau}(y - \tau \ln \lambda)\right)\right)$$

$$= H_{0;\tau \ln \lambda, \tau}(y).$$

Since

$$\tau \lim_{\xi \to 0} \frac{\lambda^\xi - 1}{\xi} = \tau \ln \lambda,$$

this ends the proof.

Let us now apply this result to study the distribution of the highest age at death in life insurance and to the determination of the probable maximum loss in property and casualty insurance.

9.6.2 Probable Maximal Loss

Broadly speaking, the probable maximum loss (PML) is the worst loss likely to happen. It is defined as a high quantile of the maximum of all claim severities hitting the portfolio during a reference period (generally, one year). Since this maximum will exceed the so-defined PML with a small, controlled probability, it is very unlikely that an individual claim amount recorded in this portfolio assumes a value larger than the PML.

Let us now make the link with the Generalized Pareto approximation derived from the Pickands-Balkema-de Haan theorem for the excesses over a sufficiently high threshold. Denote as $Y_{[N_u]}$ the maximum of the N_u claims above u. Clearly, $N_u \sim \mathcal{B}in(n, \overline{F}(u))$ is approximately Poisson for n and u sufficiently large.

Assume that $N_u \sim \mathcal{P}oi(\lambda_u)$ and, given $N_u \geq 1$, $Y_1 - u, \ldots, Y_{N_u} - u$ are independent and obey the $\mathcal{G}\mathcal{P}ar(\xi, \tau)$ distribution. Then, for $y > u$, Property 9.6.1 gives

$$P[Y_{(N_u)} \leq y] = P[Y_{(N_u)} - u \leq y - u]$$

$$= \exp\left(-\lambda_u \left(1 + \xi \frac{y - u}{\tau}\right)_+^{-1/\xi}\right).$$

Defining the PML as the quantile of $Y_{(N_u)}$ corresponding to probability level $1 - \epsilon$, for some ϵ small enough, that is,

$$\Pr[Y_{(N_u)} \leq \text{PML}] = 1 - \epsilon$$

we get

$$\text{PML} = F_{Y_{(N_u)}}^{-1}(1 - \epsilon)$$

$$= u + \frac{\tau}{\xi}\left(\left(-\frac{\lambda_u}{\ln(1 - \epsilon)}\right)^{\xi} - 1\right).$$

As an example, consider the motor third-party liability insurance portfolio studied in this chapter. With $\epsilon = 5\%$, we get

$$\widehat{\text{PML}} = 2,471,312 + \frac{7,655,438}{0.2718819}\left(\left(-\frac{29}{\ln 0.95}\right)^{0.2718819} - 1\right) = 132,039,070$$

which indeed exceeds the observed maximum of 80,258,970 Belgian francs. For calculations with this portfolio, the actuary can set the working upper bound to the claim severity distribution at $\widehat{\text{PML}}$.

9.6.3   Highest Age at Death

The highest age at death can be modeled statistically as the maximum M_N of a random sample of size N. In order to study the behavior of the highest age at death M_N, we use the Generalized Pareto approximation to the remaining lifetimes at ages $x \geq x^\star$. Under the conditions leading to the Generalized Pareto approximation $G_{\xi;\tau}$ for the remaining lifetime distribution at age $x^\star$, the number $L_{x^\star}$ of survivors at threshold age $x^\star$ is roughly Poisson with mean $\ell_{x^\star} = \text{E}[L_{x^\star}]$ as an approximation to the Binomial distribution valid for large sizes and small success probabilities. As a consequence, the distribution function of the maximum age at death $M_{L_{x^\star}}$ can be obtained from Property 9.6.1. This result shows that the distribution of the maximum $M_{L_{x^\star}}$ of the lifetimes of the $L_{x^\star}$ individuals reaching age $x^\star$ can be approached by a GEV distribution $H_{\xi;\mu,\psi}$ where

$$\mu = \tau\xi^{-1}(\ell_{x^\star}^{\xi} - 1) \text{ and } \psi = \tau\ell_{x^\star}^{\xi}.$$

Hence,

$$P[M_{L_{x^\star}} \leq s] = \exp\left(-\ell_{x^\star}\left(1 + \xi\frac{s - x^\star}{\tau}\right)_{+}^{-1/\xi}\right)$$

by (9.10). As an application, we find the following approximation for the quantile at probability level $1 - \epsilon$ of $M_{L_{x^\star}}$:

$$F_{M_{L_{x^*}}}(1 - \epsilon) \approx x^* + \frac{\tau}{\xi} \left(\left(-\frac{\ell_{x^*}}{\ln(1 - \epsilon)} \right)^{\xi} - 1 \right).$$

9.6.4 Stability Property with Increasing Thresholds

If the number of excesses over a threshold u is Poisson distributed with mean λ_u and the corresponding excesses follow the $\mathcal{G}\mathcal{P}ar(\xi, \tau)$ distribution then the number of excesses over a larger threshold $v > u$ is Poisson with expected value

$$\lambda_v = \lambda_u \left(1 + \xi \frac{v - u}{\tau} \right)^{-1/\xi}.$$

This is an application of Property 2.3.2. This expression yields

$$E[N_v] = E[N_u] \left(1 + \xi \frac{v - u}{\tau} \right)^{-1/\xi}. \tag{9.11}$$

As an application, let us consider the sum S_v of the excesses over v, with $v > u$, that is

$$S_v = \sum_{i|Y_i > u} (Y_u - v)_+ = \sum_{i|Y_i > v} (Y_i - v).$$

According to (9.11), the expected value of S_v is given by

$$E[S_v] = E[N_v] E[Y - v | Y > v]$$
$$= E[N_v] \left(\tau + (v - u) \frac{\xi}{1 - \xi} \right).$$

This expected value represents the average reinsurance benefits paid under an excess-of-loss treaty. It can be estimated by substituting point estimates for the corresponding unknown quantities.

9.7 Bibliographic Notes and Further Reading

This chapter only gives a short, non-technical description of the main aspects of EVT. The key references for EVT applied to solve risk management problems related to insurance, reinsurance and finance remain Embrechts et al. (1997) and Beirlant et al. (2005). The book devoted to quantitative risk management by McNeil et al. (2015) also contains a detailed treatment of EVT.

The application of EVT to claim severities follows Cebrian et al. (2003) who proposed a detailed analysis of the SOA large claims database. The application to the modeling of remaining lifetime at the oldest ages is taken from Gbari et al. (2017a), following Watts et al. (2006). It is interesting to notice that EVT is strongly connected to survival analysis; see e.g. Aarssen and de Haan (1994), Balkema and de Haan (1974), not to mention Gumbel (1937). The impact of working with aggregated mortality statistics, as opposed to individual lifetimes is thoroughly discussed in Gbari et al. (2017b).

We must acknowledge here that there is no consensus about the existence of a finite endpoint to the human life span and, if so, whether this upper limit changes over time. Einmahl et al. (2019) used ages at death of about 285,000 Dutch residents, born in the Netherlands, who died in the years 1986–2015 at a minimum age of 92 years. Based on extreme value theory, these authors found statistical evidence that there is indeed an upper limit to the life span for both genders, without trends in time over the last 30 years. The estimated endpoints ω obtained by these authors appear to be close to those reported in Gbari et al. (2017a), which is fortunate because Belgium and The Netherlands are neighboring countries.

This is in contrast with Rootzen and Zholud (2017) who concluded that the life span was not bounded from above. Several demographic studies have also suggested that age-specific death probabilities may level off at some point above 100 years, rather than continuing to increase with age. Notice that if the force of mortality tends to flatten at oldest ages then remaining lifetimes become ultimately Negative Exponential.

It is worth to mention here that actuaries are more interested in closing the life table in an appropriate manner, than to provide a final answer to the question of maximum lifespan. Indeed, the problem posed by the mortality pattern at oldest ages, with a still increasing, or plateauing, or even decreasing force of mortality after a maximum level has been reached, goes far beyond the actuarial expertise as almost no data is available at ages above 115. Actuaries nevertheless need an appropriate model, supported by empirical evidence, to close the life table so that actuarial calculations can be performed, typically in life annuity or reverse mortgage portfolios. Therefore, assuming that the force of mortality becomes ultimately constant, i.e that the remaining lifetime tends to the Negative Exponential distribution as the attained age grows is a conservative strategy for managing life annuities.

The International Database on Longevity (IDL) has been launched to complement the Human Mortality Database (HMD) at older ages. This is because many observations in the HMD are right-censored, particularly at the age of 110 so that HMD alone is not suitable for EVT analysis. In contrast, the IDL contains the exact age and time of death of so-called super-centenarians, that is, individuals who at least reached their 110th birthday. IDL contains data for 15 industrialized countries around the world.

The determination of the threshold beyond which the Generalized Pareto behavior becomes apparent is not an easy task in practice. We refer the reader to Scarrott et al. (2012) for a detailed review of threshold selection methods. In this chapter, we have mainly used basic graphical tools based on the mean excess function and Pareto index plots. Besides the Gertensgarbe plot used by Cebrian et al. (2003), there are also more formal selection procedures. Gbari et al. (2017a) resorted to the techniques proposed by Pickands (1975), Reiss and Thomas (1997) and Scarrott et al. (2012) to select the threshold age, because the graphical tools did not produce a clear answer. Danielsson et al. (2016) and Bader et al. (2017, 2018) proposed data-driven procedures to select the Generalized Pareto threshold. Some of these tools are implemented in the R package eva.

Instead of first isolating the extreme observations and then modeling them separately, another approach was proposed by Cooray and Ananda (2005) who combined a LogNormal probability density function together with a Pareto one. Specifically, these authors introduced a two-parameter smooth continuous composite LogNormal-Pareto model, that is, a two-parameter LogNormal density up to an unknown threshold value and a two-parameter Pareto density beyond it.

Continuity and differentiability are imposed at the unknown threshold to ensure that the resulting probability density function is smooth, reducing the number of parameters from 4 to 2. The resulting two-parameter probability density function is similar in shape to the LogNormal density, yet its upper tail is thicker than the LogNormal density (and accommodates for the large losses often observed in liability insurance).

As noted by Scollnik (2007), this model is very restrictive because whatever the data set under study, exactly 39.215% of the observations are expected to fall below the threshold separating the LogNormal and Pareto segments. This may lead to poor adjustment to data. This is why Scollnik (2007) developed a second composite Lognormal–Pareto model. Besides, Scollnik (2007) also introduced an alternative composite model in which the Generalized Pareto distribution is used above the threshold.

Pigeon and Denuit (2011) proposed the mixed composite LogNormal-Pareto model based on Scollnik's model, assuming that each observation may have its own threshold. These claim-specific thresholds are then considered as realizations of some non-negative random variable and a mixed model is proposed. Such composite models are very appealing for insurance applications. Nadarajah and Bakar (2014) proposed another new composite model based on the LogNormal distribution and demonstrated its superior fit to the Danish fire insurance data. A R package has been contributed by Nadarajah and Bakar (2013).

Of course, other distributions can be combined with the Pareto tail. Scollnik and Sun (2012) and Bakar et al. (2015) proposed a Weibull distribution for the body of the data. See also Calderin-Ojeda and Kwok (2016).

There are also several proposals to combine a nonparametric density estimate below the threshold with a Generalized Pareto tail. Let us mention the work by Tancredi et al. (2006) who modelled data with a distribution composed of a piecewise constant density from a low threshold up to an unknown end point and a Generalized Pareto distribution for the remaining tail part. MacDonald et al. (2011) suggested to approximate the upper tail of the distribution with the Generalized Pareto distribution without specifying any parametric form for the bulk of the distribution in order to avoid a specification risk. The bulk of the distribution is simultaneously estimated non-parametrically. See also Hong and Martin (2018).

References

Aarssen K, de Haan L (1994) On the maximal life span of humans. Math Popul Stud 4:259–281

Bader B, Yan J, Zhang X (2017) Automated selection of r for the r largest order statistics approach with adjustment for sequential testing. Stat Comput 27:1435–1451

Bader B, Yan J, Zhang X (2018) Automated threshold selection for extreme value analysis via ordered goodness-of-fit tests with adjustment for false discovery rate. Ann Appl Stat 12:310–329

Bakar SA, Hamzah NA, Maghsoudi M, Nadarajah S (2015) Modeling loss data using composite models. Insur Math Econ 61, 146–154

Balkema A, de Haan L (1974) Residual life time at great age. Ann Probab 2:792–804

Beirlant J, Goegebeur J, Teugels J, Segers J, Waal DD, Ferro C (2005) Statistics of extremes: theory and applications. Wiley, New York

Calderin-Ojeda E, Kwok CF (2016) Modeling claims data with composite Stoppa models. Scand Actuar J 2016:817–836

Cebrian AC, Denuit M, Lambert P (2003) Generalized Pareto fit to the society of actuaries' large claims database. N Amn Actuar J 7:18–36

Cooray K, Ananda MMA (2005) Modeling actuarial data with a composite Lognormal-Pareto model. Scand Actuar J 2005:321–334

Danielsson J, Ergun L, de Haan L, de Vries C (2016) Tail index estimation: quantile driven threshold selection. Available at SSRN: https://ssrn.com/abstract=2717478 or http://dx.doi.org/10.2139/ssrn.2717478

Einmahl JJ, Einmahl JH, de Haan L (2019) Limits to human life span through extreme value theory. J Am Stat Assoc (in press)

Embrechts P, Kluppelberg C, Mikosch T (1997) Modelling extremal events for insurance and finance. Springer, Berlin

Gbari S, Poulain M, Dal L, Denuit M (2017a) Extreme value analysis of mortality at the oldest ages: a case study based on individual ages at death. N Amn Actuar J 21:397–416

Gbari S, Poulain M, Dal L, Denuit M (2017b) Generalised Pareto modeling of older ages mortality in Belgium using extreme value techniques. ISBA Discussion Paper, UC Louvain

Gumbel EJ (1937) La Durée Extrême de la Vie Humaine. Hermann, Paris

Hong L, Martin R (2018) Dirichlet process mixture models for insurance loss data. Scandinavian Actuarial Journal 2018:545–554

MacDonald A, Scarrott CJ, Lee D, Darlow B, Reale M, Russell G (2011) A flexible extreme value mixture model. Comput Stat Data Anal 55:2137–2157

MacNeil AJ, Frey R, Embrechts P (2015) Quantitative risk management. Princeton University Press, Concepts, Techniques and Tools

Nadarajah S, Bakar S (2013) CompLognormal: an R package for composite lognormal distributions. R J 5:98–104

Nadarajah S, Bakar SA (2014) New composite models for the Danish fire insurance data. Scand Actuar J 2014:180–187

Pickands J (1975) Statistical inference using extreme order statistics. Ann Stat 3:119–131

Pigeon M, Denuit M (2011) Composite Lognormal-Pareto model with random threshold. Scand Actuar J 2011:177–192

Reiss R-D, Thomas M (1997) Statistical analysis of extreme values, with applications to insurance, finance, hydrology and other fields. Birkhauser, Basel

Rootzen H, Zholud D (2017) Human life is unlimited-but short. Extremes 20:713–728

Scarrott C, MacDonald A (2012) A review of extreme value threshold estimation and uncertainty quantification. REVSTAT-Stat J 10:33–60

Scollnik DPM (2007) On composite Lognormal-Pareto models. Scand Actuar J 20–33

Scollnik DP, Sun C (2012) Modeling with Weibull-Pareto models. N Amn Actuar J 16:260–272

Tancredi A, Anderson C, O'Hagan A (2006) Accounting for threshold uncertainty in extreme value estimation. Extremes 9:87–106

Watts K, Dupuis D, Jones B (2006) An extreme value analysis of advance age mortality data. N Amn Actuar J 10:162–178